# MEDICAL DEVICE TECHNOLOGIES

## A SYSTEMS BASED OVERVIEW USING ENGINEERING STANDARDS

Gail D. Baura

AMSTERDAM • BOSTON • HEIDELBERG • LONDON
NEW YORK • OXFORD • PARIS • SAN DIEGO
SAN FRANCISCO • SINGAPORE • SYDNEY • TOKYO

Academic Press is an imprint of Elsevier

Academic Press is an imprint of Elsevier
225 Wyman Street, Waltham, MA 02451, USA
The Boulevard, Langford Lane, Kidlington, Oxford, OX5 1GB, UK

**Notices**
Knowledge and best practice in this field are constantly changing. As new research and experience broaden our understanding, changes in research methods, professional practices, or medical treatment may become necessary.

Practitioners and researchers must always rely on their own experience and knowledge in evaluating and using any information, methods, compounds, or experiments described herein. In using such information or methods they should be mindful of their own safety and the safety of others, including parties for whom they have a professional responsibility.

To the fullest extent of the law, neither the Publisher nor the authors, contributors, or editors, assume any liability for any injury and/or damage to persons or property as a matter of products liability, negligence or otherwise, or from any use or operation of any methods, products, instructions, or ideas contained in the material herein.

**Library of Congress Cataloging-in-Publication Data**
Application submitted

**British Library Cataloguing-in-Publication Data**
A catalogue record for this book is available from the British Library

ISBN: 978-0-12-374976-5

For information on all Academic Press publications
visit our website at: www.elsevierdirect.com

Printed and bound by CPI Group (UK) Ltd, Croydon, CR0 4YY

Working together to grow
libraries in developing countries

www.elsevier.com | www.bookaid.org | www.sabre.org

ELSEVIER    BOOK AID International    Sabre Foundation

To Larry Spiro, my *bon vivant*,
without whose infinite patience and love
this book could not have been written.


Teachers open the door, but you must enter by yourself.
—Chinese Proverb

# Contents

# II
# LAB EXPERIMENTS

# Preface

In 2006, I returned to academia from the medical device industry and was asked to teach medical devices. To prepare for the course, I first listed appropriate topics. Electrical medical devices are traditionally taught in a bioinstrumentation course. I decided to construct a course with important electrical *and* mechanical medical devices. Then I searched for a textbook incorporating these topics, and found ... nothing. I created my lectures from primary literature, constructed unusual lab experiments, and endured a grueling spring semester. After teaching this course twice, I formalized the general structure of each device lecture as: relevant physiology, mathematical modeling or biocompatibility, clinical need, historical devices, technology. I improved the technology section by discussing each medical device as a system. To encourage professors to provide evidence for ABET Criterion 5, I included requirements from applicable engineering standards in the technology section. I also strengthened the three FDA recall case studies I discuss, which provide a glimpse of the working conditions of medical device engineers. Five years later, these five teaching semesters have become this textbook.

The textbook has three audiences. First, because the only prerequisites are free body diagrams, introductory circuits, and introductory differential equations, it is appropriate for bioengineering students who are second semester sophomores and first semester juniors. Many bioengineering programs purport to prepare their students to enter the medical device industry, but most do not introduce their students to basic, modern medical devices. Second, because learning in context is known to result in a deeper understanding of subject matter and increased student engagement, the book is appropriate for electrical and mechanical engineering students, as well as bioengineering students. Institute of Electrical and Electronics Engineers (IEEE) President Moshe Kam recently stated that "the intersection between our traditional fields—electrical engineering, computer engineer, and computer science—and the life sciences is 'hot.' If we fail to capture this growing interest area, others will fill the gap" (Kam, 2011). Third, for medical device engineers working on one type of medical device, the systems approach of this book provides connections to other devices. Common medical instruments are electrocardiographs, blood pressure monitors, pulse oximeters, thermometers, and respiration monitors. Common electrical stimulators are pacemakers, implantable cardioverter defibrillators (ICDs), cochlear implants, and deep brain stimulators.

Industries such as the medical device industry wish to hire graduates who are industry ready. The American Society of Mechanical Engineers determined from Vision 2030 surveys of 590 industrial managers that they believe current graduates are weak in overall systems perspective, "new technical fundamentals (new ME applications — bio, nano, info, etc.)" and engineering codes and standards (Kirkpatrick, 2010). The Institute of Electrical and Electronic Engineers Board of Directors recently approved a position paper that declares "introducing standards in the classroom will augment the learning experience by pointing students to available design tools, and to best industry practices" (IEEE, 2009). In one of its workshop summaries for its Moving Forward to Improve Engineering Education initiative, the National Science Board stated that "companies want engineers with ... systems thinking; ... an ability to understand the business context of engineering; ... and an ability to change. The public sector especially needs engineers with a sophisticated understanding of the social environment within which their activity takes place, a system understanding, and an ability to communicate with stakeholders" (National Science Board, 2007).

This textbook addresses all these issues, from the perspective of an author with thirteen years of medical device industry experience who is also an ABET bioengineering program evaluator. The text is a necessary first step for bioengineering graduates entering the medical device industry. In 2013, a second textbook by the author will describe the medical device design process in detail, as well as other aspects of medical device market release and postmarketing surveillance (Baura, 2013).

## Additional Instructional Materials

For instructors using this text in their course, an accompanying website includes support materials such as physiologic waveform files for chapter exercises, electronic images from the text, and an instructor's manual. Register at www.textbooks.elsevier .com for access to these learning resources.

## Acknowledgments

I would like to thank all the students who took my course and taught me how to teach this material. My colleagues who reviewed chapters ensured that the material is covered in an appropriate clinical and industrial context: Dr. Suhail Ahmad, Dr. Paul Benkeser, Dr. John Enderle, Stuart Gallant, Dr. Arlene Gwon, Dennis Hepp, Dr. Claude Jolly, Dr. Hugues LaFrance, Dr. James Oberhauser, Dr. Shrirang Ranade, Alex Stenzler, and an anonymous reviewer. In particular, Dennis introduced me to pacemakers years ago, reviewed (and volunteered to read!) numerous chapters, and continues to be my mentor. My industrial colleagues who generously donated or loaned medical devices were critical to the development of the lab experiments in Part II: Anne Bugge, Dr. Joe Chinn, and Dr. Sergio Shkurovich. I reprinted the figures from several Elsevier textbooks, and I am grateful to their original authors, especially Drs. Arthur Guyton and John Hall, for their use. The team at Elsevier and Macmillan Publishing Solutions ushered this manuscript through the publication process: Ed Dionne, Jeff Freeland, Joe Hayton, Mike Joyce, Becky Pease, and Jonathan Simpson. Special thanks go to Jonathan for signing my book and to Ed for keeping me sane.

The following people were not directly involved but influenced this text: My undergraduate professors at Loyola Marymount University provided an incredible engineering foundation that I use daily: Cliff d'Autremont, Dr. Joe Callanan, Dr. Tai-Wu Kao, Dr. John Page, Bob Ritter, and Dr. Paul Rude. Bob Ward gave me my first taste of medical devices when he hired me as a student intern in the biomedical engineering department of St. Mary Medical Center. My Ph.D. advisors, Dr. David Foster and Dr. Dan Porte Jr., taught me by example the importance of mathematical modeling and clinical need.

Most of all, I would like to thank my long-suffering husband Larry for his continuing encouragement and emotional support. Larry does not understand my obsession with writing textbooks, but he does endure my habit. I am so sorry that my first book wasn't my last. And I am even sorrier that I still have two books to go.

I welcome comments to this text at www.gailbaura.com.

*Gail D. Baura*
Claremont, 2011

# References

Baura, G. D. (2013). *U.S. Medical Device Regulation: An Introduction to Biomedical Product Development, Market Release, and Postmarket Surveillance*, Manuscript in preparation.

IEEE (2009). *IEEE Position Paper on the Role of Technical Standards in the Curriculum of Academic Programs in Engineering, Technology and Computing*. Piscataway, NJ:IEEE.

Kam, M. (2011). President's column: life sciences spark IEEE interest. *IEEE Institute* (March).

Kirkpatrick, A. (2010). Vision 2030 Progress Report and Curricula Recommendations. *2010 International Mechanical Engineering Education Conference*. Newport Beach, CA: ASME.

National Science Board (2007). *Moving Forward To Improve Engineering Education, NSB-07-122*. Arlington, VA: NSF.

# About the Author

Gail Dawn Baura received a BSEE from Loyola Marymount University in 1984, and an MSEE and MSBME from Drexel University in 1987. She received a PhD in Bioengineering from the University of Washington in 1993. Between these graduate degrees, Dr. Baura worked as a loop transmission systems engineer at AT&T Bell Laboratories. Since graduation, she has served in a variety of research and development positions at IVAC Corporation, Cardiotronics Systems, Alaris Medical Systems, and VitalWave Corporation (now Tensys Medical). Her most recent industrial position was Vice President of Research and Chief Scientist at CardioDynamics. Dr. Baura has conducted research on insulin kinetics and on the following medical devices: blood pressure monitor, electronic thermometer, external defibrillator, impedance cardiograph, pacemaker, syringe, and large volume infusion pumps. In 2006, she returned to academia as a Professor at Keck Graduate Institute of Applied Life Sciences, which is a member of the Claremont Colleges.

Dr. Baura is a senior member of IEEE, associate editor of *IEEE Pulse* (formerly *EMB Magazine*), a member of the Biomedical Engineering Society Accreditation Activities Committee, and an ABET program evaluator for bioengineering and biomedical engineering. She holds 20 issued U.S. patents. She serves as an intellectual property expert witness, and as an invited juror for the Medical Design Excellence Awards. Her research interests are the application of system theory to patient monitoring and other devices. She is the author of four engineering textbooks.

# Nomenclature

$u(k)$ scalar

$\mathbf{u}(k)$ vector

$\underline{\mathbf{U}}(k)$ matrix

$U(jw)$ Fourier Transform

[ ] concentration

$\cdot$ derivative superscript

$\wedge$ approximation superscript

$*$ complex conjugate superscript

$a$ arterial subscript, ambient subscript

$\alpha$ real part of quadratic root, compartmental model transfer function denominator polynomial coefficient

$A$ area, alveolar subscript, absorbance, optical coefficient

$A_i$ thermistor resistance coefficients

$\underline{\mathbf{A}}(\mathbf{k})$ feedback matrix containing rate constants

$AC$ pulsatile subscript

$b(t)$ net hepatic glucose balance

$B$ blood subscript, bispectrum

$\underline{\mathbf{B}}(\mathbf{k})$ feedforward matrix containing rate constants

$B_o$ net balance

$BI$ blood inlet subscript

$BO$ blood outlet subscript

$\beta$ imaginary part of quadratic root, compartmental model transfer function numerator polynomial coefficient

$\beta_m$ material calibration temperature

$c$ calibration subscript, chronaxie subscript, constant value, concentration, controller subscript, sound propagation velocity in tissue

$\mathbf{c}(t)$ concentration vector

$C$ capacitance

$\underline{\mathbf{C}}(\mathbf{k})$ relationship between the concentration and desired output vectors

$CM$ common-mode subscript

$d$ delivered subscript, derivative subscript, deoxyhemoglobin subscript, diffusion subscript, dialysis subscript

$d/du(k)$ derivative

$dif$ differential subscript

$\delta$ instantaneous subscript

$D$ dead space subscript, downsampling constant, diffusion coefficient, dialysate subscript, diameter

$DC$ nonpulsatile subscript, direct current subscript

$\Delta$ change

| | | | |
|---|---|---|---|
| $e$ | error, continuous envelope signal | **k** | admissible parameter space of rate constants |
| $e^-$ | electron | $K$ | rate of urea clearance, PID controller gain |
| $\varepsilon$ | strain, emissivity, molar absorptivity, membrane porosity | $l$ | length, leakage current subscript, load subscript, loss subscript |
| $E$ | photonic energy, defibrillator energy, elastance, elastance subscript | $\lambda$ | wavelength |
| $E\{\ \}$ | expected value | $L$ | inductance, inductance subscript |
| $E(t)$ | glucose effectiveness | $m$ | quadratic root, mass |
| $EO$ | effective orifice subscript | $m_i$ | measured value for Bland-Altman analysis |
| $f$ | frequency, refraction subscript, feedback subscript, filter | $\mu$ | gain constant |
| $F$ | force, factor | $M$ | moment, membrane subscript |
| $\Phi$ | infrared flux | $n$ | number of samples, summation index, refractive index, number of loading cycles |
| $g(t)$ | glucose concentration | | |
| $G$ | gain subscript | $n(k)$ | noise sequence |
| $\underline{\mathbf{G}}(\mathbf{k})$ | Jacobian matrix | $\mathbf{n}(t)$ | noise vector |
| $h$ | Planck's constant, pulse | $N$ | number of moles per unit volume, number of turns in coil |
| $\eta$ | fluid viscosity | | |
| $H(s)$ | transfer function | $N1$ | negative peak in neural response waveform |
| $\underline{\mathbf{H}}(s,\mathbf{k})$ | compartmental model transfer function | $o$ | output subscript, incident subscript, peak subscript, initial loading subscript, threshold subscript, oxyhemoglobin subscript |
| $i$ | current, general index, input subscript, integral subscript, incidence subscript | | |
| $i(t)$ | plasma insulin concentration | $p$ | photocurrent subscript, patient subscript, pulse subscript, proportionality subscript, parameter |
| $I$ | light intensity | | |
| $\underline{\mathbf{I}}(s)$ | identity matrix | | |
| $j$ | imaginary number, scale | $\mathbf{p}$ | uniquely identifiable parameter vector |
| $J$ | flux | $P$ | pressure, pressure subscript |
| $k$ | discrete time (samples), rate constant, kinematic position | $P_h$ | hydrostatic pressure gradient |

| | | | |
|---|---|---|---|
| $P_i$ | pressure at inlet of tube | $T$ | absolute temperature, tidal subscript, surface temperature, temperature subscript, thermistor subscript, transmission coefficient |
| $P_o$ | pressure at outlet of tube | | |
| P2 | positive peak in neural response waveform | | |
| $\theta$ | angle | $TR$ | transrespiratory subscript |
| $Q$ | flow | $u$ | ultrafiltration subscript |
| $r$ | correlation coefficient, radius, rheobase subscript, reflection subscript | $u(k)$ | input sequence |
| | | $\mathbf{u}(t)$ | exogenous input vector, reconstruction vector |
| $r(t)$ | insulin in remote compartment | $u_p(t)$ | glucose utilization in peripheral tissues |
| $R$ | specific gas constant, resistance, resistance subscript, ratio, response, reflection coefficient | $\upsilon$ | optical frequency |
| | | $v$ | velocity, venous subscript |
| $\rho$ | density | $visc$ | viscosity subscript |
| $s$ | signal, stored subscript, sensor subscript, saturation subscript, Laplace frequency | $V$ | voltage, volume, urea volume distribution |
| $s_i$ | reference standard value for Bland-Altman analysis | $W(n)$ | gravimetric wear |
| | | $W(z,t)$ | longitudinal particle displacement |
| $sq$ | square wave subscript | $WT(k,2^j)$ | wavelet transform |
| $S$ | sieving coefficient | $\Omega$ | continuous frequency |
| $S_{cr}$ | serum creatinine | $\Omega_c$ | continuous cutoff frequency |
| $S_I$ | insulin sensitivity | $\Omega_N$ | Nyquist frequency |
| $SS$ | steady state subscript | $\Omega_{sampl}$ | sampling frequency |
| $\sigma$ | Stefan-Boltzmann constant | $x(t)$ | state vector |
| $t$ | continuous time, pulse duration time, total subscript, torsional subscript, trial number | $y(k)$ | output sequence |
| | | $\mathbf{y}(t)$ | desired output vector |
| | | $\psi(t)$ | wavelet |
| | | $z$ | longitudinal distance |
| $\tau$ | shear stress, membrane tortuosity | $Z$ | impedance |

# MEDICAL DEVICES

A *medical device* is an apparatus used in the diagnosis, mitigation, therapy, or prevention of disease, which does not attain its primary purpose through chemical action. In Part I of this textbook, we provide an overview of medical devices (Chapter 1), followed by a discussion of 19 types of medical devices (Chapters 2–20). Because many students have minimal exposure to medical devices, Chapter 1 provides a framework for subsequent discussions. Basic concepts like instruments, stimulators, and sensors are considered.

In an introductory survey course, choosing only 19 types of medical devices for discussion is difficult. Devices were chosen on the basis of the following criteria:

1. Exclusion of imaging devices because many good textbooks on medical imaging exist.
2. Medical devices (except for imaging) that have won a Nobel Prize in Physiology or Medicine or a Lasker Clinical Medical Research Award, because these devices have met clinical needs and have saved lives.
3. Medical devices that measure the vital signs, because it is an ABET requirement that bioengineer graduates have "the ability to make measurements on and interpret data from living systems."
4. Medical devices in the four high-growth areas of cardiovascular devices, neural devices, orthopedics, and combination products, in order to provide student training for the medical device industry.

With these criteria in mind, we describe the following technologies: electrocardiographs; pacemakers; external defibrillators; implantable cardioverter defibrillators; heart valves; blood pressure monitors; catheters, bare metal stents, and vascular grafts; hemodialysis delivery systems; mechanical ventilators; pulse oximeters; thermometers; electroencephalographs; neurostimulators; cochlear implants; functional electrical stimulators; intraocular lens implants; total hip prostheses; drug-eluting stents; and the artificial pancreas.

Each chapter begins with a discussion of relevant physiology and clinical need. Historic devices are included because they provide insight into the design-improvement process. System description and system diagrams provide details on technology function and on the administration of diagnosis and/or therapy. The systems approach enables students to quickly identify the relationships between devices. Each chapter concludes with five key device features, which are requirements in an applicable consensus standard.

In three chapters, case studies of significant Food and Drug Administration recalls are included. The Bjork Shiley heart valve (Chapter 6), Guidant endovascular graft (Chapter 8), and Guidant implantable cardioverter defibrillator (Chapter 5) recalls had significant effects on how medical devices are designed and monitored after market release. These case studies provide a glimpse into medical device business practices such as design control, verification testing, postmarket surveillance (reporting of adverse events), and sales.

Exercises at the end of each chapter include traditional homework problems, analysis exercises, and four questions from assigned primary literature. In many homework problems, the students download physiologic waveforms and process them using software, such as Matlab. Because this is a textbook and not a reference book, the students are asked to analyze differences between various devices and device components after a group of devices, such as implantable stimulators, is discussed. This is a more effective teaching strategy than having these issues readily compared in an existing table.

It is recommended that the primary literature readings in the exercises be assigned before each chapter is covered. Reading about the first successful implementation of a medical device provides context for the topics covered in the corresponding chapter. This enables the questions to be discussed at the start of lecture, and facilitates active learning.

# 1

# Diagnosis and Therapy

In this chapter, we discuss foundational material for medical devices. A medical device is specifically defined in the Federal Food Drug & Cosmetic Act, and it is generally an apparatus for diagnosis and/or therapy that does not attain its primary purpose through chemical action. Many medical devices are included in the American Institute for Medical and Biological Engineering (AIMBE) Hall of Fame Innovations (Figure 1.1).

Upon completion of this chapter, each student shall be able to:

1. Understand the difference between devices, medical devices, medical instruments, and medical electrical stimulators.
2. Identify the basic building blocks of medical instruments.
3. Describe four types of sensors.
4. Identify the basic building blocks of electrical stimulators.
5. Define the terms "system" and "systems engineering."

## MEDICAL DEVICE DEFINITIONS

In 1938, as part of the Federal Food Drug and Cosmetic Act (FDC Act), a medical device was defined as:

> An instrument, apparatus, implement, machine, contrivance, implant, in vitro reagent, or other similar or related article, including a component part, or accessory which is—
> (1) recognized in the official National Formulary, or the United States Pharmacopoeia, or any supplement to them,
> (2) intended for use in the diagnosis of disease or other conditions, or in the cure, mitigation, treatment, or prevention of disease, in man or other animals, or
> (3) intended to affect the structure or any function of the body of man or other animals, and which does not achieve any of its primary intended purposes through chemical action within or on the body of man or other animals and which is not dependent upon being metabolized for the achievement of any of its primary intended purposes (Food Drug and Cosmetic Act, 1938).

AIMBE HALL OF FAME

*1950s and earlier*

➤ Artificial kidney
➤ X-ray
➤ Electrocardiogram
➤ Cardiac pacemaker
➤ Cardiopulmonary bypass
➤ Antibiotic production technology
➤ Defibrillator

*1970s*

➤Computer assisted tomography (CT)
➤Artificial hip and knee replacement
➤Balloon catheter
➤Endoscopy
➤Biological plant/food engineering
➤The cochlear implant and stimulators

*1990s until today*

➤ Genomic sequencing and
   micro-arrays
➤ Positron emission
   tomography
➤ Image-guided surgery

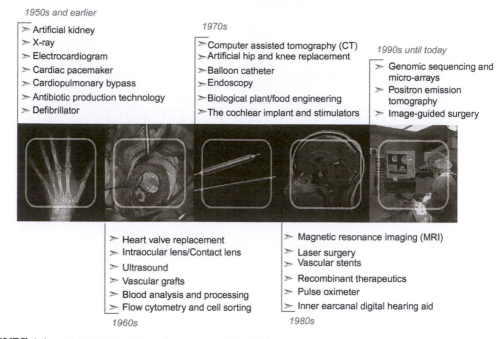

➤ Heart valve replacement
➤ Intraocular lens/Contact lens
➤ Ultrasound
➤ Vascular grafts
➤ Blood analysis and processing
➤ Flow cytometry and cell sorting

*1960s*

➤ Magnetic resonance imaging (MRI)
➤ Laser surgery
➤ Vascular stents
➤ Recombinant therapeutics
➤ Pulse oximeter
➤ Inner earcanal digital hearing aid

*1980s*

**FIGURE 1.1**    AIMBE Hall of Fame Innovations (AIMBE, Washington, D.C.).

In this definition, the National Formulary and United States Pharmacopeia refer to the two parts of the annually issued *United States Pharmacopeia and National Formulary*, which gives the composition, description, method of preparation, and dosage for drugs.

As stated, a medical device is an apparatus used in the diagnosis, mitigation, therapy, or prevention of disease. *Diagnosis* is the identification of the nature and cause of a disease. *Mitigation* is the alleviation of the course of a disease. *Therapy* is the treatment of a disease. *Prevention* is the interposition of an obstacle to a disease. For example, a blood pressure monitor, an ablation catheter that destroys Barrett's esophagus precancerous cells, a cochlear implant, and a condom are used for diagnosis, mitigation, therapy, and prevention, respectively.

A medical device is distinguished from a drug or biologic in that it does not achieve its primary intended purpose through chemical action. A drug, such as the antihistamine loratadine (also known as Claritin), consists of pure chemical substances that are easily analyzed after manufacture. In contrast, a biologic, such as blood or a vaccine, is derived from living organisms and is easily susceptible to contamination. The FDC Act definition states that an in vitro reagent, which is a chemical agent used for an in vitro diagnostic reaction, is a medical device.

## CLINICAL NEED

Recently, it has become fashionable to speak about "translational research." Taking research from the bench to the bedside and back again is the National Institutes of Health's (NIH) mandate in recently launching a consortium of 12 academic health centers for multidisciplinary investigation. NIH believes that a "broad re-engineering effort" for research investigation is needed to "synergize multidisciplinary and interdisciplinary clinical and translational research and researchers to catalyze the application of new knowledge and techniques to clinical practice at the front lines of patient care" (Translational Research, 2008a).

Perhaps because of her vantage point as a longtime medical device researcher, the author believes that medical device researchers have *always* engaged in translational research. We develop medical devices to meet clinical needs, usually through diagnosis and therapy. In the early days–and you will read about these in subsequent book chapters–researchers like cardiologist Paul Zoll raced to invent a fully implantable pacemaker. Patients with complete heart block require pacing, as atrial stimulation is not conducted to the ventricles. In 1960, bioengineer Wilson Greatbatch and heart surgeon William Chardack achieved this milestone with their first successful long-term human implant (Chardack et al., 1960). More recently, as entrepreneurs attempt to start their device companies, the first question venture capitalists ask before funding is, "What is the market size?" While market size translates into potential profits for the venture capitalists, it also translates into clinical need for bioengineers (Baura, 2008b).

## MEDICAL DEVICES VS. MEDICAL INSTRUMENTS

Traditionally, in the bioengineering curriculum, students are introduced to medical devices through a course in bioinstrumentation. Although medical instruments are an important subset of medical devices, one bioinstrumentation course is insufficient preparation for a career in the medical device industry. In this textbook, we highlight both medical devices and medical instruments.

### Instruments Make Measurements

A *medical instrument* is a medical device that makes measurements, often for the diagnosis of disease. The physiologic quantity, property, or condition that the system measures is the *measurand*. The energy or information from a measurand is converted to another form by a *transducer*. If the transducer output is an electrical signal, then the transducer is a sensor (Webster, 1998). Many textbooks have different definitions of sensors and transducers. Throughout this book, we consider a *sensor* to be a device that transforms biologic, chemical, electrical, magnetic, mechanical, optical, or other stimuli input into an electrical signal output.

An electronic instrument requires a power source, such as standard line 60-Hz/120-V, or alkaline or other batteries. In an electronic instrument, patient isolation is positioned inline before or after the sensor to prevent the patient from receiving an electric shock. The isolated sensor output may be optionally amplified. In a digital electronic instrument, the amplified sensor output then undergoes data acquisition. *Data acquisition* consists of analog filtering, analog-to-digital conversion (ADC), digital filtering, and optional down-sampling. The data acquisition output in a digital electronic instrument or amplified sensor output in an analog electronic instrument may receive additional processing from a processor module. Here, a *processor module* is a microcontroller, microprocessor, or digital signal processor, with the required memory and peripherals. A clinical user may provide input, such as height or weight, which aids in the additional processing. The processing output in an electronic instrument or transducer output in a simple instrument is then displayed. Typical displays include the liquid crystal display (LCD) and thermal printer. An optional transceiver may transmit data to another medical device, or it may receive data from another device for processing and display. The communications protocol may involve the internet, telemetry, or other means (Figure 1.2).

Let us illustrate these concepts with a few examples. Two simple medical devices are an intravenous (IV) administration set and a galinstan glass thermometer (Figure 1.3). A standard IV administration set for gravity feed enables drug therapy. It contains a spike, drip chamber, polyvinyl chloride (PVC) tubing, and a roller clamp. When the spike penetrates an IV container hanging from an IV pole, gravity can be used to feed a prescribed drug solution from the container, through the administration set, to a catheter into a patient. A clinician uses the roller clamp to adjust the observed flow rate through the drip chamber. A galinstan thermometer enables the diagnosis of fever. Galinstan is a gallium-indium-tin alloy. We use a galinstan, rather than mercury, thermometer, because hospitals have been replacing mercury thermometers with less harmful alternatives for over a decade.

Of these two medical devices, only the galinstan thermometer is a simple medical instrument. The measurand is patient temperature. The transducer is the galinstan contained in the glass tube, which expands with temperature. Temperature scale markings on the glass tube enable the patient's temperature to be displayed.

Another substitute for the mercury thermometer is the electronic thermometer, which may be analog or digital. The original analog electronic thermometer was invented by IVAC Corporation in 1970 (Georgi, 1972) and required 30 s to display the temperature. After being inserted into a patient's mouth, the temperature probe increased from ambient temperature toward a steady-state temperature at five minutes. Within the probe, a temperature sensor called a thermistor, in contact with other circuitry, provided a pulse train to a counter. Within 30 s, the counter would count to the predicted steady-state temperature. The analog thermometer was powered by alkaline batteries; so high voltage was not involved. Further, a plastic probe cover isolated the patient from the thermometer probe.

More recently, Welch Allyn (formerly Diatek) preheats the probe of its SureTemp digital electronic thermometer in order to decrease the time needed to reach the steady-state temperature after inserting the probe (Figure 1.4). After thermistor values are digitized, they are used to estimate the steady-state oral temperature within 4–6 s (Gregory and Stevenson, 1997). This digital thermometer is powered by three AA 1.5-V alkaline batteries; so, again, high voltage is not involved. A plastic probe cover isolates the patient from the thermometer probe (SureTemp, 2003).

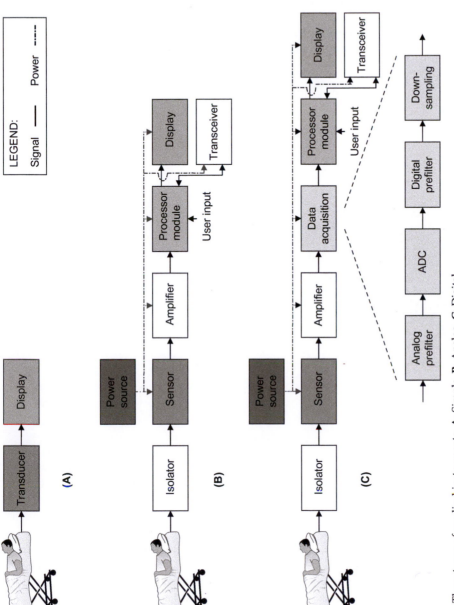

**FIGURE 1.2** Three types of medical instruments. **A:** Simple. **B:** Analog. **C:** Digital.

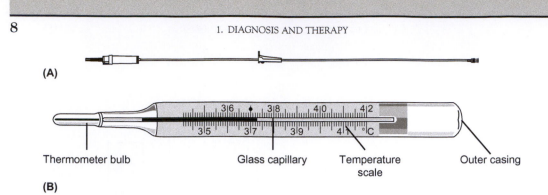

**(A)**

**(B)**

**FIGURE 1.3**   Two simple medical devices. **A**: Carefusion Model 42000 IV administration set (Carefusion, San Diego, California). **B**: Geratherm Classic galinstan thermometer (Geratherm, Geschwenda, Germany).

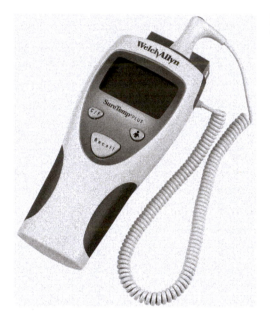

**FIGURE 1.4**   Welch Allyn SureTemp Plus 690 digital electronic thermometer (Welch Allyn, Skaneateles, New York).

## Input Dynamic Range and Frequency Response

Each medical instrument expects its measurand to be within a certain range of amplitude, amplitude variation, and frequency. An *input dynamic range* requirement specifies the mean signal level (when not zero), the range of differential signal level, and the fastest acceptable amplitude rate of change (also known as slew rate). A *frequency response* requirement typically specifies the frequency bandwidth in terms of the cutoff frequency at which the magnitude of harmonics has fallen to a significant fraction of the fundamental frequency magnitude. The specific fraction is a medical device design choice. When the frequency response requirement includes a lower frequency bound, a large direct current (DC) component is being filtered out (Figure 1.5).

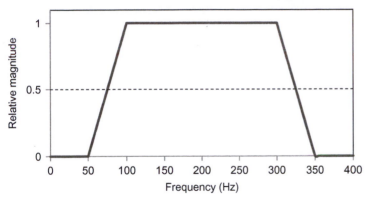

FIGURE 1.5 Sample frequency response specification.

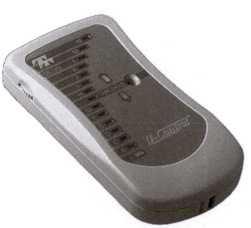

FIGURE 1.6 Thought Technology U-Control electromyograph (Thought Technology U-Control, Plattsburgh, New York).

For example, Thought Technology's U-Control electromyograph enables a urinary-incontinent patient to self-monitor and strengthen pelvic muscle contraction (Figure 1.6). An *electromyogram* refers to the biopotential associated with muscle contraction; an *electromyograph* records electromyograms. The U-Control system can be switched between two specified signal ranges of 0.2–37.5 μV or 0.6–112.5 μV. The slew rate is not specified. The frequency range is 90–300 Hz. With this lower bound of 90 Hz, the mean voltage is 0 V (U-Control, 2008b).

## Accuracy, Bias, and Precision

In a medical instrument, measured values from a transducer or sensor are subject to accuracy constraints. As defined by the National Bureau of Standards, accuracy "usually denotes in some sense the closeness of the measured values to the true value, taking into consideration both precision and bias. Bias, defined as the limiting mean and the true value, is a constant, and does not behave in the same way as the index of precision, the standard deviation" (Ku, 1988). However, in practice, accuracy is usually specified as the percentage difference between the measured and true values, based on a full-scale reading. For example, the accuracy of the

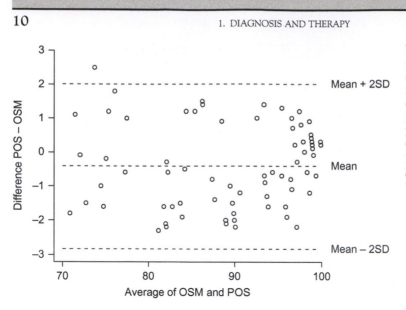

**FIGURE 1.7** Bland-Altman plot of oxygen saturation measured by oxygen saturation monitor (OSM) and pulsed oxygen oximeter (POS). After publication of this graph, a pulsed oxygen oximeter became commonly known as a pulse oximeter [*Reproduced by permission from Bland & Altman (1986)*].

Welch Allyn SureTemp Plus digital thermometer is $\pm 0.2°$F, for the patient temperature range of $80-110°$F (SureTemp Plus, 2003).

When the true values are unavailable for calculations because no recognized standard exists from the National Institute of Standards and Technology, then a reference standard may be substituted. This situation often occurs when a new noninvasive monitoring technology is compared to its invasive reference standard. Bland-Altman analysis is used to calculate the bias and precision. First, the difference between each measured, $m_i$, and reference standard, $s_i$, value is calculated for all $n$ measurement pairs. Bias and precision are calculated as:

$$\text{bias} = \frac{\sum_{i=1}^{n}(m_i - s_i)}{n},\tag{1.1}$$

$$\text{precision} = 2 \times \sqrt{\frac{\sum_{i=1}^{n}(m_i - \text{bias})^2}{n-1}}.\tag{1.2}$$

Bias and precision are then plotted on a graph of the differences vs. mean of each measurement pair (Figure 1.7).

Bland-Altman analysis is preferable to linear correlation analysis for the comparison of a new measurement to its reference standard measurement. The correlation coefficient, $r$, can be artificially increased by increasing the measurement range; it determines whether the relationship between the new and reference measurements is linear according to *any* line, not specifically the line of identity (Bland & Altman, 1986).

## Noise Sources

Up to this point, we have assumed that we have measured only the physiologic signal of interest with our sensor. Often, we are not so fortunate. Even though we have grounded and shielded our circuitry, used low noise amplification, and used the lowest DC power supply potentials, noise may seep into our system. In general, *system noise* refers to any artifact we would like to minimize. The hospital environment, in particular, is an infinite source of signal distortion. In older monitors, 60-Hz interference from power lines may distort the signal of interest. During surgery, electromagnetic interference is generated by the electrosurgical unit used for cautery. In unanesthetized patients, patient motion is a significant source of distortion. Even respiration and blood pressure may obscure the signal of interest (Baura, 2002).

Although not necessarily true, it is often assumed that the system noise is additive, so that the total digitized signal, $\mathbf{y}(k)$, is the sum of the physiologic signal, $\mathbf{u}(k)$, and noise, $\mathbf{n}(k)$:

$$\mathbf{y}(k) = \mathbf{u}(k) + \mathbf{n}(k) \tag{1.3}$$

Eq. (1.3) is thus a sum of vectors.

## SENSORS

Having introduced the basic medical instrumentation system, let us discuss each component in detail. We discuss sensors before discussing isolation, so that we understand what signals must be isolated from the patient. Because so many sensors can be used for medical instruments, we limit our discussion to four typical types: surface electrodes, pressure sensors, thermistors, and photodiodes. For each of these sensors, the output is a resistance, analog voltage, or current.

### Surface Electrodes

Electrodes are used to monitor *biopotential voltages*, which reflect the electrical stimulation that precedes the mechanical action of muscle contraction. Most muscles are stimulated by an action potential current that has been transmitted from the brain through a descending motor nerve. The biopotential reflecting electrical brain activity is called the *electroencephalogram* (*EEG*). The biopotential that corresponds to the action potential at the muscle target is an *electromyogram* (*EMG*). For cardiac muscles only, the action potential originates in the cardiac sinoatrial node. The corresponding biopotential is called an *electrocardiogram* (*ECG*). For our discussion, we assume that each biopotential is recorded *only* at the skin surface. Thus, each biopotential, on the order of microvolts to millivolts, corresponds to the summation of several potentials (Figure 1.8).

A *surface electrode* is a sensor that converts ionic current in the body to electrical current. This occurs through the creation of a double layer of charge at the electrode-electrolyte interface. Most commonly, potassium chloride, KCl, in an electrode gel that is at least 25% water by weight enables the creation of an electrode-electrolyte interface within the surface electrode (3M Material Safety Data Sheet, 2006) (Figure 1.9).

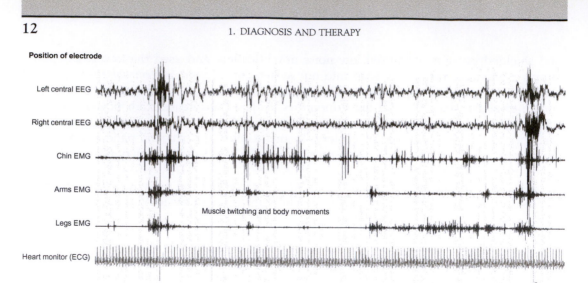

Position of electrode

Left central EEG

Right central EEG

Chin EMG

Arms EMG

Muscle twitching and body movements

Legs EMG

Heart monitor (ECG)

**FIGURE 1.8**  Examples of biopotentials in a patient with a sleep disorder *[Reproduced by permission from Mahowald & Schenck (2005)].*

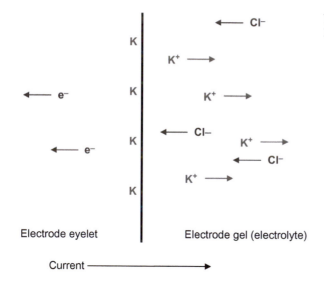

**FIGURE 1.9**  Electrode-electrolyte interface for KCl.

At the interface, the following chemical reactions occur:

$$K \rightleftharpoons K^+ + e^- \tag{1.4}$$

and

$$Cl^- \rightleftharpoons Cl + e^-. \tag{1.5}$$

As illustrated in Figure 1.9, electrons moving in a direction opposite to that of the current in the electrode, $K^+$ cations moving in the same direction as the current, and

Cl⁻ anions moving in a direction opposite to that of the current in the electrolyte all contribute to a net current that crosses the interface and to a DC offset voltage. Manufactured electrodes are expected to meet the Association for the Advancement of Medical Instrumentation (AAMI) DC offset voltage requirement of an electrode pair gel-to-gel offset voltage within 100 mV (AAMI, 2005b). Silver and silver chloride, Ag/AgCl, may substitute for KCl to generate electrical current.

These days, surface electrodes are generally not the metal plate or suction electrodes discussed in older textbooks (Webster, 1998; Carr & Brown, 2001; Webster, 2010). Of three surface electrode applications, ECG electrodes are the most widely used. *ECG electrodes* are typically manufactured from a cloth, vinyl, or foam basepad; a polystyrene label; a nickel-plated brass stud; a carbon eyelet coated with Ag/AgCl (the stud and eyelet mate to form a snap); and wet or solid gel. First, the basepad material is cut into the general electrode shape, with a center circle removed. The general shape may be a circle or rectangle with rounded corners. Next, a polystyrene label with a larger area than the center circular cut is adhered, creating a well in the base. A stud and eyelet are mated through the label. Solid hydrogel or wet gel is placed in the well, in contact with the eyelet. Either gel contains approximately 5% KCl or Ag/AgCl. The adhesive on the basepad and stickiness of the gel act together to stabilize the electrode on the skin (Figure 1.10A).

Alternatively, instead of the typical ECG electrode, an *ECG tab electrode* may be used. For a tab electrode, a base of polyester film is covered with a Ag/AgCl coating. The coating may or may not be continuous (Figure 1.10B). The coating does not cover the portion of the film acting as the tab; it is next covered with a layer of solid hydrogel. The tab electrode is used specifically for resting (i.e., no movement, unlike during stress testing), short-term diagnostic applications. Surface EEG electrodes for monitoring the bispectral index and surface EMG electrodes are constructed similarly to the typical ECG electrode.

For both the typical ECG and ECG tab electrodes, the solid or wet gel performs an important function of minimizing skin impedance. Webster and colleagues estimated skin

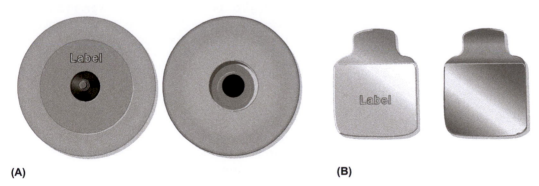

**(A)**                                                                       **(B)**

FIGURE 1.10   Top and bottom of two ECG electrodes: **A**: Typical ECG electrode; **B**: ECG tab electrode.

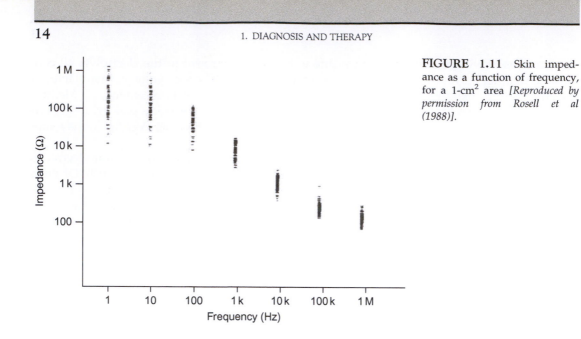

**FIGURE 1.11** Skin impedance as a function of frequency, for a 1-cm² area *[Reproduced by permission from Rosell et al (1988)].*

impedance along the thorax, forehead, or leg of 10 human subjects as varying from $1 \times 10^6$ to 100 Ω, with increasing frequency (Rosell et al., 1988) (Figure 1.11).

When we position two electrodes over the skin, we wish to measure the voltage drop within a defined tissue area beneath the skin. The skin contains three layers, each having its own sublayers. The layers are the epidermis, dermis, and subcutaneous. Within the outermost epidermis are the stratum corneum, stratum germinativum, and stratum granulosum. Within the outermost stratum corneum are layers of dead material on the skin's surface that act as a high impedance to measurement. Ideally, the skin should be prepared for measurement by (1) shaving excessive hair at the electrode site, (2) cleansing the skin with an alcohol pad and letting it dry completely, and (3) rubbing the skin briskly. This preparation is usually detailed in the electrode directions for use. However, in actual clinical use, skin is rarely prepared. Instead, clinicians rely on the electrode gel to penetrate the stratum corneum and decrease the overall measurement impedance.

## Pressure Sensors

As noted, electrical stimulation precedes mechanical action in the body. Often, we wish to measure these mechanical action changes as pressure, $P$, in units of millimeters of mercury (mm Hg). Note that 760 mm Hg = 1 atmosphere (atm) = 101,325 Pascals (Pa). In the lungs, according to the kinetic theory of gases, pressure is a measure of the total kinetic energy of the molecules:

$$P = NRT, \tag{1.6}$$

where $N$ = number of moles per unit volume, $R$ = a specific gas constant, and $T$ = absolute temperature.

For a liquid, or fluid, such as blood at rest, pressure is defined as the force, $F$, exerted perpendicularly on the unit area, $A$, of a boundary surface:

$$P = \frac{dF}{dA}. \tag{1.7}$$

Pressure in a blood vessel is greatly influenced by its radius, $r$. Making the assumption that a blood vessel can be represented by a rigid tube, we can rearrange Hagen-Poiseuille's equation for laminar flow of a Newtonian fluid to obtain pressure at the inlet, $P_i$, and the outlet, $P_o$, of the tube as:

$$P_i - P_o = \frac{8\eta l Q}{\pi r^4}, \tag{1.8}$$

where $\eta$ = fluid viscosity, $l$ = tube length, and $Q$ = flow.

When pumping saline into the veins, we must take into account the hydrostatic pressure gradient, $P_h$, with respect to a patient's right atrium. The hydrostatic pressure gradient adds 1.34 mm Hg per cm of elevation.

A pressure sensor contains three components: a diaphragm or plate of known area, $A$; a detector that responds to the applied force, such as a metal wire strain gauge; and an interface circuit, such as a Wheatstone bridge, that outputs voltage (Figure 1.12).

Strain is the fractional change, $\Delta l$, in length:

$$\varepsilon = \frac{\Delta l}{l}. \tag{1.9}$$

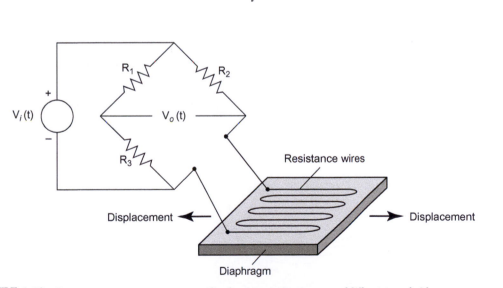

FIGURE 1.12    Pressure sensor components: a diaphragm, strain gauge, and Wheatstone bridge.

As a diaphragm is stretched, the wire of a metal wire strain gauge, in contact with the diaphragm, is lengthened. The lengthening of the wire causes a corresponding resistance change, $\Delta R$:

$$\Delta R \approx \frac{2R\Delta l}{l}, \tag{1.10}$$

assuming $\Delta l \ll l$.

The resistance change is converted to a voltage change by a Wheatstone bridge. A Wheatstone bridge consists of an input voltage, $V_i$; three resistors, and an operational amplifier. Under normal balanced conditions, the output voltage, $Vo$, equals zero and assumes that $R_1 = R_2$, and $R_3 =$ initial strain gauge resistance. When strain occurs, the resistance change changes the output voltage to:

$$V_o(t) = V_i \left( \frac{\Delta R(t) - 1}{2} \right). \tag{1.11}$$

Two common sensors provide an example. In the 1980s and 1990s, IVAC was able to compete successfully against its larger competitors Abbott Laboratories and Baxter Healthcare Corporation because IVAC volumetric pumps could sense catheter pressure in a vein to within ±2 mm Hg (Philip & Philip, 1985). At the time, nurses generally believed that a large change in catheter pressure was predictive of infiltration. Infiltration is a drug infusion complication that results from puncture of a blood vessel wall by a catheter needle and subsequent infusion into tissue (Philip, 1989).

To sense pressure, an administration set contained a pressure diaphragm in contact with the saline solution. The diaphragm interfaced to a pressure sensor (Cunningham, 1983) in the infusion pump (Figure 1.13). The pressure diaphragm consists of a thin, flexible PVC membrane, overlying an upper surface of a disk. The membrane is sealed around

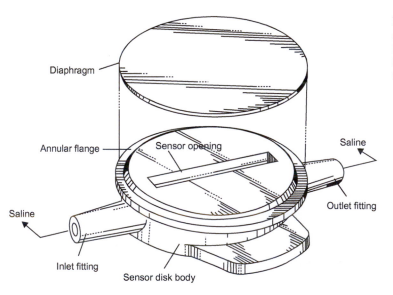

FIGURE 1.13    IVAC pressure sensor with pressure diaphragm. Both ends of the diaphragm disk interface with administration set tubing [*From Cunningham (1983)*].

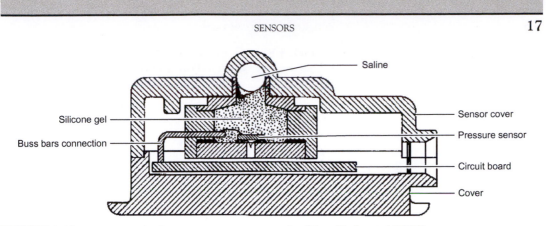

**FIGURE 1.14** Deseret Medical pressure sensor cross section *[From Hanlon et al (1987)]*.

the periphery of the upper surface, but not to the upper surface. Positive pressure in the administration set is transmitted through the opening in the upper surface to the membrane. When the pressure diaphragm is seated in its mating infusion pump (the door is closed against the front panel), the pressure from the membrane is transmitted to an adjacent pressure sensor in the pump. The pressure sensor consists of a stainless steel diaphragm, a strain gauge, and a Wheatstone bridge. As the steel diaphragm is stretched, the strain gauge outputs a proportional resistance change, which is converted to a voltage by the Wheatstone bridge.

Our second example is a disposable intraarterial blood pressure (IAP) sensor. Intraarterial blood pressure, measured in the radial artery, is the gold standard of reference blood pressure measurements. Historically, a catheter inserted into the radial artery was connected through saline-filled tubing to an external, reusable pressure sensor for IAP measurements.

In the late 1980s, through innovations in micromachining, new disposable IAP sensors began to be used in the United States. The Deseret Medical pressure sensor (Hanlon et al., 1987) is shown in Figure 1.14. Deseret Medical was acquired by Becton Dickinson in 1986. In this pressure sensor, fluid that transmits pressure is in contact with a silicone gel, which transmits the pressure to a pressure diaphragm. The pressure diaphragm is connected to a resistive strain gauge, whose output goes to a ceramic circuit board for circuitry impedance matching and temperature compensation. The diaphragm and strain gauge are shock-mounted in the sensor case so that direct loadings on the case are not directly absorbed by the diaphragm. Similarly, the circuit board is fully suspended to eliminate the effects of impact on the board. In this way, a voltage proportional to pressure is transmitted from the circuit board through a standard pressure outlet connector.

## Thermistors

A thermistor is a thermal resistor that is used to measure temperature. Although many types of thermistors exist, we discuss a *negative thermal coefficient (NTC)* bead thermistor (Figure 1.15). NTC refers to the phenomenon of decreasing resistance with increasing

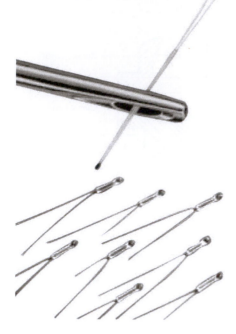

**FIGURE 1.15**  GE Sensing NTC Miniature Series thermistors (GE Sensing, St. Marys, Pennsylvania).

temperature. A bead thermistor consists of a mixed metal oxide ceramic body, applied onto parallel lead wires, encased in glass. It is the best thermistor for measuring body temperature because it is stable (the resistance does not change with age) and possesses a fast response time. For hospital thermometers, which require greater accuracy than consumer products (tolerance from the nominal value within ±0.3°F), manufacturers purchase thermistors that have been individually calibrated over the entire operating temperature range of typically 80–110°F. Individual calibration increases the thermistor cost, but, without it, the tolerances of nominal resistance may reach ±20%.

Based on experimental data, a thermistor's resistance, $R_T(T)$, may be approximated as

$$R_T(T) = exp\left(A_0 + \frac{A_1}{T} + \frac{A_2}{T^2} + \frac{A_3}{T^3}\right), \tag{1.12}$$

where $T$ = temperature and $A_0$, $A_1$, $A_2$, and $A_3$ are specified by the manufacturer. A typical resistance curve is shown in Figure 1.16. Often, this resistance curve is specified by using a thermistor parameter called the *material characteristic temperature*, $\beta_m$, and a single calibrating temperature/resistance pair, $(T_c, R_c)$:

$$R_T(T) \approx R_c \, exp\left[\beta_m\left(\frac{1}{T} - \frac{1}{T_c}\right)\right]. \tag{1.13}$$

However, since simplified models lead to decreased accuracy, these simplifications are not used for hospital thermometers. In either case, a thermistor is connected to interface circuitry so that a voltage proportional to temperature is output.

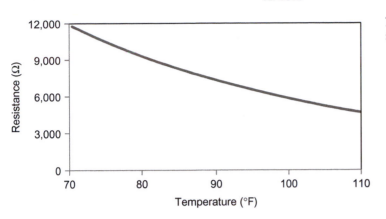

**FIGURE 1.16** Resistance as a function of temperature.

## Photodiodes

The sensors previously discussed have measured energy that originates in the body. In contrast, light detectors, such as photodiodes, are used to measure changes in light waves that have interacted with the body. *Light* is electromagnetic radiation in the spectral range from ultraviolet to infrared. An example of such a light change is the light transmitted after incident light travels through the blood. This change is described by the Beer-Lambert law (discussed in Chapter 11).

Typically, a light emitting diode (LED), rather than a more expensive laser, is used as the light source. LEDs emit more than one wavelength of light; a laser emits a single wavelength. The product of wavelength, $\lambda$, and optical frequency, $v$, is the speed of light in a vacuum, which equals $3 \times 10^8$ m/s. The energy, $E$, of a photon is related to the optical frequency by Planck's constant, $h = 6.626 \times 10^{-34}$ J-s:

$$E = hv. \tag{1.14}$$

After light is transmitted through the blood and/or other media, it is commonly detected by a *photodiode*, which consists of p-type and n-type semiconductor materials. A semiconductor, such as silicon or germanium, possesses four valence electrons, which are arranged in crystalline lattices. Because of this periodic structure, substituted atoms can dramatically change the electrical properties. Adding impurity atoms with five valence electrons, such as antimony or arsenic, produces an n-type semiconductor, which freely contributes electrons. Adding impurity atoms with three valence electrons, such as boron or aluminum, produces a p-type semiconductor, creating deficiencies in electrons, or "holes." When the p-type and n-type materials are placed in direct contact, some free electrons in the n-region diffuse across the p-n junction and combine with holes in the p-region to form negative ions. Thus, these free electrons leave behind positive ions. A space charge builds, creating a depletion region that inhibits any further electron transfer.

For the operation of a photodiode, the p-n junction is reverse biased; that is, the positive side of a battery is connected to the n-side. When a photon strikes the photodiode's surface, part of its energy detaches an electron from the surface. The remaining energy is given as kinetic energy to the electron. Electron-hole pairs on both sides of the junction are created. If the transferred energy is sufficiently high, the electron may become mobile and flow toward the positive side of the battery. Correspondingly, each created hole flows to the negative terminal, such that photocurrent, $i_p$, flows in the circuit. Even without light, a leakage current, $i_l$, flows, mainly due to the thermal generation of charge carriers (Fraden, 2004). The photodiode structure and equivalent circuit are shown in Figure 1.17.

As an alternative to the PN photodiode, an extra-high-resistance intrinsic layer may be placed between the $p$ and $n$ layers to improve the response time. This photodiode is called a PIN photodiode. In actual operation, a photodiode is connected, without a bias voltage, to a circuit such as Figure 1.18. The circuit used with a photodiode contains an operational amplifier and feedback resistor, $R_f$. We describe the operational amplifier in more detail in the Amplification section. The circuit provides a constant bias voltage of 0 V across the photodiode, and converts the photocurrent into an output voltage (Fraden, 2004):

$$V_o(t) = -R_f[i_p(t) + i_l]. \tag{1.15}$$

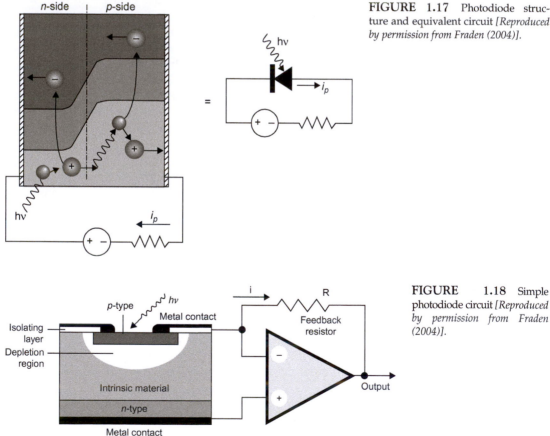

FIGURE 1.17 Photodiode structure and equivalent circuit [Reproduced by permission from Fraden (2004)].

FIGURE 1.18 Simple photodiode circuit [Reproduced by permission from Fraden (2004)].

# PATIENT AND OPERATOR SAFETY

A patient must not be harmed while connected to a medical device. Similarly, a clinician or operator must not be harmed while using a device. As discussed, electronic instruments are powered by voltages, and their sensors output voltage or current. Significant instrument current must not be transmitted to the patient and operator. When 60-Hz current travels through the body for 1 s, human subjects begin to "let go" at 16 mA—the pain threshold. As current through the trunk increases and approaches 100 mA in experimental animals, increasing numbers exhibit ventricular fibrillation (Bruner, 1967). The pain threshold increases as the frequency of current stimulation increases.

Requirements for patient and operator safety are specified in *International Electrotechnical Commission (IEC) 60601-1:2005 (MOD) Medical Electrical Equipment—Part 1: General Requirements for Basic Safety and Performance.* This 321-page international specification was modified slightly by AAMI in six places to reflect agreement with other national specifications (AAMI, 2005a). In this section, we briefly discuss two of numerous IEC 60601-1 safety requirements: leakage current and defibrillation protection. In terms of medical instrument components, Figure 1.2 illustrates that isolation and patient safety exist specifically *between* the sensor and amplifier. In actual practice, many parts of the medical instrument, in numerous locations, have been designed or modified to maximize safety. However, design requirements such as creepage distance and air clearance are beyond the scope of this discussion.

## Leakage Current

Leakage current is measured to ensure that direct contact between a patient or operator and a medical device does not result in electrical shock. Under normal conditions, total patient leakage current between the same types of an applied part, connected together, may not exceed 50 μA DC and 500 μA AC. An applied part, such as an ECG electrode cable connector, normally comes in contact with a patient when a medical device performs its function (AAMI, 2005a).

A design decision that significantly contributes to low leakage current is the addition of an isolator between the main part of a medical device and its secondary circuit. We have already discussed two "isolators" of sorts. In electronic analog and digital electronic thermometers, the plastic probe cover minimizes contamination and prevents the patient's tongue from contacting the electronic probe. In the disposable IAP sensor, fluid in contact with the patient is separated from an electrical strain gauge by a silicone gel.

A typical patient isolator is an *isolation transformer*, a simple form of which is constructed from two neighboring coils surrounding an iron core. By Faraday's law, the voltage in a coil is equal to the time rate of change of the total magnetic flux, which is equal to the inductance times the first time derivative of the current:

$$V_L(t) = L\frac{di(t)}{dt}.$$

(1.16)

Primary
$N_1$    $N_2$    Secondary

$V_{L1}(t)$    $V_{L2}(t)$

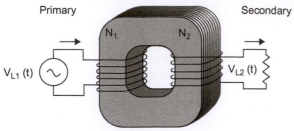

**FIGURE 1.19**  Isolation transformer *[Adapted from Nave (2010)].*

Because the coils are colocated, the changing current in the first coil, or primary winding, produces an induced voltage in the secondary winding. Assuming that the transformer is linear and that the coils are mutually coupled, the ratio of the two voltages is equal to the ratio of the number of turns, $N_i$, in each coil:

$$\frac{V_{L1}(t)}{V_{L2}(t)} = \frac{N_1}{N_2}. \tag{1.17}$$

By using an isolation transformer, the main part of the circuit, associated with the primary winding, is isolated from its secondary circuit, associated with the secondary winding (Figure 1.19).

### Defibrillation Protection

There is one exception to patient isolation from electrical shock. If a patient requires emergency defibrillation, any attached equipment cannot interfere with the delivery of a defibrillation shock. *Defibrillation* is an attempt to stop uncoordinated myocardial contractions, called *fibrillation*, that do not pump blood to the circulation. During external defibrillation with a biphasic truncated exponential waveform, a peak voltage and current of 1750 V and 35 A, respectively (assuming 50-Ω thoracic resistance) may be delivered to the patient.

According to *IEC 60601-1*, during external defibrillation, hazardous peak voltages exceeding 1 V cannot appear on another medical device enclosure. The enclosure includes connectors in patient leads and cables that are connected to the medical device and that are not delivering a defibrillation shock (AAMI, 2005a). In other words, the medical device under test cannot prevent a shock delivered from an external defibrillator from reaching the patient. This requirement is met by carefully considering the numerical value, wattage rating, and quality of the drop resistors chosen for protection circuitry.

## AMPLIFICATION

As with the photodiode, the output of a sensor may need amplification. In most cases, an operational amplifier (op amp) is sufficient to obtain a voltage output in the right range for data acquisition, such as $0 - +5$ V. However, in the case of surface electrodes, a special biopotential amplifier is required.

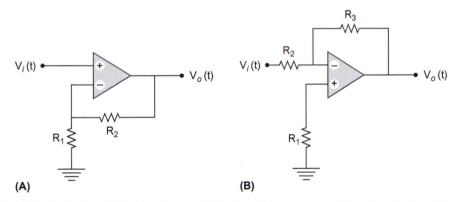

**(A)**  **(B)**

FIGURE 1.20 Typical amplification circuits. **A**: Noninverting op amp configuration. **B**: Inverting amplifier configuration.

## Operational Amplifiers

An *op amp* is an integrated circuit containing discrete components such as transistors, resistors, capacitors, and inductors. It is designed with the following properties:

- One noninverting (+) input, one inverting (−) input 180° out of phase with the noninverting input
- A high input resistance (100 M$\Omega$)
- A low output resistance (<1 $\Omega$)
- The ability to drive capacitive loads
- A low input offset voltage, ($\mu$V to mV range)
- A low input bias current (pA)
- A very high open-loop gain ($10^4$–$10^6$ times)
- A high common-mode rejection ratio (CMMR)
- Low intrinsic noise
- A broad operating frequency range
- A low sensitivity to power supply voltage variations
- A high environmental stability of its own characteristics (Fraden, 2004).

The common-mode signals are the parts of the two input signals that are in phase and of equal magnitude. The *common-mode rejection ratio (CMRR)* is the ability of an amplifier to suppress the common-mode signals.

Two typical amplification circuits, among many possible circuits, are shown in Figure 1.20. We leave it up to the reader to calculate the gain, $V_o(t)/V_i(t)$, of each circuit in Exercise 1.3.

## Biopotential Amplifiers

Unlike other sensors, surface electrodes are not interfaced to simple amplifier circuits. Environmental noise present in the surface electrode signals is so large that common-mode

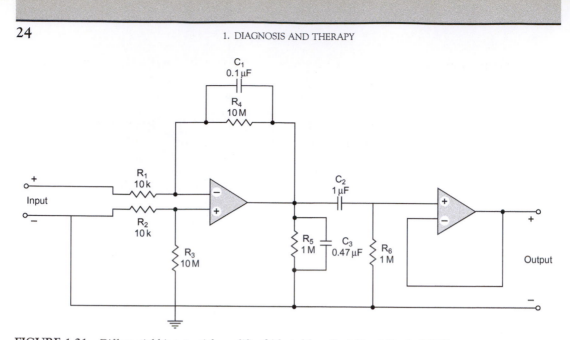

**FIGURE 1.21**    Differential biopotential amplifier *[Adapted from Prutchi and Norris (2005)].*

potentials swamp the weak biopotentials. Instead of amplifying the electrode signals directly, the difference between the voltages at each electrode is amplified in order to cancel out the environmental noise. A differential amplifier is designed to specifications of gain, frequency response, common-mode rejection, and noise minimization.

A differential amplifier may be constructed from two op amps (Figure 1.21). Here, the first stage of the circuit includes an op amp acting as a differential amplifier, the first capacitor acting as a lowpass filter, and the first and third capacitors acting as oscillation attenuators. The gain of this stage is 1000. The second capacitor and sixth resistor act as a highpass filter between the first and second stages. During the second stage, the second op amp buffers the signal with unity gain.

For this circuit, amplifier specifications can be determined by observation and measurement. The differential gain, $G_{dif}$, is the ratio between the resistor pairs:

$$G_{dif} = \frac{R_4}{R_1} = \frac{R_3}{R_2}. \tag{1.18}$$

Let us input a 5-$V_{p-p}$ sine wave of 60 Hz into the differential inputs of the biopotential amplifier, and consider this sine to be the common-mode input voltage, $V_{i,CM}$. By measuring the common-mode output voltage, $V_{o,CM}$, at the output of the circuit, we can determine the common-mode gain, $G_{CM}$:

$$G_{CM} = \frac{V_{o,CM}}{V_{i,CM}}. \tag{1.19}$$

The common-mode rejection ratio is then the ratio of differential to common-mode gains (Prutchi and Norris, 2005):

$$CMRR = \frac{G_{dif}}{G_{CM}}. \tag{1.20}$$

This differential amplifier is a simple example of a biopotential amplifier, which is useful for discussion purposes only. Two of several specific ECG biopotential amplifier circuit designs are the driven-right-leg (Winter & Webster, 1983) and driven-ground (Kim et al., 2005) circuits, which have been optimized to minimize noise. Analog Devices manufactures an AD8295 integrated circuit of two op amps and an instrumentation amplifier, which can be used as a biopotential amplifier. An instrumentation amplifier is a differential amplifier with finite gain ($G < 100$); we use an instrumentation amplifier in the lab experiment in Chapter 21. Integrated circuit AD8295 is the basis of Exercise 1.4.

## DATA ACQUISITION

When an amplified signal is digitized, it experiences data acquisition. As shown in Figure 1.22, *data acquisition* contains four components: an analog frequency-selective filter, an analog-to-digital converter (ADC), a digital frequency-selective filter, and optional downsampling. An analog lowpass filter restricts, or bandlimits, the signal frequencies that are sent to the ADC. During analog-to-digital conversion, an analog signal is sampled at a specified sampling frequency, resulting in digital samples that are equally spaced in time. Typically, digital samples contain 12 bits of information. If a small-amplitude signal has a large DC component, 16 bits may be required to obtain sufficient resolution of the AC component. A subsequent digital lowpass filter prevents aliasing (discussed in the next subsection). Because all digitized samples may not be required in further processing, optional downsampling may occur. Downsampling by a factor of $D$ can be accomplished either by taking every $D^{th}$ sample or by taking the mean of every $D$ samples.

Generally, the filter characteristics of the analog and digital prefilters are matched to minimize signal distortion. A common medical device design choice is to use second-order analog and digital Butterworth filters. The frequency response of an analog second-order Butterworth filter, with a lowpass cutoff frequency of 100 Hz, is shown in Figure 1.23.

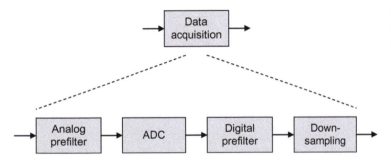

FIGURE 1.22 Data acquisition components: analog frequency-selective filter, analog-to-digital converter (ADC), digital frequency-selective filter, and downsampling.

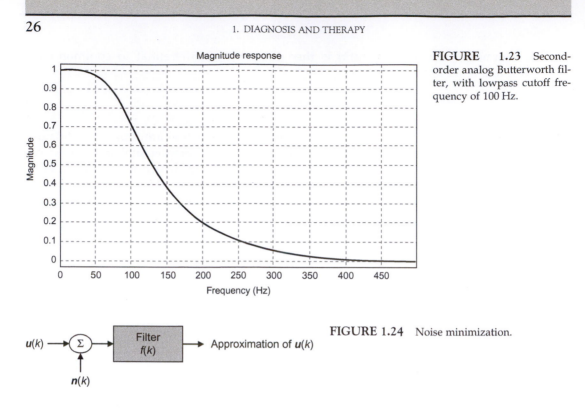

FIGURE 1.23 Second-order analog Butterworth filter, with lowpass cutoff frequency of 100 Hz.

FIGURE 1.24   Noise minimization.

After digital filtering, if noise is still present in a specific frequency band that does not overlap with signal content, $u(k)$, the digitized signal may be further filtered (Figure 1.24). The design details of these filters are not necessary in this discussion. When the signal and noise do not overlap in frequency, filter design is a very small portion of an industrial medical instrument development project. For more information on frequency-selective filters, please see (Oppenheim & Schafer, 1989). When the signal and noise do overlap in frequency, filter design must be accomplished and verified by a research or advanced development engineer, before the filter can be incorporated into an industrial medical instrument development project. For example, the inline pressure from each heartbeat may distort catheter pressure measurements being used to estimate catheter impedance, as a means for detecting drug infusion infiltration. For the pseudorandom binary sequence solution to this problem and other examples of noise minimization techniques, see Baura (2002; 2008a).

## Nyquist Sampling Theorem

As already stated, an analog frequency-selective filter is the first step in data acquisition because this filter bandlimits the range of frequencies in the analog signal that are passed to the ADC. According to the Nyquist sampling theorem, "if a function $f(t)$ contains no frequencies higher than $W$ cycles per second, it is completely determined by giving its ordinates at a series of points spaced $\frac{1}{2} W$ seconds apart" (Nyquist, 2002; Shannon, 1998). In simpler language, this theorem states that, if we bandlimit a signal at $W$ Hz and then digitize the signal at a sampling rate greater than $2 \cdot W$ Hz, the signal is uniquely determined by its samples (Oppenheim & Schafer, 1989).

Let $\Omega$ represent continuous frequency in radians/sec and $\Omega_c$ represent the continuous cutoff frequency. The Nyquist frequency, $\Omega_N$, is then the bandlimited frequency, and the Nyquist rate, $2\,\Omega_N$, is double the Nyquist frequency. If the sampling frequency, $\Omega_{sample}$, is insufficient because it is at or below the Nyquist frequency, then aliasing occurs. *Aliasing* is the resulting distortion phenomenon.

For example, look at the frequency content of a 10-Hz square wave (Figure 1.25A). From studying Fourier series, we know that this 10-Hz square wave, $u_{sq}(t)$, can be represented as:

$$u_{sq}(t) = \sum_{\substack{n=1 \\ n\ odd}}^{\infty} \frac{1}{n} sin\,(n10t). \tag{1.21}$$

The squared magnitudes of the power spectral density squared magnitudes of this square wave are shown in Figure 1.25B. The squared magnitudes do not exactly match Eq. (1.21) because the power spectrum calculation requires prewindowing, which adds some distortion. The illustrated power spectrum uses a Kaiser window.

If we digitize the square wave with a sampling frequency of 1000 Hz and lowpass filter the digitized signal at 20 Hz, we isolate the fundamental frequency, 10 Hz. The result, which is a sine wave, is shown in Figure 1.25C. If we increase the passband to 100 Hz and add four nonzero-amplitude harmonics to the fundamental frequency, the summation of sine waves begins to resemble a square wave (Figure 1.25D).

To see how aliasing affects an ECG, let us look at the one period of a real ECG shown in Figure 1.26. The fundamental frequency of this ECG can be calculated from its heart rate: 60 beats per minute = 1.0 Hz. From the power spectral density (Figure 1.26B), the original signal appears to die off around 15 Hz. By sampling the original signal at 100 Hz and lowpass filtering it at 6 Hz, we isolate the fundamental frequency and five harmonics. The result is a digitized signal retaining gross approximations of the original ECG features (Figure 1.26C). In contrast, if we lowpass filter the signal at 40 Hz, the ECG is basically recreated (Figure 1.26D). Therefore, to preserve more harmonic content, it is best to sample at a rate higher than the *minimum* Nyquist rate. In practice, engineers use sampling rates that are much higher (five to 10 times higher) than the Nyquist rate to ensure that the original signal is preserved. Because a higher sampling rate is used, the extra samples that are obtained may be downsampled after analog-to-digital conversion.

## MEDICAL ELECTRICAL STIMULATORS

Another subset of medical devices consists of *medical electrical stimulators*, which administer current to electrically excitable cells as therapy to manage chronic disease. Though early stimulators such as the pacemaker and defibrillator applied current externally, the majority of recent stimulators are implanted.

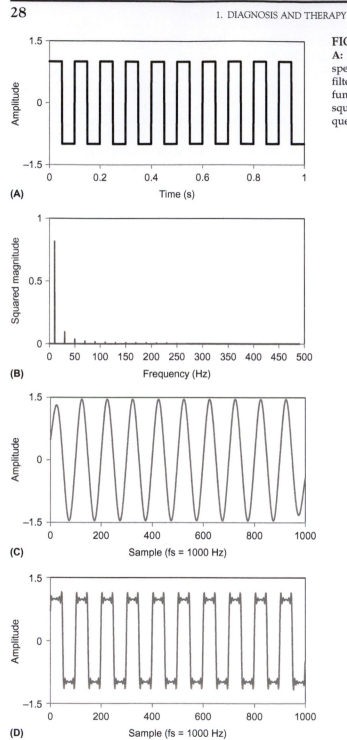

**FIGURE 1.25** A square wave simulation. **A:** Original 10-Hz square wave. **B:** Power spectral density of square wave, using Kaiser filter. **C:** Recreated square wave, based on fundamental frequency only. **D:** Recreated square wave, based on fundamental frequency and four nonzero harmonics.

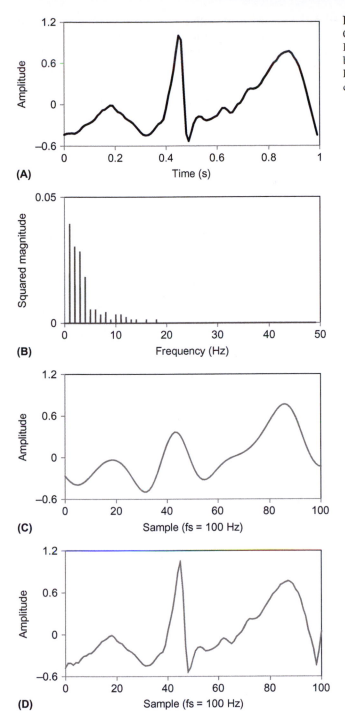

(A)

(B)

(C)

(D)

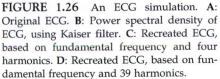

FIGURE 1.26 An ECG simulation. A: Original ECG. B: Power spectral density of ECG, using Kaiser filter. C: Recreated ECG, based on fundamental frequency and four harmonics. D: Recreated ECG, based on fundamental frequency and 39 harmonics.

The components of an external stimulator are illustrated in Figure 1.27. An external electrical stimulator requires a power source such as standard line 60-Hz/120-V or lithium batteries. Based on user input, the processing module activates a waveform circuit that discharges an appropriate waveform through electrodes to the patient. If the waveform is discharged over the skin, then surface electrodes are used, with electrode cables connecting the stimulator to the electrodes. Here, electrical current is converted to ionic current. If the electrodes are discharged internally, such as with an ablation catheter, then the cables connecting the stimulator to the electrodes are lead wires that are specially designed for the application. Lead wires are often referred to as leads. The processing module also transmits data to the display and optionally to a transceiver. A typical display is an LCD; a transceiver, such as a data downloader, may communicate using an ethernet or other protocol. Voice prompts may assist the user in performing therapy. The first biphasic truncated exponential automated external defibrillator, the Philips HeartStream Forerunner, is a good example of an external electrical stimulator (Figure 1.28).

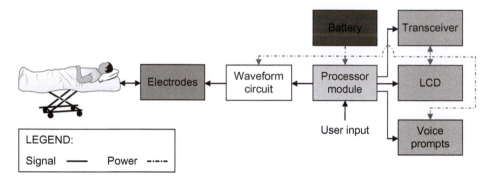

**FIGURE 1.27**    External electrical stimulator system diagram.

**FIGURE 1.28**  Philips HeartStream Forerunner automated external defibrillator (Philips Medical, Andover, Massachusetts).

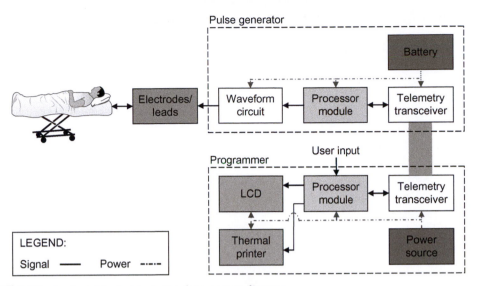

**FIGURE 1.29**   Implantable electrical stimulator system diagram.

An implantable electrical stimulator is necessarily more complicated because it is used for chronic disease management. The three parts of an implantable electrical stimulator are the leads, pulse generator, and programmer (Figure 1.29). Implanted leads, which include the electrodes, enable current to be delivered to the tissue to which they are affixed. In an implanted pulse generator, the battery powers the processor module, which enables a waveform circuit to output current through its leads and a telemetry transceiver to transmit and receive data. In the external programmer, a telemetry transceiver (which may include a wand) transmits and receives data, which are then processed and displayed. Typical displays include an LCD and thermal printer. The external programmer is powered by standard line 60-Hz/120-V or lithium batteries. A user, either the clinician or the patient, may input parameters to the programmer for processing. For example, a clinician may program a new atrioventricular delay for a pacemaker.

Medical electrical stimulators, whether external or implantable, may also be medical instruments. For example, as discussed in Chapter 3, early implantable pacemakers always paced the right ventricle at a steady rate. In contrast, modern demand pacemakers sense whether an R wave is present. If the R wave, which represents the onset of ventricular depolarization, is not present, then the pacemaker paces the right ventricle. Therefore, a demand pacemaker paces *on demand*.

## Stimulator Batteries

Although the waveform circuitry and leads vary between implantable stimulators, the battery and telemetry transceiver are common. A stimulator battery provides a voltage that results from the movement of cations to a negative electrode and of anions to a positive electrode within an electrolyte within a cell.

**FIGURE 1.30** Greatbatch Medical primary cell battery (Greatbatch Medical, Clarence, New York).

The theoretical voltage of a cell may be calculated as the difference between the half-cell potentials of the ions. The theoretical capacity may be calculated by the weight of the ions. An *equivalent weight*, which is equal to the atomic weight divided by the valence, produces 1 faraday (F), or 26.8 amp-hours (A-h), during reaction. Using this relationship, lithium (Li) has an atomic weight of 6.941 and theoretical capacity of $26.8/6.941 = 3861$ A-h/g.

Special requirements for medical batteries include high-energy-density, end-of-life indication, performance predictability, and reliability. These requirements are met by using lithium batteries encased in titanium (Figure 1.30). Lithium is extremely dense and has a low half-cell potential of $-3.05$ V. When companied with any cathode material, lithium has a high potential.

For a low-current application such as a pacemaker, a $Li/I_2$ battery is used, which has the cell reaction:

$$Li + \frac{1}{2}I_2 \rightarrow Li\ I. \tag{1.22}$$

Here, the cathode contains a mixture of iodine and poly-2-vinylpyridine, which have been reacted together at high temperature to form a conductive charge transfer complex. Because the mixture has been poured into the cell while molten, a layer of LiI forms at the anode, producing a separator layer *in situ*. During discharge, the LiI layer grows in thickness, resulting in a rise in cell impedance, which can be measured. Because the electrolyte is solid, $Li/I_2$ cells supply low currents in the µA range that have been proven to be reliable for over a 10-year period (Figure 1.31).

For a high-current application, such as an implantable cardioverter defibrillator, a lithium/silver vanadium oxide (Li/SVO) battery is used, which has the cell reaction:

$$7\,Li + Ag_2V_4O_{11} \rightarrow Li_7Ag_2V_4O_{11}. \tag{1.23}$$

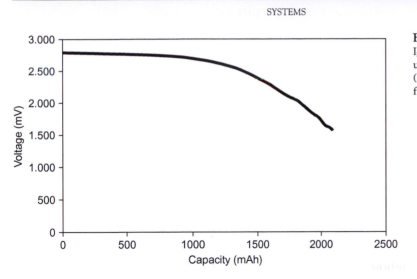

**FIGURE 1.31** Typical Li/$I_2$ battery discharge plot, under a 100-k$\Omega$ load at 37°C (Reprinted by permission from Takeuchi et al., 2004).

A lithium salt is dissolved in an electrolyte, such as the nonaqueous propylene carbonate, with a polypropylene separator. Due to the several different oxidations states of vanadium, as well as silver reduction, a stepped discharge curve occurs over time. This predictable discharge curve indicates the battery's state of the charge. The formation of metallic silver greatly increases the conductivity of the cathode material, thereby increasing the high-current-carrying capability of the system. Good long-life characteristics over at least 9 years have been obtained (Takeuchi et al., 2004).

## SYSTEMS

The discussion of medical instruments and electrical stimulators necessarily occurs in the context of systems and systems engineering. According to the International Council on Systems Engineering (INCOSE), a *system* is "a construct or collection of different elements that together produce results not obtainable by the elements alone. The elements, or parts, can include people, hardware, software, facilities, policies, and documents; that is, all things required to produce system-level results" (INCOSE, 2010). In the next 19 chapters, we demonstrate how each medical device contains subsystems of components, such as the isolators and sensors discussed in this chapter. The component connections are illustrated through a type of system diagram called the *functional flow block diagram*, such as those in Figures 1.2, 1.27, and 1.29.

INCOSE defines *systems engineering* as "an interdisciplinary approach and means to enable the realization of successful systems. It focuses on defining customer needs and required functionality early in the development cycle, documenting requirements, and then proceeding with design synthesis and system validation while considering the complete problem. Systems engineering considers both the business and technical needs of all customers with the goal of providing a quality product that meets the user needs" (INCOSE, 2010).

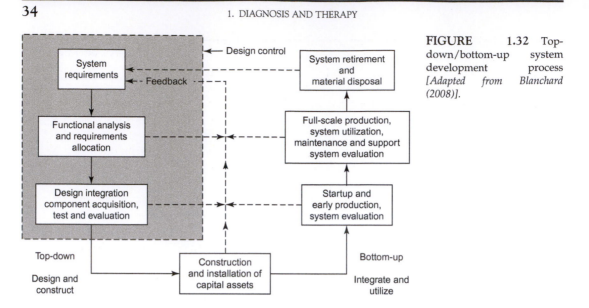

FIGURE 1.32 Top-down/bottom-up system development process [*Adapted from Blanchard (2008)*].

Systems engineers focus on the top-down/bottom-up system development process (Blanchard, 2008), which is shown in Figure 1.32. A product begins as systems requirements, progresses from early prototypes to a mass-produced product, with identified issues fed back through the process. The shaded area in this figure is consistent with the design control process, which is part of the Food and Drug Administration's Good Manufacturing Practices.

In the following 19 chapters, we address aspects of systems engineering, specifically clinical (customer) need and engineering standards, which are the basis of product requirements. We revisit design control as engineering design in Chapter 19, after we have discussed various medical device systems in Chapters 2—18. A future second textbook by the author will describe the design process in detail, as well as other aspects of medical device market release and postmarketing surveillance (Baura, 2013).

## SUMMARY

A medical device is an apparatus that is used in the diagnosis, mitigation, therapy, or prevention of disease and that does not attain its primary purpose through chemical action. In this chapter, we discussed two types of electrical medical devices: medical instruments and electrical stimulators.

A medical instrument is a medical device that makes measurements, often for the diagnosis of disease. The physiologic quantity, property, or condition that the system measures is the measurand. The energy or information from a measurand is converted to another form by a transducer. If the transducer output is an electrical signal, then the transducer is a sensor. Commonly used sensors are the surface electrode, pressure sensor, thermistor, and photodiode. An electronic instrument requires a power source such as a standard 60-Hz/120-V line or alkaline or other batteries. In an electronic instrument, patient isolation is positioned inline before or after the sensor to prevent the patient from receiving an electric shock. The isolated sensor output may be optionally amplified.

In a digital electronic instrument, the amplified sensor output undergoes data acquisition. Data acquisition consists of analog filtering, analog-to-digital conversion, digital filtering, and optional downsampling. The data acquisition output in a digital electronic instrument or the amplified sensor output in an analog electronic instrument may receive additional processing from a processor module. Here, a processor module is taken to be a microcontroller, microprocessor, or digital signal processor with the required memory and peripherals. A clinical user may provide input, such as height or weight, which aids in the additional processing. The processing output from an electronic instrument or simple instrument transducer is then displayed. Typical displays are the liquid crystal display and thermal printer. An optional transceiver may transmit data to another medical device, or it may receive data from another device for processing and display. The communications protocol may involve the Internet, telemetry, or other means.

An external electrical stimulator requires a power source such as standard line 60-Hz/120-V or lithium batteries. Based on user input, the processing module activates a waveform circuit that discharges an appropriate waveform through electrodes to the patient. If the waveform is discharged over the skin, then surface electrodes are used, with electrode cables connecting the stimulator to the electrodes. Here, electrical current is converted to ionic current. If the electrodes are discharged internally, such as with an ablation catheter, then the cables connecting the stimulator to the electrodes are lead wires (referred to as leads) that are specially designed for the application. The processing module also transmits data to the display and optionally to a transceiver. A typical display is an LCD; a transceiver may communicate using an ethernet or other protocol.

An implantable electrical stimulator is necessarily more complicated because it is used for chronic disease management. The three parts of an implantable electrical stimulator are the leads, pulse generator, and programmer. Implanted leads, which include the electrodes, enable current to be delivered to the tissue to which they are affixed. In an implanted pulse generator, the battery powers the processor module, which enables a waveform circuit to output current through its leads and a telemetry transceiver to transmit and receive data. In the external programmer, a telemetry transceiver transmits and receives data, which are then processed and displayed. Typical displays include an LCD and thermal printer. The external programmer is powered by standard 60-Hz/120-V line or lithium batteries. A user, either the clinician or patient, may input parameters to the programmer for processing. For example, a clinician may program a new atrioventricular delay for a pacemaker.

Medical electrical stimulators are often also electronic instruments. For example, early implantable pacemakers always paced the right ventricle at a steady rate. In contrast, modern demand pacemakers sense whether an R wave is present, which represents the onset of ventricular depolarization. If the R wave is not present, then the pacemaker paces the right ventricle. Therefore, a demand pacemaker paces on demand.

## Acknowledgment

The author wishes to thank Dr. Paul Benkeser for his review of this chapter. Dr. Benkeser is Associate Chair for Undergraduate Studies and Professor, Biomedical Engineering, Georgia Institute of Technology.

## Exercises

**1.1** Referring to Figure 1.1, which AIMBE Hall of Fame Innovations are *not* medical devices? Give a reason for each.

**1.2** Finer control of the drug flow rate through an administration set can be obtained using an infusion controller, which was invented in 1970 (Georgi, 1974). The administration set tubing interfaces with the controller directly, for the control and adjustment of fluid drops. An external drop sensor is also clamped to the administration set drip chamber for direct measurement of the flow rate through the counting of fluid drops. The optical drop sensor has a reference beam that is transmitted and received through the drip chamber. An amplified pulse train, with each pulse representing a drop, is output from the drop sensor. This sensor pulse train is compared to a pulse train representing the desired flow rate, which has been input by the user. This comparison is used to adjust the actual flow rate through the tubing until it matches the desired flow rate.

   **(a)** Is an infusion controller a medical instrument? Explain your answer.

   **(b)** Over time, infusion controllers were replaced by volumetric infusion pumps to overcome the controller's pressure limitation. What is the specific pressure limitation of an infusion controller?

   A typical infusion pump uses a special administration set, in which the section of tubing in contact with the infusion pump mechanism is made of silicone, rather than traditional PVC. The more pliable silicone enables the pump mechanism to output much more accurate flow rates, so that a drop sensor is no longer needed. In the Carefusion Alaris SE infusion pump, the catheter pressure is measured as an input for estimation of catheter impedance (Voss et al., 1997). Catheter impedance changes have been linked to the onset of the patient drug infusion complications (Scott et al., 1996).

   **(c)** Is the Alaris SE pump a medical instrument? Explain your answer.

**1.3** Calculate the gain, $V_o(t)/V_i(t)$, of each circuit in Figure 1.20, using the listed op amp properties.

**1.4** In the driven-right-leg circuit, the common-mode voltage on the body is sensed by two averaging resistors, inverted, amplified, and fed back to the right leg (Winter & Webster, 1983) (Figure 1.33). Download the data sheet for Analog Devices AD8295 IC: http://www.analog.com/static/imported-files/data_sheets/AD8295.pdf. Draw the circuit for a driven-right-leg circuit that utilizes the AD8295 IC.

**1.5** Plot pulse oximetry waveform 1_1 that has been provided by your instructor. Waveform characteristics are documented in the associated header1_1.txt ascii file. Take the first beat of the waveform, and calculate its frequency response. What do you estimate is the bandwidth of this signal? Filter the waveform with a second-order Butterworth filter with a corner frequency that is equivalent to your chosen bandwidth. Filter the waveform with a second-order Butterworth filter with a corner frequency that is half of your chosen bandwidth. Plot the original waveform and your two filtered waveforms in one figure. Comment on how the results were affected by your choice of bandwidth.

Read Bland & Altman (1986), which is the origin of Bland-Altman analysis. Then work Exercises 1.6–1.9.

**1.6** In linear correlation analysis, what does the squared correlation coefficient, $r^2$, statistically represent? When a paper includes a figure with $r$, rather than $r^2$, in the figure legend, what

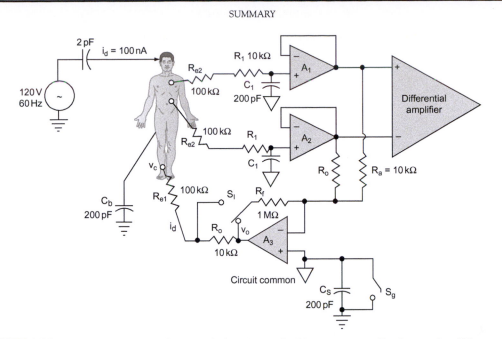

**FIGURE 1.33** Winter & Webster's driven-right leg circuit. $A_3$ drives $v_c$ to a small voltage value. When switch $S_g$ is open, this circuit is an isolated amplifier with a stray capacitance to earth ground of $C_s$. $C_b$ is the capacitance between the body and earth ground [*Adapted from Winter & Webster (1983).*].

do you suspect? What do the authors suggest are the reasons for the popularity of linear correlation analysis?

**1.7** How would you define clinical agreement? What does Figure 2 tell you about peak expiratory flow rate data that Figure 1 does not?

**1.8** How is repeatability related to precision? How do Bland and Altman derive their calculation of precision for repeated measurements?

**1.9** When is it appropriate to use a logarithmic transformation? What is the result of taking this approach?

## References

AAMI (2005a). *ANSI/AAMI ES60601-2-12 Medical Electrical Equipment—Part 1: General Requirements for Basic Safety and Performance* (3rd ed.). Arlington, VA: AAMI.

AAMI (2005b). *ANSI/AAMI EC12:2000/(R)2005 Disposable ECG Electrodes.* Arlington, VA: AAMI.

Baura, G. D. (2002). *System theory and practical applications of biomedical signals.* Hoboken, NJ: Wiley-IEEE Press.

Baura, G. D. (2008a). *A Biosystems Approach to Industrial Patient Monitoring and Diagnostic Devices.* San Rafael, CA: Morgan Claypool.

Baura, G. D. (2008b). Translational research. In G. Madhavan, & L. Kun (Eds.), *Career Development in Bioengineering and Biotechnology.* New York: Springer.

Baura, G. D. (2013). *U.S. Medical Device Regulation: An Introduction to Biomedical Product Development, Market Release, and Postmarket Surveillance.* Manuscript in preparation.

Blanchard, B. S. (2008). *System Engineering Management.* Hoboken, NJ: John Wiley.

Bland, J. M., & Altman, D. G. (1986). Statistical methods for assessing agreement between two methods of clinical measurement. *Lancet, 1*, 307–310.

Bruner, J. M. (1967). Hazards of electrical apparatus. *Anesthesiology, 28*, 396–425.

Carr, J. J., & Brown, J. M. (2001). *Introduction to Biomedical Equipment Technology* (4th ed.). Upper Saddle River, NJ: Prentice Hall.

Chardack, W. M., Gage, A. A., & Greatbatch, W. (1960). A transistorized, self-contained, implantable pacemaker for the long-term correction of complete heart block. *Surgery, 48*, 643–654.

Cunningham, J. N. (1983). Pressure diaphragm. US 4,398,542.

Food Drug and Cosmetic Act (1938). U.S. Code. Title 21, Chapter 9, Subchapter II, §321(h).

Fraden, J. (2004). *Handbook of Modern Sensors: Physics, Designs, and Applications* (3rd ed.). New York: AIP Press.

Georgi, H. (1972) Electronic thermometer. US 3,072,076.

Georgi, H. (1974). Method and apparatus for fluid flow control. US 3,800,794.

Gregory, T. K., & Stevenson, J. W. (1997). Medical thermometer. US 5,632,555.

Hanlon, S. P., Kerby, W. L., Purdy, E. R., & Strom, J. (1987), Pressure Transducer. US 4,679,567.

INCOSE (2010). *A Consensus of INCOSE Fellows.* San Diego, CA: INCOSE.

Kim, K. K., Lim, Y. K., & Park, K. S. (2005). Common mode noise cancellation for electrically non-contact ECG measurement system on a chair. *Proceedings of the Twenty-Seventh Annual International Conference of the Ieee Engineering in Medicine and Biology Society, Vols. 1–7*(27), 5881–5883.

Ku, H. H. (1988). *Statistical Concepts in Metrology—with a Postscript on Statistical Graphics. NBS Special Publication 747.* Gaithersburg, MD: National Bureau of Standards.

Mahowald, M. W., & Schenck, C. H. (2005). Insights from studying human sleep disorders. *Nature, 437*, 1279–1285.

Nave, C. R. (2010). *HyperPhysics.* Atlanta: Georgia State University.

Nyquist, H. (2002). Certain topics in telegraph transmission theory. *Proceedings of the IEEE, 90*, 280–305 Reprinted from (1928). *Transactions of the AIEE* (February), 617–44.

Oppenheim, A. V., & Schafer, R. W. (1989). *Discrete-Time Signal Processing.* Englewood Cliffs, NJ: Prentice Hall.

Philip, J. H. (1989). Model for the physics and physiology of fluid administration. *Journal of Clinical Monitoring, 5*, 123–134.

Philip, J. H., & Philip, B. K. (1985). Hydrostatic and central venous pressure measurement by the IVAC 560 infusion pump. *Medical instrumentation, 19*, 232–235.

Prutchi, D., & Norris, M. (2005). *Design and Development of Medical Electronic Instrumentation: A Practical Perspective of the Design, Construction, and Test of Medical Devices.* Hoboken, NJ: Wiley-Interscience.

Rosell, J., Colominas, J., Riu, P., Pallas-Areny, R., & Webster, J. G. (1988). Skin impedance from 1 Hz to 1 MHz. *IEEE Transactions on Bio-Medical Engineering, 35*, 649–651.

Scott, D. A., Fox, J. A., Philip, B. K., Lind, L. J., Cnaan, A., & Palleiko, M. A., et al. (1996). Detection of intravenous fluid extravasation using resistance measurements. *J. Clin. Monit., 12*, 325–330.

Shannon, C. E. (1998). Communication theory of secrecy systems. 1945. *MD Computing, 15*, 57–64.

SureTemp Plus (2003). *User Manual.* Skaneateles, NY: Welch Allyn.

Takeuchi, E. S., Leising, R. A., Spillman, D. M., Rubino, R., Gan, H., & Takeuchi, K. J., et al. (2004). Lithium batteries for medical applications. In G. Nazri, & G. Pistoia (Eds.), *Lithium Batteries: Science and Technology.* Norwell, MA: Kluwer.

3M Material Safety Data Sheet: 3M Red Dot Monitoring Electrodes 2660, 2670, 2360, 2364, 9651. (2006). St. Paul, MN: 3M Medical Division.

Translational Research: Overview. (2008a). Bethesda, MD: National Institutes of Health, http://www.nihroadmap.nih.gov/clinicalresearch/overview-translational.asp.

U-Control (2008b). *User's Manual.* Montreal: Thought Technology.

Voss, G. I., Butterfield, R. D., Baura, G. D., & Barnes, C. W. (1997). Fluid flow impedance monitoring system. US 5,609,576.

Webster, J. G. (Ed.), (1998). *Medical Instrumentation: Application and Design* (3rd ed.). Hoboken, NJ: Wiley.

Webster, J. G. (Ed.), (2010). *Medical Instrumentation: Application and Design* (4th ed.). Hoboken, NJ: Wiley.

Winter, B. B., & Webster, J. G. (1983). Driven-right-leg circuit design. *IEEE Transactions on Bio-Medical Engineering, 30*, 62–66.

# 2

# Electrocardiographs

In this chapter, we discuss the relevant physiology, history, and key features of electrocardiographs. An electrocardiograph is an instrument that enables cardiac biopotentials to be measured, displayed, and analyzed (Figure 2.1).

Upon completion of this chapter, each student shall be able to:

1. Understand the mechanisms underlying the electrocardiogram.
2. Identify two classes of arrhythmias.
3. Describe five key features of electrocardiographs.

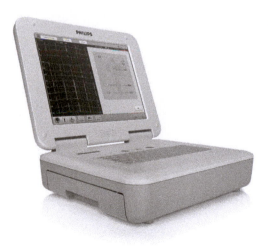

**FIGURE 2.1**  Philips PageWriter TC70 Electrocardiograph (Philips Medical Systems, Andover, Massachusetts).

## CARDIAC ELECTRICAL CONDUCTION

Each beat of the cardiac cycle is a series of complex, synchronized physiologic events that enables oxygen to be delivered throughout the body. In approximately 1 minute, 5–6 L of blood are pumped from the heart to tissues and back again. A heartbeat begins with electrical stimulation and ends with mechanical valve closure.

### Cardiac Anatomy

With each heartbeat, blood moves through the systemic and pulmonic circulations (Figure 2.2). The *systemic circulation* delivers oxygenated blood from the left atrium to the left ventricle, then through the aorta, arteries, arterioles, and capillaries, and finally to the tissues. From the tissues, deoxygenated blood is delivered through capillaries, venules, veins, and vena cava to the right atrium. Similarly, the *pulmonic circulation* delivers deoxygenated blood from the right atrium to the right ventricle, through the pulmonary artery to the lungs, and back through the pulmonary vein to the left atrium. Blood vessels that are *arteries* move blood away from the heart. Blood vessels that are *veins* move blood toward the heart.

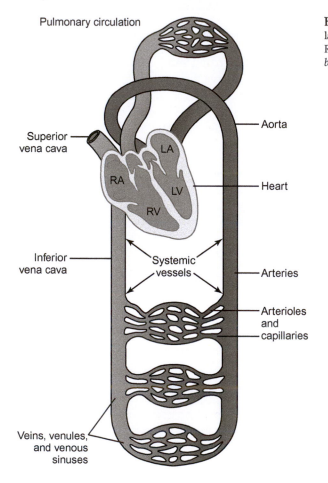

**FIGURE 2.2** The systemic and pulmonic circulations. LA = left atrium, LV = left ventricle, RA = right atrium, RV = right ventricle *[Reproduced by permission from (Guyton & Hall (2006)].*

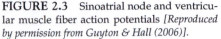

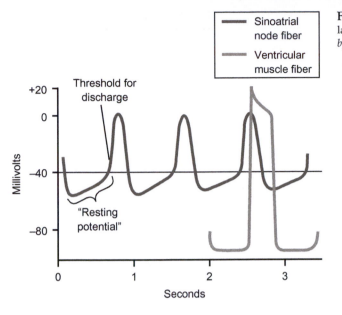

FIGURE 2.3 Sinoatrial node and ventricular muscle fiber action potentials [*Reproduced by permission from Guyton & Hall (2006)*].

## The Spread of Action Potentials

Blood circulation begins when cardiac fibers of the dominant pacemaker and conducting system generate the action potentials that result in cardiac contraction. Sinoatrial (SA) node fibers possess the ability to depolarize spontaneously until a threshold potential of about −40 mV is reached, which generates a new action potential (Figure 2.3). These pacemaker action potentials spread to working myocardial fibers, resulting in working myocardium action potentials (Figure 2.3). In the ventricle, a working myocardium action potential begins with a rapid reversal of the myocardial cell membrane potential, from a resting potential of about −90 mV to the initial peak of about +20 mV. This rapid phase of depolarization lasts less than 1 ms and is followed by a prolonged plateau and then repolarization to the resting potential. Lasting 200−400 ms, this action potential is generated through a combination of membrane potential changes, changes in ionic conductivity, and ion currents.

The dominant pacemaker fibers of the SA node are located in the wall of the right atrium at the opening of the superior vena cava (Figure 2.4). At rest, the SA node paces the heart at about 70 beats per minute (bpm). From the SA node, the excitation is conducted over the working myocardia of both atria, initiating atrial contraction. The excitation is then briefly delayed at the atrioventricular (AV) node, which allows the atrial blood to enter the ventricles. From the AV node, the excitation quickly moves at a velocity of about 2 m/s through the remainder of the system, from the AV bundle, through the left and right bundle branches of the Purkinje fibers, to their ends. Because the ends of the Purkinje fibers penetrate the muscle mass, an action potential quickly spreads to the entire ventricular muscle mass. Only about 30 ms elapse between conduction from the AV bundle to the ventricular muscle mass (Guyton & Hall, 2006).

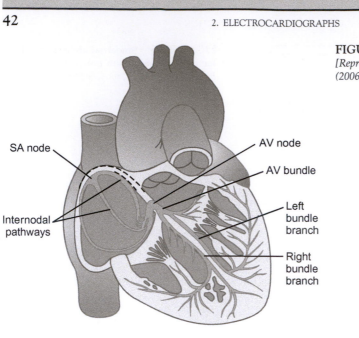

**FIGURE 2.4** Cardiac conduction system [*Reproduced by permission from Guyton & Hall (2006)*].

## STANDARD LEADS

The spread of excitation can be observed by measuring cardiac biopotentials. Cardiac biopotentials are measured as voltages on the skin surface by means of electrodes that have been placed in standard positions. Because a voltage is the potential difference between two points, these cardiac biopotentials are measured as the difference between a positive electrode and a reference electrode or reference electrode network. The measurements are called *electrocardiograms* (ECGs).

As shown in Figure 2.5, the standard 10-electrode positions are the left arm (LA), right arm (RA), left leg (LL), right leg (RL), and six chest positions (V1–V6). The electrodes may not be actually positioned on an arm or leg. Rather, because the body is assumed to be purely resistive at ECG frequencies, arm and leg electrodes may be positioned along the edges of the chest (thorax). The specific locations for the chest leads are as follows:

- V1: Fourth intercostal (between the ribs) space, right of sternum (breastbone)
- V2: Fourth intercostal space, left of sternum.
- V4: Fifth intercostal space, in the same vertical line as the clavicle (collarbone).
- V3: Midway between V2 and V4.
- V6: Fifth intercostal space, in same vertical line as armpit fold.
- V5: Between V4 and V6.

The 10 electrode positions are used for voltage measurements. Each standard electrode configuration for voltage measurement is referred to as a lead. The first three leads—Leads I, II, and III—were defined by Willem Einthoven, based on LA, LL, and RA

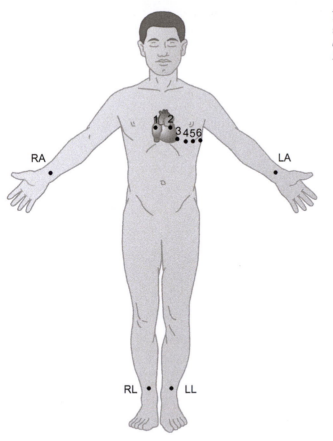

**FIGURE 2.5** Ten electrode locations for a standard 12-lead electrocardiogram [*Modified and reproduced by permission from Guyton & Hall (2006)*].

(Figure 2.6A). Each of these limb leads is based on potential differences between two single electrodes. Notably, Lead III is the difference between Lead II and Lead I:

$$\text{Lead III} = \text{Lead II} - \text{Lead I}. \tag{2.1}$$

Rather than using a single electrode as the reference, several electrodes may be connected through similarly valued resistors to a common node, with a fourth right-leg electrode connected to ground. The voltage at this common node is referred to as *Wilson's central terminal*, after Frank Wilson. Emanuel Goldberger defined augmented, or amplified, limb leads, based on this common node reference voltage. His three augmented limb leads are augmented voltage left (AVL), augmented voltage right (AVR), and augmented voltage foot (AVF) (Figure 2.6B). The six standard chest leads use a chest electrode as the positive electrode and Wilson's central terminal as the reference. Each chest lead is oriented through the AV node, to the patient's back. In total, 12 standard leads are measured from 10 electrodes. These 12 leads provide the clinician with distinct frontal plane (limb and augmented limb leads) and horizontal plane (chest leads) views of cardiac stimulation.

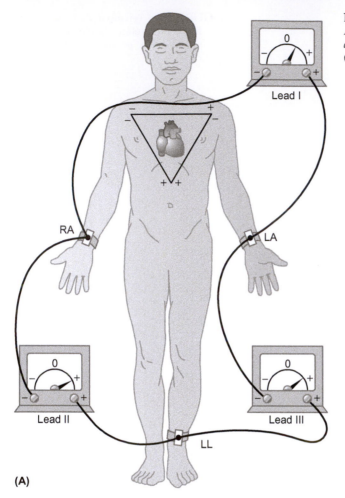

**FIGURE 2.6 A:** Limb leads. **B:** Augmented limb leads *[Modified and reproduced by permission from Guyton & Hall (2006)].*

(A)

Clinically, a 12-lead ECG is used to diagnose *cardiac arrhythmias* (atypical heartbeats). An atypical heartbeat is a deviation from the "normal" sinus rhythm, which is illustrated in Figure 2.7. Sinus rhythm is generated by the SA node and maintains a rate of 60–100 bpm. During the P wave, the atria are depolarized, leading to the simultaneous contraction of the atria. During the QRS complex, the ventricles are depolarized, leading to the beginning of ventricular contraction. An R-wave orientation of a positive or negative deflection is determined by the chosen lead orientation. During the ST segment, which occurs between the S and T waves, the initial phase of ventricular repolarization occurs. Finally, during the T wave, the rapid phase of ventricular repolarization occurs. The peak-to-peak amplitude is on the order of a millivolt.

On standard ECG tracings, which are recorded at a paper speed of 25 mm/s, rate is easily determined from the thin and thick squares in the *x*-axis. Each thin square, or

FIGURE 2.6 (*Continued*)

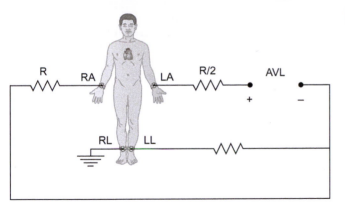

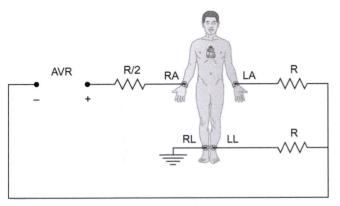

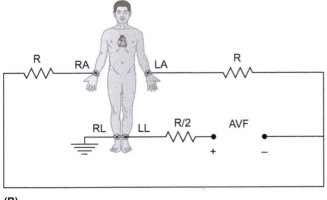

(B)

minor division, is 1 mm; each thick square, or major division, is 5 mm. For a rate of 60 bpm, you should be able to observe five major divisions (or 25 minor divisions) between adjacent R waves. Similarly, an R-to-R distance of one major division represents 200 ms and 300 bpm, two major divisions represent 150 bpm, and three major divisions represent 100 bpm.

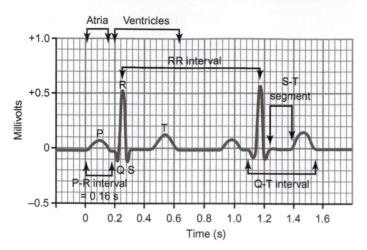

**FIGURE 2.7** Sinus rhythm measured by Lead II *[Reproduced by permission from Guyton & Hall (2006)].*

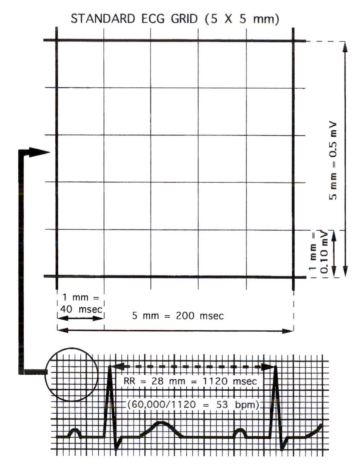

**FIGURE 2.8** Heart rate determination *[Reproduced by permission from Atlee (1996)].*

Further, four major divisions represent 75 bpm, and six major divisions represent 50 bpm. In Figure 2.8, an R-to-R distance of 28 minor divisions represents 1,120 ms and 53 bpm. Note that each minor division in the *y*-axis represents 1 mm or 0.10 mV.

A heart rate lower than 60 bpm is referred to as *sinus bradycardia*. A heart rate higher than 100 bpm is referred to as *sinus tachycardia*.

## TWO ARRHYTHMIA CLASSES

Electrocardiograms enable us to observe arrhythmias. Arrhythmias can be categorized into five classes:

1. Irregular rhythms.
2. Escape.
3. Premature beats.
4. Heart blocks.
5. Tachyarrhythmias.

In this section, we discuss the first two classes of arrhythmias. Premature beats and heart blocks are discussed in Chapter 3; tachyarrhythmias are discussed in Chapter 4.

### Irregular Rhythms

Fundamentally, the heart "wants" to beat, with the SA node its dominant pacemaker. During emergency situations, a potential pacemaker may replace a failing SA node. These potential pacemakers in cardiac tissue are known as *ectopic foci*. If normal SA node pacing fails, they begin to pace at their inherent rate, according to a physiologic hierarchy. The fastest available pacemaker paces and suppresses slower pacemaking activities, in what is known as *overdrive suppression*. Should the SA node fail, an atrial focus begins to pace at an inherent rate of 60–80 bpm. If atrial foci fail to pace, an AV junctional focus begins to pace at an inherent rate of 40–60 bpm. The AV junction is the portion of the AV node possessing ectopic foci. If AV junctional foci fail to pace, a ventricular focus paces at an inherent rate of 20–40 bpm. Ventricular foci are composed of Purkinje cells within the Purkinje fibers.

Figure 2.9 illustrates several irregular rhythms caused by multiple, active ectopic foci. Wandering pacemaker occurs when pacemaker activity "wanders" from the SA node to nearby atrial ectopic foci. This results in rate variation within the normal range and a shape variation in the P wave. In Figure 2.9A, the heart rate varies from 51 to 111 bpm, along with varying P-wave morphology. Should this rate increase into the tachycardia range, it becomes *multifocal atrial tachycardia (MAT)*. Now the rate is greater than 100 bpm, and the P-wave shape varies greatly, as three or more atrial foci pace. MAT is a common rhythm in chronic obstructive pulmonary disease patients, whose airways are inflamed. In Figure 2.9B, MAT, at 104 bpm, has at least five different P-wave morphologies in Lead II. When multiple atrial ectopic foci continuously fire, complete depolarization of the atria may be prevented, with only occasional depolarization through the AV node to stimulate the ventricles. *Atrial fibrillation (Afib)* is present—chaotic atrial spikes and irregular ventricular rhythm. In Figure 2.9C, Afib results in a ventricular rate of 100 bpm (Dubin, 2000). Afib may also be caused by reentry, discussed in Chapter 4.

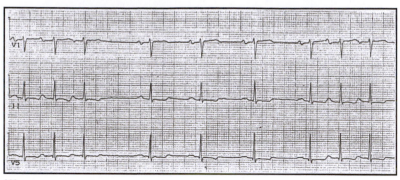

**(A)**

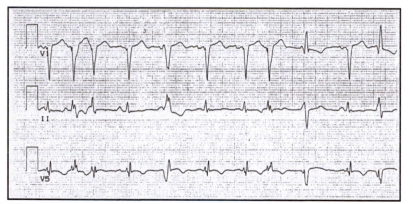

**(B)**

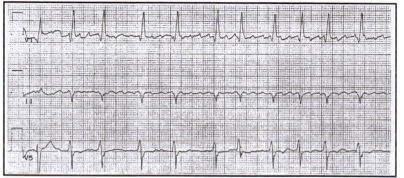

**(C)**

FIGURE 2.9 Irregular rhythm arrhythmias. **A:** Wandering pacemaker. **B:** Multifocal atrial tachycardia. **C:** Atrial fibrillation *[Reproduced by permission from Atlee (1996)].*

## Escape

A pause in SA node pacing activity may lead to an *escape arrhythmia*, which refers to the ability of an ectopic focus to "escape" overdrive suppression. If SA node pacing has completely stopped, an escape rhythm may occur. If SA node pacing pauses briefly, an escape beat may occur.

Escape rhythms are not unwelcome. During sinus arrest, a very sick SA node completely stops pacing. In response to sinus arrest or transient sinus block, an ectopic focus is no longer suppressed and initiates pacing activity. If this occurs at the atrial level, an atrial escape rhythm or beat occurs (Figure 2.10A). The resulting rate is 60–80 bpm. Because this rhythm or beat originates in an ectopic focus, the P-wave shape changes. If an atrial ectopic focus does not pace, an ectopic focus in the AV junction may escape overdrive suppression to produce a junctional escape rhythm or beat. The resulting rate is 40–60 bpm. Usually, a junctional escape rhythm or beat conducts mainly to the ventricles, producing a series of lone QRS complexes, without P waves (Figure 2.10B). However, if the rhythm or beat also causes the atria to depolarize from below, retrograde (inverted)

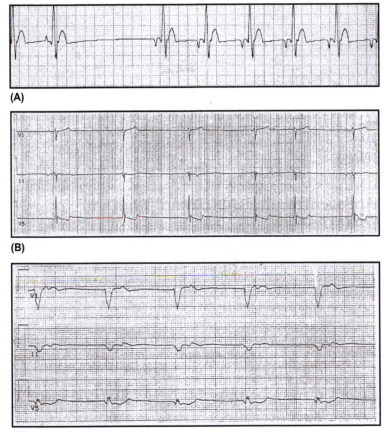

(A)

(B)

(C)

**FIGURE 2.10** Escape arrhythmias. **A:** Sinus rhythm, followed by atrial escape beats, printed with Rivertek RSIM-100 Simulator (Rivertek, St. Paul, Minnesota). **B:** Sinus beat, followed by a junctional escape beat and then a junctional escape beat with retrograde P wave superimposed on the T wave; **C:** Ventricular escape rhythm [**B** and **C** reproduced by permission from Atlee (1996)].

P waves are observed in the ECG. These retrograde P waves may occur before or after each QRS complex or be buried within it (Figure 2.10B). Finally, without atrial or junctional ectopic focus pacing, an ectopic focus in the ventricles may escape overdrive suppression to produce a ventricular escape rhythm or beat (Figure 2.10C). This rhythm is also known as an "ideoventricular rhythm." The resulting rate is 20–40 bpm. Enormous ventricular complexes, without P waves, are observed in the ECG (Dubin, 2000).

## CLINICAL NEED

As illustrated by our discussion of arrhythmias, electrocardiograms permit noninvasive diagnosis of cardiac conduction abnormalities. Whenever possible, clinicians opt for noninvasive rather than surgical diagnoses.

## HISTORIC DEVICES

As physiologists began to understand the importance of cardiac biopotentials, they attempted to measure them. Because cardiac potential amplitudes are on the order of millivolts, the history of electrocardiography is inextricably linked to the recognition and accurate measurement of these small voltages.

### Early Devices

In 1856, Rudolph von Koelliker and Heinrich Muller confirmed, by placing the nerve of a Galvani nerve-muscle preparation on a beating heart, that an electrical current accompanies each heartbeat. They observed the synchronous contraction of the muscle and the heart. Later, in 1872, Gabriel Lippmann invented the capillary manometer, which enabled cardiac biopotentials to be recorded. The capillary manometer consisted of a thin glass tube with a column of mercury beneath sulphuric acid. The mercury level changed proportionally with the potential difference between the mercury and sulfuric acid.

British physiologist Augustus Waller used the capillary manometer to record and publish the first human electrocardiogram. Waller recorded his cardiograms across the limbs, making electrical contact by dipping his subjects' hands and feet into saline. Waller often demonstrated his technique to other physiologists by using his bulldog Jimmy as a subject. Jimmy is reported to have enjoyed being a subject and patiently stood with paws in glass jars of saline (Figure 2.11).

### Enabling Technology: String Galvanometer

Dutch physiologist Willem Einthoven witnessed one of Wallers' demonstrations. Einthoven began investigating this new electrocardiogram curve, developing the necessary mathematical corrections for capillary electrometer properties to obtain an accurate electrocardiogram in absolute units. Using an improved manometer in 1895, he observed deflections, which he named P, Q, R, S, and T. Later, to avoid the need for these corrections, he began developing an instrument with a faster frequency response for measuring electrocardiograms.

By 1901, Einthoven had invented a new "string galvanometer" (Figure 2.12). The string galvanometer weighed 600 lbs and required five people to operate. Its mechanism consisted of a thin thread stretched in a magnetic field. When current passed through the string, the current strength displaced the string proportionally. Galvanometer sensitivity

FIGURE 2.11   Waller's bulldog Jimmy.

FIGURE 2.12   Einthoven's string galvanometer.

was increased by using a very thin wire, using a strong magnetic field of 20,000 gauss, and placing the wire in a vacuum. An optical illumination system allowed the deflections to be photographed.

With his instrument, Einthoven was able to conduct thorough studies of the electrocardiogram with his students. Einthoven received the Nobel Prize in Physiology or Medicine in 1924 "for his discovery of the mechanism of the electrocardiogram" (Einthoven, 1965; Fisch, 2000).

## SYSTEM DESCRIPTION AND DIAGRAM

Electrocardiographs no longer weigh 600 lbs. Instrumentation sensitivity is obtained through the use of a biopotential amplifier (discussed in Chapter 1). The modern electrocardiograph is an excellent example of a digital instrument and can be described by the general diagram in Figure 1.2.

In Figure 2.13, we have modified the general diagram for cardiograph-specific components. Here, a patient interacts with ECG electrodes, which output analog cardiac biopotentials. These small voltages are isolated with a transformer, amplified using a biopotential amplifier, and then they undergo data acquisition. During data acquisition, the analog voltages are analog filtered, digitized with an analog-to-digital converter (ADC), digitally filtered, and downsampled. A typical sampling frequency is 120 Hz. The resulting downsampled, digital voltages are analyzed by a processor module for the detection of arrhythmias and the identification of parameters like heart rate. The same module formats the user interface, enabling lead ECGs and parameters to be displayed on a liquid crystal display user screen and thermal printer.

## KEY FEATURES FROM ENGINEERING STANDARDS

Electrocardiographs cannot be marketed in the United States without meeting specific requirements. The Food and Drug Administration (FDA) Center for Devices and Radiologic Health (CDRH) recommends consensus standards for certain medical devices. In the case of the electrocardiograph, as of 2010, the consensus standard was developed by the Association for the Advancement of Medical Instrumentation (AAMI) and approved by the American National Standards Institute (ANSI). The specific standard is *ANSI/AAMI EC11:2007 Diagnostic Electrocardiographic Devices* (AAMI, 2007).

### Input Dynamic Range

Because ECGs are taken from specific anatomic locations, the general range of peak-to-peak (p-p) amplitudes can be determined from large ECG databases. At a minimum, an electrocardiograph must be able to respond to and display a ±5-mV differential signal, varying at a rate within 320 mV/sec, from a direct current offset voltage in the range of −300−+300 mV. Although larger p-p amplitudes have been observed, particularly in pediatric patients, AAMI and the American Heart Association are in agreement that a 10-mV p-p range is sufficient (AAMI, 2007).

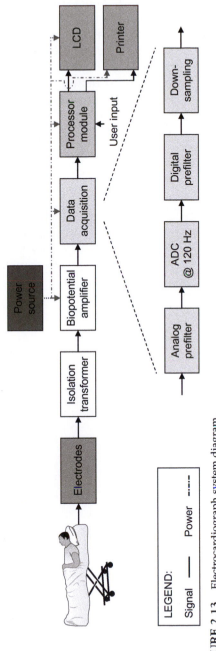

FIGURE 2.13  Electrocardiograph system diagram.

## Frequency Response

In Chapter 1, we discussed frequency response in terms of harmonics. For ECGs that do not contain pacemaker spikes or for ECGs obtained from a cardiograph with special analog circuitry for pace artifact detection, a general rule of thumb is to preserve 30 harmonics. Therefore, for a heartbeat with a typical rate of 60–100 bpm, or 1 to about 2 Hz, the Nyquist frequency and rate are about 60 Hz and 120 Hz, respectively. A sinus rhythm heartbeat and its power spectrum are shown in Figure 2.14.

Instead of using harmonics, AAMI chose to define frequency response in terms of the output amplitudes for various input sinusoid amplitudes and frequencies. For this measurement procedure, the output gain setting is assumed to be 10 mm/mV; all lead settings are to be tested.

The appropriate patient cable connectors are first given a sine wave, with the amplitude adjusted to obtain a 1.0-mV p-p test input.

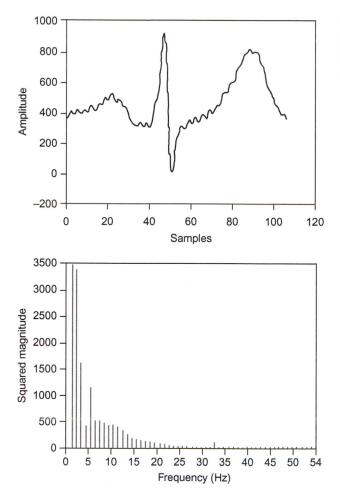

**FIGURE 2.14** Sinus rhythm heartbeat and power spectrum. The noisy ECG was acquired with a sampling frequency of 200 Hz. The power spectrum was calculated after the mean ECG was subtracted.

- The sine wave frequency is then varied from 0.67 Hz to 40 Hz. It must be verified that the output amplitude remains within ±10% of 1.0-mV p-p.
- Next, a sine wave is given, with the amplitude adjusted to obtain a 0.5-mV p-p test input. The sine wave frequency is then varied from 40 Hz to 100 Hz. It must be verified that the output amplitude remains within −30% to +10% of 0.5-mV p-p.
- Similarly, the sine wave is varied from 150 Hz to 500 Hz. It must be verified that the output amplitude remains within −100% to +10% of 0.5 mV p-p.
- Finally, a sine wave is given with the amplitude adjusted to obtain a 0.25-mV p-p test input. The sine wave frequency is then varied from 100 to 150 Hz. It must be verified that the output amplitude remains within − 30% to +10% of 0.25 mV p-p (AAMI, 2007).

## System Noise

Because very-low-amplitude ECG features such as P waves can have diagnostic implications, system noise needs to be minimized. AAMI requires that noise due to patient cables, all internal circuits, and output displays not exceed 30 μV p-p over a 10-s period. This voltage measurement should be made with a manufacturer's recommended cable, with all inputs connected through a special circuit. The special circuit consists of a 51-kΩ resistor in parallel with a 47-nF capacitor, in series with each patient-electrode connection (AAMI, 2007).

## Arrhythmia Detection

Although presumably arrhythmia detection performance is also specified, AAMI chose to specify the testing procedures only for arrhythmias. The FDA recommends that *ANSI/ AAMI EC57:2003 Testing and Reporting Performance Results of Cardiac Rhythm and ST-Segment Measurement Algorithms* (AAMI, 2003) be followed before market release of a electrocardiograph. In this standard, specific bench testing and reporting procedures are detailed, using waveforms from databases such as the American Heart Association (AHA) and Massachusetts Institute of Technology—Beth Israel Hospital (MIT-BIH) databases. For example, if a cardiograph being tested detects atrial fibrillation at the same time intervals as is referenced in a database waveform, this interval is considered a *true positive*. If the cardiograph continues to detect Afib after the database Afib reference notation stops, this time interval is considered *false negative*. Using this AAMI standard, the performance between various cardiographs can be directly compared (AAMI, 2003).

According to a member of the AAMI Committee responsible for the 1998 version of this standard, representatives from smaller companies were willing to agree to minimum performance standards, whereas representatives from larger companies forced only testing and reporting procedures to be standardized.

## Leads-Off Detection

The requirements for input dynamic range and frequency response apply to diagnostic cardiographs. According to the AAMI definition, a diagnostic electrocardiograph is used to obtain a set of conventional ECG signatures that accurately represents detailed wave-forms and beat-to-beat variability. Typically, a patient is measured at the bedside or in a physician's office within, say, 1 hr.

Other electrocardiograph-like instruments are also used in patient care, such as those defined by AAMI to be "cardiac monitors." Cardiac monitors are used to obtain a heart rate indication, to display heart rate and/or ECG waveforms, and to provide alarms when cardiac standstill, bradycardia, and/or tachycardia are detected. Cardiac monitors may be used in the operating room and intensive care units (ICUs) of hospitals. A critically ill patient is continuously monitored by the electrocardiograph. In a typical ICU module, a central console, where an ICU nurse sits, is surrounded by six or more single-patient rooms. The equipment in each room interfaces with the central console.

The FDA recommends that cardiac monitor manufacturers follow *ANSI/AAMI EC13:2002 Cardiac Monitors, Heart Rate Meters, and Alarms* (AAMI, 2002), rather than ANSI/AAMI EC11:2007. Cardiographs from these manufacturers possess features that are useful for con-tinuous patient monitoring. One such feature is *leads-off detection*: Less than 1 μA of current is transmitted systematically to various pairs of patient cable electrodes in order to determine whether each electrode is properly connected to the patient. If an open circuit is detected, an alarm in the cardiac monitor sounds and is transmitted to the central console (AAMI, 2002).

## SUMMARY

An electrocardiograph is a digital instrument that enables cardiac biopotentials to be measured, displayed, and analyzed. A diagnostic electrocardiograph is used to obtain a set of conventional ECG signatures that accurately represents detailed waveforms and beat-to-beat variability. A cardiac monitor is an electrocardiograph used to obtain a heart rate indication, display heart rate and/or ECG waveforms, and provide alarms when car-diac standstill, bradycardia, and/or tachycardia are detected.

Cardiac biopotentials originate in the SA node. There, the dominant pacemaker cells gener-ate pacemaker action potentials that spread to working myocardial cells, resulting in working myocardium action potentials. From the SA node, an excitation is conducted over the working myocardium of both atria, initiating atrial contraction. The excitation is then briefly delayed at the AV node, allowing the atrial blood to enter the ventricles. From the AV node, the excita-tion quickly moves at a velocity of about 2 m/s through the remainder of the system, from the left and right bundle branches of the Purkinje fibers, to the subendocardial endings of the Purkinje fibers. From the subendocardial endings, the excitation is conducted at a velocity of about 1 m/s over the ventricular musculature, initiating ventricular contraction.

This spread of excitation can be observed by measuring the cardiac biopotentials as electrocardiograms. During the P wave, the atria are depolarized, leading to the simulta-neous contraction of the atria. During the QRS complex, the ventricles are depolarized, leading to the beginning of ventricular contraction. An R-wave orientation of a positive or negative deflection is determined by the chosen lead orientation. During the ST segment, which occurs between the S- and T waves, the initial phase of ventricular repolarization occurs. Finally, during the T wave, the rapid phase of ventricular repolarization occurs. Using 10 standard electrode positions, 12 standard leads may be measured. A diagnostic

12-lead ECG is used to diagnose cardiac arrhythmias. The five classes of arrhythmias are irregular rhythms, escape, premature beats, heart blocks, and tachyarrhythmias.

Willem Einthoven invented the first electrocardiograph, which he called a string galvanometer. The modern electrocardiograph uses a biopotential amplifier for increased sensitivity. Key features include its high-input dynamic range, high-frequency response, low system noise, some type of arrhythmia detection, and leads-off detection.

## Exercises

**2.1** Plot ECG waveforms 2_1, 2_2, and 2_3 that have been provided by your instructor. Waveform characteristics are documented in the associated header2_1.txt ascii file. Identify the source of noise artifact in each waveform.

**2.2** With respect to the same waveforms, calculate and plot the power spectral density for each waveform. Noise artifact from two of the waveforms can be minimized through frequency selective filtering. To which waveforms does this apply?

**2.3** Minimize the noise artifact from the two waveforms identified in Exercise 2.2. For each Butterworth filter, state the order and cutoff frequencies used. Plot the filtered waveforms.

**2.4** How can the noise artifact of the remaining third waveform be minimized? Provide a theoretical basis for this strategy.

Read (Einthoven et al., 1950), which is actually the translation of two papers by Einthoven and collaborators from 1912–1913:

**2.5** What measurement phenomenon is Einthoven describing on pp. 196–197 (Figures 1, 2, 4)? In modern terms, how do you prevent this phenomenon from occurring?

**2.6** How does respiration affect specific features of the electrocardiogram? What phenomenon causes these effects?

**2.7** How do you measure Lead I, Lead II, and Lead III? Summarize the evidence that Lead III = Lead II–Lead I is a valid equation.

**2.8** Which is the optimum lead for measurement? Why?

## References

AAMI (2002). *ANSI/AAMI EC13:2002 cardiac monitors, heart rate meters, and alarms*. Arlington, VA: AAMI.

AAMI (2003). *ANSI/AAMI EC57:2003 testing and reporting performance results of cardiac rhythm and ST-Segment measurement algorithms*. Arlington, VA: AAMI.

AAMI (2007). *ANSI/AAMI EC11:2007 diagnostic electrocardiographic devices*. Arlington, VA: AAMI.

Atlee, J. L. (1996). *Arrhythmias and pacemakers: Practical management for anesthesia and critical care medicine*. Philadelphia: W.B. Saunders.

Dubin, D. (2000). *Rapid interpretation of EKG's: An interactive course* (6th ed.). Tampa, FL: Cover Pub.

Einthoven, W. (1965). The string galvanometer and the measurement of action currents in the heart. In: *Nobel lectures: Physiology or medicine 1922–1941*. Amsterdam: Elsevier.

Einthoven, W., Fahr, G., & De Waart, A. (1950). On the direction and manifest size of the variations of potential in the human heart and on the influence of the position of the heart on the form of the electrocardiogram. *American Heart Journal, 40*, 163–211.

Fisch, C. (2000). Centennial of the string galvanometer and the electrocardiogram. *Journal of the American College of Cardiology, 36*, 1737–1745.

Guyton, A. C., & Hall, J. E. (2006). *Textbook of Medical Physiology* (11th ed.). Philadelphia: Elsevier Saunders.

CHAPTER

# 3

# Pacemakers

In this chapter, we discuss the relevant physiology, history, and key features of implantable cardiac pacemakers. A pacemaker is an electrical stimulator that discharges electrical current within one or more cardiac chambers, as a treatment for a cardiac arrhythmia or heart failure. A modern pacemaker is also a medical instrument that enables cardiac electrograms to be measured, displayed, recorded, and analyzed (Figure 3.1).

Upon completion of this chapter, each student shall be able to:

1. Identify two classes of arrhythmias.
2. Understand basic pulse generator parameters and modes.
3. Describe five key features of implantable cardiac pacemakers.

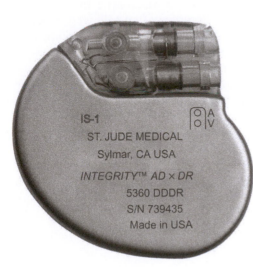

**FIGURE 3.1** The St. Jude Medical Integrity Pacemaker is $43 \times 44 \times 6$ mm and weighs 18 g (St. Jude Medical, St. Paul, Minnesota).

# TWO ARRHYTHMIA CLASSES

In the last chapter, we discussed two classes of arrhythmias, irregular and escape rhythms, as a means of introducing the utility of electrocardiographs. In this section, we discuss two other classes of arrhythmias: premature beats and heart blocks. In particular, the need to treat severe heart blocks led to the development of the implantable pacemaker.

## Premature Beats

When an ectopic focus in the atria, AV junction, or ventricles becomes irritable, it may spontaneously fire, leading to one or more premature beats. Atrial and junctional foci may become irritable from excess epinephrine or norepinephrine, which stimulates the adrenergic receptors of foci, releases adrenergic substances that mimic this effect, or creates other conditions that increase the release of epinephrine or norepinephrine. Ventricular foci may become irritable when they experience either hypoxemia (low oxygen) or hypokalemia (low serum potassium). When one premature beat follows and is coupled to a sinus beat, the group of two beats is called *bigeminy*. When two premature beats follow the sinus beat, the group of three beats is called *trigeminy*.

A *premature atrial beat* can be identified by the change in heart rate and the change in P-wave shape. The active ectopic center resets after the premature beat; so the original heart rate continues. A normal QRS complex often follows the premature atrial beat because the ventricular conduction system is usually receptive to being depolarized (Figure 3.2A). However, if one bundle branch is not completely repolarized when the other is receptive, then a slightly widened QRS complex may be observed following the premature beat. If the AV node is completely unreceptive because it is still in the refractory period of its repolarization, then no QRS complex is conducted.

Similarly, a *premature junctional beat* can be identified by the change in heart rate and the lack of an associated P wave. Again, an active ectopic center resets after the premature beat; so the original heart rate continues. A normal QRS complex often follows the premature junctional beat because the ventricular conduction system is usually receptive to being depolarized. However, if one bundle branch repolarizes faster than the other, a slightly widened QRS complex may also result (Figure 3.2B).

A *premature ventricular contraction* (PVC) can be identified by the change in heart rate and significantly larger beat height and duration (Figure 3.2C). Because a region of the ventricular wall where the irritable focus is located depolarizes before the rest of this ventricle and then the second ventricle, a very wide ventricular complex is produced. After the PVC, a pause is observed because the SA node is not reset.

As already stated, PVCs may result from low serum potassium. But numerous PVCs from the same ventricular focus can be a warning of hypoxia. Hypoxia may due to airway obstruction, suffocation, or another emergency. Six or more PVCs in 1 min is considered pathological (Dubin, 2000).

## Heart Block

As discussed with escape rhythms in Chapter 2, a sick sinus (SA) node may temporarily stop pacing for at least one beat. This pause, known as sinus block, may lead to an escape

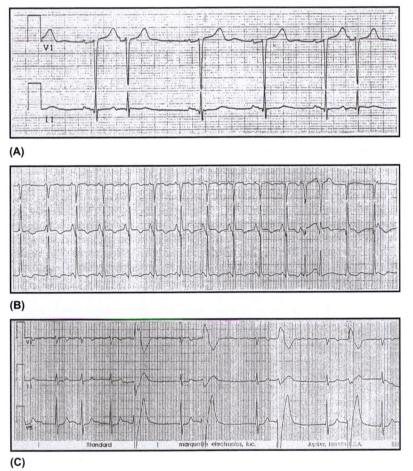

(A)

(B)

(C)

FIGURE 3.2 Premature beats. **A**: Sinus rhythm with premature atrial beats 2 and 6 (see Lead II). **B**: Two junctional beats, followed by nine sinus beats and then two premature junctional beats. **C**: Ventricular bigeminy in patient with atrial fibrillation [*Reproduced by permission from Atlee (1996)*].

beat from an ectopic focus in the atria, AV junction, or ventricles. Sinus block is characterized by identical P waves before and after the pause (Figure 3.3A).

Heart block may also occur in the AV node. With minimal first-degree (1°) AV block, the impulse from the atria is delayed, resulting in a longer than normal pause before ventricular stimulation. Typically, the P-R interval, measured from the beginning of the P wave to the beginning of the QRS complex, is less than 0.2 s, or one major division on an ECG tracing. In 1° AV block, normal sinus rhythm is present, but the P-R interval exceeds 0.2 s (Figure 3.3B).

With second-degree (2°) AV block, the ECG continues to degrade. With Type I (Wenckebach) 2° AV block, the P-R interval becomes progressively longer with each beat, until the AV node no longer conducts an impulse from the atria (Figure 3.4A). Type I 2° AV block is caused by parasympathetic excess or by drugs that mimic these parasympathetic effects. In both cases, AV conduction is slowed.

With Type II (Mobitz II) 2° AV block, the P-R interval remains consistent, but a QRS complex does not follow every P wave. Two to four P waves may occur before ventricular

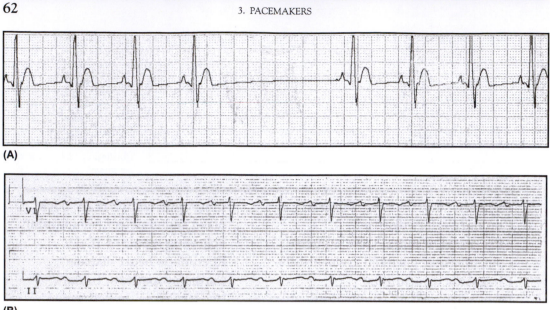

**(A)**

**(B)**

**FIGURE 3.3** Heart blocks. **A:** Sinus block printed from Rivertek RSIM-1500 Simulator (Rivertek, St. Paul, Minnesota). **B:** 1° AV block with heart rate of 88 bpm and P-R interval of 0.28 s *[Reproduced by permission by Atlee (1996)].*

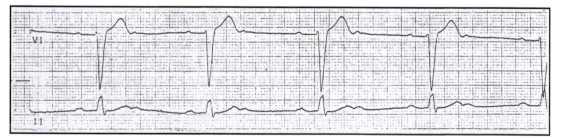

**(A)**

**(B)**

**FIGURE 3.4** Second-degree AV block. **A:** Type I 2° AV block. P-R intervals for beats 3–6 are 0.23, 0.33, 0.37, and 0.41 s. **B:** Type II 2° AV block, with a 2:1 conduction ratio. *[Reproduced by permission by Atlee (1996)].*

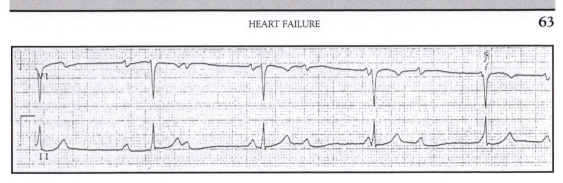

**FIGURE 3.5** Ventricular third-degree AV block *[Reproduced by permission by Atlee (1996)].*

depolarization and a QRS complex are observed. The number of P waves and resulting conduction ratio of P waves to QRS complex relate to the increasing severity of the block in the AV bundle or bundle branches. The atrial rate, or P-P interval, also remains consistent (Figure 3.4B). Type II 2° AV block usually progresses to complete AV block.

With third-degree (3°), or complete, AV block, the atrial rate becomes completely dissociated from the ventricular rate, or R-R interval. No atrial depolarization is conducted to the ventricles; so an ectopic focus below the block escapes overdrive suppression. If the ventricular rate is 40−60 bpm, then the focus is pacing in the AV junction. If the ventricular rate is 20−40 bpm, then no active AV junctional focus is present, and the focus is pacing in the ventricles. When a slower ventricular rate occurs in ventricular 3° AV block, blood flow to the brain becomes inadequate, which causes the patient to lose consciousness. This is known as *Stokes-Adams syndrome* (Figure 3.5).

Heart block may also occur in one of the bundle branches. Typically, both ventricles are depolarized simultaneously. However, block in one of these branches delays depolarization to the ventricle it supplies. As a result, the R wave degrades from one major division to two minor divisions, with a QRS duration of at least 0.12 s, which is equivalent to three minor divisions. The wider R wave of left bundle branch block (LBBB) can be observed in the left chest leads V4, V5, or V6 (Figure 3.6A). The wider R wave of right bundle branch block (RBBB) can be observed in the right chest leads V1, V2, or V3 (Figure 3.6B) (Dubin, 2000).

## HEART FAILURE

Bundle branch block, with a corresponding increased QRS duration, may occur as a result of heart failure (HF). Heart failure refers to the reduced pumping capacity of the heart. It is typically caused by decreased myocardial contractility after coronary artery disease, high blood pressure, or diabetes. The body compensates for acute reduced pumping and resulting lower cardiac output (blood flow) by increasing sympathetic nervous stimulation. This stimulation results in increased pumping of functional heart muscle and venous return to the heart. When this compensation is insufficient to balance outgoing versus returning blood flow, chronic heart failure occurs.

Approximately 5.7 million people in the United States have heart failure, with an additional 670,000 cases diagnosed each year (AHA, 2009). Heart failure patients generally present to their healthcare provider with decreased exercise tolerance, increased fluid

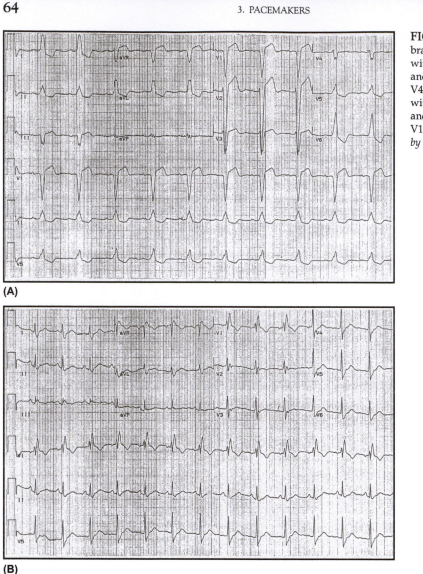

(A)

(B)

FIGURE    3.6 Bundle branch blocks. **A**: Left BBB, with QRS duration of 0.16 s and two R peaks visible in V4 and V5. **B**: Right BBB, with QRS duration of 0.15 s and two R peaks visible in V1 [*Reproduced by permission by Atlee )1996)*].

retention, or symptoms of another cardiac or noncardiac disorder. For diagnosis, an echo-cardiogram is obtained to assess left ventricular (LV) ejection fraction, LV structure, and possible structural abnormalities. *Ejection fraction* (*EF*) refers to the percentage of total blood ejected from either ventricle. Patients are classified according to New York Heart Association (NYHA) functional status (Table 3.1).

Significant predictors of HF prognosis include decreased LVEF, worsening NYHA functional status, widened QRS, chronic low blood pressure, and resting tachycardia. A QRS duration greater than 0.13 s is associated with dyssynchronous contraction between the ventricles (Jessup et al., 2009). The left ventricle remodels; that is, it experiences dilation,

**TABLE 3.1** New York Heart Association Heart Failure Functional Classification

| Class | Patient Reaction During Physical Activity |
| --- | --- |
| I | No symptoms and no limitation in ordinary physical activity |
| II | Mild symptoms and slight limitation during ordinary activity. Comfortable at rest. |
| III | Marked limitation in activity due to symptoms, even during less-than-ordinary activity. Comfortable only at rest. |
| IV | Severe limitations. Experiences symptoms even while at rest. |

changes in sphericity, wall thinning, functional mitral regurgitation (backflow of blood through the mitral valve), and increased wall stress (Salazar & Abraham, 2009).

## TISSUE RESPONSE TO STIMULATION VOLTAGE

As a result of heart failure or an arrhythmia, a ventricle may either not contract efficiently or not contract at all. Direct electrical stimulation of the ventricle may act as therapy to elicit contraction. The response to electrical stimulation in frog gastrocnemius muscle experiments was modeled by physiologist Louis Lapicque as a resistor and capacitor in parallel (Brunel & van Rossum, 2007). Lapique found that the general tissue response could be modeled as:

$$i(t) = i_r\left(1 + \frac{t_c}{t}\right),\tag{3.1}$$

where $t$ = pulse duration time, $t_c$ = chronaxie pulse duration, $i(t)$ = stimulation threshold current, and $i_r$ = constant rheobase current. *Rheobase* is the lowest stimulation current at any pulse duration. *Chronaxie* is the pulse duration corresponding to twice the rheobase current (Lapicque, 1931); it varies with tissue type. For example, the chronaxie of mammalian cardiac muscle is 2 ms. Because we administer constant voltage pulses, rather than constant current pulses, this relationship is usually discussed in terms of voltage. An idealized strength-duration curve based on Eq. (3.1) is given in Figure 3.7.

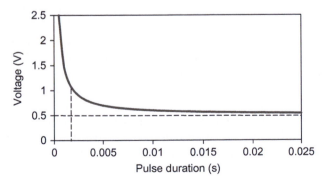

FIGURE 3.7 Strength-duration curve for constant voltage stimulation. The rheobase is 0.5 V; the chronaxie is 0.002 s = 2 ms.

## CLINICAL NEED

As discussed, the ventricular rate of 3° AV block cannot sustain activity. Patients with ventricular 3° AV block require a faster ventricular rate, which can be triggered by a cardiac pacemaker. Cardiac patients with other diagnoses that result in slowed rates (bradycardia) can also be treated with a cardiac pacemaker.

About one-third of Class III/IV heart failure patients with low EF manifest a QRS duration greater than 0.12 s. Associated ventricular dyssynchrony can be improved through timed stimulation of each ventricle by using a pacemaker therapy that we will discuss, called cardiac resynchronization therapy (CRT) (Jessup et al., 2009).

## HISTORIC DEVICES

The search for a viable pacemaker lasted several decades. Initially, an external pacemaker solution was investigated. Once cardiac patients were successfully externally paced, it became clear that they would not be able to resume daily activities unless their pacemakers were completely implanted.

### Early Devices

The concept of a pacemaker originated in 1933 when Albert Hyman patented his invention for an "artificial pace maker for the heart." Hyman sought to restart the heart after it "ceased to function as a result of accident, electrocution, gas poisoning, ether anesthesia." The invention consisted of a needle that would stimulate the heart with current from a generator (Hyman & Hyman, 1933); it did not revive the patient. Years later, in 1952, cardiologist Paul Zoll demonstrated how external stimulation of the heart, with electrodes attached to needles subcutaneously across the chest, could revive a patient during asystole. This procedure required 130-V pulses of unspecified duration, at 90 bpm, to be administered (Zoll, 1952). Zoll's invention was commercialized by Electrodyne and became a standard of care in hospitals during the 1950s (Figure 3.8). Zoll received the 1973 Lasker Clinical Medical Research Award "for his development of the life-saving closed-chest defibrillator and the pacemaker" (Lasker Awards, 2009).

Over time, the pacemaker evolved into a portable design. Earl Bakken, an engineer who serviced equipment for the University of Minnesota Hospital's Department of Surgery as cofounder of Medtronic, was commissioned by surgeon Walton Lillehei to build a lightweight, battery-powered pacemaker that could be worn around the neck. Bakken used a 9.4-V zinc/mercuric oxide battery and a transistor circuit in his design. This pacemaker stimulated the heart with a 2-ms square wave amplitude that varied from 1 to 20 mA; the rate was set between 100 and 120 bpm. Stimulation occurred through two braided tantalum lead wires, or leads, percutaneously inserted into the myocardium (Lillehei et al., 1960). Although an improvement over the Zoll design, the leads introduced risks for mechanical failure and infection. Several groups, including Paul Zoll's, began research on a fully implantable pacemaker.

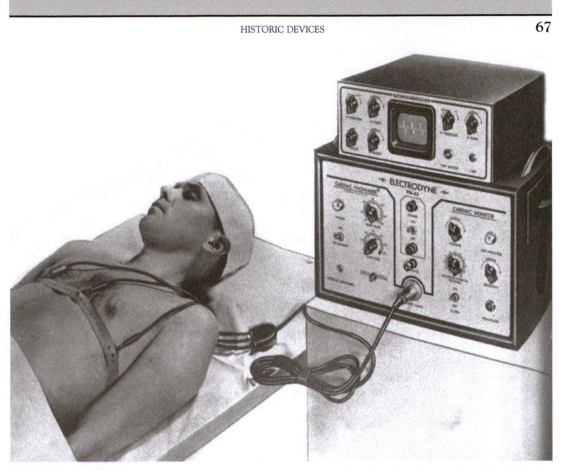

FIGURE 3.8 A patient being monitored by his Electrodyne pacemaker *[Reproduced by permission from Nicholson et al. (1959)].*

## Enabling Technology: Ruben-Mallory Zinc Mercuric Oxide Battery

Engineer Wilson Greatbatch and surgeon William Chardack achieved this goal in 1960 with the first successful long-term implant of a fixed-rate ventricular pacemaker. In this discussion, *successful long-term implant* is defined as an implant that results in patient survival for at least 12 months without a serious wound infection, electrode failure, or electronics failure (Chardack et al., 1960; Chardack et al., 1961). A pacemaker designed by Rune Elmqvist was implanted in 1960 in Uruguay. It was explanted after 8 months when the patient died of sepsis due to an infection from the thoracic incision (Fiandra, 1988).

The issue of a stable, nonelevating electrode voltage had already been solved by Hunter and Roth. Greatbatch and Chardack adopted the Hunter-Roth electrode, which was composed of two stainless steel prongs, spaced 1 cm apart, and a flat silicone patch. To these electrodes, Greatbatch and Chardack added a simple oscillator circuit and 10 Ruben-Mallory zinc/mercuric oxide (Zn/HgO) button batteries. These batteries were chosen for their small size and high energy density. Because the circuit drew only 11 mA for a 50-bpm, 2-ms, 4-mA amplitude pulse, the 450-mAh batteries were estimated to last at least five years. The entire assembly was potted in epoxy resin and then coated in a thin layer of silicone rubber (Figure 3.9) (Chardack et al., 1960; Greatbatch & Holmes, 1991). (We leave it to the reader to analyze oscillator behavior in Exercise 3.3.)

Each Zn/HgO cell conducted the following reactions in its aqueous potassium hydroxide environment:

$$Zn + 2OH^- \rightarrow ZnO + H_2O + 2e^- \tag{3.2}$$

$$HgO + H_2O + 2e^- \rightarrow Hg + 2OH^- \tag{3.3}$$

Net reaction:            $$Zn + HgO \rightarrow ZnO + Hg. \tag{3.4}$$

When the HgO was exhausted, zinc combined with $H_2O$ to form ZnO and hydrogen gas, $H_2$. The design choices of epoxy resin and silicon rubber coating enabled hydrogen gas from the batteries to escape.

Implantation required open chest surgery to place the electrode patch on the right ventricular surface and to have the prongs penetrate the myocardium. The electrode was then tunneled subcutaneously to reach the pacemaker pulse generator, which was subcutaneously implanted in the lower abdomen (Chardack et al., 1960).

Earl Bakken of Medtronic licensed the Greatbatch design and began to manufacture this pacemaker (Greatbatch & Holmes, 1991). Through the ingenuity of numerous engineers and physicians, the basic design evolved to encompass demand pacing (i.e., pacing only when the heart stopped), transvenous pacing (the threading of electrodes into the endocardium through a major vein which eliminated open chest surgery), pacing in both

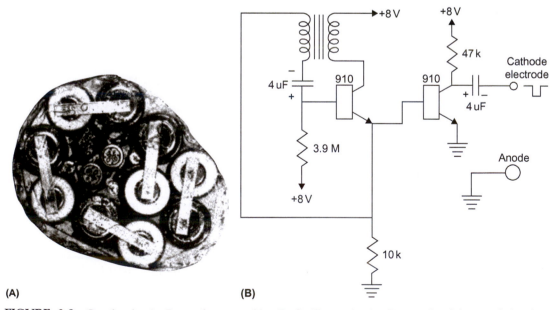

**(A)**                                        **(B)**

FIGURE 3.9 Greatbatch. A: Pacemaker assembly. B: Oscillator circuit. *[A reproduced by permission from Chardack et al. (1961); B reproduced by permission from Chardack et al. (1960)].*

ventricles and the right atrium, sturdy leads, a longer-life lithium battery, telemetry, programmability, and sensors for motion and other parameters. For more of the history of the implantable pacemaker, see the excellent website compiled by the Heart Rhythm Society: http://www.hrsonline.org/News/ep-history/.

## SYSTEM DESCRIPTION AND DIAGRAM

The Greatbatch pacemaker was an early version of the electrical stimulator shown in Figure 1.29. A modern pacemaker system, based on the Greatbatch design, is also an analog medical instrument (Figure 1.2) because intracardiac voltages (electrograms) are sensed for pacing on demand (pacing only when a specified heart chamber stops beating at the desired rate). Sensor-based enhancements, which include an accelerometer, can detect motion and cause the heart rate to be adjusted during physical activity. Other optional sensors can detect respiration, the presence of thoracic fluid, or pressure. Each modern pacemaker system contains three components: the leads, which incorporate electrodes; the pulse generator, which is informally called the can; and the pacemaker programmer. A system diagram for the modern pacemaker system is given in Figure 3.10.

### Leads

A lead connects the pulse generator to the heart. Modern pacemaker leads are manufactured from an outer jacket of polyurethane or a polyurethane hybrid, silicone rubber or polyurethane insulation, and one or two metal coils of titanium or platinum/iridium that

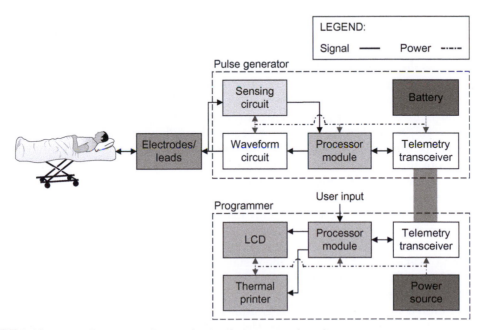

FIGURE 3.10 Pacemaker system diagram (optional sensors not shown).

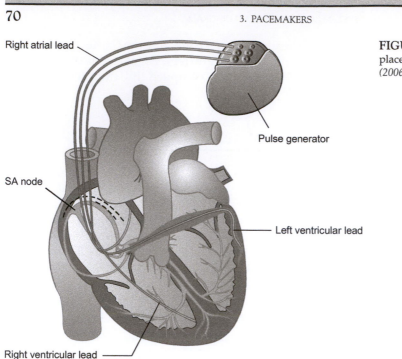

Right atrial lead

Pulse generator

SA node

Left ventricular lead

Right ventricular lead

FIGURE 3.11  Pacemaker lead placement [*Based on Guyton & Hall (2006). Reproduced by permission*].

terminate as textured-surface electrodes. A *unipolar lead* administers a voltage between one electrode as the catheter and a portion of the pulse generator can as the anode. The more common *bipolar lead* administers a voltage between a so-called tip electrode (cathode) and a ring electrode (anode), spaced about 10–30 mm apart at the end of the lead; it is necessarily larger in diameter and less flexible than a unipolar lead. The reduced spacing between electrodes in the bipolar configuration is minimally influenced by signals near the pulse generator, such as electromyograms from skeletal muscle. Furthermore, a bipolar lead positioned in the atrium records the atrial P wave with greater amplitude than a far-field R wave from the ventricles. In contrast, a unipolar lead positioned in the atrium may record a far-field R wave of greater amplitude than the atrial P wave. At the proximal end, a lead connects to the pulse generator header through a standard IS-1 connector.

Each lead is implanted by an electrophysiologist or interventional cardiologist, who threads it from the subclavian or cephalic vein (near the clavicle) to the right atrium, right ventricle, or left ventricle (through the coronary sinus) (Figure 3.11).

At the distal end, a lead is affixed to the endocardium with either a passive fixation mechanism such as a tine or fin, which extends perpendicularly from the lead, or with an active fixation mechanism such as a screw helix. With the screw helix, the actual helical-shaped tip electrode is screwed into the tissue (Figure 3.12).

Over time, scar tissue may form at the distal electrode, which increases impedance and the required stimulation current from a constant voltage source. We discuss the host reactions underlying this scar tissue (fibrous capsule) in more detail in Chapter 5. This impedance is monitored by the pacemaker programmer.

To minimize this impedance increase, steroid-eluting electrodes have been developed. In an early design, each electrode consists of a silicone core, impregnated with a small

FIGURE 3.12 St. Jude Medical Optisense lead. The screw helix can be seen at the end of the lead (St. Jude Medical, St. Paul, Minnesota).

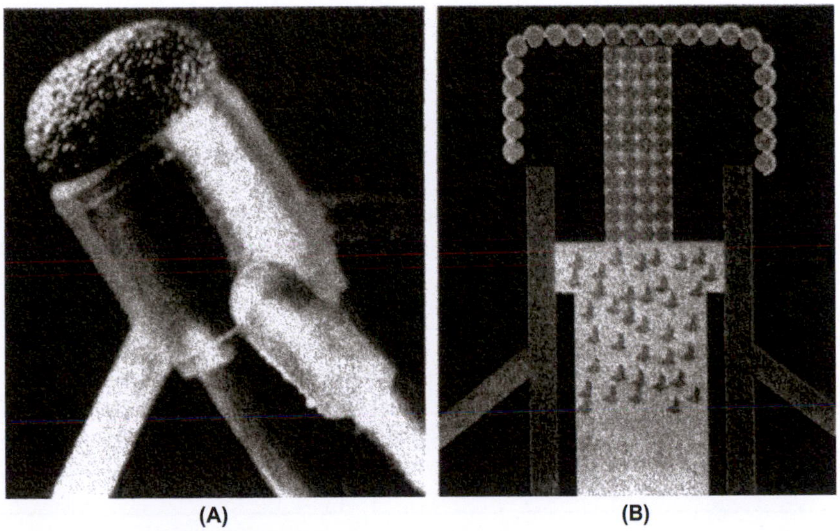

**(A)** **(B)**

**FIGURE 3.13** Medtronic CapSure electrode. **A:** Surface. **B:** Cross-sectional diagram. A silicone rubber plug compounded with dexamethasone sodium phosphate sits behind the platinum-coated titanium electrode *[Reproduced by permission from Mond & Stokes (1992)]*.

dose of dexamethasone sodium phosphate, in contact with porous titanium, coated with platinum (Stokes, 1987). The mechanism by which dexamethasone sodium phosphate prevents an impedance increase is not fully understood. The steroid-eluting electrode is an early example of a drug-eluting combination product (Figure 3.13).

## Pulse Generator

A pulse generator is a hermetically sealed titanium "can" containing a battery and circuitry, with an external connector module called a header, which interfaces with the leads. The pulse generator size is defined mostly by the battery. As discussed in Chapter 1, a $Li/I_2$ battery is used for pacing. An electrophysiologist implants the pulse generator in a small-incision, subcutaneous pocket in the pectoralis fascia below the clavicle, on the patient's nondominant side.

As shown in Figure 3.10, the processor module determines when the waveform circuit, for timing and output functions, should charge a capacitor and discharge a pulse to the patient through the leads. Intracardiac electrograms received from these leads are processed through an analog sensing circuit, in which specific frequency bands are bandpass-filtered and amplified to isolate heart waves (Figure 3.14).

Amplified filter outputs are compared to preset amplitude thresholds (sensitivities) to determine whether a wave is to be sensed as an input to demand pacing. An optional accelerometer senses motion and provides an input to the control logic for rate-responsive pacing. A telemetry transceiver provides external communication to the pacemaker programmer, which is called *interrogation* or *programming*. Older programmers may use a wand to communicate to the pulse generator.

The tank capacitor, which is charged to a preset pulse amplitude voltage around 5 V, is discharged for a preset pulse width duration. Discharge can be described by a first-order differential equation. At the end of discharge, charge voltage still exists between the bipolar electrodes. This excess charge can cause distortions in intracardiac sensing in the short term and electrode corrosion and tissue damage in the long term. To prevent these complications, the excess charge is eliminated through active discharge. In one implementation of active discharge, a coupling capacitor between the tip and ring, which has been connected to the tank capacitor during discharge, slowly discharges through the electrodes and tissue (Figure 3.15).

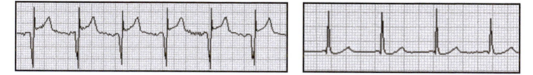

**FIGURE 3.14** Electrograms from two patients, measured from tip of a pacemaker lead placed at right ventricle apex, referenced to the can. These patients were at high risk for acute coronary syndromes [*Reproduced by permission from Fischell et al. (2010)*].

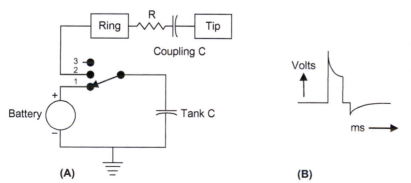

**FIGURE 3.15** Idealized schematic. **A**: Circuit for discharge and active discharge. **B**: Pulse. The tank capacitor charges at switch connection 1 and discharges at connection 2. The coupling capacitor discharges at connection 3.

### Capture Tests

For these pulses to be therapeutic, each discharged voltage must be sufficient to "capture" the heart, that is, to generate depolarization. During implantation, a capture test is conducted for each lead. For each test, assuming a constant preselected pulse width, the pulse amplitude voltage is decreased from a maximum setting to the point at which depolarization no longer occurs. This is equivalent to searching for the strength-duration curve voltage at a given pulse duration. The last pulse amplitude of capture is known as the *capture threshold*. Typically, the pacing threshold is then set at twice the capture threshold. This safety margin of two takes into account natural circadian changes in the pacing threshold, as well as long-term changes in impedance. The safety margin is set at only a factor of two to maximize battery life.

### Pacemaker Sensing Parameters and Modes

Pacemaker *sensing parameters* must also be set during implantation. Although many implanting physicians choose to use default settings, others set at least atrial sensitivity, ventricular sensitivity, AV delay, PV delay, and postventricular atrial refractory period (PVARP).

- *Atrial and ventricular sensitivities* are the minimum voltage thresholds for sensing a P wave and R wave, respectively. These waves are sensed from the bandpass-filtered electrogram.
- *AV delay* refers to the time interval after an atrial pacing pulse (as opposed to a true P wave), A, during which an R wave may be detected. If an R wave is not detected during this time, a ventricular pulse, V, is fired.
- Similarly, *PV delay* refers to the time interval after a P wave, during which an R wave may be detected. If an R wave is not detected during this time, a ventricular pulse is fired.
- *PVARP* is the time interval after R wave or ventricular stimulus during which atrial sensing is disabled. PVARP prevents retrograde P waves (which often occur following paced beats) from being sensed by the atrial amplifier and causing a pacemaker-mediated tachycardia. The related ventricular refractory period (VRP) is the time after R wave or ventricular stimulus during which ventricular sensing is disabled. VRP prevents T waves from being sensed by the ventricular amplifier (Figure 3.16).

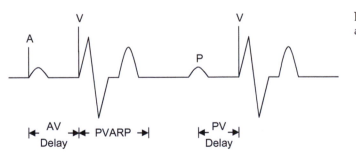

**FIGURE 3.16** Pacemaker parameters and their effect on the electrocardiogram.

**TABLE 3.2** Revised NBG Pacemaker Code (Bernstein et al., 2002)

| Position | I | II | III | IV | V |
|---|---|---|---|---|---|
| Category | Chamber(s) Paced | Chamber(s) Sensed | Response to Sensing | Rate Modulation | Multisite |
| | O = None | O = None | O = None | O = None | O = None |
| | A = Atrium | A = Atrium | T = Triggered | R = Rate Modulation | A = Atrium |
| | V = Ventricle | V = Ventricle | I = Inhibited | | V = Ventricle |
| | D = Dual (A + V) | D = Dual (A + V) | D = Dual (I + T) | | D = Dual (A + V) |

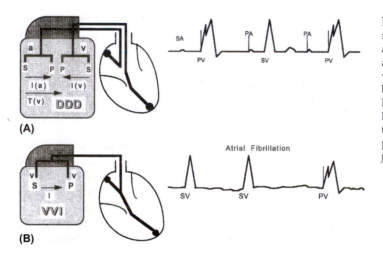

(A)

(B)

FIGURE 3.17 Two pacemaker modes. **A**: DDDxx. Both the right atrium and right ventricle are sensed and paced. **B**: VVIxx. Only the right ventricle is sensed and paced. In these beats, which occur during atrial fibrillation, the first two ventricular depolarizations are spontaneous (S), but the third ventricular depolarization is paced (P) [*Reproduced by permission from Atlee (1996)*].

Pacemaker *mode*, in terms of pacing, sensing, and rate modulation, is classified according to the Revised North American Society of Pacing and Electrophysiology/British Pacing and Electrophysiology Group (NBG) Pacemaker Code (Table 3.2).

A pulse generator is typically programmed in DDDxx mode, with the capabilities of rate modulation and left ventricular pacing dependent on the pulse generator model. Both the right atrium and ventricle are sensed and paced, with a programmable sensing response of inhibiting or triggering pacing stimuli. The emergency mode used during pacemaker programming is VVIxx. With VVI, only the right ventricle is paced and sensed, with a sensed beat inhibiting pacing stimulation (Figure 3.17).

### Cardiac Resynchronization Therapy

*Cardiac resynchronization therapy (CRT)* is the biventricular pacing of the left and right ventricles, in which the AV delay, ventricular pulse order, and ventricular intrapulse delay (also known as VV skew) may be set to a default setting or customized per patient. This biventricular pacing methodology treats the dyssynchrony between contractions from both ventricles.

The first pacemaker to receive FDA approval for CRT was developed by Medtronic. In the first published CRT multicenter, randomized, double-blind study, Medtronic investigators demonstrated that NYHA class, left ventricular ejection fraction, and QRS duration

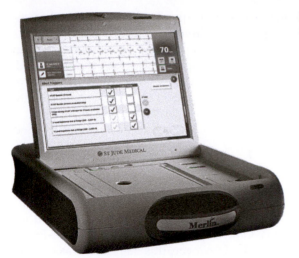

**FIGURE 3.18** St. Jude Medical Merlin programmer (St. Jude Medical, St. Paul, Minnesota).

could be improved with 6 months of biventricular pacing. The 453 patients enrolled in this study were initially NYHA Class III or IV (Abraham et al., 2002). In a subset of 323/453 of these patients, doppler echocardiograms were obtained at baseline and at 3 and 6 months. The patients who received CRT demonstrated reverse left ventricular remodeling, improved systolic and diastolic function, and decreased mitral regurgitation. In a meta-analysis of several CRT studies, heart failure patient hospitalizations and all-cause mortality were reduced by 32% and 25%, respectively (Jessup et al., 2009; McAlister et al., 2004).

Although these are promising results, only one-third of heart failure patients with low LVEF and Class III to IV symptoms of heart failure manifest a QRS duration greater than 0.12 s; these symptoms were all eligibility requirements in these CRT studies (Jessup et al., 2009). Approximately one-third of the patients receiving CRT are estimated to be nonresponders (Salazar & Abraham, 2009), which may be related to noncustomization of pacemaker settings.

## Pacemaker Programmer

As discussed in Chapter 1, the pacemaker programmer is a computer that enables pulse generators to be programmed and pacemaker data to be uploaded and displayed on an LCD or thermal printer. Pacemaker data include programmed parameters, battery voltage, lead impedance, historical changes in parameters, and arrhythmic events. These data are accessed by the follow-up physician during a patient's quarterly or semiannual visit. Each manufacturer has its own telemetry protocol for radiofrequency communication between the pulse generator and programmer (Figure 3.18).

## KEY FEATURES FROM ENGINEERING STANDARDS

Based on the discussion of electrocardiograph consensus standards that are recognized by FDA in Chapter 2, you might assume that FDA-recognized consensus standards exist for most types of medical devices. This is unfortunately not true. As of 2010, only 27 cardiology standards are recognized by FDA; we discussed three of them in Chapter 2. As of June, 2009, the only standard recognized by FDA that is related to an implantable pacemaker is the ISO 5841-3 standard for a low-profile IS-1 connector (CDRH, 2009). Even when recognized, the inclusion of a consensus standard in a product development requirement specification is not mandatory. According to FDA, "Conformance with recognized consensus standards is strictly voluntary" (CDRH, 2007).

In this section, we discuss key features from *ISO 5841-3* and a European standard used by pacemaker manufacturers.

## Lead Connection

At first glance, it may appear odd that the only FDA-recognized consensus standard for pacing is *ISO 5841-3:2000 Impants for Surgery-Cardiac Pacemakers—Part 3: Low-Profile Connectors [IS-1] for Implantable Pacemakers* (ISO, 2000). However, this standard was created to address the problem of incompatible lead connectors and pulse generator headers from different manufacturers during pacemaker implantation. With faulty or damaged sealing between the lead and pulse generator, pacemaker functions such as sensing, output pulse amplitude, and battery discharge may be compromised.

An IS-1 lead connector pin must be at least 18 mm in length and 3.1 ±0.3 mm in diameter. This connector must provide a sealing ring, or circumferential barrier, for maintaining electrical insulation between electrically isolated parts of the connector assembly. When the connector is inserted into or removed from its corresponding connector cavity, no more than 14 N of insertion or withdrawal force shall be required. The minimum electrical impedance between conductive elements must be 50 kΩ. Set screw forces that lock the connector in the header must not deform the lead connector (ISO, 2000).

## Mechanical Integrity of Leads

Leads must be able to withstand the tensile forces and flexural stresses that might occur during and after implantation because they are generally not replaced during the lifetime of a patient. Three known pacemaker lead failure mechanisms are environmental stress cracking (ESC), metal ion oxidation (MIO), and crush fracture. *Environmental stress cracking* is the cracking or crazing of a material in the presence of stress and a chemical environment. To minimize ESC, lead manufacturing techniques have been changed to remove residual stress in each device when shipped. *Metal ion oxidation* is the release of metal ions in conductor coils, resulting from interactions with hydrogen peroxide released from encapsulating fibrous tissue, which catalyzes auto-oxidative degradation of the polymer inside the lead's inner insulation. The choice of polymer composition can minimize MIO. *Crush fracture* may occur during surgical implantation via the subclavian vein; the subclavian stick procedure has been modified to minimize bone clamping of the lead (Stokes, 2006).

To address unknown lead failure mechanisms, leads may undergo three tests, according to British Standard *EN 45502-2-1:2003 Active Implantable Medical Devices—Part 2-1: Particular*

*Requirements for Active Implantable Medical Devices Intended to Treat Bradyarrhythmia (Cardiac Pacemakers)* (BSI, 2003). In the first tensile stress test, a lead is preconditioned for at least 10 days by soaking in a preconditioning bath of 9-g/l saline at 37 ±5°C. It is then subjected to a 1-min tensile load that causes 20% elongation. After loading, it must be verified that permanent lead elongation in excess of 5% has not occurred and that electrical continuity (direct current resistance measurement) has been maintained within the manufacturer's specifications. The lead is then placed in a test bath of 9-g/L saline at 37 ±5°C for 1 hr. This test bath contains an electrode plate having a noble metal surface with a minimum area of 500 mm$^2$. With the lead within 50−200 mm from the reference electrode plate, each conductor is then subjected to a 100-V direct current for at least 15 s. It must be verified that leakage current is within 2 mA while the voltage is applied.

In a second flexural stress test, each unique uniform flexible part of a lead body is placed in a holding fixture that generates a 6-mm bending radius. The fixture causes the bending angle to oscillate from 85° to 90°, at a rate of approximately 2 Hz for at least 47,000 cycles. After the oscillations, it must be verified that the lead conductor is intact and that the measured resistance is within manufacturer's specification.

In a third flexural stress test, a lead connector pin is connected to a holder that contains the lead's mating pulse generator header. The holder is positioned to hang the lead vertically, subject to gravity, with a 100-g load attached to the lead, 10 cm from the holder. The holder oscillates from −45 to +45 degrees from vertical, at a rate of approximately 2 Hz for at least 82,000 cycles. After the oscillations, it must be verified that the lead conductor is intact and that the measured resistance is within manufacturer's specification (BSI, 2003).

We discuss similar tests for implantable defibrillator leads in Chapter 5.

## Battery

A pacemaker system is only as good as its battery, which typically lasts about 7−10 years. The battery is replaced by replacing the pulse generator.

Per EN 45502-2-1:2003, the life of a pulse generator need only be 3 months, under conditions specified by the manufacturer. However, at least one power source indicator must warn of the onset of recommended replacement time (BSI, 2003). The remaining battery life for a patient's implant is assessed during testing with the programmer.

## Sensitivity

Accurate sensing threshold, or sensitivity, measurement is critical to pulse generator performance. According to British Standard EN 45502-2-1:2003, sensitivity error must be within ±10%. The error is measured by connecting the pulse generator to a test signal generator through two resistors and then observing the output on an oscilloscope (Figure 3.19).

The first test signal is a positive polarity triangular pulse, with a pulse width of 15 ms and time-to-peak of 2 ms. The pulse interval is adjusted so that it is at least 50 ms less than the default pulse interval of the pulse generator. For a particular pulse generator

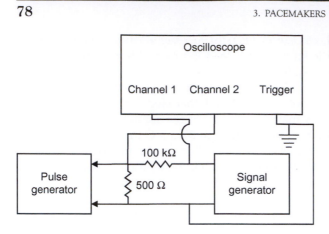

FIGURE    3.19   Sensitivity     measurement
apparatus *[Based on BSI (2003)]*.

sensitivity setting, the pulse amplitude is adjusted up from 0 mV until it consistently inhibits pulse generator stimulation. The positive sensitivity is calculated by dividing the measured test signal amplitude by 201. The procedure is then repeated with a negative polarity triangular pulse and adjustment to the amplitude; negative sensitivity is also calculated by dividing the measured test signal amplitude by 201 (BSI, 2003).

## Minimum Susceptibility to Electromagnetic Interference

Early pulse generators were susceptible to electromagnetic interference (EMI) from microwaves and other equipment. Pulse generators would react to EMI by reversion to a fixed rate or even complete cutoff (Walter et al., 1973). In general, pulse generator exposure to an electromagnetic field may induce currents from a lead into the heart, causing fibrillation or local heating. Exposure may also induce voltage in a lead that damages the pulse generator or prevents the pulse generator from accurate sensing. Further, magnetic control components in a pulse generator may be activated or damaged by magnetic fields.

Modern pulse generators are designed to minimize the effect of EMI radiating from 16.6 Hz, which is present on some European railways, to 3 GHz, the radiation field from cell phones. EN 45502-2-1:2003 specifies seven tests to assess the effects of electromagnetic nonionizing radiation. We describe two of these tests.

The first test assesses the effect of ambient continuous wave electromagnetic fields on induced voltages in leads. For a tissue-equivalent interface circuit composed of 13 resistors and seven capacitors, a sinusoidal input is applied at four well spaced frequencies per decade between 16.6 Hz and 167 kHz. At each frequency, the signal amplitude is slowly increased from 0 to 1 $V_{p-p}$. The tissue-equivalent interface circuit is also connected to a pulse generator, set at its highest sensitivity in pacing mode. It must be verified that, as each signal amplitude is increased, the pulse generator either continues to operate as set or changes to its interference mode.

The second test assesses the effect of strong static magnetic fields on magnetically sensitive components. A pulse generator is placed in a field coil, centered in the field, and aligned so that the most sensitive axis of the pulse generator is parallel to the axis of the coil. The magnetic field flux density is set to a strength of 1 mT, slowly increased to 10 mT, held at 10 mT

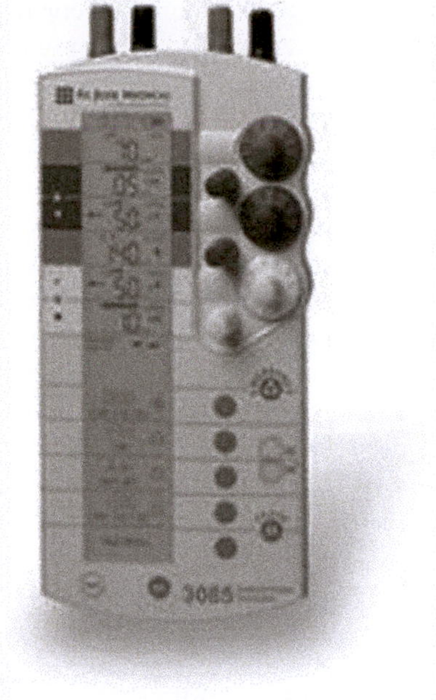

FIGURE 3.20 St. Jude Medical Model 3085 dual chamber temporary pulse generator (St. Jude Medical, St. Paul, Minnesota).

for at least 1 min, and slowly reduced to zero. As a reference for this measurement, 1 tesla (T) of flux density is achieved when 1 Newton (N) of force acts on a wire of length 1 m, carrying 1 A of current. After the field flux density has been set to zero for 5 s, it must be verified that the pulse generator functions as it did prior to the test without adjustment (BSI, 2003).

## TEMPORARY CARDIAC PACING

In addition to permanent cardiac pacing, temporary pacing may be administered to treat acute bradyarrhythmias and certain tachyarrhythmias, either mechanically, transcutaneously, or tranvenously. In *mechanical pacing*, a clinician administers sharp repetitive blows to the patient's sternum to stimulate myocardial depolarization during ventricular fibrillation. In *transcutaneous pacing*, a clinician paces the chest with an external pacemaker and defibrillator-style electrodes. The larger electrode reduces current density at the skin and prevents localized burning. The pacemaker, set to demand mode, administers up to 200 mA of a 30–180 paces per minute in monophasic, truncated waveform. We discuss defibrillation waveforms extensively in Chapter 4. In *transvenous pacing*, transvenous leads are connected to an external pulse generator (Figure 3.20), which functions as an implanted pulse generator.

## SUMMARY

A pacemaker is an electrical stimulator that discharges electrical current within two or three cardiac chambers, as a treatment for a cardiac arrhythmia or heart failure. It is also a medical instrument that enables cardiac electrograms to be measured, displayed, recorded, and analyzed for pacing on demand. Sensor-based enhancements, which include an accelerometer, can detect motion and cause the heart rate to be adjusted during physical activity. Other optional sensors can detect respiration, the presence of thoracic fluid, or pressure. The majority of pacemakers are implanted. Temporary external pacemakers may be used in a transcutaneous or transvenous manner.

Two classes of arrhythmias are premature beats and heart blocks. Premature beats occur when an ectopic focus in the atria, AV junction, or ventricles becomes irritable and spontaneously fires. When one premature beat follows a sinus beat, the group of two beats is called bigeminy. When two premature beats follow and are coupled to the sinus beat, the group of three beats is called trigeminy. Heart blocks occur when the transmission of a beat in the SA node, AV node, or a bundle branch is delayed. The most serious block is 3° AV block, when the atrial rate becomes completely dissociated from the ventricular rate, or R-R interval. When a slower ventricular rate occurs in this block, blood flow to the brain becomes inadequate, causing the patient to lose consciousness. This effect is known as Stokes-Adams syndrome.

Pacemakers were developed initially to treat Stokes-Adams syndrome, but they currently provide treatment for various arrhythmias associated with bradycardia and heart failure. Direct electrical stimulation of the right atrium or a ventricle may act as therapy to elicit contraction, with the general tissue response modeled as $i(t) = i_r\left(1 + \frac{t_c}{t}\right)$.

A pacemaker system contains three components: the leads, which incorporate electrodes; the pulse generator, which is informally called the can; and the pacemaker programmer. While many implanting physicians choose to use default settings, others set at least atrial sensitivity, ventricular sensitivity, AV delay, PV delay, and PVARP. The pulse generator is typically programmed in DDDxx mode, with the capabilities of rate modulation and left ventricular pacing dependent on the pulse generator model.

Key pacemaker features include its standardized lead connection, lead mechanical integrity, long-term battery, accurate sensitivity, and minimum susceptibility to electromagnetic interference.

## Acknowledgment

The author thanks Dennis Hepp for his detailed review of this chapter. Mr. Hepp is Managing Director at Rivertek Medical Systems.

## Exercises

**3.1** The chronaxie of human motor nerve is about 0.01 ms. How will this affect the strength-duration curve of Figure 3.7? For the same pulse duration, does human cardiac muscle or motor nerve require a higher stimulation voltage?

**3.2** Your instructor will provide the short film, *Tony's Cardiac Pacemaker*, about an early external pacemaker patient (courtesy of the Heart Rhythm Foundation). Describe the electrical safety and pain issues to which Tony was exposed. What is the function of the lower left dial that

Tony operates (on the right side of the Electrodyne pacemaker)? In a modern pacemaker, what parameters would have to be monitored to provide the same function?

**3.3** Build the circuit in Figure 3.9B, and observe the oscillating pulse. Describe how the circuit design enables oscillation.

**3.4** Write a software program to detect end-of-life in a typical $Li/I_2$ battery. Your code should base detection on the input voltage. For input data, use a copy machine to enlarge the battery discharge plot in Figure 1.31 sufficiently so that you can read off data pairs (time, voltage). (When modelers wish to verify the results of an older journal article, they resort to the copy machine trick.) How often should this detection code run? Would you store this code in the pacemaker or in the pacemaker programmer?

Read Chardack et al. (1960):

**3.5** How is the pacemaker heart rate set? What heart rate range is desirable? Why?

**3.6** Is unipolar or bipolar stimulation preferable? Why? How does this affect mercury-zinc battery life?

Read Stertzer et al. (1978):

**3.7** Explain the mechanism underlying recharging of the nickel-cadmium battery. How does a pacemaker patient physically recharge the battery? How long does recharging take?

**3.8** From a patient's perspective during the time frame of the Sterzer paper, is it preferable to power a pacemaker with mercury-zinc or nickel-cadmium batteries? Give reasons.

# References

Abraham, W. T., Fisher, W. G., Smith, A. L., Delurgio, D. B., Leon, A. R., & Loh, E., et al. (2002). Cardiac resynchronization in chronic heart failure. *The New England Journal of Medicine, 346*, 1845–1853.

AHA. (2009). Heart failure. *Diseases and conditions*. Dallas, AHA.

Atlee, J. L. (1996). *Arrhythmias and pacemakers: Practical management for anesthesia and critical care medicine*. Philadelphia: W.B. Saunders.

Bernstein, A. D., Daubert, J. C., Fletcher, R. D., Hayes, D. L., Luderitz, B., & Reynolds, D. W., et al. (2002). The revised NASPE/BPEG generic code for antibradycardia, adaptive-rate, and multisite pacing. North American Society of Pacing and Electrophysiology/British Pacing and Electrophysiology Group. *Pacing and Clinical Electrophysiology, 25*, 260–264.

Brunel, N., & Van Rossum, M. C. (2007). Lapicque's 1907 paper: From frogs to integrate-and-fire. *Biological Cybernetics, 97*, 337–339.

BSI (2003). *EN 45502-2-1:2003 Active implantable medical devices—Part 2-1: Particular requirements for active implantable medical devices intended to treat bradyarrhythmia (cardiac pacemakers)*. Brussels: BSI.

CDRH (2007). *Guidance for industry and FDA staff: Recognition and use of consensus standards*. Rockville, MD: FDA.

CDRH (2009). *Recognized consensus standards: June, 2009*. Silver Spring, MD: FDA.

Chardack, W. M., Gage, A. A., & Greatbatch, W. (1960). A transistorized, self-contained, implantable pacemaker for the long-term correction of complete heart block. *Surgery, 48*, 643–654.

Chardack, W. M., Gage, A. A., & Greatbatch, W. (1961). Correction of complete heart block by a self-contained and subcutaneously implanted pacemaker. Clinical experience with 15 patients. *The Journal of Thoracic and Cardiovascular Surgery, 42*, 814–830.

Dubin, D. (2000). *Rapid interpretation of EKG's: An interactive course* (6th ed.). Tampa, FL: Cover Pub.

Fiandra, O. (1988). The first pacemaker implant in America. *Pacing and Clinical Electrophysiology, 11*, 1234–1238.

Fischell, T. A., Fischell, D. R., Avezum, A., John, M. S., Holmes, D., & Foster, M., IIIrd, et al. (2010). Initial clinical results using intracardiac electrogram monitoring to detect and alert patients during coronary plaque rupture and ischemia. *Journal of the American College of Cardiology, 56,* 1089–1098.

Greatbatch, W., & Holmes, C. F. (1991). History of implantable devices. *IEEE Engineering in Medicine and Biology Magazine, 10,* 38–41.

Guyton, A. C., & Hall, J. E. (2006). *Textbook of medical physiology* (11th ed.). Philadelphia: Elsevier Saunders.

Hyman, C. H., & Hyman, A. (1933). Artificial pace maker for the heart. *US 1,913,595.*

ISO (2000). *ISO 5841-3:2000 implants for surgery-cardiac pacemakers—Part 3: Low-profile connectors [IS-1] for implantable pacemakers.* Geneva: ISO.

Jessup, M., Abraham, W. T., Casey, D. E., Feldman, A. M., Francis, G. S., & Ganiats, T. G., et al. (2009). 2009 focused update: ACCF/AHA Guidelines for the Diagnosis and Management of Heart Failure in Adults: A report of the American College of Cardiology Foundation/American Heart Association Task Force on Practice Guidelines: Developed in collaboration with the International Society for Heart and Lung Transplantation. *Circulation, 119,* 1977–2016.

Lapicque, L. (1931). Has the muscular substance a longer chronaxie than the nervous substance? *Journal of Physiology, 73,* 189–214.

Lasker Awards: Former Winners. (2009). Lasker Foundation.

Lillehei, C. W., Gott, V. L., Hodges, P. C., Jr., Long, D. M., & Bakken, E. E. (1960). Transistor pacemaker for treatment of complete atrioventricular dissociation. *Journal of the American Medical Association, 172,* 2006–2010.

McAlister, F. A., Stewart, S., Ferrua, S., & McMurray, J. J. (2004). Multidisciplinary strategies for the management of heart failure patients at high risk for admission: A systematic review of randomized trials. *Journal of the American College of Cardiology, 44,* 810–819.

Mond, H. G., & Stokes, K. B. (1992). The electrode-tissue interface: The revolutionary role of steroid elution. *Pacing and Clinical Electrophysiology, 15,* 95–107.

Nicholson, M. J., Eversole, U. H., Orr, R. B., & Crehan, J. P. (1959). A cardiac monitor–pacemaker: Use during and after anesthesia. *Anesthesia and Analgesia, 38,* 335–347.

Salazar, C., & Abraham, W. T. (2009). Biventricular and novel pacing mechanisms in heart failure. *Current Heart Failure Reports, 6,* 14–18.

Stertzer, S. H., Depasquale, N. P., Cohn, L. J., & Bruno, M. S. (1978). Evaluation of a rechargeable pacemaker system. *Pacing and Clinical Electrophysiology, 1,* 186–188.

Stokes, K. B. (1987). Body implantable lead. *US 4,711,251.*

Stokes, K. B. (2006). Polyurethane pacemaker leads: The contribution of clinical expertise to the elucidation of failure modes and biodegradation mechanisms. In K. M. Becker, & J. J. Whyte (Eds.), *Clinical evaluation of medical devices: Principles and case studies* (2nd ed.). Totawa, NJ: Humana Press.

Walter, W. H., Mitchell, J. C., Rustan, P. L., Frazer, J. W., & Hurt, W. D. (1973). Cardiac pulse generators and electromagnetic interference. *JAMA, 224,* 1628–1631.

Zoll, P. M. (1952). Resuscitation of the heart in ventricular standstill by external electric stimulation. *The New England Journal of Medicine, 247,* 768–771.

CHAPTER

# 4

# External Defibrillators

In this chapter, we discuss the relevant physiology, history, and key features of external defibrillators. As originally conceived, an external defibrillator is an electrical stimulator that discharges electrical current across the thorax, as a treatment for fibrillation. A modern external defibrillator is also a digital instrument that enables cardiac electrocardiograms to be measured, displayed, and analyzed (Figure 4.1).

Upon completion of this chapter, each student shall be able to:

1. Identify tachyarrhythmias.
2. Understand how cardiopulmonary resuscitation and external defibrillators work together to treat sudden cardiac arrest.
3. Describe five key features of external defibrillators.

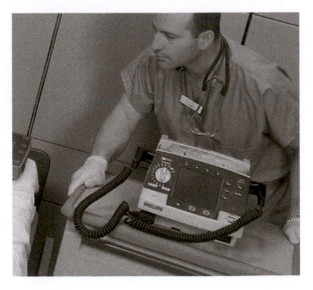

FIGURE 4.1 Philips M4735A HeartStart XL Defibrillator/Monitor (Philips Medical Systems, Andover, Massachusetts).

# TACHYARRHYTHMIAS

In the last chapter, we discussed how the need to treat severe heart blocks led to the development of the implantable pacemaker. Here, we discuss tacchyarrhythmias, which include ventricular fibrillation. The need to treat ventricular fibrillation led to the development of the external defibrillator.

We begin our tachyarrhythmia discussion by describing the phenomenon of *reentry*, which is the mechanism behind several tachyarrhythmias. Normally, an action potential travels unidirectionally, with an impulse traveling down through cells and creating a series of depolarizations and repolarizations. However, during reentry, an initiating premature stimulus, or *impulse*, gets caught in a branch of tissue that splits and reconnects, causing the impulse to retrigger, rather than travel downward through the tissue.

Tachyarrhythmias are distinguished by their high rate. With this arrhythmia class, one or more irritable foci begin to pace rapidly or reentry occurs. Using rate, we can identify subclasses of tachyarrhythmias immediately. In order of increasing rate, these subclasses are paroxysmal tachycardia, flutter, and fibrillation.

## Tachycardia

*Paroxysmal tachycardia* is identified by its general rate range of 150–250 bpm. It may be caused by irritable ectopic focus firing, with rapid sudden pacing, or by reentry. In contrast, sinus tachycardia is the SA node's gradual, not sudden, response to exercise or excitement. As before, the irritable focus or reentry circuit may lie in the atria, junction, or ventricles. Paroxysmal tachycardia due to an irritable atrial or junction focus or reentry circuit is also known as *paroxysmal supraventricular tachycardia*.

When paroxysmal atrial tachycardia is observed, the ventricular rate is high, and the P wave is present (Figure 4.2A). When paroxysmal junctional tachycardia is observed, the ventricular rate is high, and the P wave is either inverted or missing. An irritable focus in the AV junction or AV nodal reentry may initiate tachycardia pacing (Figure 4.2B).

When paroxysmal ventricular tachycardia (Vtach) is observed, the ventricular rate is high and the beats resemble PVC-like complexes. Most commonly, reentry triggers this arrhythmia. Although the atria are still paced, individual P waves are hidden in the large complexes. This serious condition usually indicates coronary artery insufficiency, with inadequate oxygen reaching the heart (Figure 4.2C) (Dubin, 2000).

## Flutter

*Flutter* is identified by its general rate range of 250–350 bpm. With *atrial flutter*, one extremely irritable atrial ectopic focus rapidly fires in this rate range; so identical "flutter" waves are observed. Reentry causes the impulse to continuously circle the crista terminalis

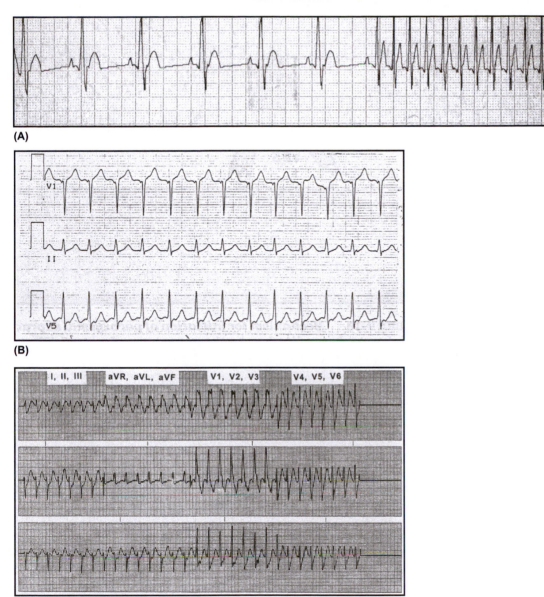

**(A)**

**(B)**

**(C)**

**FIGURE 4.2** Types of paroxysmal tachycardia. **A**: Paroxysmal atrial tachycardia printed from Rivertek RSIM-100 Simulator (Rivertek RSIM-100 Simulator, St. Paul, Minnesota). **B**: Paroxysmal junction tachycardia due to AV nodal reentry. **C**: Paroxysmal ventricular tachycardia with right BBB [*B and C reproduced by permission from Atlee (1996)*].

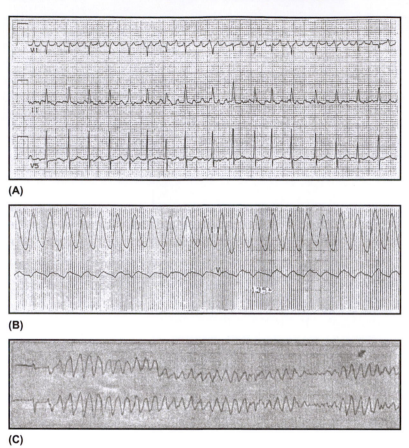

(A)

(B)

(C)

FIGURE 4.3 Flutter and torsades de pointes. **A**: Atrial flutter, with a ventricular rate of 152 bpm. **B**: Ventricular flutter at 175 bpm. **C**: Torsades de pointes *[Reproduced by permission from Atlee (1996)].*

region of the right atrium. These flutter waves are somewhat sawtooth in nature. Since every atrial depolarization does not reach the ventricles, a slower ventricular rate is observed (Figure 4.3A) (Dubin, 2000; Waldo & Feld, 2008).

Similarly, with *ventricular flutter*, one extremely irritable ventricular ectopic focus rapidly fires in the general rate range of 250–350 bpm. These flutter waves are sinusoidal in nature. Because the rate is so rapid, the ventricles hardly have time to fill with blood before contraction (Figure 4.3B). The resulting lower cardiac output and hypoxia rapidly deteriorate into deadly ventricular fibrillation.

A peculiar arrhythmia in the general rate range of flutter, 250–350 bpm, is *torsades de pointes*. Translating into "twisting of points," this arrhythmia is caused by low potassium, medications that block potassium channels, or congenital abnormalities that length the QT segment.

Torsades de pointes results from two competitive, irritable ventricular foci, which cause an undulating amplitude of sinusoids (Figure 4.3C) (Dubin, 2000).

## Fibrillation

*Fibrillation* is identified by its general rate range of 350–450 bpm. We already discussed atrial fibrillation (AF or Afib) in the context of irregular arrhythmias (Chapter 2). One or

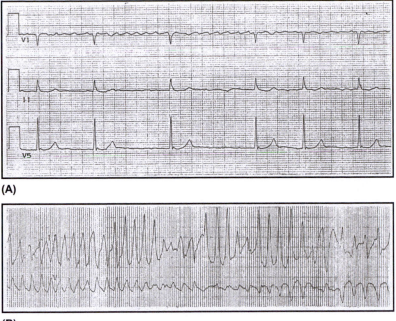

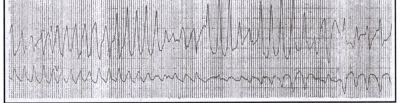

**(A)**

**(B)**

**FIGURE 4.4** Types of fibrillation. **A**: Atrial fibrillation, with a ventricular rate of 53 bpm. **B**: Ventricular fibrillation [*Reproduced by permission from Atlee (1996)*].

more atrial ectopic foci continuously fire, with reentry. Thus, the complete depolarization of the atria is prevented, with only occasional depolarization through the AV node to stimulate the ventricles (Figure 4.4A) (Dubin, 2000; Waldo and Feld, 2008).

Similarly, in *ventricular fibrillation* (VF or Vfib), one or more ventricular ectopic foci continuously fire with reentry, producing an erratic, rapid twitching of the ventricles (Figure 4.4B). Although the rate range of Vfib is described as 350–450 bpm, in reality, it may be difficult to even calculate such a high observed range. Because no blood is being pumped by the heart, a Vfib patient requires immediate attention (Dubin, 2000).

## SUDDEN CARDIAC ARREST AND CARDIOPULMONARY RESUSCITATION

Ventricular fibrillation is the most common arrhythmic cause of out-of-hospital *sudden cardiac arrest* (*SCA*), which occurs when the heart stops beating. Other causes include coronary heart disease, myocardial infarction (a "heart attack"), electrocution, drowning, or choking. SCA occurs in approximately 350,000 persons annually in the United States (AAMI, 2003).

Emergency medical services (EMS) may be dispatched to help an SCA victim. The incidence of emergency medical services (EMS)–treated all-rhythm SCA is approximately 55/100,000 person-years; the incidence of EMS–treated Vfib SCA is approximately 21/100,000 person-years. In an analysis of 31,919 SCA events in the United States from 1980 to 2003, 8.4% of all-rhythm and 17.7% of Vfib patients survived (Rea et al., 2004).

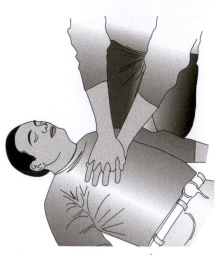

**FIGURE 4.5**  Chest compression at the sternum during cardio-pulmonary resuscitation *(Adapted from A.D.A.M., Atlanta, Georgia).*

When SCA occurs in an out-of-hospital setting, lay rescuers may perform cardiopulmonary resuscitation (CPR) until EMS arrives. This combination of chest compressions and rescue breaths enables oxygenated blood to circulate to vital organs. First, an unconscious person who is not breathing is given two rescue breaths. With both hands on the lower half of the sternum, the rescuer compresses the victim's chest about $1\frac{1}{2}$ in., 30 times within 18 s. The rescuer then administers two rescue breaths within 2 s. This cycle of 30 compressions and two rescue breaths is repeated until help arrives or the rescuer is too exhausted to continue (Figure 4.5) (ARC, 2006).

## DEFIBRILLATION MECHANISM AND THRESHOLD

When EMS arrives, a large current shock, or electrical countershock, is administered across the victim's thorax as a means of stopping fibrillation, also known as *defibrillation*. Typically, defibrillation electrode pads are placed in the anteriolateral position. The sternum electrode is placed on the patient's upper right chest, below the clavicle and to the right of the sternum. The apex electrode is placed on the patient's lower left chest, over the cardiac apex and to the left of the nipple in the midaxillary line (Figure 4.6).

The mechanism behind defibrillation is still unproven. However, it is believed that there are at least three types of Vfib. One important type is *long-duration Vfib*, which is characterized by a duration of greater than 1 min. Long-duration Vfib is the Vfib type presented by the majority of patients receiving initial therapy.

Recent studies suggest that Purkinje fibers appear to be active throughout long-duration Vfib and may either be reentrant or a focal source of impulse waves during Vfib (Ideker, 2007; Tabereaux et al., 2009). When a countershock is weak compared to the defibrillation threshold (DFT), reentry occurs, leading to renewed Vfib. As the shock strength

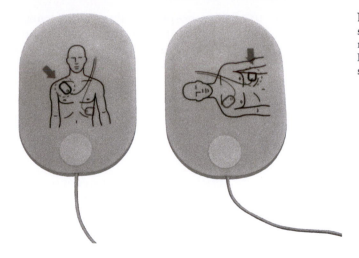

**FIGURE 4.6** Philips M3713A Heart-start defibrillation pad illustration of ante-riolateral electrode pad placement (Philips Medical Systems, Andover, Massachu-setts).

approaches the DFT, a focal epicardial activation pattern becomes responsible for failed defibrillation (Chattipakorn et al., 2004).

The defibrillation threshold refers to a threshold for successful countershock. Typically, defibrillation studies estimate defibrillation threshold by determining the peak shock voltage, $V_{50}$, and total delivered energy, $E_{50}$, required to produce a 50% likelihood that the shock will defibrillate.

## CLINICAL NEED

The countershock administered by EMS may enable a victim's heart to convert from Vfib to a perfusing rhythm. Because the time between a 911 emergency phone call and EMS arrival varies considerably among U.S. cities, smaller, user-friendly defibrillators, called *automated external defibrillators* (*AEDs*), have been developed and placed in airports, airlines, and casinos. Lay rescuers have the option of using AEDs on SCA victims and are protected in many states by Good Samaritan Laws from liability for rendering emergency AED treatment.

With quicker access to defibrillation, the probability of survival increases. In general, the American Red Cross states that "each minute that defibrillation is delayed reduces the chance of survival by about 10% (ARC, 2006)." Similarly, the Emergency Medical Services Division of the King County (Seattle) Department of Public Health modeled the survival rate of 1667 patients who underwent a witnessed cardiac arrest due to heart disease, who were in Vfib, and whose arrest occurred prior to arrival of EMS. They determined that survival, defined as hospital discharge while alive, was initially 67%, with 5.5% lost for each minute until CPR, defibrillation and definite care were completed (Larsen et al., 1993). More recently, early defibrillation programs in aircraft, airports, and a casino demonstrated a higher rate of survival, from 40% to 61% (Page et al., 2000; Caffrey et al., 2002; Valenzuela et al., 2000).

Alternatively, external defibrillators may be used in a hospital setting to treat atrial fibrillation or unstable atrial flutter. Transchest cardioversion with an external defibrillator induces coordinated changes in the critical mass of the atria to terminate reentry. Prior to the procedure, a patient may be given anticoagulation medication to reduce the risk of stroke from blood clots. Because the shock is painful, the patient is sedated before cardioversion is performed.

## HISTORIC DEVICES

The search for a viable external defibrillator was relatively easy. Investigation of an optimum defibrillation waveform spanned several decades. In particular, an optimum waveform requiring minimum energy was needed before smaller AEDs could be manufactured.

### Early External Defibrillator Devices

In 1947, surgeon Claude Beck was able to revive his patient from Vfib during a procedure when the chest cavity was already open. After massaging the heart by hand to continue circulation for 35 min., 110-V/1.5-A alternating current (AC) was administered from an experimental defibrillator. The patient's cardiac rhythm returned (Beck et al., 1947). Beck began to advocate that knowledge of this resuscitation procedure be certified by the American College of Surgeons (Beck & Rand, 1949).

Cardiologist Paul Zoll, who had already demonstrated the utility of external pacing, extended this technique to external defibrillation in 1956. Using an AC 60-Hz/240–720-V sine wave, he administered this waveform for 0.1–0.5 s through copper electrodes smeared with electrode paste against the chest wall. In the first four published patient cases, this external countershock converted Vfib to either asystole or some ventricular beats. In the three patients with ventricular beats, the rhythm eventually converted back to Vfib (Zoll et al., 1956). After retirement from clinical practice, Zoll founded ZMI Corporation in 1983, which later, as Zoll Corporation, became the first manufacturer to market a combined external pacemaker and external defibrillator in a compact unit.

### Enabling Technology: Lown-Edmark Waveform

Cardiologist Bernard Lown challenged the use of alternating current as the most efficient means of external defibrillation in 1962. His group compared the use of alternating current versus direct current countershock in 85 mongrel dogs, after fibrillation induction with a 60-Hz sine wave. DC countershock was administered by discharging energy stored across a capacitor, which was shaped by an inductor and the patient's resistance. In these experiments, DC countershock was demonstrated to be more effective, less lethal, and less likely to induce Vfib than AC countershock (Lown et al., 1962).

Around the same timeframe, cardiovascular surgeon Karl Edmark was also experimenting with the optimum capacitor/inductor combination for a DC waveform (Edmark and

Harkins, 1957; Edmark et al., 1966). In 1955, Edmark founded Physio-Control Corporation, whose Lifepak series of defibrillators was the standard of care in external defibrillation for many years. The popular 1980s Physio-Control Lifepak 5 portable defibrillator for EMS (Figure 4.7), which had rechargeable nickel cadmium batteries, even appeared in the movie *E.T.: The Extra-Terrestrial*.

The simplified circuit for a DC waveform is shown in Figure 4.8. The resulting damped sinusoid waveform is known as the *Lown-Edmark waveform*. Here, the capacitor, $C$, is charged with a DC voltage, $V_{DC}$, before $t = 0$, when the capacitor then switches to discharge. The voltage across the capacitor is $V_i(t)$. The current in the circuit is shaped by the capacitor, by the inductor, $L$, by the inductor resistance, $R_l$, and by the patient resistance, $R_p$. Typically, we assume a patient resistance of 50 Ω.

In writing the equations for this circuit, a Kirchhoff voltage loop for $t \geq 0$ sec results in:

$$\frac{-1}{C} \int i(t)dt = L\frac{di(t)}{dt} + (R_l + R_p)i(t), \qquad (4.1)$$

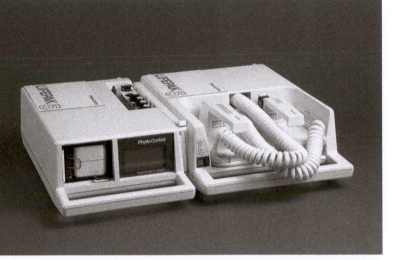

FIGURE 4.7 Physio-Control Lifepak 5 portable defibrillator (Medtronic Physio-Control, Redmond, Washington).

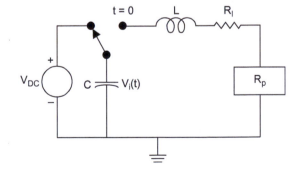

FIGURE 4.8 Simplified external defibrillation circuit for a DC waveform.

$$L\frac{di(t)}{dt} + (R_l + R_p)i(t) + \frac{1}{C}\int i(t)dt = 0. \tag{4.2}$$

If we take the first derivative with respect to time on both sides of Eq. (4.2), we obtain

$$L\frac{d^2i(t)}{dt^2} + (R_l + R_p)\frac{di(t)}{dt} + \frac{1}{C}i(t) = 0. \tag{4.3}$$

Assuming complex roots for this second-order differential equation, the form of the solution is

$$i(t) = e^{\alpha t}[c_1\cos\beta t + c_2\sin\beta t], \tag{4.4}$$

where the roots are

$$m = \alpha \pm j\beta. \tag{4.5}$$

To determine the values of the constants, $c_x$, Eq. (4.4) is considered when $t = 0$ s and current has not yet flowed through the circuit:

$$i(0) = 0 = e^{\alpha(0)}[c_1\cos\beta(0) + c_2\sin\beta(0)], \tag{4.6}$$

$$0 = c_1. \tag{4.7}$$

Substituting Eq. (4.7) into Eq. (4.4) yields

$$i(t) = c_2 e^{\alpha t}\sin\beta t, \tag{4.8}$$

with a first-time derivative of

$$\frac{di(t)}{dt} = c_2\left[\beta e^{\alpha t}\cos\beta t + \alpha e^{\alpha t}\sin\beta t\right]. \tag{4.9}$$

The voltage in the circuit at $t = 0$ is

$$V_i(0) = L\frac{di(0)}{dt} + (R_l + R_p)i(0). \tag{4.10}$$

Substituting Eqs. (4.8) and (4.9) into (4.10) yields

$$v_{DC} = L\{c_2[\beta e^{\alpha(0)}\cos\beta(0) + \alpha e^{\alpha t}\sin\beta(0)]\} + (R_l + R_p)c_2 e^{\alpha(0)}\sin\beta(0), \tag{4.11}$$

$$c_2 = \frac{V_{DC}}{L\beta}. \tag{4.12}$$

Although we have just examined defibrillation voltages and currents, the countershock selected by a clinician is based on energy. Specifically, the delivered energy, $E_d(t)$, which is selected from the user interface, is related to the stored energy, $E_s(t)$, as

$$E_d(t) = E_s(t)\frac{R_p}{R_p + R_l}. \tag{4.13}$$

At $t = 0$, the stored energy is related to the initial capacitor voltage, which is $V_{DC}$, as

$$E_s(t) = \frac{1}{2}C[V_i(0)]^2. \tag{4.14}$$

We leave it to the reader to work with these equations in Exercise 4.1. To illustrate the damped sinusoid voltage a patient receives, we assume a maximum delivered energy setting of 360 J; circuit elements of 16 µF, 0.1 H, and 11 Ω; and a patient resistance of 50 Ω. The resulting damped sinusoid has a peak amplitude of 2795 V and width of approximately 6 ms (Figure 4.9A).

## Early Automated External Defibrillator Devices

External defibrillators based on the damped sinusoid waveform weighed over 20 lbs due to the size of the dense capacitor and inductor for high voltages, and they required a clinician to recognize an electrocardiogram that should be defibrillated. Surgeon Arch Diack attempted to simplify defibrillation through a minimally invasive electrode monitoring system, whereby Vfib could be automatically detected in an unconscious patient. The system relied on a combination sensor inserted into the pharynx for sensing ECG and respiratory sounds. When both signals were not sensed, the device defibrillated (Diack et al., 1978). In 1979, Diack founded Cardiac Resuscitator Corporation, which marketed this first automated external defibrillator, called the Heart Aid.

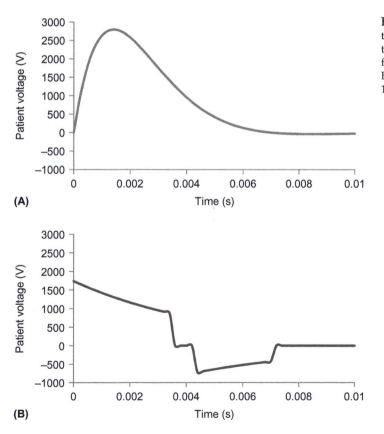

FIGURE 4.9 External defibrillation waveforms for a patient resistance of 50 Ω. A: Damped sinusoid for 360 J, 16 µF, 0.1 H, and 11 Ω. B: Biphasic truncated exponential for 150 J, 100 µF, 51% tilt.

(A)

(B)

## Enabling Technology: Biphasic Truncated Exponential Waveform

In 1992, a start-up company named HeartStream began its goal of developing a truly useful AED—one that was small, inexpensive, and easy to use. The founders were five engineers from Physio-Control, under the leadership of Carl Morgan, who had been Physio-Control's research director. To achieve their goal of smaller size and lower cost, HeartStream moved from using a large capacitor and inductor to deliver a high-energy, damped sinusoid waveform to a smaller capacitor only to deliver a low-energy biphasic truncated exponential (BTE) waveform. The BTE waveform is commonly used in implantable defibrillators. In an early study in 1995, HeartStream demonstrated in swine that a BTE waveform of lower voltage and energy was required to reach $V_{50}$ and $E_{50}$, compared to a damped sinusoid (Gliner et al., 1995).

Ease of use required the development of an accurate algorithm for analyzing shockable ECG rhythms, an efficient disposable battery, and other human factors improvements such as voice prompts. The resulting HeartStream ForeRunner (Figure 1.28), which weighed 4 lb and measured approximately $2.5 \times 8 \times 8.8$ in., received FDA notification that it could be marketed on September 10, 1996 (CDRH, 1996).

To increase its market, HeartStream initially marketed its AED to airlines and encouraged the American Heart Association (AHA) to include defibrillation training in CPR courses. In 1996, American Airlines equipped each of its planes with a ForeRunner; in 1998, AHA changed CPR guidelines to include AEDs (Davis, 2003). In the wake of the ForeRunner's success, many other similarly designed AEDs have been marketed.

In 1998, HeartStream was acquired by Hewlett Packard, which spun it out as part of Agilent Technologies in 1999. Royal Philips Electronics acquired Agilent in 2001.

The simplified H-bridge circuit for a BTE waveform is shown in Figure 4.10. When the capacitor is charged (initialization), the circuit is not connected to the patient. During the first phase of countershock, switches B and D are closed, enabling a positive voltage to be delivered to the patient. With all switches open, the patient voltage goes to zero. With switches A and C closed, the second phase of countershock is delivered as a negative voltage. To terminate the shock, all switches are again opened.

In the ForeRunner, the durations of both phases are determined from a preset table, based on the measured patient resistance. This ensures that the total *tilt*—that is, the fraction of the initial voltage delivered—is 50–60%. The low-voltage tails of excessively high total tilts may cause refibrillation (Gliner et al., 1998). The resulting BTE waveform is

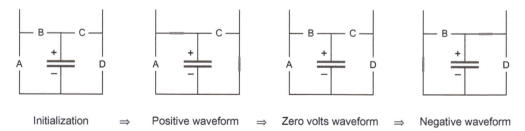

Initialization ⇒ Positive waveform ⇒ Zero volts waveform ⇒ Negative waveform

FIGURE 4.10 Simplified external defibrillation circuit for a BTE waveform, with switching. The patient is not shown but is connected to the top of the circuit.

shown in Figure 4.9B for a maximum delivered energy of 150-J and 100-µF capacitor, which the ForeRunner uses, and an assumed patient resistance of 50 Ω.

For the first phase, the current can be described as

$$i(t) = \frac{V_{DC}}{R_p} e^{-t/(R_p C)}. \tag{4.15}$$

We leave it up to the reader to derive Eq. (4.15) in Exercise 4.2.

## SYSTEM DESCRIPTION AND DIAGRAM

These days, AEDS are ubiquitous and can be found in schools, airports, and even private homes. For prehospital and hospital users, such as emergency medical technicians and nurses, defibrillators are often partly automated and partly manual external versions.

The basic system diagram for an AED is shown in Figure 4.11. An AED is both an external electrical stimulator and digital instrument. The user powers on the device, which utilizes a nonrechargeable lithium ion battery. Voice prompts and an LCD screen enable the user to attach electrodes to the victim's thorax. The electrocardiogram from the electrodes is digitized and analyzed by the Vfib detection module, which sends detection signals and the ECG waveform back to the processor module. If a shockable rhythm, such as Vfib or Vtach is detected, the processor module prompts the user to depress the defibrillate button, which causes the waveform circuit, holding with a charged capacitor, to discharge across the electrodes. Separate from therapy, the user can download ECGs and detection data through a data downloader module.

The electrode pads used are larger versions of the surface electrodes used for monitoring ECG, discussed in Chapter 1. Here, electrical current from the AED is converted to ionic current through the creation of a double layer of charge at the electrode-electrolyte interface. Similarly, ionic current in the body is converted to electrical current for an ECG measurement (Figure 4.12).

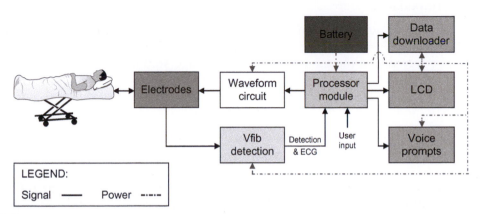

**FIGURE 4.11** Automated external defibrillator system diagram.

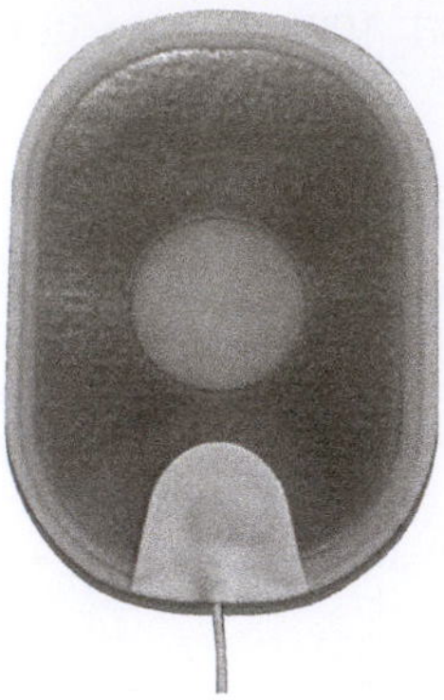

**FIGURE 4.12** Philips M3713A Heartstart defibrillation pad (Philips Medical Systems, Andover, Massachusetts).

With a combination external defibrillator, more functionality is added to the device. The defibrillator is powered by line voltage or by a rechargeable lithium ion battery. Additionally, the user may use electrodes or old-fashioned paddles, which require electrode paste, to defibrillate. The Vfib detection module not only detects arrhythmias, but also detects the timing of the QRS complex, in order to optionally synchronize countershock with the cardiac cycle. A pacing waveform circuit enables monophasic transcutaneous pulses to be administered to the patient, typically from 0 to 200 mA, at 40–180 bpm.

Optional sensor modules enable measurement of noninvasive blood pressure, arterial saturation of oxygen (pulse oximetry), or end-tidal carbon dioxide. Each sensor module contains an isolator, sensor, amplifier, data acquisition submodule, and processor submodule. In AED mode, the user is prompted to defibrillate; in manual mode, users can make this decision themselves, based on visual observation of the ECG waveform on an LCD screen or strip chart recorder (which is a thermal printer) and sensor measurements (Figure 4.13).

## Thoracic Impedance

Up to this point, we have assumed that the thoracic impedance is a purely resistive 50 $\Omega$. However, this assumption conflicts with Lapicque's work, described in Chapter 3. Lapique discovered that the tissue response to voltage stimulation may be modeled as a resistor, $R$, and capacitor, $C$, in parallel, such that the tissue impedance is:

$$Z(j\Omega) = R || \frac{1}{j\Omega C}. \tag{4.16}$$

$$Z(j\Omega) = \frac{R}{j\Omega RC + 1}. \tag{4.17}$$

Here, $\Omega$ is continuous frequency. Because it is part of the defibrillation circuit, the true patient impedance value affects the defibrillation waveform shape during discharge.

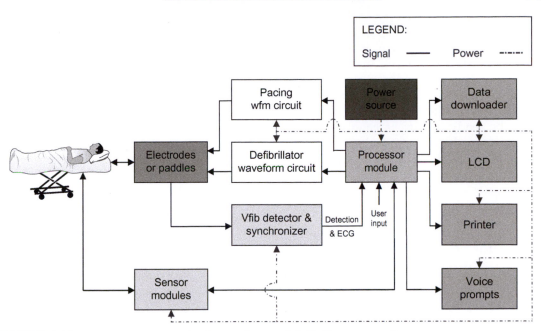

**FIGURE 4.13** Combined automated and manual external defibrillator system diagram.

To simplify estimation of thoracic impedance, Leslie Geddes' group proposed that the ratio of the peak voltage to peak current obtained from transthoracic application of a sinusoidal current with a 30-kHz frequency is approximately the ratio of peak voltage to peak current in response to a transthoracic damped sinusoid waveform (Geddes et al., 1976). More recently, Gail Baura demonstrated in swine that transthoracic impedance can be estimated as a resistor and capacitor in series. A series *RC* circuit is equivalent to a parallel *RC* circuit. The resistance was estimated by fitting the observed transthoracic voltage to its transfer function in the z-domain. Once the resistance was estimated, the capacitance was estimated through Powell's method (Baura, 2000a; Baura, 2000b; Baura, 2001). This model of transthoracic impedance enables different defibrillation waveforms to be compared, with peak currents estimated between waveforms. This peak current value may be constant between $V_{50}$ values of various waveform shapes (Baura, 2002).

## KEY FEATURES FROM ENGINEERING STANDARDS

As of 2010, the FDA recommends the AAMI standard for external defibrillators: *ANSI/ AAMI DF80:2003 Medical Electrical Equipment—Part2-4: Particular Requirements for the Safety of Cardiac Defibrillators (Including Automated External Defibrillators)* (AAMI, 2003). This consensus standard is the basis for the key features discussed here.

## Battery Charging Time

The time until a patient is defibrillated is critical for survival. For this reason, ANSI/ AAMI DF80:2003 requires that manual defibrillators be ready for discharge at maximum energy within 25–35 s of being powered on in rechargeable battery mode, based on whether the defibrillator is frequently or infrequently used, respectively. According to this standard, the threshold for frequent use is more than 2500 discharges. Similarly, AEDs must be ready for discharge at maximum energy within 40–50 s of being powered on in rechargeable battery mode, based on whether the defibrillator is frequent or infrequently used, respectively. The 40-s requirement is based on the assumptions of a 10-s self-test + a 15-s ECG analysis + a 15-s charge time. For both manual and automated defibrillators, this test is conducted after the batteries have been depleted by the delivery of 15 discharges at maximum energy (AAMI, 2003).

## Capacitor Discharge Accuracy

Delivered energy is affected by patient impedance. A defibrillator energy setting assumes a patient impedance of 50 Ω. ANSI/AAMI DF80:2003 requires that the delivered energy at all settings, for patient impedance of 25, 50, 75, 100, 125, 150, and 175 Ω, be specified. Delivered energy accuracy is then tested according to this specification and must be within ±3 J or ±15%, whichever is greater (AAMI, 2003).

## Synchronization

A defibrillator may be used for external cardioversion to treat atrial flutter or atrial fibrillation. When this shock is administered, the defibrillator is placed in synchronization mode, to ensure that the shock is timed after the peak of the QRS. Without synchronization, the shock could convert the rhythm to ventricular fibrillation or asystole.

ANSI/AAMI DF80:2003 specifies that the defibrillator not power on in synchronization mode, and that, at most, 60 ms occur between the peak of the QRS and the peak of the defibrillator waveform. The 60-ms delay assumes that the ECG is derived from the defibrillator electrodes or paddles, rather than from a signal input to the defibrillator (AAMI, 2003).

## Rhythm Recognition Detection Accuracy

For AEDS, it is obviously important that ventricular fibrillation, ventricular tachycardia, and other waveforms be accurately detected. ANSI/AAMI DF80:2003 requires minimum sensitivities and specificities of detection. *Sensitivity* refers to correct classification percentage of a shockable rhythm. *Specificity* refers to the correct classification percentage of a nonshockable rhythm.

Rather than using a standard waveform database for assessing detection accuracy, ANSI/AAMI DF80:2003 requires that an ECG waveform database be collected, using an electrode system and ECG signal processing characteristics similar to the device being tested. This database must include at least Vfib rhythms of varying amplitudes, Vtach rhythms of varying rates and QRS width, various sinus rhythms including supraventricular tachycardias, atrial fibrillation, and atrial flutter. The database must also include sinus rhythm with premature ventricular contraction, asystole, and pacemaker rhythms. Because Vfib and Vtach are the shockable rhythms, sensitivity and specificity can be calculated by testing this waveform database.

According to the standard, the sensitivity for recognizing Vfib at maximum peak-to-peak amplitude of 200 mV or greater must exceed 90% in the absence of artifacts, such as those induced by cardiopulmonary resuscitation. Similarly, for devices detecting Vtach, the sensitivity must exceed 75%. In the absence of artifacts, the specificity for detecting nonshockable rhythms must exceed 95% (AAMI, 2003).

## Recovery after Defibrillation

After a shock occurs, ANSI/AAMI DF80:2003 requires that at most 10 s pass before an ECG is visible on the monitor display and that the ECG peak-to-valley amplitude not deviate from the original amplitude by greater than 50%. If the device is an AED, then the rhythm recognition detector must be able to detect a shockable rhythm within 20 s of the shock. In this way, both clinicians and lay rescuers are able to make informed decisions about whether to administer another shock.

Testing is conducted by attaching two defibrillator electrodes to the metal plates of a test fixture. The test fixture consists of a container filled with saline, with two metal plates in contact with two sponges sitting at the bottom of the fixture. A 10-Hz signal generator is connected by a double-pole switch to two silver electrodes in contact with the saline. The device is set so that monitor sensitivity is adjusted to 10 mm/mV and that monitor frequency response is the widest possible.

With both switch poles closed, the signal generator output is adjusted so that the monitor displays 10 mm peak-to-valley. If the device has a rhythm recognition detector, the amplitude of the shockable rhythm signal is adjusted so that the device can detect a shockable rhythm. One switch pole is then opened, and the maximum energy pulse is delivered to the test fixture. The switch pole is immediately opened, to determine whether more than 10 s passes before the signal appears on the display. If the device is an AED, then the time until a shockable rhythm is also observed (AAMI, 2003).

## SUMMARY

An external defibrillator is an electrical stimulator that discharges electrical current across the thorax, as a treatment for ventricular fibrillation, atrial fibrillation, or atrial flutter. It is also a digital instrument that enables cardiac electrocardiograms to be measured, displayed, and analyzed. There are two types of external defibrillators: an automated external defibrillator (AED) and combination external defibrillator. AEDs are designed to be used by untrained users. Arrhythmias are detected, with the user notified to defibrillator if a shockable rhythm is present. A combination defibrillator is used by clinicians and can be used in its AED mode or in its manual mode for greater control.

External defibrillators are used to treat specific tachyarrhythmias, which are distinguished by their high rate, due to an irritable focus and/or reentry. Paroxysmal tachycardia is identified by its general rate range of 150–250 bpm. The arrhythmias may originate or reenter the atrium, AV junction, or ventricle.

Flutter is identified by its general rate range of 250–350 bpm. With atrial flutter, one extremely irritable atrial ectopic focus rapidly fires in this rate range; so identical "flutter" waves are observed. Reentry causes the impulse to continuously circle the crista terminalis region of the right atrium. Similarly, with ventricular flutter, one extremely irritable ventricular ectopic focus rapidly fires in the general rate range of 250–350 bpm. These flutter waves are sinusoidal in nature. A peculiar arrhythmia in the general rate range of flutter, 250–350 bpm, is torsades de pointes. Torsades de pointes results from two competitive, irritable ventricular foci, which cause an undulating amplitude of sinusoids.

Fibrillation is identified by its general rate range of 350–450 bpm. In atrial fibrillation, one or more atrial ectopic foci continuously fire, with reentry. Similarly, in ventricular fibrillation, one or more ventricular ectopic foci continuously fire with reentry, producing an erratic, rapid twitching of the ventricles. While the rate range of Vfib is described as 350–450 bpm, in reality, even calculating such a high observed range could be difficult. Because the heart is pumping no blood, a Vfib patient requires immediate defibrillation.

Early defibrillators used the Lown-Edwards waveform; modern defibrillators use the biphasic truncated exponential waveform. The BTE waveform enables defibrillators to be smaller in size and lower cost. Key features include its battery charging time, capacitor discharge accuracy, synchronization, rhythm recognition detection accuracy, and recovery after defibrillation.

## Exercises

**4.1** For the circuit used to plot Figure 4.10A, what is the equation for $i(t)$? Recalculate the patient voltage over time, in 0.0002 s increments, for a patient resistance of 25 Ω, 50 Ω, 100 Ω, and 175 Ω. Plot these voltages on the same graph. What is the peak voltage for each patient resistance?

**4.2** Using Figure 4.11 as a foundation, derive Eq. (4.15). What is the relationship between the stored and delivered energy in a BTE circuit?

**4.3** For the circuit used to plot Figure 4.10B, recalculate the patient voltage over time, in 0.0002 s increments, for a patient resistance of 25 Ω, 50 Ω, 100 Ω, and 175 Ω. Plot these voltages on the same graph. Is the Lown-Edwards or BTE circuit more sensitive to patient resistance?

**4.4** Assume an AED is not properly maintained in its workplace environment. It is discharged on a date beyond the defibrillation electrode packaging expiration date. What may be the state of the electrodes before countershock? What may be the state of the patient after countershock? Provide reasons.

Read Edmark et al. (1966):

**4.5** Describe the process by which Edmark found his preferred waveform shape. What criteria did he use?

**4.6** Describe the process by which Edmark translated his dog experiments into clinical practice. How does this differ from today's environment of translating medical device research into practice?

Read Lown et al. (1962):

**4.7** What were the three specific protocols used?
**4.8** How did AC and DC results differ when these protocols were conducted? Are there any advantages of AC versus DC stimulation? If so, please explain.

# References

AAMI (2003). *ANSI/AAMI DF80:2003 Medical electrical equipment—Part 2–4: Particular requirements for the safety of cardiac defibrillators (Including Automated External Defibrillators)*. Arlington, VA: AAMI.

ARC (2006). *First aid/CPR/AED for the workplace: Participant's workbook*. Yardley, PA: American Red Cross.

Atlee, J. L. (1996). *Arrhythmias and pacemakers: Practical management for anesthesia and critical care medicine*. Philadelphia: W.B. Saunders.

Baura, G. D. (2000a). Method and apparatus for electrode and transthoracic impedance estimation. *US 6,016,445*.

Baura, G. D. (2000b). Method and apparatus for high current electrode, transthoracic and transmyocardial impedance estimation. *US 6,058,325*.

Baura, G. D. (2001). Method and apparatus for high current electrode, transthoracic and transmyocardial impedance estimation. *US 6,253,103*.

Baura, G. D. (2002). *System theory and practical applications of biomedical signals*. Hoboken, NJ: Wiley-IEEE Press.

Beck, C. S., Pritchard, W. H., & Feil, H. S. (1947). Ventricular fibrillation of long duration abolished by electric shock. *Journal of the American Medical Association, 135*, 985–986.

Beck, C. S., & Rand, H. J., 3rd (1949). Cardiac arrest during anesthesia and surgery. *Journal of the American Medical Association, 141*, 1230–1233.

Caffrey, S. L., Willoughby, P. J., Pepe, P. E., & Becker, L. B. (2002). Public use of automated external defibrillators. *New England Journal of Medicine, 347*, 1242–1247.

CDRH (1996). *510(k) Summary of safety and effectiveness K955628*. Rockville, MD: FDA.

Chattipakorn, N., Banville, I., Gray, R. A., & Ideker, R. E. (2004). Effects of shock strengths on ventricular defibrillation failure. *Cardiovascular Research, 61*, 39–44.

Davis, R. (2003). To save lives, inventors had to change minds. *USA Today* (July 30).

Diack, A. W., Welborn, W. S., & Rullman, R. G. (1978). Cardiac resuscitator and monitoring apparatus. *US 4,088,138*.

Dubin, D. (2000). *Rapid interpretation of EKG's: An interactive course* (6th ed.). Tampa, FL: Cover Pub.

Edmark, K. W., & Harkins, H. N. (1957). Rapid detection of cardiac arrest with electrical resuscitation. *Surgical Forum, 7*, 290–293.

Edmark, K. W., Thomas, G. I., & Jones, T. W. (1966). DC pulse defibrillation. *Journal of Thoracic and Cardiovascular Surgery, 51*, 326–333.

Geddes, L. A., Tacker, W. A., Jr., Schoenlein, W., Minton, M., Grubbs, S., & Wilcox, P. (1976). The prediction of the impedance of the thorax to defibrillating current. *Medical Instrumentation, 10*, 159–162.

Gliner, B. E., Jorgenson, D. B., Poole, J. E., White, R. D., Kanz, K. G., & Lyster, T. D., et al. (1998). Treatment of out-of-hospital cardiac arrest with a low-energy impedance-compensating biphasic waveform automatic external defibrillator. The LIFE Investigators. *Biomedical Instrumentation & Technology, 32*, 631–644.

Gliner, B. E., Lyster, T. E., Dillion, S. M., & Bardy, G. H. (1995). Transthoracic defibrillation of swine with monophasic and biphasic waveforms. *Circulation, 92*, 1634–1643.

Ideker, R. E. (2007). Ventricular fibrillation: How do we put the genie back in the bottle? *Heart Rhythm, 4*, 665–674.

Larsen, M. P., Eisenberg, M. S., Cummins, R. O., & Hallstrom, A. P. (1993). Predicting survival from out-of-hospital cardiac arrest: A graphic model. *Annals of Emergency Medicine, 22*, 1652–1658.

Lown, B., Neuman, J., Amarasingham, R., & Berkovits, B. V. (1962). Comparison of alternating current with direct electroshock across the closed chest. *American Journal of Cardiology, 10*, 223–233.

Page, R. L., Joglar, J. A., Koval, R. C., Zagrodzky, J. D., Nelson, L. L., & Ramaswamy, K., et al. (2000). Use of automated external defibrillators by a U.S. airline. *New England Journal of Medicine, 343*, 1210–1216.

Rea, T. D., Eisenberg, M. S., Sinibaldi, G., & White, R. D. (2004). Incidence of EMS-treated out-of-hospital cardiac arrest in the United States. *Resuscitation, 63*, 17–24.

Tabereaux, P. B., Dosdall, D. J., & Ideker, R. E. (2009). Mechanisms of VF maintenance: wandering wavelets, mother rotors, or foci. *Heart Rhythm, 6*, 405–415.

Valenzuela, T. D., Roe, D. J., Nichol, G., Clark, L. L., Spaite, D. W., & Hardman, R. G. (2000). Outcomes of rapid defibrillation by security officers after cardiac arrest in casinos. *New England Journal of Medicine, 343*, 1206–1209.

Waldo, A. L., & Feld, G. K. (2008). Inter-relationships of atrial fibrillation and atrial flutter mechanisms and clinical implications. *Journal of the American College of Cardiology, 51*, 779–786.

Zoll, P. M., Lilenthal, A. J., Gibson, W., Paul, M. H., & Norman, L. R. (1956). Termination of ventricular fibrillation in man by externally applied electric countershock. *New England Journal of Medicine, 254*, 727–732.

CHAPTER

# 5

# Implantable Cardioverter Defibrillators

In this chapter, we discuss the relevant physiology, history, and key features of internal defibrillators, which are more commonly known as *implantable cardioverter defibrillators* (*ICDs*). Modern ICDs are combination electrical stimulators and analog instruments, with defibrillator (tachycardia) and pacemaker (bradycardia) functionality (Figure 5.1).

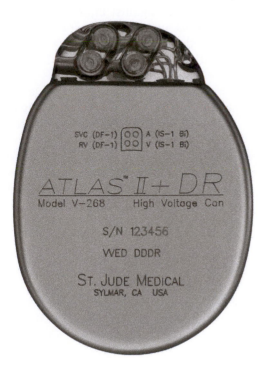

**FIGURE 5.1**   St. Jude Medical Atlas II ICD Pacemaker is 39 cc, is 14 mm long, and weighs 78 g (St. Jude Medical, St. Paul, Minnesota).

Upon completion of this chapter, each student shall be able to:

1. Identify the host reactions involved in the wound-healing response.
2. Describe five key features of ICDs.
3. Understand the impact of the Guidant ICD recall on U.S. medical device regulation.

## WOUND-HEALING RESPONSE

An ICD, like a pacemaker, contains implantable parts. After implantation, the pulse generator and leads initiate host reactions. In this chapter, we discuss the wound-healing response to pulse generator implantation. In Chapter 6, we will discuss blood coagulation resulting from contact between artificial surfaces, such as lead or heart valve surfaces, and blood.

In response to an injury, the body seeks to heal itself. Surgical implantation initiates a series of host reactions referred to as the *wound-healing response*: blood-material interactions, provisional matrix formation, acute inflammation, chronic inflammation, granulation tissue, foreign-body reaction, fibrosis/fibrous capsule development. The extent to which these reactions occur is determined by the implanted material's *biocompatibility*, which is "the ability of a material to perform with an appropriate host response in a specific application" (Williams, 1987; Ratner et al., 2004).

Many types of blood cells participate in the wound-healing response. All blood cells involved in the wound-healing response originate in the bone marrow. *Platelets* are nonnucleated, disk-shaped blood cells with a diameter of $3-4\,\mu m$ and an average volume of $10^{-8}\,mm^3$. *Leucocytes* are white blood cells that defend the body against injurious agents. *Neutrophils* are leukocytes whose main function is *phagocytosis*, which is the cellular ingestion of an injurious agent. During phagocytosis, a neutrophil recognizes and attaches to an injurious agent. It then engulfs the agent, killing or degrading it. *Monocytes* are leukocytes that become macrophages, which also engage in phagocytosis (Figure 5.2).

Blood—material interactions are the result of injury to vascularized connective tissue. At the site of blood vessel injury, platelets adhere to the implant surface. Following platelet adhesion, a complex series of reactions is initiated, causing more platelets to be recruited into a growing platelet aggregate. Within minutes to hours of implantation, platelet aggregation leads to the development of the provisional matrix, consisting of fibrin, activated platelets, inflammatory cells, and endothelial cells. *Fibrin* is an insoluble protein essential to the clotting process; endothelial cells line the inner surface of blood vessels. The fibrin network's three-dimensional structure with attached adhesive proteins provides a substrate for cell adhesion and migration.

Now acute inflammation occurs, for minutes to hours or to days, depending on the extent of injury. Fluid and plasma (the noncellular portion of the blood) exude, and leukocytes, which are predominantly neutrophils, emigrate from the blood vessels to the implant site. These neutrophils attempt to phagocytose microorganisms and foreign materials. However, because of the size difference, implant phagocytosis cannot occur. Acute inflammation normally resolves within one week; a later presence suggests the presence of infection. As inflammation becomes chronic, leucocytes called monocytes move from the blood vessels to the tissue and become macrophages.

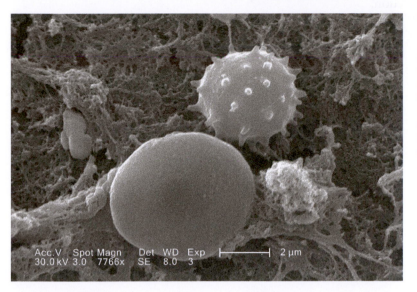

FIGURE 5.2 A number of erythrocytes enmeshed in a fibrinous matrix on the luminal surface of an indwelling vascular catheter; Magnified 7766×. The knobby-surfaced cell in the center is a leucocyte (CDC Public Health Image Library scanning electron micrograph by Janice Carr, 2005).

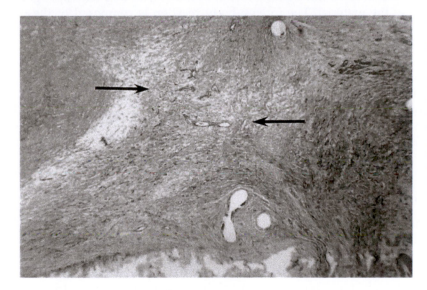

FIGURE 5.3 Granulation tissue in the anastomotic hyperplasia at the anastomosis of an ePTFE vascular graft. Capillary development (slits) and fibroblast infiltration with collagen deposition (darker grey) from the artery form the granulation tissue (arrows). Masson's Trichrome stain. Original magnification 4× [Reproduced by permission from Anderson (2004)].

The action of monocytes and macrophages stimulates vascular endothelial cells and fibroblasts in the implant site to proliferate and begin to form granulation tissue, which is the key characteristic of healing inflammation. A *fibroblast* is a connective-tissue cell that secretes collagen. Granulation appears on the surface of healing wounds as pink, soft granular tissue. Histologically, it is characterized by the proliferation of new small blood vessels and fibroblasts. The new blood vessels are formed by the budding or sprouting of preexisting vessels, which is called *angiogenesis* (Figure 5.3).

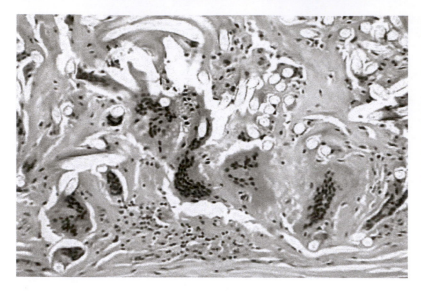

**FIGURE 5.4** Foreign-body reaction with multinucleated foreign body giant cells and macrophages at the periadventitial (outer) surface of a Dacron vascular graft. Fibers from the Dacron vascular graft are identified as clear oval voids. Hematoxylin and eosin stain. Original magnification 20× [Reproduced by permission from Anderson (2004)].

*Foreign-body giant cells* (*FBGCs*) form through the fusion of monocytes and macrophages and attempt to phagocytose the material. The foreign-body reaction is composed of these FBGCs and the components of granulation tissue, that is, the macrophages, fibroblasts, and capillaries. For a relatively flat and smooth surface, such as a breast prosthesis, the foreign-body reaction is composed of a layer of macrophages one to two cells in thickness. For a relatively rough surface such as a Dacron vascular prosthesis, a foreign-body reaction is composed of macrophages and FBGCs at the surface (Figure 5.4).

Eventually, fibrosis or fibrous encapsulation occurs. The host surrounds the implant and isolates it from the surrounding environment. The injured tissue is replaced with connective tissue that constitutes the fibrous capsule (Anderson, 2004) (Figure 5.5).

## Pulse Generator Biocompatibility

Both ICDs and pacemaker pulse generators are typically implanted in a subcutaneous pocket for at least 5 yr. The pulse generator case is made of titanium (Ti), which is known for its biocompatibility. Titanium is naturally covered by a surface layer of titanium dioxide ($TiO_2$), which reduces the inflammatory response and fibrous capsule formation. $TiO_2$ may inhibit peroxynitrite, which is produced during the inflammatory response (Suzuki et al., 2003).

For example, a Ti disk and copper (Cu) disk were subcutaneously implanted separately into 12 female Spraque-Dawley rats (200–250 g). Each Ti disk was 10 mm in diameter and 1 mm thick. Each Cu disk was created by evaporating a 0.2- to 0.5-μm thickness of Cu onto a Ti disk through physical vapor deposition. The disks were implanted for 28 or 56 days. After explant, it was observed that the Cu implants were surrounded by a significantly thicker and denser fibrous capsule than the Ti implants at 28 days ($p < 0.03$). The Cu capsule thickness did not change markedly by 56 days. While the Ti capsule thickness

**FIGURE 5.5** Fibrous capsule with a focal foreign-body reaction to silicone gel from a silicone gel-filled silicone-rubber breast prosthesis. The breast prosthesis-tissue interface is at the top of the photomicrograph. Oval void spaces lined by macrophages and a few giant cells are identified. Also identified is a focal area of foamy macrophages (arrows) indicating macrophage phagocytosis of silicone gel. Hematoxylin and eosin stain. [Reproduced by permission from Anderson (2004)].

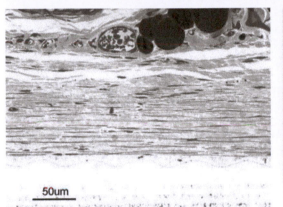

**(A)**

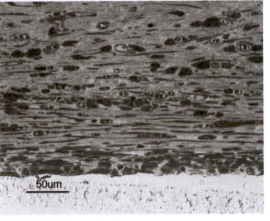

**(B)**

**FIGURE 5.6 A**: Light micrograph of electrochemically prepared section showing the tissue surrounding Ti disk 28 days after implantation in vivo, methylene blue and Azur II staining. The label shows Ti-tissue interface. Elongated profiles of fibroblasts, macrophages, and collagen fibers are arranged parallel at the material surface (top). Magnification 400×. **B**: Light micrograph of electrochemically prepared section showing tissue surrounding Cu disk 28 days after implantation in vivo, methylene blue and Azur II staining. The label shows Cu-tissue interface. Dense thick capsule is infiltrated by cells, mainly macrophages and fibroblasts. Accumulation of macrophages is observed close to the implant surface (top). Magnification 400× [Reproduced by permission from Suska et al. (2008)].

did somewhat increase by 56 days, the difference between Ti and Cu capsule thickness was still significant ($p < 0.04$). Only graphs, but not actual thickness values, were given in the reported study. Histologically, fewer macrophages, fibroblasts, and foreign-body giant cells were present in the Ti capsules (Suska et al., 2008) (Figure 5.6).

# CLINICAL NEED

As discussed in Chapter 4, the time interval between sudden cardiac arrest and CPR/defibrillation/definite care is inversely predictive of survival. For patients predisposed to ventricular fibrillation or ventricular tachycardia, an implanted defibrillator is an efficient treatment for tachycardia, much as an implanted pacemaker is an efficient treatment for bradycardia.

According to the American College of Cardiology/American Heart Association/Heart Rhythm Society 2008 guidelines for the implantation of cardiac pacemakers and antiarrhythmia devices, an ICD *should* be implanted in patients with the following conditions:

1. Cardiac arrest survival due to Vfib or hemodynamically unstable sustained Vtach after evaluation to define the cause of the event and to exclude any completely reversible causes
2. Structural heart disease and spontaneous sustained Vtach, whether hemodynamically stable or unstable
3. Syncope of undetermined origin with clinical relevant, hemodynamically significant sustained Vtach or Vfib induced at electrophysiological study
4. Left ventricular ejection fraction (LVEF) $\leq$ 35% due to prior myocardial infarction (MI), at least 40 days post-MI and in NYHA Class II or III
5. Nonischemic dilated cardiomyopathy, LVEF $\leq$ 35%, and in NYHA Class II or III
6. LV dysfunction due to prior MI, at least 40 days post-MI, LVEF $\leq$ 30%, and in NYHA Class I
7. Nonsustained Vtach due to prior MI, LVEF $\leq$ 40%, and inducible Vfib or sustained Vtach at electrophysiological study (Epstein et al., 2008).

In these Class I implantation recommendations, *syncope* is sudden loss of consciousness. *Nonischemic dilated cardiomyopathy* refers to an enlarged, dilated left ventricular muscle, resulting in decreased pumping ability and lower ejection fraction.

# HISTORIC DEVICES

Looking at the similarities between modern pacemakers and implantable defibrillators, you might assume that pacemaker manufacturers systematically planned to incorporate defibrillator and cardioverter features into their pulse generators. Nothing could be further from the truth.

## Early Devices

In 1968, cardiologist Michel Mirowski and his group began working toward their goal of an implantable defibrillator. When the National Institutes of Health (NIH) refused to fund their work, they funded themselves. Popular cardiologist opinion was against them, as evidenced by Bernard Lown's searing editorial in which Lown noted the lack of clinical need for this device. Lown compared the development of an implantable defibrillator to

Edmund Hillary's reasons for climbing Mt. Everest "because it was there" (Lown & Axelrod, 1972). As explained in Chapter 4, Lown was the leading expert at the time in external defibrillation.

Yet, by 1980, working with a small company called Medrad, the group established that their automatic implantable defibrillator (AID) could treat patients. Each AID consisted of lithium batteries and circuitry in a hermetically sealed titanium case, weighing 260 g and occupying a 145-cc volume. The batteries had a projected monitoring life of about 3 yr or a discharge capacity of about 100 shocks. The pulse generator was implanted subcutaneously in the abdomen. The Ti and silicone rubber electrodes were delivered via catheter to the superior vena cava, near the atrial junction, or positioned extrapericardially on the apex of the heart. Vfib was detected by monitoring the sampled probability density function of ventricular electrical activity. Upon detection, a truncated exponential pulse of 25 J was delivered. An external analyzer enabled battery depletion to be checked. An external recorder enabled 22.5 s of ECG preceding a shock and 67.5 s of ECG following a shock to be printed (Figure 5.7).

Patients were eligible for AID implantation if they had survived at least two episodes of cardiac arrest not associated with acute MI, with ECG-recorded Vfib at least once (Mirowski et al., 1980a). In a first patient, ventricular flutter was induced, which caused the AID to automatically defibrillate and terminate the arrhythmia. In a second patient, five episodes of spontaneous Vtach were automatically defibrillated, with the arrhythmia returning to sinus rhythm. In a third patient, who had both a pacemaker and AID implanted, pacing for a short time at 130 bpm triggered AID defibrillation "that did not evoke any abnormal rhythm and was well tolerated by the conscious patient" (Mirowski et al., 1980b).

Following initial clinical studies, Medrad's subsidiary Intec (Mower & Hauser, 1993) improved the original AID design by adding detection of ventricular tachycardia, adding cardioverter capability (including shock synchronization), and improving lead strength

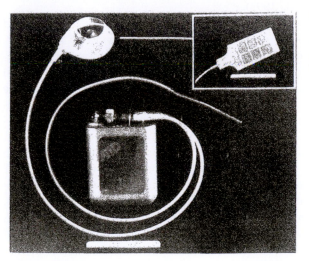

FIGURE 5.7 Mirowski's automatic implantable defibrillator with two electrodes. The cup electrode could optionally be replaced with a patch electrode shown in the insert [Reproduced by permission from Mirowski et al. (1980a)].

(Kolenik et al., 1993). Further clinical trials were conducted, resulting in FDA's approval to market the improved AID-B/BR cardioverter defibrillator in October 1985.

Even with FDA approval, the AID-B/BR ICD was not market-ready. Each AID-B/BR unit required over 400 hr to build because each was assembled from discrete components. After Eli Lilly purchased the AID technology from Intec in 1985, Lilly's wholly owned subsidiary Cardiac Pacemakers, Inc., (CPI) further improved this first ICD. CPI kept the major AID-B/BR circuit design intact but converted the design into a more reliable hybrid circuit. It created a quality assurance (QA) group that monitored AID performance in the field. This QA group included three former Intec clinical specialists who were involved in the original AID clinical trial. Additionally, CPI automated the production, designing, and building of more than 200 new tools and fixtures, and the firm implemented just-in-time manufacturing. Eventually, the 400-hr build time dropped to 17 hr for the renamed Ventak 1500 ICD (Mower & Hauser, 1993; Greatbatch & Holmes, 1991).

During the 1980s, Eli Lilly purchased several key medical device technologies, including the AID. In the mid-1990s, Lilly's device subsidiaries were spun or sold off to raise capital. In particular, CPI was merged with four other Lilly subsidiaries and spun off as Guidant Corporation.

## Enabling Technologies: Transvenous Defibrillation and Integrated Leads

ICD implantation was limited by the need for a thoracotomy. When transvenous defibrillation leads became available, they were used in some patients but required substantial technical expertise to insert. To combat this, electrophysiologist Gust Bardy and his group optimized the transvenous defibrillation lead implant procedure by combining it with a single incision, a unipolar configuration, and a biphasic waveform to mimic pacemaker implantation. The procedure used Medtronic's Model 6966 right ventricular lead, which could pace, sense, and defibrillate. In 40 patients, a defibrillation threshold of $9.3 \pm 6.0$ J was obtained, with 93% having a DFT less than 20 J. Additionally, the procedure took only $100 \pm 28$ min to implement (Bardy et al., 1993).

Model 6966 used a historically important coaxial design, which consisted of a layered arrangement of conductors coiled over each other with intervening layers of insulation. Because of high failure rates in the original design, ICD RV leads have become multiluminal within a single-insulation frame. Each lumen contains an electrode conductor in coil or cable form, with an additional protective secondary insulation sheath. One or two cable conductors enable bipolar or unipolar pacing and sensing. Two defibrillator catheter conductors are coiled and may contain a central lumen for stylet passage. The removable stylet provides extra stiffness during implantation. The primary insulator in the main lead body is silicone rubber. The main lead body may also contain compression or crush passageways. The outer insulation jacket uses a polyurethane hybrid material (Figure 5.8).

To minimize voltage drops and heat losses during high-voltage shocks with low electrical resistivity, a composite wire conductor of silver and MP-35N alloy (SPS Technologies, Cleveland, Ohio) is used in the defibrillator lead. MP-35 N is composed of nickel, cobalt, chromium, and molybdenum. As with pacing leads, titanium or platinum/iridium is used for the defibrillator electrode, with steroid elution to minimize host reactions.

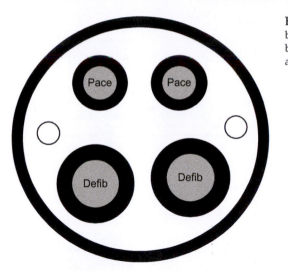

**FIGURE 5.8** Cross section drawing of an ICD lead body, with two pacing cable conductors and two defibrillator coils. Each conductor is insulated. There are also two compression passageways.

Manufacturers have attempted to minimize the diameter of ICD leads, but they must consider the rough handling each lead endures during implant. Some implant issues that can impact lead failure are subclavian venous access, trauma during endovascular passaging, position of a stiff lead with a stylet fully inserted at the right ventricular apex, overtorquing of active-fixation leads during helix extension, and kinking of the lead and connector in the subcutaneous pocket. The high rate of failure in the narrow diameter (6.6 French $\cong$ 2.2 mm) Medtronic Sprint Fidelis lead was traced to conductor fracture (Haqqani & Mond, 2009).

## SYSTEM DESCRIPTION AND DIAGRAM

As with the pacemaker, a modern ICD is an electrical stimulator and analog instrument. In the pulse generator, electrogram features are detected in sensing circuits and may trigger pacing on demand, cardioversion (low-energy shocks), or defibrillation (high-energy shocks) from the appropriate waveform circuit. Because a larger battery is required for high-energy shocks, the $Li/I_2$ pacemaker battery is replaced with the Li/SVO battery discussed in Chapter 1. Up to five leads connect to the pulse generator header: a right atrial pacing lead, a right ventricular pacing lead, a left ventricular pacing lead, right ventricular defib (+) shock coil, and a right ventricular defib (−) shock coil. An optional accelerometer detects motion and may cause the heart rate to be adjusted during physical activity. Other optional sensors may detect minute ventilation, the presence of thoracic fluid, or pressure.

The pulse generator and programmer communicate via telemetry. Through interrogation, electrograms, ECGs, and other data may be displayed and printed. A system diagram for the modern ICD system is given in Figure 5.9.

In Chapter 4, we discussed the switching circuit that shapes the capacitor discharge of the biphasic truncated exponential waveform. A typical maximum delivered energy

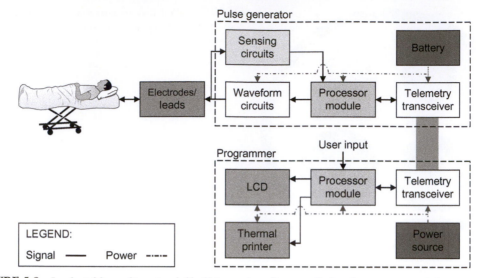

**FIGURE 5.9**  Implantable cardioverter defibrillator system diagram. Optional sensors are not shown.

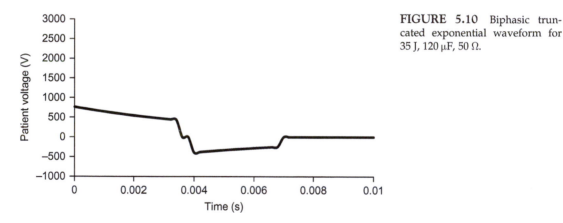

FIGURE 5.10  Biphasic truncated exponential waveform for 35 J, 120 μF, 50 Ω.

setting for an ICD is 35–41 J. The clinician may either program an ICD to output waveforms with fixed tilt at 40–65% or fixed pulse width. A typical ICD waveform is shown in Figure 5.10.

## Defibrillation Threshold Tests

To ensure effective defibrillation, defibrillation threshold (DFT) tests are conducted during implantation. After fibrillation is induced, several shocks are delivered to determine the appropriate energy setting for future shocks. One method for inducing fibrillation is a direct shock of about 8 V DC through the defibrillation lead. Unlike pacemaker capture tests, many shocks need not be delivered via a step-down protocol to find the optimum setting that successfully defibrillates with the lowest energy. This is a painful procedure for the patient, so the number of total shocks used in testing is minimized. Further, the DFT represents only a specific probability, such as 20–50%, for successful defibrillation. A test practice that is gaining in popularity involves testing only one episode of Vfib at an energy of 15–20 J below the maximum output and then programming the ICD to maximum output (Cuoco & Gold, 2008).

## Arrhythmia Detection

After implantation, an ICD should shock only when Vtach or Vfib is detected. Although arrhythmia detection has certainly improved since Mirowski's day, inaccurate detection may still result in inappropriate shocks because detection algorithms are known to have "limitations" (Cuoco & Gold, 2008). In an early algorithm disclosed by Medtronic, Vtach is detected when 12 consecutive R-to-R wave intervals of less than 400 ms are observed. Similarly, Vfib is detected when 12 out of 16 consecutive R-to-R intervals of less than 360 ms are observed. This less stringent 12/16 criterion accounts for occasional undersensing of R waves (Cuoco and Gold, 2008).

# KEY FEATURES FROM ENGINEERING STANDARDS

As with implantable pacemakers, the only standard recognized by FDA that is specific to an implantable defibrillator is that of a lead connector: *ISO 11318:2002 Cardiac Defibrillators—Connector Assembly DF-1 for Implantable Defibrillators—Dimensions and Test Requirements* (CDRH, 2009). This is an industry standard lead connector for the shock coil (ISO, 2002). In this section, we discuss key features from a standard of the European Committee for Electrotechnical Standardization Joint Working Group on Active Implantable Medical Devices (CEN/CLC JWG AIMD), which ICD manufacturers use.

## Biologic Effects

After implantation, an ICD must not release significant particulate matter within the body. This requirement is tested by removing an ICD aseptically from its sterilized package and immersing it in a saline solution bath of approximately 9 g/L in a neutral glass container, with at least five times the volume of the ICD surface area. The container is covered with a glass lid and maintained in an agitated state at $37\pm2°C$ for 8–18 hr. A reference saline bath sample of similar volume is also prepared. After 8–18 hr has elapsed, the excess average count of particles in the test bath, compared to the reference bath, must not exceed 100/mL for particles greater than 5.0 μm and must not exceed 5/mL greater than 25 μm (BSI, 2008).

## Mechanical Integrity of Leads

As with pacemakers, leads must be able to withstand the tensile forces and flexural stresses that might occur during and after implantation. In a first tensile stress test, a lead is preconditioned for at least 10 days by soaking in a preconditioning bath of 9 g/L saline at 37±5°C. It is then subjected to a 1-min tensile load that causes 20% elongation. After loading, it must be verified that permanent lead elongation in excess of 5% has not occurred and that electrical continuity (direct current resistance measurement) has been maintained within the manufacturer's specification. The lead is then placed in a test bath of 9 g/L saline at 37±5°C for 1 hr. This test bath contains an electrode plate having a noble metal surface with a minimum area of 500 mm$^2$, with the lead within 50–200 mm from the reference electrode plate. Insulation between each electrical conductor carrying a defibrillation shock and every other conductor and between each electrical conductor carry a defibrillation shock, and the reference electrode is then subjected to 2000±50 $V_{DC}$ test voltage for 15 s, and 100 V direct current for at least 15 s. Similarly, insulation between each pacing/sensing electrical conductor and every other conductor not exposed to 2000 V, as well as between each pacing/sensing electrical and the reference electrode, is then subjected to 100±5 $V_{DC}$ test voltage for 15 s. It must be verified that leakage current while each voltage is applied is within 2 mA.

The two EN 45502-2-2:2008 flexural stress tests for ICD lead oscillations between 85° and 90° and between −45 and +45 degrees from vertical are identical to the two EN 45502-2-1:2003 flexural stress tests for pacemaker lead oscillations, described in Chapter 3 (BSI, 2008).

## Delivered Voltage and Energy

The observed pulse generator peak voltages and pulse energies must match their specifications. For these measurements, the RV output is connected to a 50±1% load, and other outputs are connected to 500±5% loads. An oscilloscope is used to measure each RF output across its range of energy settings. For each acquired defibrillation pulse, the peak voltage is recorded, and the delivered pulse energy, $E_d(t)$, is calculated as:

$$E_d(t) = \int_0^{t_p} \frac{V^2(t)}{R_p} dt, \tag{5.1}$$

where $t_p$ = pulse duration, $V(t)$ = instantaneous voltage, and $R_p$ = 50 Ω (BSI, 2008).

## Battery Discharge and Warning

Per EN 45502-2-2:2008, the life of an ICD pulse generator must be at least 3 months. This requirement is tested by delivering six 30 J or maximum energy (whichever is less) pulses, spaced uniformly over a 3-month period, without any pacing, into a 50±1% load. The last pulse is delivered at the end of the test period. The ICD must also provide advanced warning when power depletion will limit future therapeutic functions. The advanced warning is tested through inspection of a design analysis provided by the manufacturer (BSI, 2008).

## Arrhythmia Detection Accuracy

It is reasonable to expect that arrhythmia detection accuracy testing is required for implantable defibrillators because arrhythmia detection accuracy testing is required for external defibrillators in ANSI/AAMI DF80:2003 (AAMI, 2003). However, according to EN 45502-2-2:2008, the ICD market and technology are

> still developing and changing rapidly. In drafting this document, the CEN/CLC JWG AIMD has tried to ensure that the requirement specified will not become obsolete, or become unnecessary limitations, as the therapy and the technology develop. For these reasons, the JWG has not included requirements for: arrhythmia detection, antitachycardia pacing, duration of high-voltage pulses (BSI, 2008).

## FDA CASE STUDY: GUIDANT ICD RECALL

By the year 2000, three pacemaker manufacturers dominated the market: Medtronic, Guidant, and St. Jude Medical. All three also manufactured ICDs. As discussed in Chapter 3, Medtronic was the first manufacturer to complete and publish its cardiac resynchronization therapy (CRT) clinical study in 2002. CRT is an effective treatment for Class III/IV heart failure patients with low ejection fraction and long QRS duration. Medtronic received FDA approval to market its biventricular pacemaker used for CRT on August 28, 2001 (CDRH, 2001); it received FDA approval for its lucrative ICD for CRT on June 26, 2002 (CDRH, 2002b).

Guidant's CRT ICD product development and clinical trial followed closely. Guidant submitted its premarket approval package to FDA in February 2001. It received approval to market its Contak Renewal CRT ICD on December 20, 2002 (CDRH, 2002a). The design of Guidant's Contak Renewal was similar to that of its earlier released Ventak Prizm 2 DR 1861 ICD and later Contak Renewal 2 ICD (Steinbrook, 2005).

On March 14, 2005, 21-yr-old college student Joshua Oukrop collapsed and died while on a spring break bicycling trip. Oukrop, who had hypertrophic cardiomyopathy, had received a Prizm 2 DR 4 yr earlier to prevent sudden cardiac death from ventricular fibrillation. Guidant analyzed his ICD and determined that the device had short-circuited internally while attempting to deliver a countershock and that the short circuit permanently disabled the device.

Two months later, four Guidant officials met with Oukrops' electrophysiologists, Barry Maron and Robert G. Hauser, in separate meetings, and disclosed that this failure had occurred 25 other times in patients with Prizm 2 DR. Guidant had fixed the short circuit in April 2002 (Steinbrook, 2005; Meier, 2005c), although about 13,900 ICDs had been implanted from the remaining original inventory of defective devices (Spitzer, 2005). The short circuits occurred over time, when polyimide wire insulation in the ICD header degraded (Steinbrook, 2005) (Figure 5.11).

When these cardiologists asked when other physicians would be informed of defective implants, Guidant replied that the risks associated with surgical replacements outweighed the risked posed by the device. Concerned about patient safety, Hauser alerted the *New York Times* of the short circuit. Hauser, a former president of CPI, had become an industry watchdog in recent years. He created a hospital database for pacemaker and ICD failures (Meier, 2005c) that was much easier to navigate than FDA's Manufacturer and User Facility Device Experience database (Hauser et al., 2001). A few hours before the

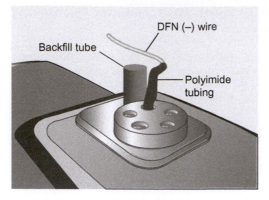

**FIGURE 5.11** When the polyimide insulation in the Ventak Prizm 2 deteriorates, the DFN wire and backfill tube could short-circuit [Adapted from Steinbrook (2005)].

newspaper published its first article on Oukrop's death on May 24, 2005, Guidant made its first public announcement about the defect identified 3 yr earlier (Steinbrook, 2005).

## Public Guidant Actions

The May 24 *New York Times* article caused FDA to open an investigation of these devices (Meier, 2006a). On June 17, 2005, Guidant issued a mandatory notification to clinicians of defective Prizm 2 DR ICDs manufactured on or before April 16, 2002, of defective Contak Renewal ICDs manufactured on or before August 26, 2004, and of defective Contak Renewal 2 ICDs manufactured on or before August 26, 2004, (CDRH, 2005). On July 1, 2005, FDA classified these notifications as Class I recalls (Steinbrook, 2005).

From a business perspective, Guidant did not want to publicly disclose ICD short-circuit problems in early 2005 because it was in the middle of being acquired. In December 2004, Johnson & Johnson (J&J) announced it planned to buy Guidant for $25.4 billion (Meier, 2005b). After the recalls were announced, J&J dropped its offer price, with Boston Scientific Corporation (BSC) then placing a competing bid. In the end, BSC acquired Guidant for $27.6 billion in April 2006 (Meier, 2006b).

## Internal Guidant Actions

According to FDA regulations, a company assesses the seriousness of each product defect and determines the necessary administrative action, including a mandatory notification or voluntary product recall. In mid-2002, an internal Guidant report estimated that the Prism 2 DR short-circuit problem was rare but "life threatening" (Meier, 2005a). After changing the design of the Prism 2DR to correct short-circuiting, Guidant did not submit the required design changes to the FDA. Instead, it reported the correction as a minor alteration in its annual report to FDA (West, 2010). If discussions and a recall of design-related Contak Renewal ICDs implanted in international and U.S. clinical trial patients had occurred, then the Contak Renewal probably would not have been approved for market in December 2002.

In June 2004, the fourth patient implanted with a Contact Renewal ICD died when his ICD short-circuited. The week before his death, his physician had observed the following pacemaker programmer screen warning during the patient's normal visit, after pacemaker interrogation: "WARNING: A shorted condition on the shocking leads has been detected. A LOW shocking lead impedance has been recorded. Please evaluate lead integrity. Select 'Reset Fault' to continue." The physician evaluated lead integrity but did not realize that the ICD, rather than leads, had short-circuited.

After investigating this patient's death in August 2004, Guidant stopped shipping uncorrected Contact Renewal ICDs (West, 2010). In October, 2004, an internal Guidant report confirmed that a small but increasing number of Contak Renewals were also shorting when polyimide degraded (Meier, 2006c).

In March 2005, Guidant sent a "Shorted Shock Lead Warning Screen" product update to all physicians using its pacemakers and ICDs. The product update did not mention the 12 Contact Renewal short-circuit failures that were known, nor did it advise that the warning screen's appearance could indicate a short circuit. This product correction was not reported to the FDA, as required (West, 2010). In May 2005, before public disclosure of Prism 2 DR problems, Guidant projected that about 0.15% of the devices were likely to short-circuit, with 12% of the patients whose units failed either experiencing a life-threatening event or death (Meier, 2005a).

## FDA Responses

In addition to merging its investigation with one being conducted by the Justice Department (Meier, 2006a), FDA began a review of its postmarket surveillance processes. This review was called the Postmarket Transformation Initiative. Although the initiative has had some effect on improving process efficiencies (CDRH, 2006), Robert Hauser stated in 2009 that "I don't think there's been much change at the FDA level" (Snowbeck, 2009).

## Aftermath

In July 2007, Boston Scientific agreed to pay $195 million to settle a class-action suit against its Guidant division. More than 1500 lawsuits representing approximately 4000 Guidant ICD patients were resolved with this settlement (Heuser, 2007). On August 20, 2007, Boston Scientific agreed to a settlement of $16.75 million with 35 states and the District of Columbia to provide ICD protection and education to consumers (Solomon, 2007).

On February 25, 2010, criminal charges were filed against Guidant LLC (owned by Boston Scientific) by the U.S. District Court of Minnesota. Guidant was charged with one count of submitting a periodic postapproval report on the Prizm 2 DR that was "materially false and misleading" and with one count of failing to submit a written report to FDA of the product update correction made to the Contak Renewal 1. A criminal fine of $253,926,251 was proposed. Guidant was also given notice of forfeiture of Contak Renewal 1 ICDs manufactured prior to June 17, 2005. The value of the property was estimated at

$42,079,675 (West, 2010). A settlement of $296 million would have been the largest payout to date ever levied for violating FDA's medical device reporting requirements.

After feedback from ICD patients and Joshua Oukrop's physicians, Drs. Hauser and Maron, federal judge Donovan Frank rejected Guidant's $296 million plea deal. In an unusual decision, Frank urged that the plea agreement be modified because the penalties are "not in the best interests of justice and do not serve the public's interest because they do not adequately address Guidant's history and criminal conduct at issue." Frank suggested that probation, establishment of a compliance and ethics program, and forfeiture payments to Medicare could be parts of a modified agreement (Frank, 2010). On January 21, 2011, Frank added 3 yr of probation to the $296 million plea agreement, to which Boston Scientific agreed. As part of probation, Boston Scientific will provide $15 million for two community programs (Frank, 2011).

For a more thorough discussion of FDA regulation, see Baura (2013).

# SUMMARY

An implantable cardioverter defibrillator (ICD) is a combination electrical stimulator and analog instrument, with defibrillator (tachycardia) and pacemaker (bradycardia) functionality. Like the pacemaker, an ICD has three components: the pulse generator, leads, and the programmer. Electrograms are sensed, with appropriate defibrillator or pacemaker therapy administered as needed. Two defibrillator leads are implanted in the right ventricle. Up to three pacemaker leads are implanted in the right atrium, right ventricle, and left ventricle. An optional accelerometer detects motion and may cause the heart rate to be adjusted during physical activity. Other optional sensors may detect minute ventilation, the presence of thoracic fluid, or pressure.

Key ICD features include biologic effects, mechanical integrity of leads, delivered voltage and energy, battery discharge and warning, and arrhythmia detection accuracy. Guidant's ICD recall in 2005 has exposed the need for improved postmarket surveillance after a medical device is approved by the FDA.

## Acknowledgment

The author thanks Dennis Hepp for his detailed review of this chapter. Mr. Hepp is Managing Director at Rivertek Medical Systems.

## Exercises

**5.1** How biocompatible are ICD leads? What host reactions take place after implantation? Is it easier to explant a pulse generator or one ICD lead?

**5.2** Plot the right atrial and right ventricular electrograms (waveforms 5_1 and 5_2) and electrocardiogram (waveform 5_3) provided by your instructor. Waveform characteristics are documented in the associated header5_1.txt ascii file. These data were acquired simultaneously.

What frequency band would you use to isolate P waves in the electrogram? What frequency band would you used to isolate R waves in the electrogram?

**5.3** How would you improve the ICD test protocols given for delivered voltage and energy, battery discharge and warning, and arrhythmia detection accuracy?

**5.4** Create a test protocol that would have identified the short circuit in the Guidant ICD header.

Read Mirowski et al. (1978):

**5.5** Describe how ventricular fibrillation is detected. Be prepared to discuss this approach for a specific fibrillation waveform your instructor provides during the discussion.

**5.6** Describe how the ventricular fibrillation detection approach would handle a sinus rhythm waveform.

Read Mirowski et al. (1980b):

**5.7** How is battery function verified? Why is this important? How is detection accuracy verified? How accurate was detection to date in these patients?

**5.8** At the time this article was written, was implantable defibrillation a viable substitute for drug therapy or external defibrillation? What are disadvantages of using this implanted device?

# References

AAMI (2003). *ANSI/AAMI DF80:2003 Medical electrical equipment—Part 2-4: Particular requirements for the safety of cardiac defibrillators (Including Automated External Defibrillators)*. Arlington, VA: AAMI.

Anderson, J. M. (2004). Inflammation, wound healing, and the foreign-body response. In B. D. Ratner, A. S. Hoffman, F. J. Schoen, & J. E. Lemons (Eds.), *Biomaterials science* (2nd ed.). San Diego, CA: Elsevier.

Bardy, G. H., Johnson, G., Poole, J. E., Dolack, G. L., Kudenchuk, P. J., & Kelso, D., et al. (1993). A simplified, single-lead unipolar transvenous cardioversion-defibrillation system. *Circulation, 88*, 543–547.

Baura, G. D. (2013). *U.S. medical device regulation: An introduction to biomedical product development, market release, and postmarket surveillance*. Manuscript in preparation.

BSI (2008). *EN 45502-2-2:2008 Active implantable medical devices—Part 2-2: Particular requirements for active implantable medical devices intended to treat tachyarrhythmia (Includes Implantable Defibrillators)*. Brussels: BSI.

CDRH (2001). *Approval order P010015*. Rockville, MD: FDA.

CDRH (2002a). *Approval order P010012 S002*. Rockville, MD: FDA.

CDRH (2002b). *Approval order P010031*. Rockville, MD: FDA.

CDRH (2005). *FDA issues nationwide notification of recall of certain guidant implantable defibrillators and cardiac resynchronization therapy defibrillators*. Rockville, MD: FDA.

CDRH (2006). *Report of the postmarket transformation leadership team: Strengthening FDA's postmarket program for medical devices*. Rockville, MD: FDA.

CDRH (2009). *Recognized consensus standards*. Silver Spring, MD: FDA.

Cuoco, F. A., & Gold, M. R. (2008). The implantable cardioverter defibrillator. In K. A. Ellenbogen, & M. A. Wood (Eds.), *Cardiac pacing and ICDs* (5th ed.). Chichester, West Sussex, UK: Wiley-Blackwell.

Epstein, A. E., Dimarco, J. P., Ellenbogen, K. A., Estes, N. A., 3rd, Freedman, R. A., & Gettes, L. S., et al. (2008). ACC/AHA/HRS 2008 guidelines for device-based therapy of cardiac rhythm abnormalities: A report of the American college of cardiology/American heart association task force on practice guidelines (Writing committee to revise the ACC/AHA/NASPE 2002 guideline update for implantation of cardiac pacemakers and antiarrhythmia devices): Developed in collaboration with the American association for thoracic surgery and society of thoracic surgeons. *Circulation, 117*, e350–e408.

Frank, D. W. (2010). *United States of America v. Guidant LLC Memorandum Opinion and Order, 10-MJ-67 DWF*. Department of Justice District of Minnesota. Minneapolis, MN: United States District Court.

Frank, D. W. (2011). *United States of America v. Guidant LLC Judgment in a Criminal Case, 10-MJ-67 DWF.* Department of Justice District of Minnesota. Minneapolis, MN: United States District Court.

Greatbatch, W., & Holmes, C. F. (1991). History of implantable devices. *IEEE Enginering in Medicine and Biology Magazine, 10,* 38–41.

Haqqani, H. M., & Mond, H. G. (2009). The implantable cardioverter-defibrillator lead: Principles, progress, and promises. *Pacing and Clinical Electrophysiology, 32,* 1336–1353.

Hauser, R. G., Hayes, D. L., Almquist, A. K., Epstein, A. E., Parsonnett, V., & Tyers, G. F., et al. (2001). Unexpected ICD pulse generator failure due to electronic circuit damage caused by electrical overstress. *Pacing and Clinical Electrophysiology, 24,* 1046–1054.

Heuser, S. (2007). Boston Scientific to end suit, pay $195m: Guidant allegedly hid defect in defibrillators. *Boston Globe* (July 14).

ISO (2002). ISO 11318:2002 Cardiac Defibrillators—Connector Assembly DF-1 for Implantable Defibrillators—Dimensional and Test Requirements. Geneva: ISO.

Kolenik, S. A., Langer, A. A., Heilman, M. S., & Staewan, W. S. (1993). Engineering considerations in the development of the automatic implantable cardioverter defibrillator. *Progress in Cardiovascular Diseases, 36,* 115–136.

Lown, B., & Axelrod, P. (1972). Implanted standby defibrillators. *Circulation, 46,* 637–639.

Meier, B. (2005a). Files show Guidant foresaw some risks. *New York Times* (December 24).

Meier, B. (2005b). Maker of heart device kept flaw from doctors. *New York Times* (May 24).

Meier, B. (2005c). Repeated defect in heart devices exposes a history of problems. *New York Times* (October 20).

Meier, B. (2006a). Internal turmoil at device maker as inquiry grew. *New York Times* (February 28).

Meier, B. (2006b). Market Place: Thousands of devices for hearts are recalled. *New York Times* (June 27).

Meier, B. (2006c). U.S. subpoena opens a new front in investigation of heart devices. *New York Times* (January 28).

Mirowski, M., Mower, M. M., Langer, A., Heilman, M. S., & Schreibman, J. (1978). A chronically implanted system for automatic defibrillation in active conscious dogs: Experimental model for treatment of sudden death from ventricular fibrillation. *Circulation, 58,* 90–94.

Mirowski, M., Mower, M. M., & Reid, P. R. (1980a). The automatic implantable defibrillator. *American Heart Journal, 100,* 1089–1092.

Mirowski, M., Reid, P. R., Mower, M. M., Watkins, L., Gott, V. L., & Schauble, J. F., et al. (1980b). Termination of malignant ventricular arrhythmias with an implanted automatic defibrillator in human beings. *New England Journal of Medicine, 303,* 322–324.

Mower, M. M., & Hauser, R. G. (1993). Developmental history, early use, and implementation of the automatic implantable cardioverter defibrillator. *Progress in Cardiovascular Diseases, 36,* 89–96.

Ratner, B. D., Hoffman, A. S., Schoen, F. J., & Lemons, J. E. (2004). Biomaterials science: A multidisciplinary endeavor. In B. D. Ratner, A. S. Hoffman, F. J. Schoen, & J. E. Lemons (Eds.), *Biomaterials science* (2nd ed.). San Diego, CA: Elsevier.

Snowbeck, C. (2009). $296M settles defibrillator investigation. *St. Paul Pioneer Press* (November 6).

Solomon, L. (2007). McGrath: state settles with defibrillator manufacturer. Helena, MT: Montana Dept. of Justice.

Spitzer, E. (2005). *State of New York v. Guidant Corporation, 403656/05.* New York: Supreme Court of the State of New York.

Steinbrook, R. (2005). The controversy over Guidant's implantable defibrillators. *New England Journal of Medicine, 353,* 221–224.

Suska, F., Emanuelsson, L., Johansson, A., Tengvall, P., & Thomsen, P. (2008). Fibrous capsule formation around titanium and copper. *Journal of Biomedical Materials Research A, 85,* 888–896.

Suzuki, R., Muyco, J., McKittrick, J., & Frangos, J. A. (2003). Reactive oxygen species inhibited by titanium oxide coatings. *Journal of Biomedical Materials Research A, 66,* 396–402.

West, T. (2010). *United States of America v. Guidant, LLC Criminal Information, 10-MJ-67 DWF.* Department of Justice District of Minnesota. Minneapolis, MN: United States District Court

Williams, D. F. (1987). Definitions in Biomaterials. *Consensus Conference of the European Society for Biomaterials.* Chester, England: Elsevier.

# 6

# Heart Valves

In this chapter, we discuss the relevant physiology, history, and key features of mechanical and tissue heart valves. A tissue valve is also referred to as a bioprosthesis or bioprosthetic valve. A prosthetic heart valve is a medical device that replaces a diseased native valve and enables blood flow within the cardiac chambers to be regulated (Figure 6.1).

Upon completion of this chapter, each student shall be able to:

1. Understand the mechanisms that generate blood pressure.
2. Identify the host reactions that affect prosthetic heart valves.
3. Grasp the impact of the Bjork-Shiley heart valve recall on U.S. medical device regulation.
4. Describe five key features of prosthetic heart valves.

## CARDIAC MECHANICS

Each beat of the cardiac cycle begins with electrical stimulation and ends with mechanical valve closure. We have already discussed how excitation spreads from the sinoatrial node to the working myocardia of both atria, atrioventricular node, AV bundle, left and

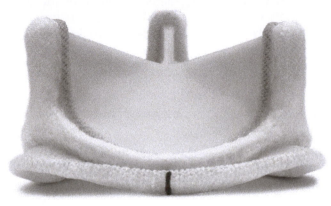

FIGURE 6.1 Edwards Lifesciences Carpentier-Edwards Perimount Magna Ease aortic heart valve (Edwards Lifesciences, Irvine, California).

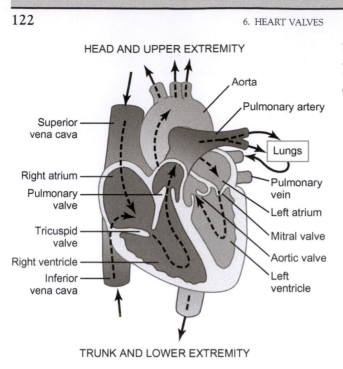

HEAD AND UPPER EXTREMITY

Aorta

Pulmonary artery

Superior vena cava

Lungs

Right atrium

Pulmonary valve

Pulmonary vein

Left atrium

Mitral valve

Tricuspid valve

Aortic valve

Right ventricle

Left ventricle

Inferior vena cava

TRUNK AND LOWER EXTREMITY

**FIGURE 6.2** The systemic (light grey) and pulmonic (dark grey) circulatory systems [*Reproduced by permission from Guyton and Hall (2006)*].

right bundle branches of the Purkinje fibers, and ventricular muscle mass. These action potentials stimulate muscle contraction. The goal of electrical stimulation and muscle contraction is to transport oxygenated blood to the tissues and deoxygenated blood to the lungs.

Deoxygenated blood flows through the superior vena cava and inferior vena cava into the right atrium and right ventricle. Flow between the two right chambers is controlled by the tricuspid valve. Flow from the right ventricle to the pulmonary artery and thereafter the lungs is controlled by the pulmonary valve. Oxygenated blood from the lungs flows through the pulmonary vein to the left atrium and left ventricle. Flow between the two left chambers is controlled by the mitral valve. Flow from the left ventricle to the aorta and thereafter the tissues is controlled by the aortic valve (Figure 6.2).

Each valve prevents backflow, or regurgitation, of blood between a ventricle and atrium [the *atrioventricular (AV)* valves] or between a large blood vessel and ventricle (the *semilunar valves*). The AV valves are the tricuspid and mitral valves; the semilunar valves are the pulmonary and aortic valves. Each valve is passively opened or closed through a pressure gradient. Anatomically, a valve consists of two or three tissue leaflets attached to an *annulus*, or ring, that separates the two cardiac regions. The AV valves also have fiber strands called *chordae tendineae* that connect their leaflets to papillary muscles on the ventricular wall (Figure 6.3).

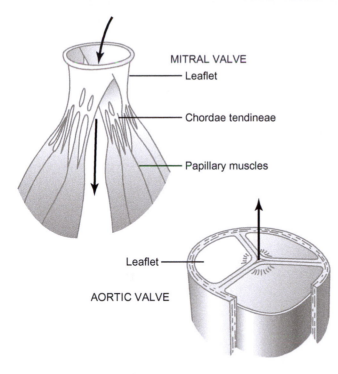

MITRAL VALVE
— Leaflet

— Chordae tendineae

— Papillary muscles

Leaflet —

AORTIC VALVE

**FIGURE 6.3** The mitral and aortic valves [*Reproduced by permission from Guyton and Hall (2006)*].

## The Cardiac Cycle

The AV and semilunar valves are critical to the *cardiac cycle*, which consists of systole and diastole. The events of the cardiac cycle are essentially the same for both ventricles, with similar ejected volumes as a result. However, the pressures driving systole and diastole in the right ventricle are much lower than the pressures in the left ventricle. For example, the semilunar valve opens in the left ventricle above 80 mmHg, but only above 8 mmHg in the right ventricle. We illustrate the events of the cardiac cycle by discussing these events for the left ventricle only.

Within systole, an isovolumetric contraction period and ejection period occur. At the onset of ventricular systole, the intraventricular pressure rises, causing the immediate closure of the mitral valve. At first, the mitral valve also remains closed, causing continual contraction around the incompressible contents. When the intraventricular pressure exceeds the diastolic pressure of about 80 mmHg, the aortic valve opens and blood begins to be expelled. Initially, the intraventricular pressure continues to rise until it reaches a maximum of about 120 mmHg. Toward the end of systole, the pressure begins to fall. As the volume curve of Figure 6.4 illustrates, under resting conditions, the ventricle ejects about half of its volume. This ejected volume of about 70 mL is the *stroke volume (SV)*, which results in an *ejection fraction (EF)* of about $70/120 \cong 60\%$. The resulting flow, or *cardiac output (CO)*, can be calculated using the *heart rate (HR)* as

$$CO = SV \cdot HR. \tag{6.1}$$

Normal cardiac output at rest is about 5 l/min.

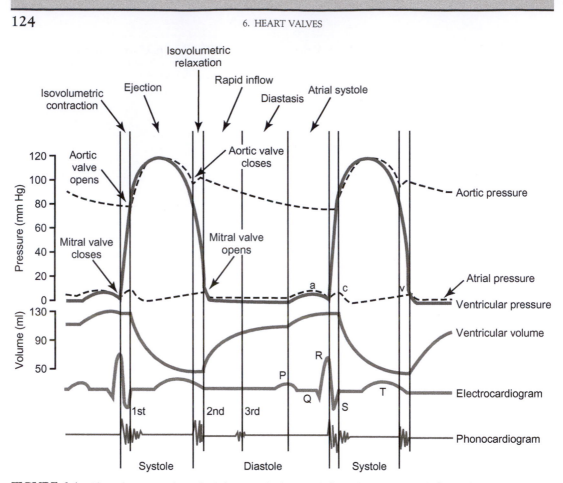

**FIGURE 6.4** Physiologic signals in the left ventricle that result from the cardiac cycle [*Reproduced by permission from Guyton and Hall (2006)*].

When the ejection phase is complete, an end-systolic volume of about 50 mL remains in the ventricle. Closure of the aortic valve marks the end of systole. At normal resting level, the duration of this contraction period, the *left ventricular ejection time (LVET)*, is about 150–300 ms in normal subjects.

During diastole, an isovolumetric relaxation period and filling period occur. Initially, for about 50 ms, both valves remain closed. The resulting relaxation is thus isovolumetric, as the intraventricular pressure falls rapidly to almost zero. When the pressure is lower than the atrial pressure, the mitral valve opens, causing the ventricle to fill in preparation for the next systole. During the filling period, the intraventricular pressure rises only slightly.

The cardiac cycle can be monitored by sound, as well as by cardiac pressures and voltages (electrocardiograms). When the AV valves close at the beginning of systole, a first

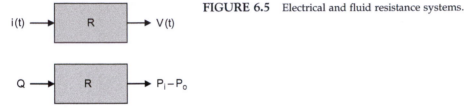

**FIGURE 6.5** Electrical and fluid resistance systems.

heart sound of low frequency and relatively long duration can be heard. When the semilunar valves close suddenly toward the end of systole, a second heart sound of high frequency and short duration can be heard. During the middle third of diastole, an occasional weak third heart sound may be heard (Figure 6.4) (Guyton & Hall, 2006). Some heart murmurs, or deviations from normal heart sounds, are linked to valvular disease.

## Fluid Mechanics

Much like Ohm's law, with

$$V(t) = i(t) \cdot R, \tag{6.2}$$

we can look at pressure as being the product of flow, $Q$, and fluid resistance, $R$, through Hagen-Poiseuille's equation (briefly discussed in Chapter 1) (Figure 6.5):

$$(P_i - P_o) = \frac{8\,\eta\,l_t Q}{\pi r_t^4}, \tag{6.3}$$

$$(P_i - P_o) = Q \cdot \frac{8\,\eta\,l_t}{\pi\,r_t^4} = Q \cdot R. \tag{6.4}$$

Here, $P_i$ = pressure at the tube inlet, $P_o$ = pressure at the tube outlet, $\eta$ = fluid viscosity, $r_t$ = total radius, and $l_t$ = total tube length. For Eq. (6.4) to be valid, this measurement is assumed to be made in a rigid cylindrical tube that is much longer than its radius. It is also assumed that the cylinder of fluid in the tube is a Newtonian, or uniform, fluid with laminar, or smooth, flow (Figure 6.6A).

At the tube outlet, the viscous force retarding fluid motion is equal and opposite to the force exerted by the pressure on the end of the cylinder of fluid. The viscous force, $F_v(r)$, is a function of radius and is equal to the product of the area of its surface, viscosity, and velocity gradient, $\frac{dv(r)}{dr}$. The force exerted by the pressure, $F_P(r)$, is the product of the pressure difference and cross-sectional area. This equality can be written as:

$$-F_v(r) = F_P(r), \tag{6.5}$$

$$-2\pi r l_t \cdot \eta \cdot \frac{dv(r)}{dr} = (P_i - P_o)\pi r^2, \tag{6.6}$$

Canceling common terms, rearranging, and integrating both sides result in

$$-2l_t \cdot \eta \cdot \frac{dv(r)}{dr} = (P_i - P_o)r, \tag{6.7}$$

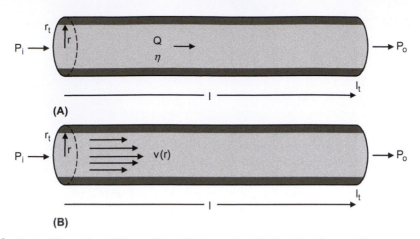

**FIGURE 6.6**   **A**: An illustration of Hagen-Poiseuille's equation. **B**: Resulting flow profile.

$$\frac{dv(r)}{dr} = \frac{-(P_i - P_o)r}{2\eta l_t}, \tag{6.8}$$

$$\int \frac{dv(r)}{dr} dr = \int \frac{-(P_i - P_o)r}{2\eta l_t} dr, \tag{6.9}$$

$$v(r) = \frac{-(P_i - P_o)r^2}{4\eta l_t} + c, \tag{6.10}$$

where $c$ is a constant. We assume the boundary condition that the velocity is 0 when $r$ equals the total radius, such that

$$v(r_t) = 0 = \frac{-(P_i - P_o)r_t^2}{4\eta l_t} + c, \tag{6.11}$$

$$c = \frac{(P_i - P_o)r_t^2}{4\eta l_t}. \tag{6.12}$$

Substituting Eq.(6.12) into Eq.(6.10) yields

$$v(r) = \frac{-(P_i - P_o)r^2}{4\eta l_t} + \frac{(P_i - P_o)r_t^2}{4\eta l_t}, \tag{6.13}$$

$$v(r) = \frac{(P_i - P_o)}{4\eta l_t}(r_t^2 - r^2). \tag{6.14}$$

Thus, the fluid velocity profile is parabolic, as shown in Figure 6.6B.

The shear stress, $\tau(r)$, for a Newtonian fluid is the stress encountered at the boundary condition of the total radius, which is equal to the product of the viscosity and velocity gradient:

$$\tau(r) = \eta \frac{dv(r)}{dr}. \tag{6.15}$$

Substituting Eq.(6.8) into Eq.(6.15) (Nichols & O'Rourke, 1998), we obtain

$$\tau(r) = \frac{-(P_i - P_o)r}{2l_t}. \tag{6.16}$$

The assumptions we have made are valid for a system such as saline solution moving through infusion pump tubing (Voss et al., 1997; Baura, 2002; Baura, 2008). But they do not approximate blood moving through cardiac chambers. For these large diameters, blood can be assumed to be a Newtonian fluid. However, blood flow does not involve a rigid cylindrical tube. Moreover, this pulsatile flow, with heart valves opening and closing, is not smooth (Nichols & O'Rourke, 1998). A more accurate model of cardiac fluid mechanics requires three-dimensional modeling based on radius, length, and angle, which is beyond the range of our discussion.

## BLOOD COAGULATION

In response to a native valve being surgically replaced by a prosthetic valve, a series of host reactions related to blood coagulation is initiated. As with the wound-healing response, blood-material interactions occur after blood contact with the implant surface. In this case, the host responses occur directly in the bloodstream, rather than in vascularized connective tissue. Platelet adhesion and aggregation lead to the platelet release reaction, coagulation activity, and consumption.

After platelets have aggregated on the implant surface, substances stored in platelet granules are secreted. This is known as the *platelet release reaction*. Alpha granule contents like platelet factor 4 are readily released. Dense granule contents like adenosine diphosphate, $Ca^{2+}$, and serotonin require platelet stimulation by the enzyme thrombin for release. These released and generated substances actively promote their incorporation into growing platelet aggregates.

Next comes platelet coagulation activity. Negatively charged membrane phosopholipids are expressed, which accelerate factor X activation and the conversion of prothrombin to thrombin in the clotting cascade. Thrombin activates platelets directly and generates threads of fibrin, which adhere to the surface of the platelet thrombus. Platelets are then consumed by the removal from circulating blood within several days (Hanson, 2004).

### Mechanical Valve Biocompatibility

The materials used in mechanical valves stimulate thrombosis and thromboembolic events (Figure 6.7A and B).

The most biocompatible valve material used is pyrolytic carbon, which is formed from the thermal decomposition of hydrocarbons, such as propane, propylene, acetylene, and

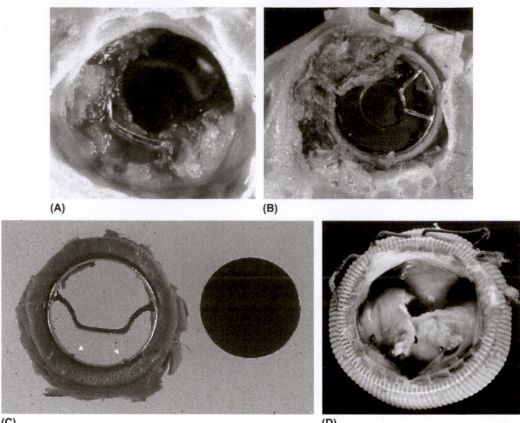

**FIGURE 6.7**  Prosthetic valve complications. **A**: Thrombosis on a Bjork-Shiley tilting disk aortic valve prosthesis, localized to outflow strut near minor orifice, a point of flow stasis. **B**: Thromboembolic infarct of the small bowel (arrow) secondary to embolus from valve prosthesis. **C**: Strut fracture of Bjork-Shiley valve, showing valve housing with single remaining strut and adjacent disk. **D**: Structural valve dysfunction (manifest as calcific degeneration with tear) of porcine valve [*B reproduced by permission from Schoen (2001); A and C reproduced by permission from Schoen et al. (1992). All four figures reproduced by permission from Padera & Schoen, 2004)*].

methane, in the absence of oxygen. The mechanism underlying pyrolytic carbon's relatively increased biocompatibility, compared to other materials such as silastic rubber, is poorly understood (More et al., 2004). All patients who choose mechanical, rather than tissue, valve replacements are placed on continued warfarin anticoagulation therapy for the lifetime of the valve (Bonow et al., 2006), which is 30–40 yr (Goldstein, 2007).

## CLINICAL NEED

In past decades, the high rate of valvular disease in the United States was caused by rheumatic fever, which is a complication of an untreated group A streptococcal infection (strep throat). Currently, valvular disease has decreased to approximately 2% of the U.S. population. This estimate is based on a recent study by Mayo Clinic researchers involving documented echocardiograph parameter estimates. An echocardiograph image of the heart is made with ultrasound measurements. In this study, echocardiograph data from a nationwide population of 11,911 patients and from a community population of 16,501 patients were examined. Prevalence of valvular disease increased with age (Nkomo et al., 2006).

According to American College of Cardiology/American Heart Association practice guidelines, an aortic valve should be replaced in symptomatic patients with severe aortic stenosis or severe aortic regurgitation. *Stenosis* refers to a narrowed valve. Similarly, a mitral valve should be replaced in symptomatic patients with moderate to severe mitral stenosis who also have moderate to severe mitral regurgitation.

Diagnosis of valve severity as mild, moderate, or severe is made using echocardiography. Stenosis classification involves threshold values of jet velocity, mean pressure gradient, and valve area. Regurgitation classification involves threshold values of regurgitant volume, regurgitant fraction, and regurgitant orifice area (Bonow et al., 2006).

## HISTORIC DEVICES

Prosthetic valve replacement would not have been possible without the development of surgeon John Gibbon's functional heart-lung machine in 1953, which enabled numerous new open heart surgical procedures to be investigated and performed. A heart-lung machine procedure is also known as *cardiopulmonary bypass*. The device removes carbon dioxide from venous blood, adds oxygen, and returns the blood to the aorta.

John Gibbon Jr. began his investigation of an extracorporeal circuit for an artificial heart and lung in 1931. Over 22 yr and countless animal experiments, he learned how to convert a stream flow of several liters of blood into a thin film exposed to oxygen and convert it back again to stream flow. He accomplished these conversions, with minimal damage to blood cells, using heparin and a screen oxygenator. The screen oxygenator consisted of several vertical screens hanging in a plastic case filled with an atmosphere of oxygen. Heparinized blood was filmed on the screens and exposed to oxygen. By maintaining a blood flow of 10 L/min, normal carbon dioxide tension was maintained (Gibbon, 1978).

In 1953, Gibbon successfully performed cardiopulmonary bypass with his then latest device, the Model II, on an 18-yr-old female patient with an atrial septal defect. The Model II consisted of a screen oxygenator and three roller pumps (Cohn, 2003). Gibbon received the 1968 Lasker Clinical Medical Research Award in 1968 for "designing and developing the heart-lung machine" (Lasker Awards, 2009).

### Early Devices

Cardiac surgeon Charles Hufnagel began experimenting with prosthetic valves in 1944. From 1952 to 1953, Hufnagel and his colleagues treated 23 patients with aortic insufficiency by rapidly inserting a ball valve into each patient's descending aorta. Each valve consisted of a methyl methacrylate ball, with an outer layer of polyethylene, held within a molded plastic tube. The valve prevented only regurgitant flow from the lower body (Hufnagel et al., 1954).

## Enabling Technology: Ball and Cage Valve

Cardiac surgeon Albert Starr, working with engineer M. Lowell Edwards, improved the ball design and performed the first successful orthotopic (in the normal position) valve replacement on September 21, 1960 (Chaikof, 2007). Starr and Edwards had experimented with other designs but settled on a silastic rubber ball, silicone-coated stainless steel cage, and Teflon sewing cuff to minimize thrombosis (Starr & Edwards, 1961a). The original design for a mitral valve was improved over the next 5 yr and manufactured for mitral, aortic, and tricuspid valve replacement by Edwards' new company, which became Edwards Lifesciences Corporation (Figure 6.8) (Starr, 2007). The Starr-Edwards valve's tenure is a testament to its design, as it was retired from manufacture only in 2007.

Mechanical valve designs contain three basic parts: an occluder, an occluder retention mechanism, and a sewing cuff. Over time, dozens of mechanical heart valve designs were developed. These designs fall into three basic classes: ball and cage valves, tilting and non-tilting disc valves, and bileaflet valves. The Bjork-Shiley valve, which was developed by cardiac surgeon Viking Bjork and engineer Don Shiley, is an example of a tilting disc valve. Here, a disc tilts within struts at an angle in the open position and totally occludes the valve orifice in the closed position. The St. Jude Medical (SJM) Masters Series bileaflet valve contains two semicircular occluders called leaflets, which pivot about a recessed hinge. The pyrolitic carbon leaflets allow blood flow through a central rectangular orifice and two semicircular lateral orifices in the open position, and some degree of leakage flow through small gaps in the closed position (Dasi et al., 2009) (Figure 6.9).

## Enabling Technology: Glutaraldehyde Treatment

In the early 1960s, after treating a patient with a mechanical valve replacement for a cerebral embolism, cardiac surgeon Alain Carpentier was inspired to conduct research on harvested tissue valves. The cerebral embolism began as a clot formed at the valve that migrated to the brain. Over several years of research, he discovered that stented porcine valves treated with glutaraldehyde could be used as tissue valves. Glutaraldehyde, which

Original
1960

Model 6000
1961–1965

Model 6120
1965–present

FIGURE 6.8 Evolution of the Starr-Edwards valve design *[Reproduced by permission from Starr, 2007)].*

is a colorless tanning agent, prevents collagen denaturation by intermolecular cross linkages and reduces thromboembolic risk. A tissue valve is composed of leaflets, an optional stent, and a sewing cuff. For this tissue valve, the leaflets are the treated porcine valve, the stent is the Teflon-covered stainless steel stent, and the sewing cuff is Dacron fabric covering the stent (Carpentier, 2007) (Figure 6.10).

After implantation of the tissue valve, an unexpected complication occurred. Tissue calcification emerged in patients younger than 60 yr of age (Figure 6.7D). These patients have less calcium absorption and greater calcium circulation in blood than older patients.

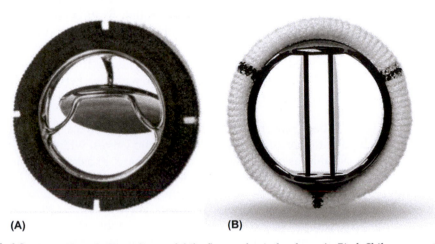

(A)  (B)

FIGURE 6.9  Examples of tilting disc and bileaflet mechanical valves: **A**: Bjork-Shiley monostrut convexo-concave valve; **B**: St. Jude Medical Masters Series valve *[A from Bjork (1985), reproduced by permission from Baura (2006); **B** from St. Jude Medical (St. Paul, Minnesota)].*

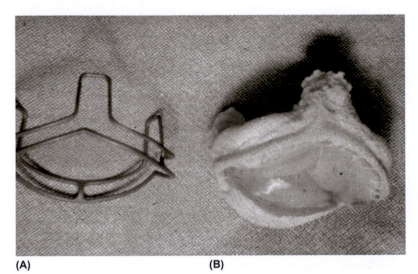

(A)  (B)

FIGURE 6.10 First glutaraldehyde-treated porcine valve implanted in human: **A**: Teflon-covered stainless steel stent used to support the valve, preventing its deformation during implantation; **B**: Stent covered with Dacron fabric to facilitate fixation and host-tissue incorporation. The porcine valve has been sutured into the stent preserving a full motion of the leaflets, avoiding fibrous incorporation *[From Carpentier et al. (1968). Reproduced by permission from Carpentier (2007)].*

Calcification in prosthetic tissue is associated with glutaraldehyde fixation processes and intrinsic tissue components (Demer, 1997; Grabenwoger et al., 1996). Calcium binding may result from major electrostatic attractions between calcium and residual glutaraldehydes and from attractions between calcium and membrane-associated phospholipids (Schoen and Levy, 2005, Zilla et al., 2000, Herrero et al., 1991).

Carpentier created a pericardial valve design to minimize flow turbulence, which contributes to calcification. The Carpentier-Edwards Perimount pericardial valve uses bovine pericardial tissue that has been shaped and stented into leaflets (Carpentier, 2007) (Figure 6.1). It is manufactured using the ThermaFix process, which targets calcium binding sites for calcium mitigation. A pH temperature-controlled fixation technique mitigates the formation of residual glutaraldehyde moieties. Ethyl alcohol and a surfactant extract phosoplipids from tissue.

These improvements increased tissue valve lifetime to about 15 yr. For 435 patients with a mean age of $60.7 \pm 11.6$ yr, who were implanted with the Perimount pericardial valve, freedom from explants due to structural valve deterioration at 14 yr was $83.4 \pm 2.3$ % (Marchand et al., 2001).

Alain Carpentier and Albert Starr received the 2007 Lasker Clinical Medical Research Award "for the development of prosthetic mitral and aortic valves, which have prolonged and enhanced the lives of millions of people with heart disease" (Lasker Awards, 2009).

The next improvement in heart valve design is the transcatheter valve. Many valve designs are being investigated. The two designs with the most patient experience are the Edwards SAPIEN and Medtronic CorValve heart valves. The SAPIEN valve is a bovine pericardium prosthesis mounted on a balloon-expandable stent. The CorValve is a porcine pericardial prosthesis mounted in a self-expanding nitinol frame. Both are deployed via catheter delivery (Zajarias and Cribier, 2009). We discuss catheter delivery systems in Chapter 8. When the transcatheter valve receives FDA approval, valve replacement will occur without open chest surgery (Figure 6.11).

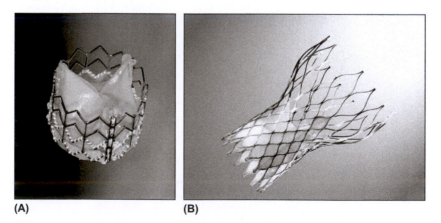

(A)                                      (B)

FIGURE 6.11  Two transcatheter heart valves: **A**: Edwards SAPIEN bovine pericardial valve; **B**: Medtronic CoreValve porcine pericardial valve *[Reproduced by permission from Zajarias & Cribier (2009)].*

## SYSTEM DESCRIPTION AND DIAGRAM

During surgical replacement, the diseased valve is excised. The tissue annulus diameter (TAD) is sized, with TAD measured as the diameter in mm of the smallest flow area within the patient's native valve annulus. A prosthetic valve of the proper size is attached to the native valve annulus. A biocompatibility system diagram for a prosthetic heart valve is given in Figure 6.12.

Ideally, a prosthetic valve produces minimal pressure drops, yields relatively small regurgitation volumes, minimizes production of turbulence and turbulent shear stresses, and does not create stagnation or flow separation regions in the immediate vicinity. A small pressure drop requires less energy to drive flow. The pressure drop is inversely proportional to the effective orifice area, $A_{EO}$, which is calculated from the Bernoulli equation as

$$A_{EO} = \frac{\text{Root mean square forward flow}}{51.6 \cdot \sqrt{\frac{(P_i - P_o)}{\eta}}}, \qquad (6.17)$$

with flow in units of ml/sec, pressure in units of mmHg, and density in units of $g/cm^3$.

Regurgitation occurs as reverse flow during valve closure and leakage (Figure 6.13), and is typically measured by stroke volume percentage. High shear stress leads to sublethal and/or lethal damage to blood cells and platelet activation. Regions of stagnation and flow separation lead to thrombus formation, tissue overgrowth, and/or calcification (Yoganathan et al., 2004).

### Comparison of Velocities and Turbulent Shear Stresses

For many years, velocity and turbulent shear stress have been measured *in vitro* by Yoganathan's group for various prosthetic valves. These measurements provide a framework for performance comparison. For example, we can examine the velocities and turbulent shear stresses downstream for the Starr-Edwards and SJM bileaflet valves at peak systole on the centerline. These measurements were made 30 mm downstream for the Starr-Edwards valve and 15 mm downstream for the SJM bileaflet valve (Figure 6.14).

The Starr-Edwards valve has low velocity reverse flow in the wake distal to the ball, whereas the SJM bileaflet ball has a more uniform velocity profile. The Starr-Edwards valve also experiences higher turbulent shear stresses than does the SJM valve. Clinically, the SJM bileaflet valve is the most commonly used mechanical valve.

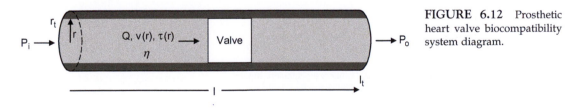

FIGURE 6.12 Prosthetic heart valve biocompatibility system diagram.

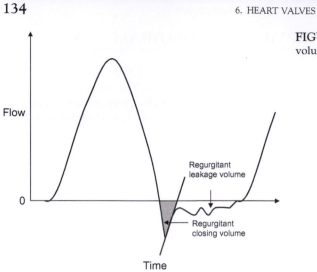

**FIGURE 6.13** Flow waveform and regurgitant volumes for one cycle [*Adapted from AAMI (2005)*].

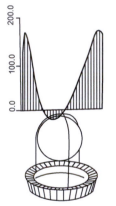

**(A)** Numbers are velocities in cm/s

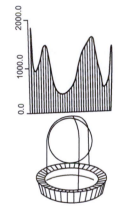

**(B)** Numbers are turbulent shear stresses in dynes/cm²

**FIGURE 6.14** Velocity and turbulent shear stress measurements for aortic Starr-Edwards and St. Jude Medical bileaflet valves. The Starr-Edwards velocity (**A**) and turbulent shear stress (**B**) measurements were made 30 mm downstream on the centerline; the St. Jude Medical velocity (**C**) and turbulent shear stress (**D**) measurements were made 13 mm downstream on the centerline across the major and minor orifices [*Reproduced by permission from Yoganathan (1995)*].

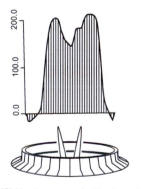

**(C)** Numbers are velocities in cm/s

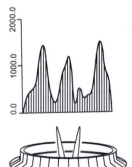

**(D)** Numbers are turbulent shear stresses in dynes/cm²

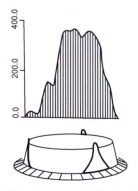 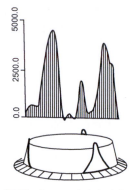

**(A)** Numbers are velocities in cm/s

**(B)** Numbers are turbulent shear
stresses in dynes/cm$^2$

FIGURE 6.15 Velocity and turbulent shear stress measurements for aortic Carpentier-Edwards 2625 porcine
and pericardial valves. The porcine
valve velocity (**A**) and turbulent shear
stress (**B**) measurements were made
15 mm downstream on the centerline;
the pericardial valve velocity (**C**) and
turbulent shear stress (**D**) measurements were made 17 mm downstream
on the centerline [*Reproduced by permission from Yoganathan (1995)*].

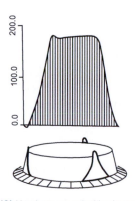 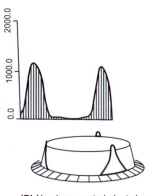

**(C)** Numbers are velocities in cm/s

**(D)** Numbers are turbulent shear
stresses in dynes/cm$^2$

Similarly, we can compare the velocity and turbulent shear stresses for the Carpentier-
Edwards 2625 porcine and pericardial tissue valves. The pericardial valve has a much
more uniform and velocity profile than does the porcine valve. It also experiences much
less turbulent sheer stresses of lower amplitude than does the porcine valve (Yoganathan
et al., 2004) (Figure 6.15).

Clinically, Albert Starr's group compared the long-term performance of these Carpentier-
Edwards pericardial and porcine valves in 1539 patients, who had similar age distribution
and clinical profiles. At 10 yr, the survival rates, freedom from major adverse cardiac events,
and endocarditis (inflammation of inner heart lining) were insignificantly different.
However, freedom from explants was higher for the pericardial versus porcine group
($97 \pm 1\%$ vs. $90 \pm 2\%$, $p = 0.04$), due to less structural valve deterioration (Gao et al., 2004).

## FDA CASE STUDY: BJORK-SHILEY HEART VALVE

After the Medical Device Amendments of 1976 were enacted, medical device manufacturers were required to request market release for their devices through 510(k) or

premarket approval submissions to the FDA. In these early days of device regulation, submissions were very short in length. For example, a 510(k) submission might be five pages long and would briefly describe its substantial equivalence to a device on the market before May 28, 1976 (Kahan, 2004).

The Bjork-Shiley valve was developed by cardiac surgeon Viking Bjork and engineer Don Shiley, and was manufactured by Shiley's company Shiley, Inc. Originally constructed with a flat disc held between two welded struts, the first implantation occurred in 1969. In 1976, the valve design was changed to a convexo-concave (C-C) disc, with the inflow strut becoming an integral part of the valve ring without welds, but the outflow strut still welded. The disc continued to open in an aortic position of 60° (Figure 6.7A) (Bjork, 1985). Shiley attempted to obtained to obtain FDA notification by submitting two 510(k)s demonstrating substantial equivalence to the original valve design. However, since Shiley made the design change to allegedly reduce thromboembolism, the FDA insisted that clinical data to support this claim be submitted.

Shiley's PMA submission for this 60° C-C Bjork-Shiley valve became the first PMA submission for a heart valve. During a premarket approval panel meeting on February 2, 1979, Shiley provided no specific clinical data on the 60° convexo-concave valve and no proof of reduced thromboembolism. Even though one strut fracture (Figure 6.7C) had occurred during clinical trials, Shiley characterized the fracture as an "isolated" incident that was being investigated. When a strut fracture occurs, the disc escapes, causing uncontrolled blood flow through the heart. Death results in about two out of three strut fractures.

The panel agreed that the application would be "approvable" upon receipt of data regarding thromboembolism and mortality. Shiley sent Bjork's data on blood, hemodynamics, mortality, and thromboembolism to the FDA on February 6, 1979, which were then distributed to panel members for comment. When no comments were received to delay approval, the valve was approved on April 27, 1979. On March 8, 1979, Shiley, Inc., was acquired by Pfizer.

As of January 11, 1990, when data were compiled for a U.S. House of Representatives investigation, 40,000 60° valves had been implanted in the United States. Outside the United States, 41,791 60° and 4030 70° valves were implanted. Shiley had reported 389 valve strut fractures, with 248 deaths (64%). A full 2.3% of implanted 70° valves had fractured, with 1.7% causing death. These fractures were probably underreported because hospitals did not report all fractures, and sudden deaths were not always recognized as related to fractures.

## Public Shiley Actions

Publicly, Shiley attempted to minimize outlet strut fracture by increasing the disc opening angle from 60° to 70°. According to the PMA Supplement it filed in October 1980, which was never approved by the FDA because no clinical study data were included, the increased angle would "significantly reduce the opening forces of the disc upon the downstream strut, with no significant changes on the closing impact" (U.S. Congress, 1990).

The 70° convexo-concave valve was seen as an interim measure before the final modification to a 70° monostrut valve (Figure 6.9A), in which the valve was machined from a single piece of metal with no welds.

Shiley also initiated six medical-device recalls and product withdrawals between 1980 and January 1990. These were all voluntary actions, as the FDA did not have the power at this time to initiate recalls and product withdrawals. Between 1980 and 1983, four recalls occurred for fractures and changes to manufacturing or testing procedures. In 1985, 29-, 31-, and 33-mm 60° valves made between March 1, 1981 and June 30, 1982 were recalled. Those made after June 30, 1982 were withdrawn from market, with production ceased. In 1986, the remaining sizes of 60° valves (19, 21, 23, 25, 27 mm) were withdrawn, with production ceased. In four of these cases, the FDA was not notified before the actions were already underway.

During the 1980s, Shiley continually provided the medical community with incomplete and misleading information on strut fractures. So-called Dear Doctor letters asked only implanting surgeons, but not patients' physicians, to monitor their patients with certain valves, but they did not provide any details of what this monitoring entailed. Even though a condition of premarket approval was reporting of an adverse reaction to the FDA within 10 days, many of the fracture reports were delayed from 3 weeks to 24 months.

## Internal Shiley Actions

Internally, Shiley conducted numerous studies which concluded that fractures were caused by a fatigue problem that could be accelerated by manufacturing deficiencies (U.S. Congress, 1990). An investigation initiated by the chief executive officer of Pfizer's Hospital Products Group in 1983 determined that welding during the manufacturing process was "out of control—not monitored," that welding patterns were nonuniform, and that final inspection of valves before shipping was inadequate (U.S. Congress, 1990). [In 1998, it was determined that <0.5-ms outlet-strut-tip impacts due to closing disc over-rotation in patients with high dP/dt could fatigue the outlet strut over time (Wieting et al., 1999).]

In response, Shiley made at least five major changes in manufacturing and quality control procedures to attempt to control strut fractures. Because it was not legally required to do so, Shiley did not notify the FDA of these changes beforehand.

In 1980, early 70° valves were fabricated from existing flanges originally machined to specification for the 60° valve. The inlet struts of these early valves were remilled to allow the additional 10° opening angle. Only in 1982 were valves fabricated from flanges initially machined to 70° specification. According to one Shiley engineer, remilling left very high levels of residual stress in the outlet strut, the area in which the fractures occurred (U.S. Congress, 1990). As shown in Figure 6.16, a good weld was consistent with homogeneous dendritic formations following the energy path from the welding torch. But some welds, such as those converted from 60° to 70°, demonstrated brittle areas within the weld, resulting in phase segregation. Cracks could form in these brittle areas in subsequent manufacturing steps or during valve implantation, and propagate during constant clinical cycling (Bjork, 1985). It was later demonstrated that 70° valves fractured at a rate seven times that of the 60° valve (U.S. Congress, 1990).

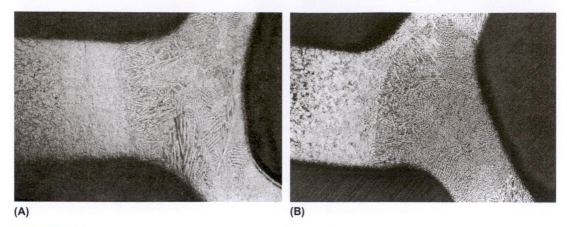

**(A)**                                                **(B)**

FIGURE 6.16    Photographs of a section through a weld. **A**: Perfect weld shows homogenous dendrite formation through the outflow strut in the drill hole and continuing 2 mm into the strut. **B**: Imperfect weld show dendrite structure has not penetrated all of the outflow strut in the upper ring-strut angle, producing an anterior area
of more brittle phase segregation *[From (Bjork (1985). Reproduced by permission from Baura, 2006)].*

Shiley's damage control tactics included efforts to silence its primary clinical investiga-
tor, Viking Bjork. Paul Morris, director of product engineering, sent a telex to Bjork in 1980
stating that "we would prefer that you did not publish the data relative to strut fractures.
We expect a few more and until the problem has been corrected, we do not feel comfortable" (U.S. Congress, 1990). When the first five 70° valve fractures occurred in Sweden
between November 1981 and June 1982, Bjork notified Shiley promptly but never notified
the FDA or the Swedish Board of Health. The Swedish Board later strongly censured Bjork
for his lack of judgment in not reporting the fractures promptly to the board, as Swedish
regulations require.

## FDA Responses

The FDA undermined its own authority during its interactions with Shiley. The agency
failed to heed the warning and reports of product failure from both the Los Angeles
District Office and internal compliance staff, and it failed to suspend manufacturing operations until a solution to the strut fracture problem could be identified. The Los Angeles
District Office conducted four manufacturing inspection visits in 1979, 1982, 1983, and
1985, and each time reported extensive manufacturing problems to the FDA. When the
FDA requested that the District Office conduct yet another inspection in 1985, the compli-
ance officer involved sent a letter asking the FDA, "When is enough, enough?" (U.S.
Congress, 1990). Although the FDA refused to allow the 70° valve to be marketed in the
United States because no clinical data had been submitted, it allowed 70° valves to be
exported outside the country until 1983.

In the absence of receiving timely and accurate reports of adverse events, the FDA
failed to aggressively monitor and demand product performance information from Shiley.
When the agency asked Shiley in September 1985 for specific documentation that would

be used to assess manufacturing and quality assurance procedures, Shiley responded with over 300 documents that required more than 6 months to analyze. The FDA engineer who analyzed these data described it as "data dump without analysis" (U.S. Congress, 1990). When the FDA determined that the data were inconclusive, it called for a planned review by the Medical Devices Advisory Panel in October 1986. Shiley recalled all remaining valves from the market before this meeting could take place.

In 1987, the FDA asked Shiley for in-tolerance, marginal, and defective recalled valves so that the FDA could conduct *in vitro* life testing. When no valves broke during testing, FDA engineers discovered that Shiley had sent all good valves. The FDA again requested recalled valves but acquiesced when Shiley would not deliver the requested valves. As of 1990, the FDA had not withdrawn the valve's premarket approval.

After the Medical Device Amendments passed in 1976, the FDA did not specify the requirement for manufacturing change notification until 1986 (U.S. Congress, 1990).

## Aftermath

In 1979, the United States filed a civil suit against Shiley and Pfizer, under the False Claims Act and common law. On July 1, 1994, Shiley agreed to pay the federal government $10.75 million, to recover payments from Medical and the Veterans Health Administration for heart valves (multiplied by a factor of two and a half times for penalties), and pay $20 million for future medical costs. Shiley admitted no liability as part of this settlement (Michaud, 1994). In 1992, the class-action suit *Arthur Ray Bowling, et al. v. Pfizer Inc., et al.* was settled. With that decision, $80 million was placed into a Consultation Fund for claimant medical and psychological consultation, and $12.5 million was placed into a Patient Benefit Fund for research on strut valve fraction research and evaluation (Spiegel, 1992).

In 1990, the U.S. House of Representatives Subcommittee on Oversight and Investigations of the Committee on Energy and Commerce conducted an investigation of the Bjork-Shiley heart valve that they subtitled "Earn as You Learn" (U.S. Congress, 1990). The Safe Medical Devices Act of 1990 is a direct result of this investigation (Baura, 2006). Some of the major provisions of this Act are as follows:

- The FDA has the authority to recall devices.
- The FDA may temporarily suspend a PMA with cause.
- Manufacturers must report device removal or correction promptly to the FDA.
- Manufacturers of high-risk devices must track each device to the patient.
- User facilities must report medical device adverse events (Safe Medical Devices Act of 1990, 1990).

For a more thorough discussion of FDA regulation, see (Baura, 2013).

## KEY FEATURES FROM ENGINEERING STANDARDS

As would be expected, design and manufacture of prosthetic heart valves have become tightly controlled in the United States. As of 2010, the FDA recognizes consensus standard *ANSI/AAMI ISO 5840:2005 Cardiovascular Implants—Cardiac Valve Prostheses* (AAMI, 2005).

## Hydrodynamic Performance

A valve design must meet minimum requirements of $A_{EO}$ and regurgitant fraction for various TAD sizes. Tests are conducted on three samples/size, in a test chamber with heart rate = 70 bpm, simulated $CO = 5.0\,l/min$, mean aortic pressure = 100 mmHg, and systolic duration = 35%.

- For aortic valve sizes of 19, 21, 23, 25, 27, 29, and 31 mm, the minimally acceptable $A_{EO}$ is 0.70, 0.85, 1.00, 1.20, 1.40, 1.60, and 1.80 cm$^2$, respectively.
- For mitral valve sizes of 25, 27, 29, and 31 mm, the minimally acceptable $A_{EO}$ is 1.20, 1.40, 1.60, and 1.80 cm$^2$, respectively.
- For aortic sizes of 10, 21, and 23 mm, the regurgitant fraction must be within 10%.
- For aortic or mitral valve sizes of 25 and 27 mm, the regurgitant fraction must be within 15%.
- For aortic or mitral valve sizes of 29 or 31 mm, the regurgitant fraction must be within 20% (AAMI, 2005).

## Component Fatigue

Each manufacturer must determine and justify an appropriate approach to fatigue assessment of a valve. Suggestions for fatigue assessment are given, but specific tests are not mandated. With this approach, the lifetime of each structural component is determined as the minimum duration for which each component can withstand anticipated repeated loadings associated with *in vivo* conditions. Initial *in vitro* test results are later strengthened through observation of structural valve deterioration rates during 1-yr and follow-up human clinical trials (AAMI, 2005).

## One-Year Clinical Study

After preclinical *in vivo* testing in animals for a new valve design has been successful, human testing must be conducted in at least 8 institutions. At least 150 aortic valve patients and 150 mitral valve patients must be evaluated. After 1 yr and at least 400 aortic valve years and at least 400 mitral valve years have passed, safety and efficacy are assessed by determining the rates of performance and other complications. If complication rates are within the rates in Tables 6.1 and 6.2, no further justification is required (AAMI, 2005).

## Clinical Study Long-Term Follow-Up

Additionally, at least 150 patients from the original cohort must be followed for at least 5 or 10 yr after implantation, depending on whether the valve is rigid or flexible, respectively. Selection of specific patients must be statistically justified to minimize selection bias. The follow-up data to be collected are rates of performance and other complications, blood studies, cardiac rhythm, reoperation reports, explant analysis, and dates/causes of death (AAMI, 2005).

**TABLE 6.1**   Acceptable Performance Complication Rates (%)

| Performance Complication | Rigid Valve | Flexible Valve |
|---|---|---|
| Thromboembolism | 3.0 | 2.5 |
| Valve thrombosis | 0.8 | 0.2 |
| All hemorrhage | 3.5 | 1.4 |
| Major hemorrhage | 1.5 | 0.9 |
| All paravalvular leak | 1.2 | 1.2 |
| Major paravalvular leak | 0.6 | 0.6 |
| Endocarditis | 1.2 | 1.2 |

*Reproduced with permission from AAMI (2005).*

**TABLE 6.2**   Acceptable Other Complication Rates. RIND = Reversible Ischemic Neurologic Deficit (%)

| Complication | Aortic | | Mitral | |
|---|---|---|---|---|
| | Rigid Valve | Flexible Valve | Rigid Valve | Flexible Valve |
| Structural valve deterioration | 0.00 | 0.030 | 0.00 | 0.12 |
| Thromboembolism (Major, RIND) | 2.78 | 2.06 | 2.63 | 2.48 |
| Valve thrombosis | 0.00 | 0.10 | 0.61 | 0.19 |
| Anticoagulant-related hemorrhage | 2.44 | 0.45 | 1.95 | 0.80 |
| Prosthetic valve endocarditis | 0.93 | 0.59 | 0.54 | 0.68 |
| Nonstructural valve dysfunction/paravalvular leak | 0.84 | 0.38 | 1.75 | 1.05 |
| Reoperation | 1.09 | 0.77 | 1.95 | 1.05 |

*Reproduced by permission from AAMI (2005).*

## Biocompatibility

*ANSI/AAMI ISO 5840:2005* also states that heart valve material biocompatibility shall be determined in accordance with *ISO 10993-1:2009 Biological Evaluation of Medical Devices— Part 1: Evaluation and Testing Within a Risk Management Process* (AAMI, 2005). In this Part 1 of the ISO standard, a test matrix for appropriate testing of surface devices, external communicating devices, and implant devices is given. This matrix points to tests in other parts of the standard that assess cytotoxicity, sensitization, irritation, toxicity, genotoxicity, implantation, hemocompatibility, and carcinogenicity (ISO, 2009).

## SUMMARY

A prosthetic heart valve is a medical device that replaces a diseased native valve and that enables blood flow in the cardiac chambers to be regulated. A mechanical valve consists of an occluder, an occluder retention mechanism, and a sewing cuff. A typical tissue valve consists of porcine or bovine tissue, a stent, and a sewing cuff. The porcine leaflets or bovine pericardial cutouts are treated with glutaraldehyde to prevent collagen denaturation.

Each beat of the cardiac cycle begins with electrical stimulation and ends with valve closure. At the onset of ventricular systole in the left heart, the intraventricular pressure rises, causing immediate closure of the mitral valve. At first, the mitral valve also remains closed, causing continual contraction around the incompressible contents. When the intraventricular pressure exceeds the diastolic pressure of about 80 mmHg, the aortic valve opens and blood begins to be expelled. Initially, the intraventricular pressure continues to rise, until it reaches a maximum of about 120 mmHg. Toward the end of systole, the pressure begins to fall. Under resting conditions, the ventricle ejects about half of the volume it contains.

At the onset of diastole, for about 50 ms, both valves remain closed. The resulting relaxation is thus isovolumetric, as the intraventricular pressure falls rapidly to almost zero. When the pressure is lower than the atrial pressure, the mitral valve opens, causing the ventricle to fill, in preparation for the next systole. During the filling period, the intraventricular pressure rises only slightly.

Replacement of a native valve is indicated when the valve experiences severe stenosis or regurgitation. Both mechanical and tissue valves possess biocompatibility issues. Patients who choose mechanical, rather than tissue, valve replacements are placed on continued warfarin anticoagulation therapy for the lifetime of the valve, which is 30–40 yr, to minimize thrombosis. Patients who choose tissue valve replacements do not require this medication but will have to replace their valves in about 15 yr due to calcification.

One of the early mechanical valves was the Bjork-Shiley valve, which could fracture due to a fatigue issue that was accelerated by manufacturing deficiencies. The U.S. House of Representatives investigation of this defective valve precipitated the Safe Medical Devices Act of 1990, which gave the FDA the authority to recall medical devices and to conduct postmarket surveillance. As would be expected, heart valves have become subject to strict design and manufacture requirements. Some of these requirements target hydrodynamic performance, component fatigue, 1-yr clinical study, clinical study long term follow-up, and biocompatibility.

## Acknowledgment

The author thanks Dr. Hugues LaFrance for his detailed review. Dr. LaFrance is Senior Manager of Heart Valve Therapy R&D Engineering at Edwards Lifesciences LLC.

## Exercises

**6.1** Calculate and plot the velocity profile for increasing radii (in 3-cm increments) in an idealized rigid artery, assuming length = 100 cm, viscosity = 0.04 dyne-s/cm$^2$, input pressure = 100 mmHg, output pressure = 80 mmHg, and radius = 24 cm. Check your units as you make calculations. Also calculate the shear stress for these increasing radii.

**6.2** Insert a prosthetic valve in the artery of Ex. 6.1, assuming that the length of the valve is 25 cm, with a root mean square forward flow of 5 L/min. Calculate the pressure at the output of the valve and the effective orifice area.

**6.3** Please respond to the following statement: "Why are we reading about a case study that only had 0.5% (389/81,791) total device failures?"

**6.4** How is the manufacturing conversion of 60° to 70° Bjork-Shiley convexo-concave valves similar to the Guidant ICD case study? What was the underlying reason that these Shiley and Guidant specific actions were taken?

Read Starr & Edwards (1961a):

**6.5** Before the Starr-Edwards valves were implanted in humans, what was the typical length of survival in patients with implanted heart valves? What three issues prevented long-term survival? Do you obtain the same implant results with canines as with humans?

**6.6** How were patients chosen for the Starr-Edwards implants? If these human studies were to take place today (assuming the same alternatives for treatment as described in the article and considering that the FDA currently oversees medical device regulation), would the same patients be chosen? Explain.

Read Starr & Edwards (1961b):

**6.7** Describe the design iterations of the caged ball valve, starting with the original Lucite cage and stainless steel fixation (look at both articles). Why was each design change implemented?

**6.8** Describe how thrombus affects an implanted caged ball valve over time. What design improvements can minimize thrombus?

# References

AAMI (2005). *ANSI/AAMI/ISO 5840:2005 cardiovascular implants—cardiac valve prostheses.* Arlington, VA: AAMI.

Baura, G. D. (2002). *System theory and practical applications of biomedical signals.* Hoboken, NJ: Wiley-IEEE Press.

Baura, G. D. (2006). *Engineering ethics: An industrial perspective.* Burlington, VT: Elsevier Academic Press.

Baura, G. D. (2008). *A biosystems approach to industrial patient monitoring and diagnostic devices.* San Rafael, CA: Morgan Claypool.

Baura, G. D. (2013). *U.S. medical device regulation: An introduction to biomedical product development, market release, and postmarket surveillance.* Manuscript in preparation.

Bjork, V. O. (1985). Metallurgic and design development in response to mechanical dysfunction of Bjork-Shiley heart valves. *Scandinavian Journal of Thoracic and Cardiovascular Surgery, 19,* 1–12.

Bonow, R. O., Carabello, B. A., Chatterjee, K., Deleon, A. C., Jr., Faxon, D. P., & Freed, M. D., et al. (2006). ACC/AHA 2006 guidelines for the management of patients with valvular heart disease: A report of the American College of Cardiology/American Heart Association Task Force on Practice Guidelines (writing Committee to revise the 1998 guidelines for the management of patients with valvular heart disease) developed in collaboration with the Society of Cardiovascular Anesthesiologists endorsed by the Society for Cardiovascular Angiography and Interventions and the Society of Thoracic Surgeons. *Journal of the American College of Cardiology, 48,* e1–e148.

Carpentier, A. (2007). Lasker Clinical Research Award. The surprising rise of nonthrombogenic valvular surgery. *Nature Medicine, 13,* 1165–1168.

Carpentier, A., Blondeau, P., Laurens, B., Hay, A., Laurent, D., & Dubost, C. (1968). Mitral and tricuspid valve replacement with frame-mounted aortic heterografts. *Journal of Thoracic and Cardiovascular Surgery, 56,* 388–394.

Chaikof, E. L. (2007). The development of prosthetic heart valves–lessons in form and function. *New England Journal of Medicine, 357,* 1368–1371.

Cohn, L. H. (2003). Fifty years of open-heart surgery. *Circulation, 107,* 2168–2170.

Dasi, L. P., Simon, H. A., Sucosky, P., & Yoganathan, A. P. (2009). Fluid mechanics of artificial heart valves. *Clinical and Experimental Pharmacology and Physiology, 36,* 225–237.

Demer, L. L. (1997). Lipid hypothesis of cardiovascular calcification. *Circulation, 95,* 297–298.

Gao, G., Wu, Y., Grunkemeier, G. L., Furnary, A. P., & Starr, A. (2004). Durability of pericardial versus porcine aortic valves. *Journal of the American College of Cardiology, 44,* 384–388.

Gibbon, J. H., Jr. (1978). The development of the heart-lung apparatus. *American Journal of Surgery, 135,* 608–619.

Goldstein, J. L. (2007). Creation and revelation: two different routes to advancement in the biomedical sciences. *Nature Medicine, 13,* 1151–1154.

Grabenwoger, M., Sider, J., Fitzal, F., Zelenka, C., Windberger, U., & Grimm, M., et al. (1996). Impact of glutaral-dehyde on calcification of pericardial bioprosthetic heart valve material. *Annals of Thoracic Surgery, 62,* 772–777.

Guyton, A. C., & Hall, J. E. (2006). *Textbook of medical physiology* (11th ed.). Philadelphia: Elsevier Saunders.

Hanson, S. R. (2004). Blood coagulation and blood-materials interactions. In B. D. Ratner, A. S. Hoffman, F. J. Schoen, & J. E. Lemons (Eds.), *Biomaterials science* (2nd ed.). San Diego, CA: Elsevier.

Herrero, E. J., Gutierrez, M. P., Escudero, C., & Castilloolivares, J. L. (1991). Inhibition of Bovine Pericardium Calcification—A Comparative-Study of Al-3+ and Lipid Removing Treatments. *Journal of Materials Science—Materials in Medicine, 2,* 86–88.

Hufnagel, C. A., Harvey, W. P., Rabil, P. J., & McDermott, T. F. (1954). Surgical correction of aortic insufficiency. *Surgery, 35,* 673–683.

ISO (2009). *ISO 10993-1:2009 biological evaluation of medical devices—Part 1: Evaluation and testing within a risk management process.* Geneva: ISO.

Kahan, J. S. (2004). Device regulation: Policies, practices, and procedures. *MD&DI, 26,* 74–82.

Lasker Awards: Former Winners. (2009). Lasker Foundation.

Marchand, M. A., Aupart, M. R., Norton, R., Goldsmith, I. R. A., Pelletier, L. C., & Pellerin, M., et al. (2001). Fifteen-year experience with the mitral Carpentier-Edwards PERIMOUNT pericardial bioprosthesis. *Annals of Thoracic Surgery, 71,* S236–S239.

Michaud, A. (1994). Shiley Inc. settles false-claims suit courses: Irvine-based firm and its parent to pay $10.75 million to U.S. in case involving flawed artificial heart valves. *LA Times* (July 1).

More, R. B., Haubold, A. D., & Bokros, J. C. (2004). Pyrolytic carbon for long-term medical implants. In B. D. Ratner, A. S. Hoffman, F. J. Schoen, & J. E. Lemons (Eds.), *Biomaterials science* (2nd ed.). San Diego, CA: Elsevier.

Nichols, W. W., & O'Rourke, M. F. (1998). *McDonald's blood flow in arteries* (4th ed.). New York: Oxford University Press.

Nkomo, V. T., Gardin, J. M., Skelton, T. N., Gottdiener, J. S., Scott, C. G., & Enriquez-Sarano, M. (2006). Burden of valvular heart diseases: A population-based study. *Lancet, 368,* 1005–1011.

Padera, R. F., Jr., & Schoen, F. J. (2004). Cardiovascular medical devices. In B. D. Ratner, A. S. Hoffman, F. J. Schoen, & J. E. Lemons (Eds.), *Biomaterials science* (2nd ed.). San Diego, CA: Elsevier.

Safe Medical Device Act of 1990 (1990). Public Law 101-629. U.S. Code amendment to the Federal Food Drug and Cosmetics Act.

Schoen, F. J. (2001). Pathology of heart valve substitution with mechanical and tissue prostheses. In M. D. Silver, A. I. Gotlieb, & F. J. Schoen (Eds.), *Cardiovascular pathology* (3rd ed.). New York: Churchill Livingstone.

Schoen, F. J., & Levy, R. J. (2005). Calcification of tissue heart valve substitutes: Progress toward understanding and prevention. *Annals of Thoracic Surgery, 79,* 1072–1080.

Schoen, F. J., Levy, R. J., & Piehler, H. R. (1992). Pathological considerations in replacement cardiac valves. *Cardiovascular Pathology, 1,* 29.

Spiegel, S. A. (1992). *Arthur Ray Bowling v. Pfizer Inc. Supplemental agreement of compromise and settlement, C-1-91-256.* Cinncinati, OH: United States District Court.

Starr, A. (2007). Lasker Clinical Medical Research Award. The artificial heart valve. *Nature Medicine, 13,* 1160–1164.

Starr, A., & Edwards, M. L. (1961a). Mitral replacement: Clinical experience with a ball-valve prosthesis. *Annals of Surgery, 154,* 726–740.

Starr, A., & Edwards, M. L. (1961b). Mitral replacement: The shielded ball valve prosthesis. *Journal of Thoracic and Cardiovascular Surgery, 42,* 673–682.

U.S. Congress. (1990). *The Bjork-Shiley Heart Valve: "Earn as You Learn," Committee Print 101-R.* Comittee on energy and commerce subcommittee on oversight and investigations. Washington, DC: U.S. Government Printing Office.

Voss, G. I., Butterfield, R. D., Baura, G. D., & Barnes, C. W. (1997). Fluid flow impedance monitoring system. *US 5,609,576.*

Wieting, D. W., Eberhardt, A. C., Reul, H., Breznock, E. M., Schreck, S. G., & Chandler, J. G. (1999). Strut fracture mechanisms of the Bjork-Shiley convexo-concave heart valve. *Journal of Heart Valve Disease, 8,* 206–217.

Yoganathan, A. P. (1995). Cardiac valve prostheses. In J. D. Bronzino (Ed.), *The biomedical engineering handbook.* Salem, MA: CRC Press.

Yoganathan, A. P., He, Z., & Casey Jones, S. (2004). Fluid mechanics of heart valves. *Annual Review of Biomedical Engineering, 6,* 331–362.

Zajarias, A., & Cribier, A. G. (2009). Outcomes and safety of percutaneous aortic valve replacement. *Journal of the American College of Cardiology, 53,* 1829–1836.

Zilla, P., Weissenstein, C., Human, P., Dower, T., & Von Oppell, U. O. (2000). High glutaraldehyde concentrations mitigate bioprosthetic root calcification in the sheep model. *Annals of Thoracic Surgery, 70,* 2091–2095.

CHAPTER

# 7

# Blood Pressure Monitors

In this chapter, we discuss the relevant physiology, history, and key features of blood pressure (BP) monitors. A blood pressure monitor is an instrument that enables systolic, diastolic, and often mean blood pressures to be measured and displayed (Figure 7.1).

Upon completion of this chapter, each student shall be able to:

1. Understand the importance of blood pressure evaluation.
2. Identify three blood pressure measurement techniques.
3. Describe five key features of blood pressure monitors.

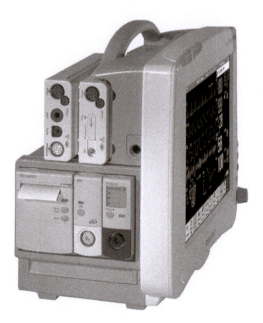

**FIGURE 7.1**    Philips Intellivue MP50 Patient Monitor with optional Multi-measurement Module (top left) that provides oscillometric and arterial blood pressure measurements: The MP50 is part of a family of patient monitors (MP20—MP70) that are optimized for various patient care environments (Philips Healthcare, Andover Massachusetts).

# BLOOD PRESSURE PROPAGATION

Oxygenated blood is ejected from the left ventricle during each cardiac beat and travels from the aorta to smaller arteries, arterioles, and capillaries. In concert, the pressure wave in the aorta propagates through the peripheral arteries. If the systemic vasculature could be modeled as a rigid cylindrical tube, then the pressure contour associated with each beat would remain constant. But, as shown in Figure 7.2, the pressure contour changes as the pressure wave travels toward smaller vessels.

Notice that the incisura in the proximal aorta waveform, which originates during aortic valve closure (see Figure 6.4), changes as the pressure wave propagates. The pressure contour changes result from traveling wave reflections. When a traveling wave experiences an impedance change, a portion of the traveling wave is reflected back through the system. Impedance changes occur at branching points, areas of alteration in arterial compliance, and high-resistance arterioles. Propagation through the upper body arteries results in different contour changes compared to propagation through the lower body arteries. These differences have been attributed to arterial differences due to age and wave velocity (Nichols & O'Rourke, 1998).

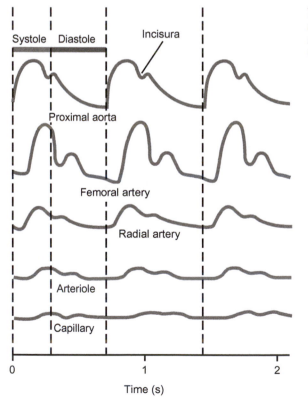

**FIGURE 7.2** The pressure contour changes as a pressure wave travels from the aorta toward smaller vessels [*Reproduced by permission from Guyton & Hall, 2006*].

At each artery, the systolic, diastolic, and mean arterial pressures can be measured. The systolic and diastolic pressures are the peak and minimum pressures measured during a cardiac beat, respectively. The *mean arterial pressure* (*MAP*) is the arithmetic mean of all pressures encountered during a cardiac beat. The mean is often grossly approximated by clinicians using the 1/3–2/3 rule, which amounts to estimating the mean as 1/3 the distance between diastolic and systolic pressures. For example, for BPs of 100 mmHg/61 mmHg (systolic/diastolic), the mean may be grossly approximated as 74 mmHg. This estimation may or may not be accurate.

In the systemic circulation, *MAP* and *cardiac output* (*CO*) are related by the *central venous pressure* (*CVP*) and *systemic vascular resistance* (*SVR*). *CVP* is the pressure measured in the thoracic vena cava, which is an approximation of the pressure in the right atrium. If we remember the fluid analog of Ohm's law that we developed from Hagen-Poiseuille's equation in Chapter 6 (Eq. (6.4)), we have

$$(P_i - P_o) = Q \cdot \frac{8\eta l_t}{\pi r_t^4} = Q \cdot R. \tag{7.1}$$

Cardiac output is blood flow; so we can substitute *CO* for *Q* in Eq. (7.1). If we assume that *MAP* stays approximately constant throughout the systemic arteries, then *MAP* is the input pressure. *CVP* is the output pressure at the end of the systemic circulation. Our resistance, *R*, for the systemic circulation is now *SVR*. These substitutions lead to

$$(MAP - CVP) = CO \cdot SVR. \tag{7.2}$$

Blood pressure is typically measured either at the brachial artery using auscultation or oscillometry, or at the radial artery using an *intraarterial pressure* (*IAP*) sensor.

## The Vital Signs

Blood pressure is one of the vital signs that indicate an individual's level of physical functioning: blood pressure, respiratory rate, heart rate, and temperature. Many clinicians, including nurses, recognize pain as a fifth vital sign. Other clinicians and medical device engineers consider arterial saturation of oxygen, measured by pulse oximetry, to be the fifth vital sign (Birnbaum, 2009). Normal adult ranges for the five vital signs are:

- Blood pressure: <120 mmHg systolic, <80 mmHg diastolic (NHLBI, 2004).
- Respiratory rate: 12–18 breaths/min.
- Heart rate: 60–80 bpm at rest.
- Temperature: 96.8–99.5°F (Guyton & Hall, 2006).
- Arterial saturation of oxygen: 93–100%.

Each vital sign can be measured discretely or continuously. A discrete measurement, which is often conducted in a physician's office, is known as a *diagnostic test*. The continuous observation of repeating events of physiologic function to guide therapy or to monitor the effectiveness of interventions is known as *patient monitoring*. Medical instruments for patient monitoring are most widely deployed in the intensive care units (ICUs) and operating rooms (ORs) of hospitals.

# CLINICAL NEED

Periodically over the last three decades, the National Heart, Lung, and Blood Institute (NHLBI) has assessed the state of American blood pressure diagnosis and treatment. The latest assessment from 2004 is called the *Seventh Report of the Joint National Committee on Prevention, Detection, Evaluation, and Treatment of High Blood Pressure* (JNC-7). According to JNC-7, BP may be classified according to the ranges in Table 7.1.

At least 50 million (one in six) Americans have hypertension warranting some form of treatment. Approximately 30% of adults are unaware of their hypertension. More than 40% of those with hypertension are not on treatment. Even for those being treated, two-thirds of hypertensive patients are not being controlled to prehypertension levels (NHLBI, 2004).

With the collapse of the world economy since JNC-7 was written, the health of newly unemployed workers is expected to significantly decline. Researchers at the Harvard School of Health analyzed U.S. job loss for 8125 workers who lost their jobs in 1999, 2001, and 2003. Workers who lost a job through no fault of their own were twice as likely to report developing a new condition like hypertension, diabetes, or heart disease over the next $1\frac{1}{2}$ yr, compared to those who were continuously employed (Strully, 2009; Baura, 2009b; Rabin, 2009).

Without blood pressure measurement, there is no blood pressure control. Control of blood pressure is critical because BP predicts cardiovascular risk. For every 20 mmHg systolic or 10 mmHg diastolic increase, the mortality from both ischemic (restricted blood flow) heart disease and stroke doubles. Further, longitudinal data from the Framingham Heart Study indicate that BP values in the range of 130–139 mmHg/85–89 mmHg are associated with more than twofold increase in relative risk from cardiovascular disease as compared with those whose BPs are classified as JNC-7 normal (NHLBI, 2004). Control may be as simple as a physician's monthly prescription and follow-up for generic atenolol (Baura, 2009a), which is available over the counter at a very low cost, such as $4 (Walmart, 2009).

During surgery, continuous BP monitoring is indicated for cardiac patients and hypertensive patients.

**TABLE 7.1**  JNC-7 Blood Pressure Classification (NHLBI, 2004)

| Classification | Systolic/Diastolic (mmHg/mmHg) |
| --- | --- |
| Normal | <120/80 |
| Prehypertension | 120–139/80–89 |
| Stage 1 hypertension | 140–159/90–99 |
| Stage 2 hypertension | 160–179/100–109 |
| Stage 3 hypertension | ≥180/110 |

# HISTORIC DEVICES

In 1616, William Harvey proposed that a finite amount of blood circulated unidirectionally through the body. Later, in 1733, Reverend Stephen Hales began to measure BP by recording the rise of blood in a brass pipe inserted above a horse's left ventricle. John Leonard Marie Poiseuille introduced the units of mmHg when he substituted a shorter column of mercury for the longer brass tube during BP measurement.

## Early Devices

Although BP could be measured directly when a pressure sensor was in contact with blood, this was not a convenient measurement. Noninvasive pressure measurement began to be investigated in the 1850s. In 1881, Austrian physician Samuel von Basch connected a mercury column manometer to a water-filled bag, which was called a *sphygmomanometer*. Fifteen yr later, Italian internist Scipione Riva-Rocci improved the sphygmomanometer design. Riva-Rocci's apparatus consisted of a mercury manometer, an arm-encircling inflatable elastic cuff, and a rubber bulb to inflate the cuff. With this device, Riva-Rocci estimated systolic pressure by determining the cuff pressure at which the radial pressure could not be felt (Roguin, 2006).

## Enabling Technology: Auscultation

Russian army surgeon Nikolai Korotkoff improved Riva-Rocci's method by discovering a more accurate pressure marker during cuff deflation. Rather than looking for the cessation of radial pressure, Korotkoff discovered that systolic pressure coincided with the beginning of sound heard with a stethoscope below the cuff and that diastolic pressure corresponded with the end of sound. He verified through canine experiments that these sounds, later called *Korotkoff sounds*, were locally generated by blood vessel compression, which he reported in 1905 (Shevchenko & Tsitlik, 1996). Auscultation is still considered the reference standard for noninvasive BP measurement.

## Enabling Technology: Oscillometry

More recently, bioengineer and physician Maynard "Mike" Ramsey III developed the first automated BP monitor, based on oscillometry. Oscillometry is the analysis of pressure pulsations as an overinflated cuff over the brachial artery is deflated. It had been known that systolic pressure could be determined with oscillometry. Ramsey and his company, Applied Medical Research, demonstrated in 1977 that the mean pressure occurred when the peak-to-peak amplitude of the pulsation was maximal. *MAP* could be estimated accurately using an automated, microprocessor-controlled monitor (Ramsey, 1979) (Figure 7.3). Ramsey sold his Dinamap technology to Johnson & Johnson Critikon in 1979. He then led a Critikon team that developed Critikon Dinamap BP monitors for measuring systolic, diastolic, and mean pressures (Szeto, 2002).

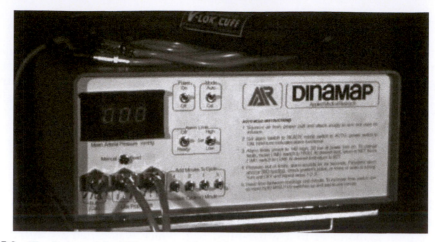

FIGURE 7.3   The first model of Dinamap marketed by Applied Research Corporation was the Model 825 (Maynard Ramsey III, Tampa, Florida).

Auscultation and oscillometry are technologies for discrete, *noninvasive blood pressure* (*NIBP*) measurement. In the OR, it is desirable to continuously monitor BP noninvasively. To this end, *continuous noninvasive blood pressure (cNIBP)* monitors are being investigated. Some of the monitoring technologies under investigation are tonometry (Belani et al., 1999), variations of tonometry (Baura, 2002a; Baura, 2003), and the method of Penaz (Molhoek et al., 1984; Imholz et al., 1998). With tonometry, BP can be measured when sufficient force is applied to the radial artery, such that the transmural pressure equals zero and the external pressure equals the internal pressure (Pressman & Newgard, 1963). With the method of Penaz, BP can be measured when the cuff pressure at which finger pulse pressure amplitudes are maximized; this point is assumed to indicate when the arterial diameter is maximal (Penaz, 1969; Baura, 2002b).

## SYSTEM DESCRIPTIONS AND DIAGRAMS

The invasive reference standard for BP measurement is the blood pressure sensor, which is commonly referred to as a blood pressure transducer. (Pressure sensors were discussed in Chapter 1.) A BP transducer has an operating range of −30 to +300 mmHg and a frequency range of 0−200 Hz (AAMI, 2006).

Typically, a disposable BP transducer (Figure 7.4) connects to the patient through a catheter sitting in the radial artery and connects to a BP monitor through a reusable electrical cable (Figure 7.5). The tubing length and the presence of air bubbles must be minimized, so that the system frequency response is not degraded.

The noninvasive reference standard for BP measurement is auscultation. As in Korotkoff's day, a sphygmomanometer cuff over the brachial artery is inflated above the systolic pressure with a rubber bulb, and it is then slowly deflated (Figure 7.6). During

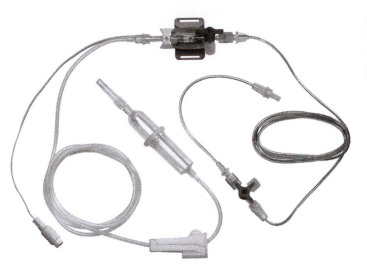

**FIGURE 7.4** Philips M1567 single-line blood pressure kit connects to the Multi-measurement module in Figure 7.1 (Philips Healthcare, Andover Massachusetts).

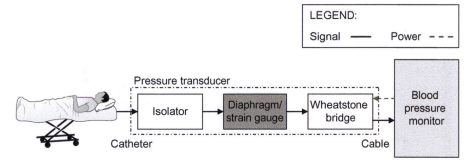

**FIGURE 7.5** Intraarterial pressure transducer system diagram.

deflation, the clinician listens with a stethoscope for Korotkoff sounds below the cuff and observes the corresponding aneroid manometer pressures. When the deflation pressure equals the systolic pressure, the sound of blood spurting under the cuff is heard. As the deflation pressure decreases and equals the diastolic pressure, the muffled sound transitions to silence (Figures 7.7 and 7.8).

To acquire accurate blood pressure readings during auscultation, the correct size cuff must be used to evenly transmit pressure to the brachial artery. The clinician must be able to hear the sounds, which means she must have good hearing and be listening in a quiet environment. For hypotensive patients, the sounds are lower frequency, and they may not be correctly heard.

Automated noninvasive BP monitors are most commonly based on oscillometry. Again, an appropriate size cuff over the brachial artery (Figure 7.9) is inflated above the systolic

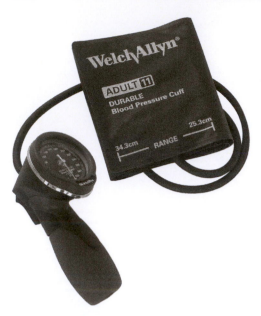

**FIGURE 7.6**   Welch Allyn Platinum Series DS48A pocket aneroid sphygmomanometer (Welch Allyn, Skaneateles, New York).

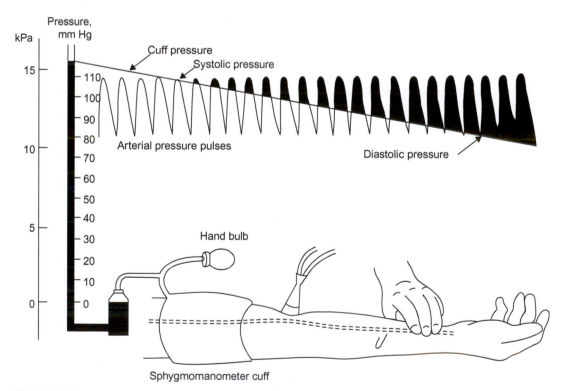

**FIGURE 7.7**   Auscultation measurement *[Reproduced by permission from Rushmer (1970)].*

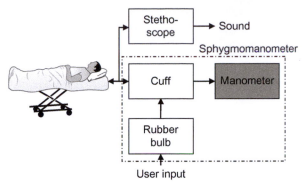

**FIGURE 7.8**  Auscultation system diagram.

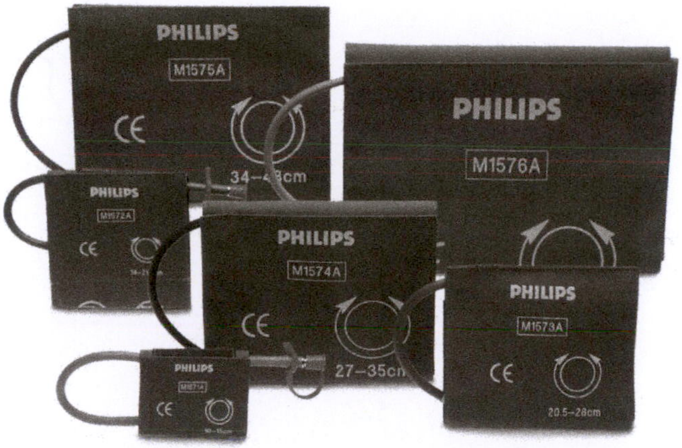

**FIGURE 7.9**  Philips M157xA cuffs connect to the Multi-measurement module in Figure 7.1 through interconnect tubing (Philips Healthcare, Andover Massachusetts).

pressure and slowly deflated. During deflation, a pressure sensor in the cuff transmits pressure pulses to an amplifier. The amplified pressures are digitized and then processed to determine systolic, mean, and diastolic pressures. The pressure at which pulsatile pressures begin is generally designated as the systolic pressure. As Ramsey discovered, the pressure at which maximal peak-to-peak pressure occurs is the *MAP*. The pressure at which pulsatile pressures disappear, or a similar pressure disappears, is designated as the diastolic pressure (Figures 7.10 and 7.11).

As with auscultation, the correct cuff size must be used to evenly distribute pressure. For accurate pressure sensor measurement, the artery marking on the cuff should be correctly centered over the inner upper arm, with no clothing between the cuff and arm. The patient should be seated with his brachial artery at the level of the right atrium. He should not move during the measurement, as motion artifact may result in an inaccurate reading. Clinicians should realize that the diastolic reading is the least accurate, subject to diastolic pressure algorithm implementation.

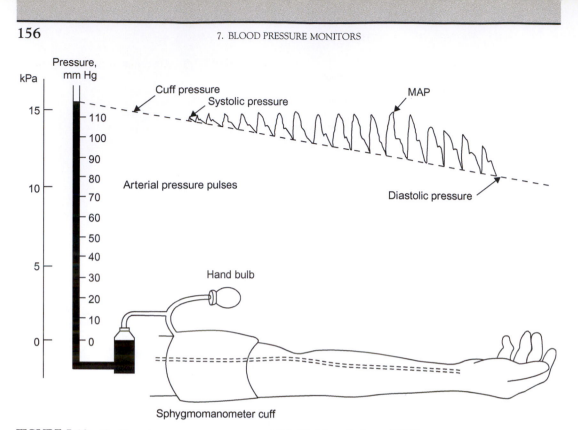

**FIGURE 7.10**   Oscillometry pressure measurement *[Modified from Rushmer (1970)].*

## Measurement Equivalency

Because IAP measurements are made in the radial artery and noninvasive measurements are made across the brachial artery, you might assume that systolic, diastolic, and mean pressures remain constant between both arteries. This assumption is not true, as demonstrated in a study of six young, healthy male volunteers. Systolic, diastolic, and mean pressures were measured invasively at the radial artery and by oscillometry at the brachial artery. For each method, three measurements were made and averaged during the following conditions: vasconstriction, vasodilation, mild sweating, intense sweating, cooldown, and general anesthesia. Cooldown was defined as the period between intense sweating through a forced-air warmer and administration of propofol-$N_2O$ general anesthesia.

The (radial − oscillometric brachial) pressure differences are shown in Table 7.2. The (radial − oscillometric brachial) *MAP* mean differences varied from −12 to −3 mmHg across conditions, suggesting that pressures between the two locations fundamentally differ. The added effect of each selected condition varies. Using the Wilcoxon rank sum test ($p < 0.05$), *MAP* differences did not significantly differ between any two conditions, while diastolic pressure differences differed only between vasoconstriction and cooldown and

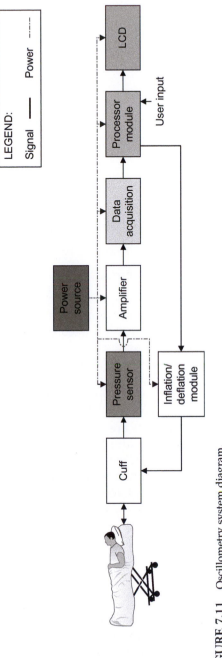

**FIGURE 7.11** Oscillometry system diagram.

**TABLE 7.2**   Differences Between Direct Radial Artery and Oscillometric Brachial Artery Blood Pressures (Radial − Oscillometric Brachial)

| Condition | Systolic | MAP | Diastolic |
|---|---|---|---|
| Vasoconstriction | 15 ± 9 | −3 ± 6 | −8 ± 9 |
| Vasodilation | 6 ± 7 | −5 ± 10 | −6 ± 9 |
| Mild sweating | −5 ± 17 | −11 ± 8 | −8 ± 10 |
| Intense sweating | −14 ± 10 | −12 ± 7 | −6 ± 10 |
| Cooldown | 10 ± 11 | −7 ± 7 | −3 ± 11 |
| General anesthesia | −6 ± 6 | −8 ± 8 | −1 ± 8 |

*All pressures have units of mmHg [Adapted from Hynson et al. (1994)].*

between vasodilation and general anesthesia. However, systolic pressure differences significantly differed between 10 of 15 paired conditions (Hynson et al., 1994).

## Consumer Blood Pressure Monitor Accuracy

Blood pressure monitors sold for home use are subject to the same accuracy standards as hospital monitors. In 2009, 39 blood pressure monitors (classification code DXN) were specifically cleared by the FDA for home use. All used oscillometry, either with an upper arm or wrist cuff, as the measurement technique. Of these, 37 of 39 monitors, according to their 510(k) summary statements, met the ANSI/AAMI SP10 accuracy requirement (CDRH, 2010). We discuss this accuracy requirement in the next section. ANSI/AAMI SP10 states that the mean error of BP measurements must be within ±5 mmHg, with a standard deviation within ±8 mmHg (AAMI, 2008). Presumably, accurate blood pressures can be obtained when the directions for use are followed.

Determining the accuracy of blood pressure monitors available for free use at pharmacies and grocery stores is more difficult (Figure 7.12). In the first half of the 1990s, the Vita-Stat 8000C was one of the most commonly used community devices, with approximately 8000 devices in use in the United States. The device uses auscultation to measure blood pressure (Whitcomb et al., 1995). The Vita-Stat 8000-C's 510(k) summary statement is not available from the FDA 510(k) database, so it is unknown if this device met any consensus standard accuracy requirement (CDRH, 2010). Three Vita-Stat 8000s, each in a local Denver grocery store, were tested against auscultation in 63 grocery shoppers. During testing, 63.2% of subjects had Vita-Stat readings greater than 5 mmHg different from their auscultation readings. The difference in systolic readings, compared to sphygmomanometer readings, varied from −60 to +58 mmHg. The Vita-Stat sensitivity and specificity in predicting hypertension, defined as ≥ 140 mmHg systolic and/or ≥ 90 mmHg diastolic, were 26% and 100%, respectively (Whitcomb et al., 1995).

A later version of the Vita-Stat, the Vita-Stat 90553, was cleared by the FDA in 1994. The Vita-Stat 90553's 510(k) summary statement is not available from the FDA 510(k) database;

**FIGURE 7.12** Free blood pressure screening at a drugstore or grocery store may not provide accurate readings.

so it is unknown whether this device met any consensus standard accuracy requirement (CDRH, 2010). A similar model, the Vita-Stat 90550, was available in about 3000 Canadian communities in 2000.

When three Vita-Stat 90553 devices, in three Canadian pharmacies, were tested against auscultation in 108 subjects, the mean systolic pressure difference was $4.4 \pm 9.4$ mmHg and the mean diastolic pressure difference was $1.0 \pm 6.2$ mmHg. In this study sample, all but two subjects had blood pressures between 100 mm Hg/60 mm Hg and 180 mm Hg/100 mm Hg. Therefore, a large range of blood pressures was not tested (Lewis et al., 2002).

More recently, the KEITO K5 was cleared by the FDA in 2000 for measurements of blood pressure using wrist oscillometry while standing. The predicate device in its 510(k) summary statement is the Vita-Stat 8000-C. Tested accuracy was "in accordance with the German's regulations for clinical testing of blood pressure systems (BGBI.IS759.771 and 1667)" (CDRH, 2000). When a KEITO K5 in a Norwegian pharmacy was tested against auscultation in 390 patients, the mean systolic pressure difference was $-5 \pm 18.5$ mmHg and the mean diastolic pressure difference was $-4 \pm 8.5$ mmHg. KEITO K5 sensitivity and specificity in predicting systolic ($\geq 140$ mmHg) and diastolic ($\geq 90$ mmHg) hypertension were 71%/32% and 100%/100%, respectively (Staal et al., 2004).

Looking at these study results, sources of community blood pressure monitor error include the use of a single cuff size and measurement in the standing position. In attempting to simplify human factors associated with these devices, accuracy may be compromised. Another source of error is the date of last calibration. For accurate blood pressure measurements, it is better to purchase a monitor whose labeling states that ANSI/AAMI SP10 was met, rather than to use a free monitor. As a recent editorial from the Mayo Clinic stated, "until manufacturers of these devices remedy the defects that cause inaccurate measurements, patients should be discouraged from using these devices, and physicians should not alter antihypertensive therapy on the basis of measurements derived from these sources" (Graves, 2005).

## KEY FEATURES FROM ENGINEERING STANDARDS

As of 2010, the FDA recommends two AAMI standards for blood pressure measurement. The first applies to BP measurements that use an occluding cuff: *ANSI/AAMI SP10:2002/(R)2008 Manual, Electronic, or Automated Sphygmomanometers* (AAMI, 2008). The second applies to BP measurements that use an indwelling catheter: *ANSI/AAMI BP22:1994/(R)2006 Blood Pressure Transducers* (AAMI, 2006). These consensus standards are the basis for the key features we discuss below.

### Monitor Life

A BP monitor should maintain its safety and performance characteristics for a reasonable length of time. Ten thousand full-scale pressure cycles approximates 5 yr of typical use by the home or physician's office user. A full-scale cycle is defined as a pressure change from 20 mmHg or less, to the full scale of the device, and then back to 20 mmHg or less. Industry experience suggests that most device failures occur within 8000 cycles. Therefore, ANSI/AAMI SP10:2002/(R)2008 requires that the device maintain safety and performance characteristics for 10,000 full-scale cycles (AAMI, 2008).

### Monitor Cuff Pressures

BP monitors with an occluding cuff must not harm the patient during use through unnecessary applied pressure. ANSI/AAMI SP10:2002/(R)2008 limits maximum cuff pressure to 300 mmHg, or 30 mmHg above the upper limit of the manufacturer-specific operating range, whichever is lower. The standard requires that cuff pressure is not maintained above 15 mmHg for longer than 3 min. During cuff deflation, the time for pressure reduction from 260 mmHg to 15 mmHg must not exceed 10 s (AAMI, 2008).

### Monitor Accuracy with Auscultatory Method as Reference Standard

BP measurements must be accurate. Reference measurements may be made using the auscultatory method. Each reference measurement is the mean of two auscultatory measurements made by two observers. It is required that 100% of measurements made by these observers agree within 10 mmHg, with 90% agreeing within 5 mmHg. This method is used to assess systolic and diastolic pressure measurement accuracy.

The accuracy test protocol requires at least 85 subjects, with at least 10% of the subjects below 100 mmHg systolic reference, 10% above 160 mmHg systolic reference, 10% below 60 mmHg diastolic reference, and 10% above 100 mmHg diastolic reference. In a first method of comparison, at least 255 observations must be obtained from these 85 subjects.

Each individual observation is then used in the error calculations. The mean error of BP measurements must be within ±5 mmHg, with a standard deviation within ±8 mmHg.

In an alternative method of comparison, three paired observations from each of the 85 subjects are averaged. After the mean error is calculated for 85 mean observations, the standard deviation must be within a certain prespecified corresponding range. For example, for a mean difference of 0 mmHg, the standard deviation must be within 6.95 mmHg. For a maximum mean difference of 5.0 mmHg, the standard deviation must be within 4.81 mmHg (AAMI, 2008).

## Monitor Accuracy with Intraarterial Method as Reference Standard

Alternatively, intraarterial pressure, preferably obtained from the brachial, axillary, or subclavian artery, may be used as the reference standard. Each reference measurement is the mean of several measurements synchronized to the noninvasive measurement. With intraarterial references, systolic, diastolic, and mean pressure measurement accuracy may be assessed.

The accuracy test protocol requires at least 15 subjects, with at least 10% of the subjects below 100 mmHg systolic reference, 10% above 160 mmHg systolic reference, 10% below 60 mmHg diastolic reference, and 10% above 100 mmHg diastolic reference. Between 5 and 10 paired measurements must be made per subject. Each individual observation is then used in the error calculations. The mean error of BP measurements must be within ±5 mmHg, with a standard deviation within ±8 mmHg (AAMI, 2008).

## Transducer Accuracy

According to ANSI/AAMI BP22:1994/(R)2006, transducer accuracy must be within ±1 mmHg plus ±1% of reading over the pressure range −30 to 50 mmHg, and within ±3 mmHg over the pressure range 50–300 mmHg. This is tested by exciting the transducer with 2.5-kHz sinewave, with 6-$V_{DC}$ baseline, and applying an accurate pressure source with an associated gauge. The pressure source and gauge must have a combined nonlinearity and hysteresis of less than 0.2% of reading to the transducer. The tested pressures in order are 0, 25, 50, 100, 200, 300, 200, 100, 50, 25, 0, −10, −30, −10, 0 mmHg. The test is repeated again at two other baseline voltages of 4 and 8 $V_{DC}$ (AAMI, 2006).

## SUMMARY

A blood pressure monitor is an instrument that enables systolic, diastolic, and often mean blood pressures to be measured and displayed. Blood pressure may be measured either noninvasively or invasively, with a pressure sensor in contact with the blood. The two proven methods of noninvasive blood pressure measurement are auscultation and oscillometry. Both methods begin with overinflation of an arm cuff, followed by slow deflation. In auscultation, the Korotkoff sounds are used as event markers for systolic (blood spurting sound) and diastolic (transition to silence) pressures. In oscillometry, the peak-to-peak amplitudes of pulse pressures are used as event markers for systolic (beginning of pulse pressures), mean (maximum pulse pressure), and diastolic (disappearance of pulse pressures) pressures.

The blood pressure waveform shapes in smaller vessels are influenced by the transmission of the original pressure wave through the vasculature. The incisura in the proximal aorta waveform, which originates during aortic valve closure, changes as the pressure wave propagates. The pressure contour changes result from traveling wave reflections. When a traveling wave experiences an impedance change, a portion of the traveling wave is reflected back through the system. Impedance changes occur at branching points, areas of alteration in arterial compliance, and high-resistance arterioles. Propagation through the upper body arteries results in different contour changes compared to propagation through the lower body arteries. These differences have been attributed to arterial differences due to age and wave velocity.

Blood pressure measurement at the brachial or radial artery is important. At least 50 million (one in six) Americans have hypertension warranting some form of treatment. Approximately 30% of adults are unaware of their hypertension. More than 40% of those with hypertension are not on treatment. Even for those being treated, two-thirds of hypertensive patients are not being controlled to prehypertension levels.

Blood predicts cardiovascular risk. For every 20 mmHg systolic or 10 mmHg diastolic increase, the mortality from both ischemic (restricted blood flow) heart disease and stroke doubles. Without blood pressure measurement, there is no blood pressure control.

Key blood pressure monitor features include monitor life; monitor cuff pressures; monitor accuracy, with auscultatory method as reference standard; monitor accuracy, with intraarterial method as reference standard; and transducer accuracy.

## Exercises

**7.1** Plot the aortic and radial blood pressure waveforms (waveforms 7_1 and 7_2) that have been provided by your instructor. Waveform characteristics are documented in the associated header7_1.txt ascii file. Model the impulse response between the aortic and radial waveforms. Use the Matlab function arx in the System Identification toolbox. What is the equation describing the model accessed by the arx function? Assume a delay of 15 samples and the use of 19 feedforward parameters. Model the impulse response using either 5, 6, or 7 feedback parameters. Which number of feedback parameters provides the best fit? Give reasons.

**7.2** Looking at Table 7.2, radial and brachial pressures appear to be fundamental different. Why is radial pressure measured in the operating room, whereas brachial pressure is measured in a physician's office? Does this fundamental difference affect the utility of either measurement?

**7.3** Take your blood pressure with an oscillometric monitor. As the overinflated cuff begins to deflate to make a measurement, move the arm being measured. How does the monitor react? Does motion artifact affect oscillometric measurements?

**7.4** Analyze the ASCII dataset (data 7_1) that has been provided by your instructor. This data set contains blood pressure readings from a new noninvasive monitor, as well as from an intraarterial monitor. Calculate the mean error and standard deviation of the error. Are these blood pressure measurements accurate? Do you believe that the ANSI/AAMI SP10 accuracy criteria of mean error of ±5 mmHg and error standard deviation of ±8 mmHg are stringent criteria? Provide reasons.

Read Ramsey (1979):

**7.5** What was the reference measurement for these studies? What was an important consideration for calibration of this reference measurement? Explain in specific terms why this is an important consideration.

**7.6** Analyze the data in Table 1 using Bland-Altman analysis. What is the bias? What is the precision? Are the data in good agreement? Explain. Analyze the data in Table 1 using linear correlation analysis. What is $r^2$?

Read Whitcomb et al. (1995):

**7.7** What was the reference measurement for these studies? What specific measures were parts of the protocol to ensure an accurate reference measurement was obtained?

**7.8** Are Vita-Stat blood pressure estimates accurate? Explain your reasoning.

# References

AAMI (2006). *ANSI/AAMI BP22:1994–2006 blood pressure transducers.* Arlington, VA: AAMI.

AAMI (2008). *ANSI/AAMI SP10:2002–2008 manual, electronic, or automated sphygmomanometers.* Arlington, VA: AAMI.

Baura, G. D. (2002a). Method and apparatus for the noninvasive determination of arterial blood pressure. *U. S.* 6,471,655.

Baura, G. D. (2002b). *System theory and practical applications of biomedical signals.* Hoboken, NJ: Wiley-IEEE Press.

Baura, G. D. (2003). Method and apparatus for the noninvasive determination of arterial blood pressure. *U.S.* 6,514,211.

Baura, G. D. (2009a). Access for all. *IEEE Engineering in Medicine and Biology Magazine, 28,* 63–64.

Baura, G. D. (2009b). Two worlds [Point of View]. *IEEE Engineering in Medicine and Biology Magazine, 28,* 102,110.

Belani, K., Ozaki, M., Hynson, J., Hartmann, T., Reyford, H., & Martino, J. M., et al. (1999). A new noninvasive method to measure blood pressure: Results of a multicenter trial. *Anesthesiology, 91,* 686–692.

Birnbaum, S. (2009). Pulse oximetry: identifying its applications, coding, and reimbursement. *Chest, 135,* 838–841.

CDRH (2000). *510(k) summary of safety and effectiveness K984083.* Rockville, MD: FDA.

CDRH (2010). *510(k) premarket notification database: January, 2010.* Silver Spring, MD: FDA.

Graves, J. W. (2005). Blood pressure measurement in public places. *American Family Physician, 71,* 851–852.

Guyton, A. C., & Hall, J. E. (2006). *Textbook of medical physiology* (11th ed.). Philadelphia: Elsevier Saunders.

Hynson, J. M., Sessler, D. I., Moayeri, A., & Katz, J. A. (1994). Thermoregulatory and anesthetic-induced alterations in the differences among femoral, radial, and oscillometric blood pressures. *Anesthesiology, 81,* 1411–1421.

Imholz, B. P., Wieling, W., Van Montfrans, G. A., & Wesseling, K. H. (1998). Fifteen years experience with finger arterial pressure monitoring: assessment of the technology. *Cardiovascular Research, 38,* 605–616.

Lewis, J. E., Boyle, E., Magharious, L., & Myers, M. G. (2002). Evaluation of a community-based automated blood pressure measuring device. *Canadian Medical Association Journal, 166,* 1145–1148.

Molhoek, G. P., Wesseling, K. H., Settels, J. J., Van Vollenhoven, E., Weeda, H. W., & De Wit, B., et al. (1984). Evaluation of the Penaz servo-plethysmo-manometer for the continuous, non-invasive measurement of finger blood pressure. *Basic Research in Cardiology, 79,* 598–609.

NHLBI. (2004). *The Seventh Report of the Joint National Committee on Prevention, Detection, Evaluation, and Treatment of High Blood Pressure* (7th ed.). Bethesda, MD: NIH.

Nichols, W. W., & O'Rourke, M. F. (1998). *McDonald's blood flow in arteries* (4th ed.). New York: Oxford University Press.

Penaz, J. (1969). Patentova Listina. *CISLO 133205.*

Pressman, G. L., & Newgard, P. M. (1963). A transducer for the continuous external measurement of arterial blood pressure. *IEEE Transactions on Biomedical Engineering, 10,* 73–81.

Rabin, R. C. (2009). Unemployment may be hazardous to your health. *New York Times* (May 9).

Ramsey, M., III (1979). Noninvasive automatic determination of mean arterial pressure. *Medical & Biological Engineering and Computing, 17,* 11–18.

Roguin, A. (2006). Scipione Riva-Rocci and the men behind the mercury sphygmomanometer. *International Journal of Clinical Practice, 60,* 73–79.

Rushmer, R. F. (1970). *Cardiovascular dynamics* (3rd ed.). Philadelphia: W. B. Saunders.

Shevchenko, Y. L., & Tsitlik, J. E. (1996). 90th anniversary of the development by Nikolai S. Korotkoff of the auscultatory method of measuring blood pressure. *Circulation, 94,* 116–118.

Staal, E. M., Nygard, O. K., Omvik, P., & Gerdts, E. (2004). Blood pressure measurements by the Keito machine: Evaluation versus office blood pressure by physicians. *Blood Pressure Monitoring, 9,* 167–172.

Strully, K. W. (2009). Job loss and health in the U.S. labor market. *Demography, 46,* 221–246.

Szeto, A. Y. J. (2002). Faces and Places: Mike Ramsey—A medical device entrepreneur. *IEEE Engineering in Medicine and Biology Magazine, 21,* 12–13.

Walmart (2009) $4 prescriptions program.

Whitcomb, B. L., Prochazka, A., Loverde, M., & Byyny, R. L. (1995). Failure of the community-based Vita-Stat automated blood pressure device to accurately measure blood pressure. *Archives of Family Medicine, 4,* 419–424.

CHAPTER

# 8

# Catheters, Bare Metal Stents, and Synthetic Grafts

In this chapter, we discuss the relevant physiology, history, and key features of cardiac catheters, bare metal stents, and grafts. Cardiac catheter systems enable the performance of minimally invasive cardiac procedures. A bare metal stent, which is an expandable tube of metallic mesh, provides a physical scaffold for keeping a blood vessel patent. A synthetic graft allows a weakened blood vessel wall to be replaced (Figure 8.1). These medical

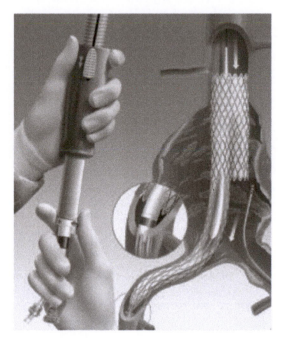

FIGURE 8.1 Sixth-generation Medtronic AneuRx system for repairing an abdominal aortic aneurysm with a graft delivered by a catheter. In this system, the graft is stented. FDA initially approved the second-generation system in 1999 (Medtronic, Santa Rosa, California).

devices have been grouped into this chapter as a basis for a discussion of the Guidant Ancure Endovascular Graft System recall. Because drug-eluting stent design has changed greatly in the last few years, this advanced stent merits its own dedicated discussion in Chapter 19.

Upon completion of this chapter, each student shall be able to:

1. Understand the mechanisms underlying atherosclerosis and aneurysms.
2. Identify the steps of the percutaneous transluminal coronary angioplasty procedure and bare metal stent implantation.
3. Describe five key features of vascular catheters, bare metal stents, and grafts.

## ATHEROSCLEROSIS

In past chapters, we have assumed that blood flows freely from the left ventricle, through the systemic circulation, and back to the right atrium. Although this is the ideal case for blood flow, blood may encounter narrowed arterial passages due to atherosclerosis.

*Atherosclerosis* is a disease in which fatty lesions develop on the inside surfaces of the arterial walls of large and intermediate-sized arteries. When atherosclerosis occurs in arteries associated with the limbs, it is called *peripheral artery disease (PAD)*. When atherosclerosis occurs in the coronary (heart) or carotid (neck) arteries (Figure 8.2), it is called *coronary artery disease (CAD)*, or carotid artery disease, respectively.

Coronary artery occlusion prevents oxygen transport to the myocardium, resulting in myocardial tissue damage. This is known as a *myocardial infarction (MI*, or a heart attack). Carotid artery occlusion prevents oxygen transport to the brain, resulting in cerebral tissue damage. This is known as a *stroke*.

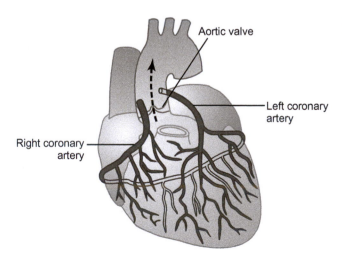

**FIGURE 8.2** The coronary arteries [*Reproduced by permission from Guyton & Hall (2006)*].

A healthy artery is composed of its inner lining endothelium, intima, smooth muscle-celled media, and outer adventitia. When the endothelium is damaged, circulating mono-cytes and lipids accumulate at the site of injury. The monocytes cross the endothelium and enter the intima. They differentiate to become macrophages and then ingest and oxidize accumulated lipoproteins. These macrophage foam cells aggregate on the blood vessel and form a visible fatty streak. Over time, fatty streaks grow larger and proliferate to form pla-ques, which occlude the artery (Guyton & Hall, 2006) (Figure 8.3).

Fibrous plaque is formed first. It consists of an amorphous central lipid core of cellular debris, cholesterol, cholesterol esters, and foam cells. A fibrous cap surrounds the fibrous plaque, consisting mainly of smooth-muscle cells in a matrix of collagen and proteoglycans (Figure 8.4A). Over time, fibrous plaque may become complicated plaque. Complicated plaque consists of calcification, hemorrhage into the plaque, ulceration of the fibrous cap, and luminal thrombosis (Figure 8.4B). Often, complicated plaque is the underlying mecha-nism behind acute coronary clinical events of ischemic (restricted blood flow) origin (Virmani et al., 1994).

## PERCUTANEOUS CORONARY INTERVENTIONS

Until 1977, coronary stenosis was treated by *coronary artery bypass graft* (*CABG*) surgery. CABG is an open heart procedure, during which a leg vein is grafted around the narrowed portion of a coronary artery. Several interventional techniques have been developed to treat coronary stenosis in a minimally invasive manner. Collectively, these techniques are known as percutaneous coronary interventions (PCI). Percutaneous refers to delivery of the intervention through a small skin puncture.

The first PCI technique that was widely used is *percutaneous transluminal coronary angio-plasty* (*PTCA*). Transluminal refers to a procedure that occurs through the inner cavity of a blood vessel. Here, a balloon catheter, which is a plastic tube with an inflatable balloon at its end, is transported to a coronary stenosis. The balloon is inflated several times for peri-ods ranging from 10 s to several minutes, to restore blood flow. During inflation, 4–12 atm of pressure may be generated (Kern, 2003) (Figure 8.5).

During balloon angioplasty, the intimal plaque is split, but endothelial injury also occurs. When a large split is present, intimal flaps form. It is believed that an expanded balloon must stretch the media and adventitia to increase the outer artery circumference and truly enlarge the lumen (Virmani et al., 1994) (Figure 8.6). Short-term failure of angio-plasty, within hours to days, may occur due to elastic recoil of the vessel wall, acute thrombosis at site of angioplasty, or acute dissection of the vessel wall beyond the area of angioplasty.

The endothelial injury initiates a wound-healing response. Typically, fibrous tissue forms in the lumen after excessive medial smooth muscle proliferation. Progressive, prolif-erative restenosis results (Padera & Schoen, 2004), unless offset by an initial larger acute lumen and smaller plaque size during angioplasty (Sangiorgi et al., 1999). Long-term fail-ure due to restenosis occurs in 30–60% of patients, most frequently within the first 6 months (Bauters & Isner, 1998, Rajagopal & Rockson, 2003).

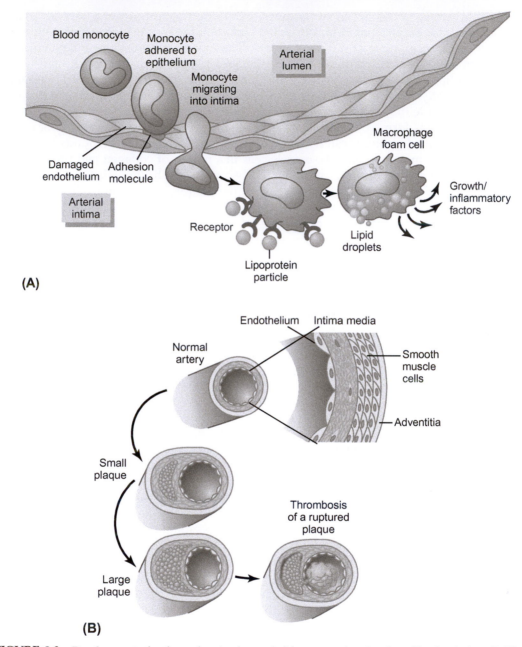

**FIGURE 8.3** Development of artherosclerotic plaque. **A:** Monocyte migration from blood to intima. **B:** Plaque formation [*Modified from Libby (2002). Figure reproduced by permission from Guyton & Hall (2006)*].

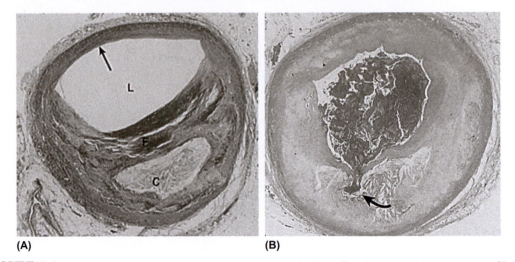

**(A)**                                                **(B)**

**FIGURE 8.4**  Atherosclerotic plaque in the coronary artery. **A:** Overall architecture demonstrating a fibrous cap (F) and a central lipid core (C) with typical cholesterol clefts. The lumen (L) has been moderately narrowed. Note the plaque-free segment of the wall (arrow). **B:** Coronary thrombosis superimposed on an atherosclerotic plaque with focal disruption of the fibrous cap (arrow), triggering fatal myocardial infarction [*A reproduced by permission from Schoen (1999). **B** reproduced by permission from Schoen (1989). Both figures reproduced by permission from (Padera and Schoen (2004)].*

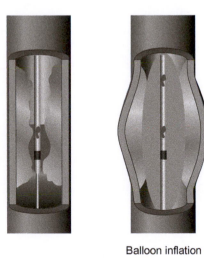

Balloon inflation

**FIGURE 8.5**  Percutaneous transluminal coronary angioplasty *(Adapted from A.D.A.M., Atlanta, Georgia).*

A stent may be deployed to the site immediately after balloon angioplasty, to support the disrupted vascular wall and minimize thrombus formation (Figure 8.7). Short-term stent failure generally involves subacute stent thrombosis that occurs in 1–3% of patients within 7–10 days (Padera & Schoen, 2004).

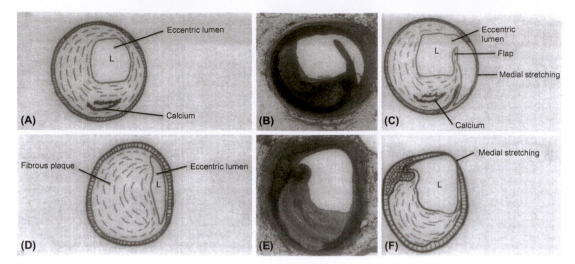

**FIGURE 8.6** Probable plaque morphologic features and lumen size (**A** and **D**) before balloon dilatation. **C**−**F**: Post-PTCA arteries drawn from histologic sections (**B**−**E**). **A**: Note eccentric, fibrous plaque with calcification. **B**−**C**: Corresponding histologic section and line drawing 14 days after successful balloon angioplasty. Note flap and marked stretching and thinning of media. **D**: Note eccentric fibrous plaque with severe luminal narrowing. **E**−**F**: Marked stretching, thinning, and disruption of media and successful angioplasty with significant increase in lumen (L) size. Movat's pentachrome stain. **B**: original magnification ×15. **E**: original magnification ×12 [*Reproduced by permission from Virmani et al. (1994)*].

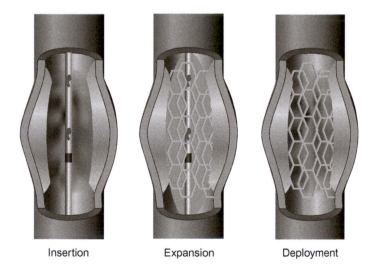

Insertion          Expansion          Deployment

**FIGURE 8.7** One example of percutaneous coronary intervention: stent deployment (*Adapted from A.D.A.M., Atlanta, Georgia*).

During stent inflation, early damage to the endothelial line and stretching of the vessel wall occur, stimulating platelet and leukocyte adherence and accumulation. Fibrous cap rupture and stent strut penetration into the plaque lipid core may increase neointimal growth, which results in restenosis (Farb et al., 2002) (Figure 8.8). This long-term failure of

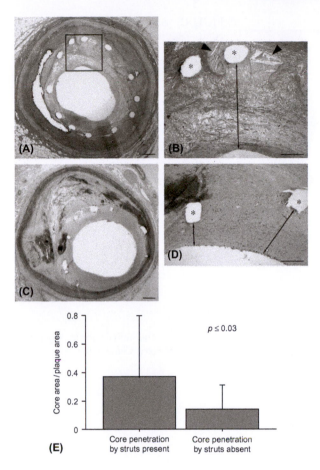

**FIGURE 8.8** Lipid core penetration by stent struts (low power in **A** and high power in **B**) was associated with increased neointimal thickness (arrows) compared with stent struts in contact with fibrous plaque (**C–D**). Cholesterol clefts are indicated (arrowheads in **B**). Plaques with large lipid cores, occupying more than one-third of the total plaque area, were at increased risk of lipid core penetration by the stent (**E**). **A–D**: Movat pentachrome stain. Scale bars = 0.37 mm in **A**, 0.22 mm in **B**, 0.47 mm in **C**, and 0.20 in **D** [*Reproduced by permission from Farb et al. (2002)*].

restenosis occurs in 20–30% of patients (Bauters and Isner, 1998, Rajagopal and Rockson, 2003). To combat this restenosis, drug-eluting stents have been developed. We discuss drug-eluting stents in Chapter 19.

## ANEURYSMS

Blood flowing through the systemic circulation may also encounter widened passages due to aneurysms. An *aneurysm* is a localized dilatation due to a weakened arterial wall. The most common aneurysms occur in the abdominal aorta; they also occur in the thoracic aorta and cerebral (brain) arteries.

An *abdominal aortic aneurysm* (*AAA*) is believed to be a complication of chronic atherosclerosis. In the abdominal aorta, mechanical or chemical insults including hypertension, smoking metabolites, and oxidized lipids initially damage the endothelial surface. As before, fibrous plaques and complicated plaques form. Then extensive structural remodeling occurs, characterized by degeneration of extracellular matrix, destruction of elastic

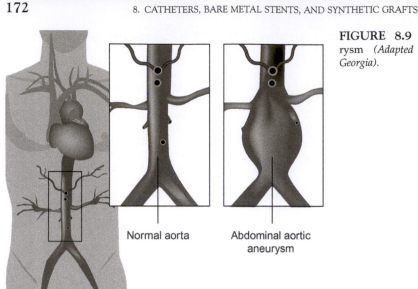

**FIGURE 8.9** Abdominal aortic aneurysm *(Adapted from A.D.A.M., Atlanta, Georgia).*

Normal aorta

Abdominal aortic aneurysm

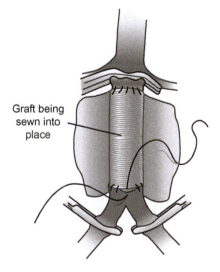

**FIGURE 8.10** Surgical repair of abdominal aortic aneurysm *(Adapted from the Society for Vascular Surgery, Chicago, Illinois).*

Graft being sewn into place

lamina, and reduction of smooth-muscle cells. These are the specific changes associated with the weakening of the aortic wall and subsequent bulge. Compared to a normal aorta, the aortic media is significantly thinner (Guo et al., 2006) (Figure 8.9).

When an AAA reaches about 5 cm in diameter, surgical intervention is recommended to prevent rupture. Traditionally, surgical repair involves open abdominal surgery, during which the aorta is reconstructed with a synthetic graft, followed by a 3-month recovery (Figure 8.10). More recently, grafts have been deployed in a minimally invasive manner using cardiac catheterization (Figure 8.1). This procedure is called *endovascular aneurysm repair* (*EVAR*).

## CLINICAL NEED

According to 2006 statistics from the American Heart Association, peripheral artery disease affects about 8 million Americans. PAD prevalence dramatically increases with age. Only about 10% of PAD patients have the classic symptoms of intermittent leg pain. Diagnosis is critical, because PAD patients possess a four to five time greater risk of MI or stroke.

Similarly, CAD affects about 17 million Americans that are age 20 yr and older. Each year, 610,000 new and 325,000 recurrent myocardial infarctions occur. Ninety percent of CAD patients have prior exposure to at least one of these risk factors: high total blood cholesterol levels, current medication with cholesterol-lowering drugs, hypertension, current medication with blood pressure-lowering drugs, current cigarette use, or clinical report of diabetes.

In 2005, stroke affected 6.5 million Americans that were age 20 yr and older. Each year, 600,000 new and 185,000 recurrent stroke attacks occur. Risk factors for stroke include current cigarette use, atrial fibrillation, and hypertension (AHA, 2009).

Cardiac catheterization underlies many of the interventional techniques performed for MI and stroke. In 2006, 1,115,000 cardiac catheterizations were performed in the United States (AHA, 2009). Cardiac catheterization is indicated for suspected or known CAD, myocardial infarction, sudden cardiovascular arrest, valvular heart disease, congenital heart disease, aortic dissection, pericardial constriction, cardiomyopathy, and heart transplant assessment (Kern, 2003). In 2006, an estimated 1,313,000 percutaneous coronary intervention procedures were performed in the United States. Less than 30% of these procedures were performed with bare metal, rather than drug eluting, stents (AHA, 2009).

Abdominal aortic aneurysms are diagnosed in about 200,000 patients in the United States annually. Of these, nearly 15,000 have AAA diameters large enough to cause death from rupture if not treated (Society for Vascular Surgery, 2009).

## HISTORIC DEVICES

A catheter is a thin plastic tube that allows passage of fluid. Surgical resident Werner Forssmann conducted the first human cardiac catheterization through self-experimentation in 1929. Disobeying his advisor's direction to conduct animal experiments first, Forssmann inserted a ureteral catheter in his left cubital vein in the upper arm and pushed it about 65 cm, the estimated distance to the right atrium. He then walked from the operating room to the x-ray room of the hospital where he worked and took x-rays. In modern-day fluoroscopy, an x-ray source under the patient's table passes x-rays through the patient to the detector above the table. In this way, a cardiac catheter may be correctly positioned.

## Early Devices

After being severely reprimanded for his experiment, Forssmann continued research experiments for a few years, including early attempts at right atrial visualization using iodide preparation injections (Forssmann-Falck, 1997). Cardiologists Dickinson Richards and Andre Cournand read about Forssmann's technique and utilized it in the 1940s for measuring cardiac output, based on the Fick principle (Cournand et al., 1945). Cournand, Forssmann, and Richards received the Nobel Prize in Physiology or Medicine in 1956 "for their discoveries concerning heart catheterization and pathological changes in the circulatory system" (Cournand, 1957).

## Enabling Technology: Percutaneous Transluminal Coronary Angioplasty

Pediatric cardiologist F. Mason Sones Jr. discovered serendipitously that iodide contrast medium could be safely injected into the coronary arteries in 1958. He developed special catheters for injecting the dye into these arteries and capturing images with x-rays, calling his technique cinecardioangiography (Sones, 1958). This technique is generally referred to as coronary angiography. Sones received the 1983 Lasker Clinical Medical Research Award "for combining the techniques of cardiac catheterization and coronary artery cinematography, thus inaugurating the modern era of diagnosis and treatment of coronary artery disease" (Lasker Awards, 2009).

Later, in 1964, vascular radiologist Charles Dotter introduced transluminal angioplasty, in which he used special catheters to open blocked peripheral arteries during angiography (Dotter and Judkins, 1964). Cardiologist Andreas Gruentzig designed a balloon for Dotter catheters and performed the first PTCA in an awake human in 1977 (Gruentzig, 1981). Radiologist Julio Palmaz and cardiologist Richard Schatz then improved PTCA by developing a balloon-expandable bare metal stent that extends the patency time of an artery after PTCA. The Palmaz-Schatz coronary stent received FDA approval in 1994 (CDRH, 1994). Representative coronary angiograms before and after stent implantation are shown in Figure 8.11.

For more history about PTCA, please see the excellent independent educational website: http://www.ptca.org/nv/timeline.html.

## Enabling Technology: Vinyon "N" Cloth

Separate from the development of percutaneous transluminal coronary angioplasty, vascular surgeon Arthur Voorhees Jr. observed during an autopsy in the late 1940s that a silk suture he neglected to remove during a canine procedure was now coated a few months later by a thrombus-free film. Voorhees postulated that a vascular graft of fine mesh cloth would terminate leakage of blood through its walls by formation of fibrin plugs (Voorhees et al., 1952). He began to experiment with a Vinyon "N" cloth tube prosthesis in dogs.

In 1953, Voorhees used the prosthesis to repair 18 aortic aneurysms. Seventeen were AAAs; one was located in the popliteal artery. As shown in Figure 8.12, the aortic segment was replaced by a Vinyon "N" cloth tube, with reinforcing cuffs above and reinforcing

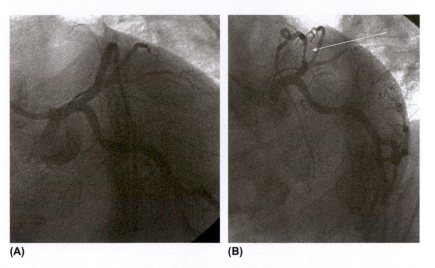

**(A)** **(B)**

**FIGURE 8.11** Angiograms. **A:** Obstruction of the proximal left anterior descending artery. **B:** The same artery after insertion of a 3.0-mm-diameter by 23-mm-long stent, which resolved the patient's angina (chest pain) *[Reproduced by permission from Ragosta (2011)].*

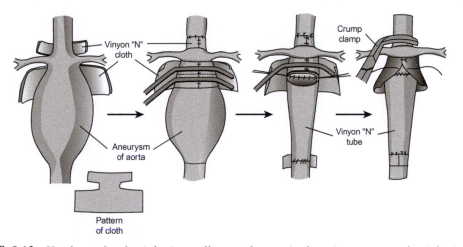

**FIGURE 8.12** Voorhees placed reinforcing cuffs over the proximal aortic segment and reinforcing strips around the line of anastomosis *[Adapted from Blakemore and Voorhees (1954)].*

strips below the graft. Ten patients survived; three patients died in the operating room. Postmortem examination of 4/5 patients who died in the postoperative period revealed structural and function integrity of the prosthesis (Blakemore & Voorhees, 1954).

When Union Carbide stopped production of Vinyon "N" cloth, other graft materials were investigated. Synthetic grafts are currently fabricated from poly(ethylene terephthalate), otherwise known as Dacron®, or expanded polytetrafluorethylene (ePTFE,

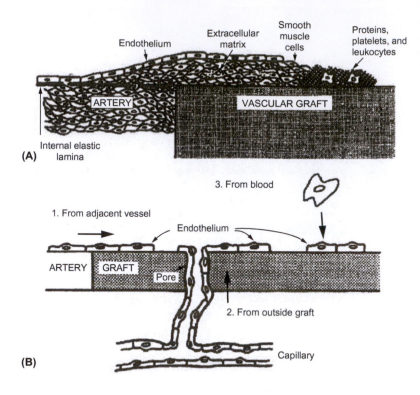

**FIGURE 8.13** Synthetic graft healing. **A:** Smooth muscle cells migrate from the media to the intima of the adjacent artery and extend over and proliferate on the graft surface; this smooth muscle cell layer is covered by a proliferating layer of endothelial cells. **B:** Possible sources of endothelium on the blood-contacting surface of the vascular graft [*Reproduced by permission from Schoen (1989). Figure reproduced by permission from Padera & Schoen, 2004)*].

Goretex®) (Padera and Schoen, 2004). The material ePTFE is formed by extrusion or sintering of PTFE, also known as Teflon®. An implanted graft heals primarily by the ingrowth of endothelium and smooth muscle cells from the cut edges of the adjacent artery or other tissue (Figure 8.13). It then becomes encapsulated in the surrounding connective tissue and elicits a typical foreign-body reaction.

## SYSTEM DESCRIPTIONS AND DIAGRAMS

Catheter systems are composed of parts that assist in delivering a catheter to the site of diagnosis and/or therapy. A large-bore stainless steel needle is used to pierce the skin and enter an artery or vein, such as the femoral artery in the thigh (Figure 8.14A and B). A stainless steel guidewire is passed through the needle into the vessel to its final location (Figure 8.14C), under the guidance of fluoroscopy. When the guidewire is well positioned in the aorta at the level of the renal arteries, the needle is removed, with firm pressure applied over the puncture site to control bleeding (Figure 8.14D). A plastic catheter sheath is then advanced over the guidewire, which increases patient comfort and limits the arterial damage by several catheter exchanges through the artery (Figure 8.14E).

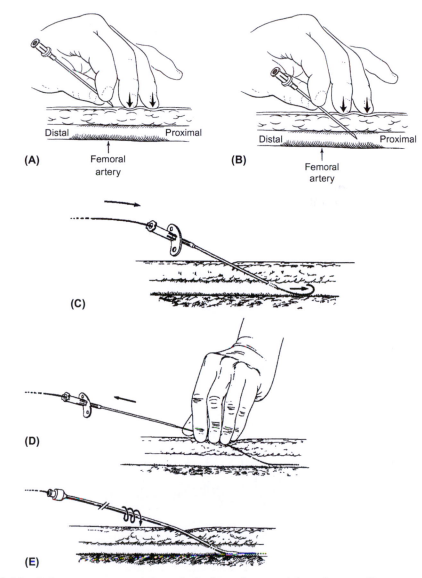

**FIGURE 8.14** Catheter system access through the femoral artery. A large-bore needle punctures (**A**) the skin and then (**B**) the artery. **C**: The flexible tip of the guidewire is passed through the needle into the artery. **D**: The needle is withdrawn, the artery is compressed, and the guidewire is pinched and fixed. **E**: The catheter sheath is advanced over the guidewire [*From Uretsky (1997). Reproduced by permission from Kern (2003)*].

## Cardiac Catheterization Procedures

With the catheter sheath in place, an appropriate catheter may be deployed. The main catheter shaft is often made of nylon. The proximal end of the catheter is a luer fitting. Catheter parts are added as needed to perform a particular function.

## Angiography

An angiography catheter enables the coronary arteries or cerebral arteries to be visualized with contrast injections for precise diagnosis of CAD or a cerebral aneurysm. These catheters are thin and flexible for small artery access. The coronary angiography catheter used for the majority of patients is the Judkins catheter, which has special preshaped curves and tapered end-hole tips (Kern, 2003) (Figure 8.15).

## Angioplasty

Angioplasty enables a balloon and stent to be delivered to and inflated at a coronary lesion. Angioplasty uses three catheters: a guiding catheter, a balloon catheter, and a stent catheter. A guiding catheter has thinner walls and larger lumens than does an angiography catheter, which allows contrast injections to be performed while the balloon or stent catheter is in place. As shown in Figure 8.16, the guiding catheter supports the balloon or

FIGURE 8.15 Judkins right (top) and left (bottom) coronary diagnostic catheters.

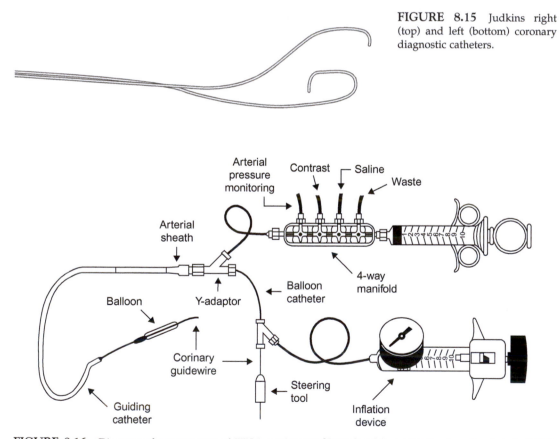

FIGURE 8.16 Diagram of components of PTCA equipment *[Reproduced by permission from Freed et al. (1996). Reproduced by permission from Kern (2003)].*

stent catheter's advancement into one lumen into the coronary artery, followed by balloon or stent inflation. A disposable syringe inflates the nylon balloon with precise measurement of the inflation pressure. Arterial pressure may be monitored from another lumen (Kern, 2003).

### Blood Pressure Monitoring

A pigtail catheter enables blood pressures to be monitored on both sides of a valve. As discussed in Chapter 7, the catheter is connected to a pressure transducer for measurement. The fluid in the catheter sheath is also connected to a pressure transducer. For example, the left ventricular pressure in contact with the catheter and aortic pressure in contact with the sheath may be measured simultaneously. Normally, no systolic pressure gradient should exist, and the aortic diastolic pressure should be about 80 mmHg. But for a patient with minimal obstructive hypertrophic cardiomyopathy (Figure 8.17), a systolic pressure gradient of 100 mmHg exists, with an aortic diastolic pressure of 60 mmHg (Kern, 2003).

### Thermodilution

A Swan-Ganz catheter enables cardiac output to be monitored in the right atrium. The catheter balloon is inflated to float the catheter tip to the pulmonary artery. A proximal port 30 cm from the catheter tip is used for rapid infusion of cold saline and pressure measurement, when this port is positioned in right atrium. A thermistor at the distal tip measures temperature in the pulmonary artery (Figure 8.18).

With thermodilution, cardiac output may be estimated in the right atrium from the change in temperature over time as the injected saline is mixed with blood. Thermodilution was a standard of care for many years, before a study published in *JAMA* demonstrated that ICU patients receiving right heart catheterization had increased

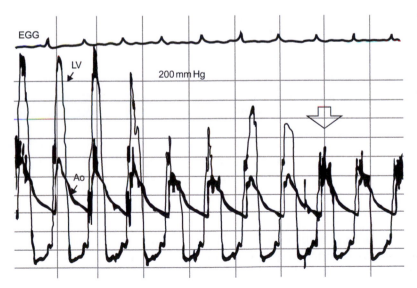

**FIGURE 8.17** Simultaneous left ventricular (LV) and aortic (Ao) pressures in a patient with obstructive hypertrophic cardiomyopathy during pullback of the catheter from the distal portion of the left ventricle. After catheter is pulled out of left ventricle (arrow), the pressure gradient is lost [*Reproduced by permission from Kern (2003)*].

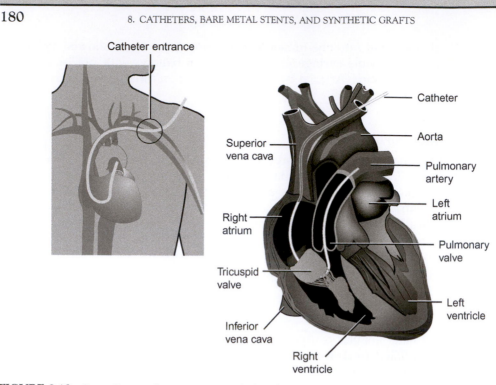

**FIGURE 8.18**  Swan-Ganz catheter positioning *(Adapted from A.D.A.M., Atlanta, Georgia).*

30-day mortality and longer length of stay in ICU, compared to their matched controls (Connors et al., 1996). Some companies believed that this study would cause CO estimation to shift from invasive to noninvasive measurement (Baura, 2001, Baura & Ng, 2003, Baura et al., 2009). Instead, much fewer CO estimates as a whole are made for diagnostic purposes.

### Electrophysiologic Studies and Ablation

*Specialty mapping catheters,* which are beyond the range of our discussion, enable electrophysiologists to acquire electrograms and localize the point of origin of ectopic beats such as those present during atrial fibrillation or atrial flutter. Once the point of origin is known, an ablation catheter applies radiofrequency energy to this site to decrease or eliminate the occurrence of ectopic beats (Figure 8.19). For example, an ectopic focus in a pulmonary vein may initiate atrial fibrillation (Haissaguerre et al., 1998).

### Biopsy

A biopsy forceps may be used to obtain a sample of right ventricular myocardial tissue. The forceps consists of a formable tip, stainless-steel cutting jaws, a wire-braided body, and a two- or three-ring plastic proximal handle controlling the operation of the jaws

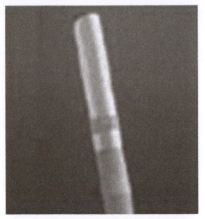

**FIGURE 8.19** Biosense Webster NaviStar DS 8-mm tip ablation catheter (Biosense Webster, Diamond Bar, California).

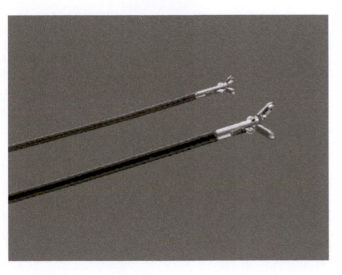

**FIGURE 8.20** Cook Flexible myocardial biopsy forceps (Cook Medical, Bloomington, Indiana).

(Figure 8.20). This procedure is indicated for monitoring cardiac transplant rejection and anthracycline cardiotoxicity.

## Bare Metal Stents

During angioplasty, a balloon is delivered by catheter to a blocked coronary or other artery and inflated to open the artery. A bare metal stent is then deployed and expanded into place. Stents are typically manufactured from stainless steel or nitinol, which is a nickel titanium alloy that self-expands with heat (i.e., after implantation). Each mesh tube generally ranges from 8 to 38 mm in length and from 2.5 to 5.0 mm in diameter (Padera & Schoen, 2004).

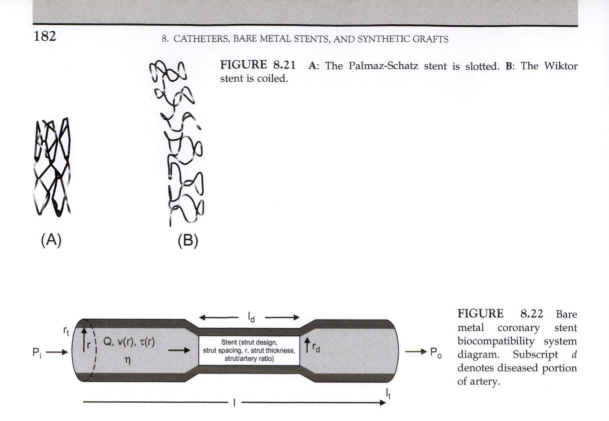

**FIGURE 8.21** **A:** The Palmaz-Schatz stent is slotted. **B:** The Wiktor stent is coiled.

(A)                    (B)

**FIGURE 8.22** Bare metal coronary stent biocompatibility system diagram. Subscript *d* denotes diseased portion of artery.

Stents are classified according to a slotted or coiled configuration. The struts of a *slotted stent* are disposed predominantly along the longitudinal axis, making the stent rigid along this axis. The struts of a *coiled stent* are disposed predominantly perpendicular to the longitudinal axis, making the stent flexible (Figure 8.21). A functional stent unit consists of struts joined to form rings. When the rings expand due to the radial force of the balloon, plastic deformation of the struts occurs (Palmaz et al., 2002).

An implanted stent is much more difficult to model than an implanted valve. The device diameter and blood flow rate are much lower, so laminar flow cannot be assumed. Even if we assume that the artery and stent are cylindrical tubes, the diseased artery has at least two radii, with a smaller radius at the localized plaque. After stent implantation, the stent reaction to the plaque is a function of strut design, strut spacing, radius of curvature, strut thickness, and stent-to-artery ratio (Timmins et al., 2008). A biocompatibility system diagram for a bare metal coronary stent is given in Figure 8.22.

Modeling such a complicated system requires *finite element analysis (FEA)*, in which solutions to partial differential equations are approximated by a grid of uniformly spaced nodes. FEA is beyond the range of this discussion. For example, arterial wall shear stress gradients were recently modeled for four stents, and used to examine the differences in clinical restenosis rates (Duraiswamy et al., 2009) (Figure 8.23). As you recall from Chapter 6, high shear stress leads to platelet activation. Because bare metal stent design has been historically conducted "by trial and error" (Zarins & Taylor, 2009), systematic

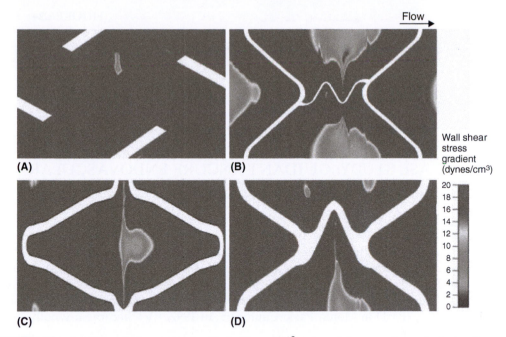

**FIGURE 8.23** Arterial wall shear stress gradients (dynes/cm³) taken at the mean flow rate during the accelerating phase of the flow cycle. **A**: The Wall stent, **B**: The Bx Velocity stent. **C**: The Aurora stent. **D**: The NIR stent. Dark pink regions near struts show wall shear stress gradients $>>$ 20 dynes/cm³ *[Reproduced by permission from Duraiswamy et al. (2009)].*

biomechanical modeling could lead to improved stent designs, which could result in lower clinical restenosis rates (Moore, 2009).

## Synthetic Grafts

A Dacron or ePTFE graft performs well in large-diameter, high-flow, low-resistance locations such as the aorta and the iliac and proximal femoral arteries. Grafts used in these locations have 5- to 10-yr patency rates of 90%. These grafts may be stented to aid in fixation. But small grafts, such as those less than 6−8 mm in diameter, have poorer performance, with 5-yr patency rates less than 50% (Padera & Schoen, 2004). This is why CABG uses biological leg vein grafts.

In the implanted stent model, we assume that the artery and stent are cylindrical tubes and that the diseased artery has at least two radii, with a larger radius at the localized dilatation. After implantation, the graft reaction to the dilatation is a function of graft design, graft fixation, radius of curvature, graft length, graft porosity (pore diameter and pore number), and graft-to-artery ratio. A successful graft reduces aneurysm wall stress so that risk of rupture is eliminated or so that the new, sealed conduit can maintain its integrity in the presence of a rupture (Moore, 2009). A biocompatibility system diagram for a synthetic graft is given in Figure 8.24.

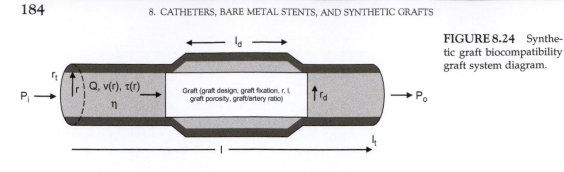

**FIGURE 8.24** Synthetic graft biocompatibility graft system diagram.

# FDA CASE STUDY: GUIDANT ANCURE ENDOVASCULAR GRAFT SYSTEM

Guidant's 2005 ICD recall, discussed in Chapter 5, was not its first recall scandal. In the 1990s, both Guidant and Medtronic were developing systems for endovascular aneurysm repair. Ideally, through a catheter delivery system, an aortic graft would be delivered to the site of an AAA. After the graft was attached and the delivery catheter was removed, blood would flow through the graft, avoiding the aneurysm, which would shrink over time. Guidant acquired its endovascular graft technology when it purchased Endovascular Technologies, Inc., (EVT) in Menlo Park, California, in 1997. EVT became a wholly owned subsidiary of Guidant.

During clinical trials, Guidant learned from physicians that they perceived the Guidant Ancure system (Figure 8.25) was more difficult to use than the Medtronic AneuRx system (Figure 8.1). A compressed, unreinforced Ancure graft was expanded at the site of aneurysm and attached with hooks, using balloon fixation. In contrast, the nitinol stent rings surrounding an AneurRx graft self-expanded at the site of aneurysm for anchoring. The Ancure and AneuRx systems were both approved by FDA on September 30, 1999. In order to not harm marketing efforts, Guidant officials believed that the Ancure system needed to be successfully deployed in a significant number of cases.

After the Ancure system began to be sold in the United States, Guidant became aware of various complications during implantation. The most severe complication was that the catheter delivery system could become lodged in a patient when the Ancure jacket sheath did not retract. When this occurred, the Ancure system could not be removed without resorting to traditional open surgical repair. A Guidant sales representative assisted in developing a technique, in which the catheter delivery handle was broken, so that catheter pieces could be removed from the patient's body. This technique became known as the handle breaking technique (HBT) and was taught to other sales representatives and physicians.

## Public Guidant Actions

Approximately a year after the Ancure system received premarket approval (PMA), in July 2000, an FDA inspector conducted a mandatory inspection of manufacturing facilities for the Ancure system. Per procedure, the inspector requested a list of all medical device reports (MDRs) related to the Ancure system. FDA requires an MDR to be submitted

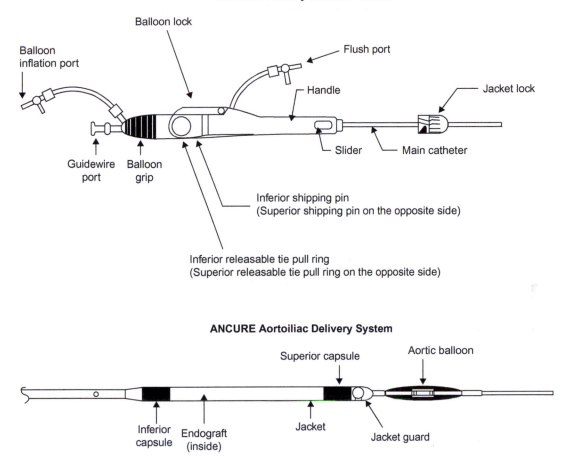

**FIGURE 8.25** Guidant Ancure System *[Reprinted from CDRH (1999). Reproduced by permission from Baura (2006)].*

within 30 days when a death or serious injury occurs in connection with medical device use. (This is a direct result of the Bjork-Shiley recall and subsequent Safe Medical Devices Act of 1990.) Guidant provided FDA with a list of 55 MDRs, even though more than 200 adverse events, as defined by FDA regulations, existed at the time (Ryan, 2003a).

## Internal Guidant Actions

On January 26, 2000, the handle breaking technique was used unsuccessfully in an operation, with the patient ultimately dying. A dozen Guidant employees (Ryan, 2003b), including one engineer (Jacobs, 2003b), concluded from this incident that the handle breaking technique required testing and validation. They also concluded that if the HBT were used, it needed to be reported to FDA. Per FDA regulations, directions for use are considered labeling. Updated labeling of an approved device requires the submission of a PMA supplement to FDA. Without this PMA supplement, the device is considered misbranded and cannot be sold legally.

After the FDA inspection in July 2000, seven anonymous employees, called the Anonymous Seven in court documents, sent a letter to FDA and to the Guidant chief compliance officer. Their letter detailed safety concerns that included:

1. [Guidant] conducted incomplete testing and analysis on currently recommended procedures.
2. [Guidant] recommended the use of the device in a manner that was outside the direction for use approved by FDA.
3. The jacket retraction failure mode, which involved the failure of the sheath of the Ancure device to retract as intended, had a corresponding complaint rate of approximately 20%.
4. [Guidant] failed to report to FDA product changes that affected safety and efficacy as legally required.
5. [Guidant] failed to submit MDRs to FDA as legally required (Ryan, 2003b).

Receipt of this letter initiated an internal Guidant investigation. It was determined that EVT had "serious quality system regulation violations, incomplete and untimely complaint handling and documentation, incomplete MDR reporting, inadequate corrective and preventative action activities, incomplete record keeping and poor traceability practices, and was significantly out of compliance with FDA regulations and its own internal policies" (Ryan, 2003b).

On March 23, 2001, Guidant disclosed to FDA the existence of 2628 additional MDRs that had not been reported. Included in these MDRs were 12 deaths and 57 conversions to open surgical repair (Ryan, 2003b). These results were known by Guidant executives by January 2001 but were not presented to FDA until Guidant learned it was under criminal investigation. Guidant suspended Ancure commercial sales on March 16, 2001 (Jacobs, 2003b).

## FDA Responses

When the FDA received the Anonymous Seven's letter in 2000, it launched its own investigation with the Office of Criminal Investigations and FBI (Jacobs, 2003a). After Guidant revised physician instructions for troubleshooting in a PMA supplement, the FDA allowed the Ancure device to be remarketed in August 2001 (Jacobs, 2003b).

## Aftermath

On June 9, 2003, criminal charges were filed against EVT by the U.S. District Court Northern District of California. On June 12, 2003, Guidant pled guilty to one felony count of false statement for not providing a full list of MDRs to an FDA inspector and guilty to nine felony counts of interstate shipment of misbranded devices. Guidant agreed to a criminal fine of $32.5 million for these felony violations. It also agreed to forfeit $10.9 million to compensate for profit made from illegal sales. Further, Guidant agreed to pay a civil settlement of $49 million (Ryan, 2003b) to settle claims that the firm's actions caused the Medicare, Medicaid, and Veteran Affairs programs to pay millions of dollars for the adulterated and misbranded devices (Jacobs, 2003a). At the time, this total settlement of $92.4 million was the largest payout ever levied for violating FDA's medical device reporting requirements. Guidant ceased Ancure system manufacture and closed its EVT facility in 2003 (Jacobs, 2003b; Baura, 2006).

The Medtronic AneuRx system continues to be sold in the United States and currently competes against several EVAR systems from other manufacturers. Endovascular aneurysm repair is becoming a standard of care. For recent generations of EVAR systems, the 5-yr survival of EVAR patients is insignificantly different from that of open surgery patients (Mani et al., 2009).

For a more thorough discussion of FDA regulation, see Baura (2013).

## KEY FEATURES FROM ENGINEERING STANDARDS

As of 2010, FDA does not recognize any consensus standards that are specific to cardiac catheterization systems. We discuss one requirement from a general FDA-recognized catheter standard: *ISO 10555-1 Sterile, Single-Use Intravascular Catheters—Part 1: General Requirements* (ISO, 1995). FDA does recognize three brief consensus standards that are specific to vascular stents for elastic recoil, dimensional attributes, and corrosion susceptibility. We discuss the most important of these: *ASTM F 2079-02 Standard Test Method for Measuring Intrinsic Elastic Recoil of Balloon-Expandable Stents* (ASTM, 2002). We also discuss requirements from the FDA-recognized standard for tubular vascular grafts: *ANSI/AAMI/ISO 7198:1998/2001(R)2004 Cardiovascular Implants—Tubular Vascular Prostheses* (AAMI, 2004).

### Freedom from Liquid Leakage Under Pressure

Catheters used for PTCA experience high pressures but must not leak. According to ISO 10555-1, leakage is tested by connecting a catheter hub to a 10-mL syringe filled with de-aerated distilled water, through a leakproof connector with 6% Luer fitting. The syringe contains a volume of water that is the nominal graduated capacity. The distal end of the catheter is occluded. To perform the test, an axial force is applied to the syringe plunger to generate a pressure of 300−320 kPa for 30 s. The catheter tube is then examined for liquid leakage, such as one or more falling drops of water. It is required that no leakage occurs (ISO, 1995).

### Elastic Recoil

As already detailed, stents began to be implanted after angioplasty to prevent restenosis. Yet stent strut penetration into the plaque lipid core may increase neointimal growth, which results in restenosis. ASTM F 2079-02 describes the technique for measuring the elastic recoil of balloon-expandable stents. If a stent diameter decreases excessively after inflation, then greater inflation pressure may be required during inflation to compensate for this elastic recoil. This greater inflation could, in turn, promote greater restenosis.

To measure elastic recoil, a stent balloon is inflated to the nominal expansion pressure for 15–30 s. The outer diameter of the stent in two approximately orthogonal rotational positions is measured. The balloon is then deflated, with the outer diameters in the same positions measured again after 10 s of deflation. Stent recoil is then calculated, at each location, as the percentage difference.

The number of required stent samples is not specified. The maximum allowable recoil is not detailed (ASTM, 2002).

## Microscopic Porosity

Synthetic grafts heal by the ingrowth of endothelium and smooth muscle cells. Although a minimum pore diameter is not specified by ANSI/AAMI/ISO 7198:1998/2001 (R)2004, this standard does require that microscopic porosity be measured and recorded.

A scanning electron micrograph of a section of the graft test specimen is prepared. The diameters of at least six representative pores from the photomicrograph are measured with a device that has an accuracy of ±1% of the manufacturer's mean pore diameter. From these measurements, the mean and standard deviation of the pore diameter are calculated in microns. The number of pores in a known area is also recorded. It must be verified that the mean porosity of the sample is within the nominal range declared by the manufacturer.

## Biocompatibility

As per the AAMI heart valve standard, ANSI/AAMI/ISO 7198:1998/2001(R)2004 requires that vascular material biocompatibility shall be determined in accordance with *ISO 10993-1:2009 Biological Evaluation of Medical Devices—Part 1: Evaluation and Testing Within a Risk Management Process* (AAMI, 2004). Part 1 of the ISO standard presents a test matrix for appropriate testing of surface devices, external communicating devices, and implant devices. This matrix points to tests in other parts of the standard that assess cytotoxicity, sensitization, irritation, toxicity, genotoxicity, implantation, hemocompatibility, and carcinogenicity (ISO, 2009).

## *In vivo* Clinical Study

Per ANSI/AAMI/ISO 7198:1998/2001(R)2004, the safety and efficacy of a vascular graft are assessed during a short-term clinical study of at least 1-yr duration. The study is conducted at a minimum of three institutions, with at least 10 implants per institution. The minimum number of patients is not specified. After implantation and follow-up, patient accountability, including rationale for the exclusion of data; adverse events, deaths, and patency rates, and adverse event rates are reported. Acceptable rates of performance and complications are not specified.

# SUMMARY

In this chapter, we have grouped three types of medical devices. Cardiac catheter systems enable the performance of minimally invasive cardiac procedures. A bare metal stent, which is an expandable tube of metallic mesh, provides a physical scaffold for keeping a blood vessel patent. A synthetic graft allows a weakened blood vessel wall to be replaced.

Cardiac catheter systems enable bare metal stents and synthetic grafts to be deployed to blood vessels that have been damaged by complications of atherosclerosis. Atherosclerotic plaque begins as fibrous plaque. This consists of an amorphous central lipid core of cellular debris, cholesterol, cholesterol esters, and foam cells. A fibrous cap surrounds the fibrous plaque, consisting mainly of smooth-muscle cells in a matrix of collagen and proteoglycans. Over time, fibrous plaque may become complicated plaque. Complicated plaque consists of calcification, hemorrhage into the plaque, ulceration of the fibrous cap, and luminal thrombosis. Often, complicated plaque is the underlying mechanism behind acute coronary clinical events of ischemic origin. When extensive structural remodeling occurs after complicated plaques form, the aortic wall may be weakened, leading to a subsequent bulge that eventually needs to be repaired.

Cardiac catheterization is a minimally invasive procedure that enables other procedures besides stent and graft implantation to be performed: angiography, angioplasty, blood pressure monitoring, thermodilution, electrophysiologic studies and ablation, and biopsy. Key features include catheter freedom from liquid leakage under pressure, stent elastic recoil, vascular graft microscopic porosity, vascular graft biocompatibility, and vascular graft *in vivo* clinical study.

Guidant's 2005 ICD recall, discussed in Chapter 5, was not its first recall scandal. Guidant's Ancure system was developed to deliver and deploy a vascular graft by catheter to the site of an abdominal aortic aneurysm. Because of competitive pressure from a Medtronic system that was approved by the FDA on the same day, Ancure complications were hidden from the FDA during a routine inspection. Eventually, due to internal company pressures, Guidant disclosed to the FDA the existence of 2628 additional MDRs that had not been reported to the FDA, as required by law. Included in these MDRs were 12 deaths and 57 conversions to open surgical repair. On June 12, 2003, Guidant pled guilty to one felony count of false statement, when it did not provide a full list of MDRs to an FDA inspector, as well as nine felony counts of interstate shipment of misbranded devices.

## Exercises

**8.1** Radiologist Charles Dotter was also an amateur filmmaker. Watch his short film on transluminal angioplasty: http://www.ptca.org/archive/bios/dotter.html. Describe the filmed procedure, including all medical devices that were used.

**8.2** Bare metal stents are fabricated from steel or nitinol. What mechanical properties of metals are optimized by choosing these materials? To answer this question, check the mechanical properties tables in a standard biomaterials textbook.

**8.3** Describe the specific host reactions occur after stent implantation. Theoretically, what type of drug would be administered to minimize these host reactions?

**8.4** Why does the FDA recommend more consensus standard requirements for vascular grafts than for vascular stents when stenting is a more utilized procedure? Is it coincidental that Guidant had two big recalls within a short time period? Give reasons.

Read Cournand et al. (1945):

**8.5** How were catheters inserted in this laboratory to reach the right atrium? What were the two types of listed catheter complications? Were these complications preventable?

Read Serruys et al. (1994):

**8.6** After initial rejection in 1993, FDA approved the Palmaz-Schatz stent for market release after reviewing these data. What were the primary clinical end points? Why are these important for the clinical study design?

**8.7** What is an angiogram? Compare the angiogram data between groups before the procedure, after it, and at follow-up (Table 3). Does the stent significantly affect results?

**8.8** What are the complications of stenting compared to balloon angioplasty alone? Do the increased cost and complications justify stenting over balloon angioplasty alone? Give reasons.

# References

AAMI (2004). *ANSI/AAMI/ISO 7198:1998/2001(R)2004 cardiovascular implants—Tubular vascular prostheses.* Arlington, VA: AAMI.

AHA (2009). *Heart disease & stroke statistics: 2009 update At-a-Glance.* Dallas: AHA.

ASTM (2002). *ASTM F2079-02 standard test method for measuring intrinsic elastic recoil of balloon-expandable stents.* West Conshohocken, PA: ASTM.

Baura, G. D. (2001). Method and apparatus for high current electrode, transthoracic and transmyocardial impedance estimation. *US 6,253,103.*

Baura, G. D. (2006). *Engineering ethics: An industrial perspective.* Burlington, VT: Elsevier Academic Press.

Baura, G. D. (2013). *U.S. medical device regulation: An introduction to biomedical product development, market release, and postmarket surveillance.* Manuscript in preparation.

Baura, G. D., Malecha, J., & Bouguerra, R. (2009). Method and apparatus for signal assessment including event rejection. *US 7,570,989.*

Baura, G. D., & Ng, S. K. (2003). Method and apparatus for hemodynamic assessment including fiducial point detection. *US 6,561,986.*

Bauters, C., & Isner, J. M. (1998). The biology of restenosis. In E. J. Topol (Ed.), *Textbook of cardiovascular medicine.* Philadelphia: Lippincott-Raven.

Blakemore, A. H., & Voorhees, A. B., Jr. (1954). The use of tubes constructed from vinyon N cloth in bridging arterial defects; experimental and clinical. *Annals of Surgery, 140,* 324–334.

CDRH (1994). *Approval order P000043.* Rockville, MD: FDA.

CDRH (1999). *Approval order P990017.* Rockville, MD: FDA.

Connors, A. F., Jr., Speroff, T., Dawson, N. V., Thomas, C., Harrell, F. E., Jr., & Wagner, D., et al. (1996). The effectiveness of right heart catheterization in the initial care of critically ill patients. SUPPORT Investigators. *Journal of the American Medical Association, 276,* 889–897.

Cournand, A. (1957). Pulmonary circulation; its control in man, with some remarks on methodology. *Science, 125,* 1231–1235.

Cournand, A., Riley, R. L., Breed, E. S., Baldwin, E. D., Richards, D. W., & Lester, M. S., et al. (1945). Measurement of cardiac output in man using the technique of catheterization of the right auricle or ventricle. *Journal of Clinical Investigation, 24,* 106–116.

Dotter, C. T., & Judkins, M. P. (1964). Transluminal treatment of arteriosclerotic obstruction. Description of a new technic and a preliminary report of its application. *Circulation, 30,* 654–670.

Duraiswamy, N., Schoephoerster, R. T., & Moore, J. E., Jr. (2009). Comparison of near-wall hemodynamic parameters in stented artery models. *Journal of Biomedical Engineering, 131,* 061006-1–061006-9.

Farb, A., Weber, D. K., Kolodgie, F. D., Burke, A. P., & Virmani, R. (2002). Morphological predictors of restenosis after coronary stenting in humans. *Circulation, 105,* 2974–2980.

Forssmann-Falck, R. (1997). Werner Forssmann: a pioneer of cardiology. *American Journal of Cardiology, 79,* 651–660.

Freed, M., Grines, C., & Safiean, R. D. (1996). *The new manual of interventional cardiology.* Birmingham, MI: Physician's Press.

Gruentzig, A. R. (1981). Percutaneous transluminal coronary angioplasty. *Seminars in Roentgenology, 16,* 152–153.

Guo, D. C., Papke, C. L., He, R., & Milewicz, D. M. (2006). Pathogenesis of thoracic and abdominal aortic aneurysms. *Annals of the New York Academy of Sciences, 1085,* 339–352.

Guyton, A. C., & Hall, J. E. (2006). *Textbook of medical physiology* (11th ed.). Philadelphia: Elsevier Saunders.

Haissaguerre, M., Jais, P., Shah, D. C., Takahashi, A., Hocini, M., & Quiniou, G., et al. (1998). Spontaneous initiation of atrial fibrillation by ectopic beats originating in the pulmonary veins. *New England Journal of Medicine, 339,* 659–666.

ISO (1995). *ISO 10555-1 sterile, single-use intravascular catheters—Part 1: General requirements.* Geneva: ISO.

ISO (2009). *ISO 10993-1:2009 biological evaluation of medical devices—Part 1: Evaluation and testing within a risk management process.* Geneva: ISO.

Jacobs, M. J. (2003a). *Press release: Endovascular Technologies, Inc.* San Francisco: United States Attorney's Office Northern District of California.

Jacobs, P. (2003b). Medical firm's dangerous secret: Device's troubles were well-known at Menlo Park company. *San Jose Mercury News* (August 3).

Lasker Awards: Former Winners. (2009). Lasker Foundation.

Libby, P. (2002). Inflammation in atherosclerosis. *Nature, 420,* 868.

Kern, M. J. (2003). *The cardiac catheterization handbook* (4th ed.). Philadelphia: PA: Mosby.

Mani, K., Bjorck, M., Lundkvist, J., & Wanhainen, A. (2009). Improved long-term survival after abdominal aortic aneurysm repair. *Circulation, 120,* 201–211.

Moore, J. E., Jr. (2009). Biomechanical issues in endovascular device design. *Journal of Endovascular Therapy, 16* (Suppl. 1), I1–I11.

Padera, R. F., Jr., & Schoen, F. J. (2004). Cardiovascular medical devices. In B. D. Ratner, A. S. Hoffman, F. J. Schoen, & J. E. Lemons (Eds.), *Biomaterials science* (2nd ed.). San Diego, CA: Elsevier.

Palmaz, J. C., Bailey, S., Marton, D., & Sprague, E. (2002). Influence of stent design and material composition on procedure outcome. *Journal of Vascular Surgery, 36,* 1031–1039.

Ragosta, M. (2011). *Cases in interventional cardiology.* Philadelphia: PA: Saunders/Elsevier.

Rajagopal, V., & Rockson, S. G. (2003). Coronary restenosis: A review of mechanisms and management. *American Journal of Medicine, 115,* 547–553.

Ryan, K. V. (2003a). *United States of America v. Endovascular Technologies, Inc. Criminal Information,* 02-0179 SI. Department of justice northern district of California. San Francisco: United States District Court.

Ryan, K. V. (2003b). *United States of America v. Endovascular Technologies, Inc. Plea Agreement,* 02-0179 SI. Department of justice northern district of California. San Francisco: United States District Court.

Sangiorgi, G., Taylor, A. J., Farb, A., Carter, A. J., Edwards, W. D., & Holmes, D. R., et al. (1999). Histopathology of postpercutaneous transluminal coronary angioplasty remodeling in human coronary arteries. *American Heart Journal, 138,* 681–687.

Schoen, F. J. (1989). *Interventional and surgical cardiovascular pathology: Clinical correlations and basic principles.* Philadelphia: W. B. Saunders.

Schoen, F. J., & Cotran, R. S. (1999). Blood vessels. In R. S. Cotran, V. Kumar, & T. Collins (Eds.), *Robbins pathologic basis of disease* (6th ed.). Philadelphia: W. B. Saunders.

Serruys, P. W., De Jaegere, P., Kiemeneij, F., Macaya, C., Rutsch, W., & Heyndrickx, G., et al. (1994). A comparison of balloon-expandable-stent implantation with balloon angioplasty in patients with coronary artery disease. Benestent Study Group. *New England Journal of Medicine, 331*, 489–495.

Society for Vascular Surgery (2009). *VascularWeb: Abdominal aortic aneurysm.* Chicago: Society for Vasculary Surgery.

Sones, F. M., Jr. (1958). Diagnosis of septal defects by the combined use of heart catheterization and selective cine-cardioangiography. *American Journal of Cardiology, 2*, 724–731.

Timmins, L. H., Meyer, C. A., Moreno, M. R., & Moore, J. E. (2008). Effects of stent design and atherosclerotic plaque composition on arterial wall biomechanics. *Journal of Endovascular Therapy, 15*, 643–654.

Uretsky, B. (Ed.), (1997). *Cardiac catheterization: Concepts, techniques, and applications.* Walden, MA: Blackwell Science.

Virmani, R., Farb, A., & Burke, A. P. (1994). Coronary angioplasty from the perspective of atherosclerotic plaque: Morphologic predictors of immediate success and restenosis. *American Heart Journal, 127*, 163–179.

Voorhees, A. B., Jr., Jaretzki, A., 3rd, & Blakemore, A. H. (1952). The use of tubes constructed from vinyon "N" cloth in bridging arterial defects. *Annals of Surgery, 135*, 332–336.

Zarins, C. K., & Taylor, C. A. (2009). Endovascular device design in the future: transformation from trial and error to computational design. *Journal of Endovascular Therapy, 16*(Suppl. 1), I12–I21.

# CHAPTER

# 9

# Hemodialysis Delivery Systems

In this chapter, we discuss the relevant physiology, history, and key features of hemo-dialyzers and hemodialysis (HD) delivery systems. A *dialyzer* is a device that separates substances in solution, or solutes, through unequal diffusion through a semipermeable membrane (Figure 9.1). *Simple diffusion* refers to the random movement of molecules or ions through intermolecular spaces in a membrane, across a concentration gradient. When the solution is blood, a dialyzer is called a *hemodialyzer*. A *hemodialysis delivery system*, or hemodialysis machine, is an instrument that enables blood to be safely dialyzed of its waste products, with a hemodialyzer and hemodialysis solution as its key components (Figure 9.2).

Dialysis may be given to acute renal failure patients in the intensive care units or general dialysis floors of hospitals or to chronic renal failure patients. The discussion in this chapter is limited to patients with *end-stage renal disease (ESRD)* who receive long-term

Dialyzer with individual capillary membranes

**FIGURE 9.1** Fresenius FX100 Polysulfone dialyzer (Fresenius SE, Homburg, Germany).

**FIGURE 9.2** Fresenius 2008T Hemodialysis Machine (Fresenius Medical Care North America, Waltham, Massachusetts).

hemodialysis, rather than *peritoneal dialysis* (PD). With hemodialysis, blood is taken from a patient's circulation, processed in a hemodialysis delivery system, and then returned through a patient's venous system. With peritoneal dialysis, dialysis fluid, or dialysate, is infused into the peritoneum, which is lined by the peritoneal membrane covering the abdominal wall and visceral organs. Because the peritoneal membrane contains blood vessels, exchange of molecules occurs between blood and the dialysate that dwells in this cavity for a prespecified period (Figure 9.3).

Upon completion of this chapter, each student shall be able to:

1. Understand the mechanisms underlying renal function.
2. Explain the mechanisms behind removal of waste products and water during dialysis.
3. Describe five key features of hemodialyzers and hemodialysis delivery systems.

## BODY FLUID COMPARTMENTS

Total body fluid is mainly distributed between extracellular fluid and intracellular fluid compartments. Extracellular fluid accounts for about 20% of body weight. About 75% of the fluid outside the cells is the interstitial fluid (ISF) bathing the cells, while the remaining

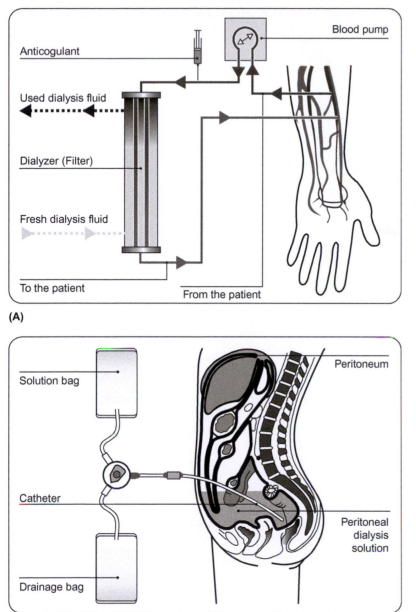

Anticogulant

Blood pump

Used dialysis fluid

Dialyzer (Filter)

Fresh dialysis fluid

To the patient

From the patient

**(A)**

Solution bag

Peritoneum

Catheter

Peritoneal dialysis solution

Drainage bag

**(B)**

**FIGURE 9.3** Hemodialysis (**A**) vs. (**B**) peritoneal dialysis (Fresenius SE, Homburg, Germany).

25% is the plasma part of the blood. Solute molecules and water are continuously exchanged between ISF and plasma through the pores of capillary membranes. About two-thirds of body fluid, the intracellular fluid, is contained within the 75 trillion cells in the body.

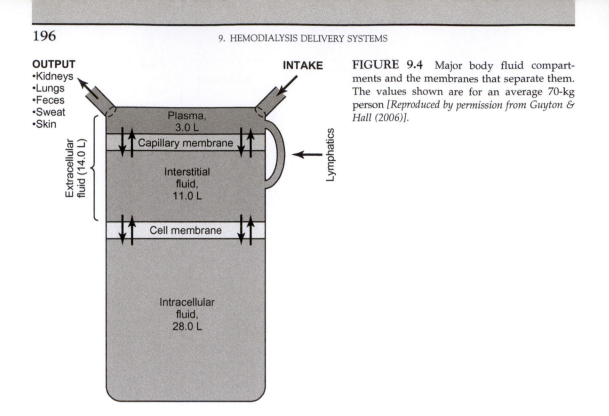

**FIGURE 9.4** Major body fluid compartments and the membranes that separate them. The values shown are for an average 70-kg person [*Reproduced by permission from Guyton & Hall (2006)*].

A relatively constant volume and stable composition and pH of the body fluids is essential for our bodies to maintain *homeostasis*, or physiologic internal balance. Daily body water loss through the skin, respiratory tract, sweat, feces, and kidneys is balanced by daily water intake through the ingestion of liquids or water in food and synthesis resulting from carbohydrate oxidation (Guyton & Hall, 2006) (Figure 9.4).

## RENAL ANATOMY AND PHYSIOLOGY

As blood flows from the left ventricle through the systemic circulation, it eventually arrives at the tissues. There, in the microcirculation, arterioles, with internal diameters of 10–15 microns, become capillaries, with internal diameters of 5 microns, which become venules, which have larger inner diameters than arterioles. At the capillaries, nutrients are transported to ISF and then tissue cells, and cell waste moves to ISF and eventually to the venous system. Since few cells are located more than 50 microns from a capillary, movement of almost any substance from the capillary to the cell occurs within a few seconds (Figure 9.5).

Within the kidney, capillary exchange is responsible for many functions, including the regulation of water and electrolyte balances and the excretion of metabolic waste products and foreign chemicals. Waste products with potential toxicity include urea, creatinine,

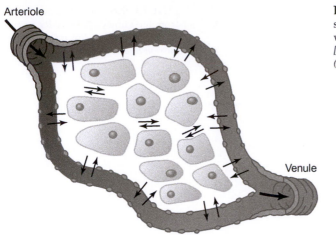

Arteriole

Venule

**FIGURE 9.5** Movement of fluid and dissolved constituents through the capillary walls and through the interstitial spaces [*Reproduced by permission from Guyton & Hall (2006)*].

excess phosphorus, phenols, and phenolic acids. Urea is a 60-Da solute produced in the liver through the combination of ammonia and carbon dioxide:

$$2\,NH_2 + CO_2 \rightarrow H_2N - \underset{\underset{O}{\|}}{C} - NH_2 + H_2O. \tag{9.1}$$

The ammonia in this reaction is released during protein breakdown. Creatinine is a waste product of muscle metabolism. Phosphates are ingested through milk products and meat.

Blood flow to the two kidneys is about one-fifth of total cardiac output. Within each kidney, the renal artery branches to form the interlobar arteries, arcuate arteries, interlobular arteries, afferent arterioles, and glomerular capillaries. These capillaries merge to form efferent arterioles, which lead to peritubular capillaries, which surround the renal tubules. The peritubular capillaries merge to eventually form the interlobular veins, arcuate veins, interlobar veins, and renal vein (Figure 9.6).

The functional unit of the kidney is the *nephron*. Each human kidney contains about 800,000 to 1 million nephrons, each capable of producing urine. Each nephron contains a *Bowman's capsule*, in which fluid is filtered from the blood, and a long tubule in which filtered fluid is converted into urine (Figure 9.7).

## Urine Formation

Each Bowman's capsule encases glomerular capillaries, and is lined by visceral epithelial cells. These capillaries freely filter most substances in plasma, except for proteins. About 20% of renal plasma flowing through the kidney is filtered into the Bowman's space. The filtered fluid, called the *glomerular filtrate*, leaves Bowman's capsule and passes through the tubules. The concentrations of glomerular filtrate constituents, including most salts and organic molecules, are similar to the concentrations in plasma. The *glomerular filtration rate* (*GFR*) in the average adult human is about 125 mL/min or 180 L/day (Guyton & Hall, 2006). GFR can be calculated from creatinine clearance, which requires 24 hour

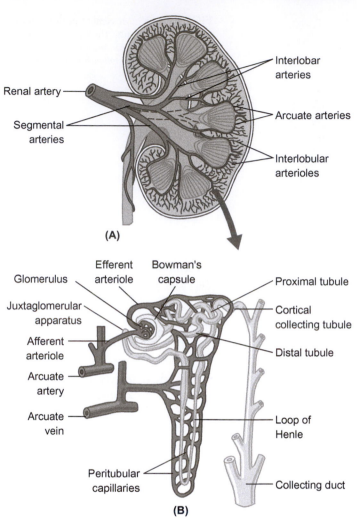

**FIGURE 9.6**   **A**: Blood flow to and from the kidney. **B**: The microcirculation of a nephron *[Reproduced by permission from Guyton & Hall (2006)].*

collection of all urine and one blood test to determine creatinine concentration. To avoid the problems of 24 hour urine collection, the National Institute of Diabetes and Digestive and Kidney Diseases (NIH, 2010) has recommended that GFR be estimated as a function of serum creatinine ($S_{cr}$), age, ethnicity, and gender (Levey et al., 2006):

$$GFR = 175 \times S_{cr}^{-1.154} \times age^{-0.203} \times 1.212 \ [if \ black] \times 0.742 \ [if \ female]. \qquad (9.2)$$

Equation (9.2) is referred to as the *modification of diet in renal disease (MDRD) study equation.*

Glomerular filtrate passes through the proximal tubule, the loop of Henle, the distal tubule, the connecting tubule, and the collecting duct. About 65% of the filtered load of sodium and water and a slightly lower percentage of filtered chloride are reabsorbed by the proximal tubule before the glomerular filtrate reaches the loop of Henle. About 20% of the filtered water and 25% of the filtered loads of sodium ($Na^+$), chloride, and potassium ($K^+$)

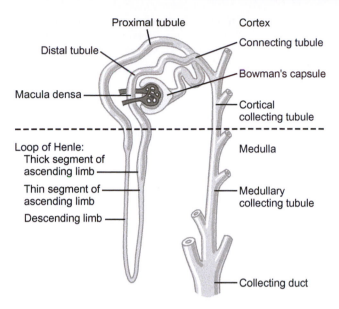

Proximal tubule

Distal tubule

Macula densa

Loop of Henle:
Thick segment of ascending limb

Thin segment of ascending limb

Descending limb

Cortex

Connecting tubule

Bowman's capsule

Cortical collecting tubule

Medulla

Medullary collecting tubule

Collecting duct

**FIGURE 9.7** Structures of the nephron. The relative lengths of the different tubular segments are not drawn to scale [*Reproduced by permission from Guyton & Hall (2006)*].

are reabsorbed in the loop of Henle. Most sodium, chloride, and potassium—but no water and urea—are reabsorbed in the first half of the distal tubule. Reabsorption occurs as the filtrate moves across the tubular epithelial membrane into renal interstitial fluid and then through the peritubular capillary membrane back into the blood. Note that most waste products are poorly reabsorbed; so a normal glomerular filtration rate effectively removes these waste products from the body.

When the glomerular filtrate passes to the second half of the distal tubule and the cortical collecting tubule, $Na^+$, water, and $K^+$ are reabsorbed, while $K^+$ and hydrogen ($H^+$) ions are secreted. Sodium reabsorption and potassium secretion depend on the activity of a sodium-potassium ATPase pump. Hydrogen ion secretion is mediated by a hydrogen-ATPase transport mechanism. The ATPase pump and ATPase transport mechanism are beyond the range of this discussion.

The filtrate moves to the medullary collecting duct, where less water and urea are reabsorbed if antidiuretic hormone (ADH) is present. The collecting ducts merge to form the renal pelvis, which is the funnel-shaped continuation of the upper end of the ureter, the tube that carries the urine to the bladder. The four processes of filtration, reabsorption, secretion, and excretion, which determine the composition of urine, are summarized in Figure 9.8.

## Regulation of Water and Electrolyte Balances

Hormones control tubular reabsorption to regulate body fluid volumes and solute concentrations. A *hormone* is a substance that is secreted from an endocrine gland or gonad and transported through the blood to the site of action. Aldosterone acts on the collecting tubule and duct cells to increase $Na^+$ reabsorption and $H^+$ and $K^+$ secretion. Angiotensin

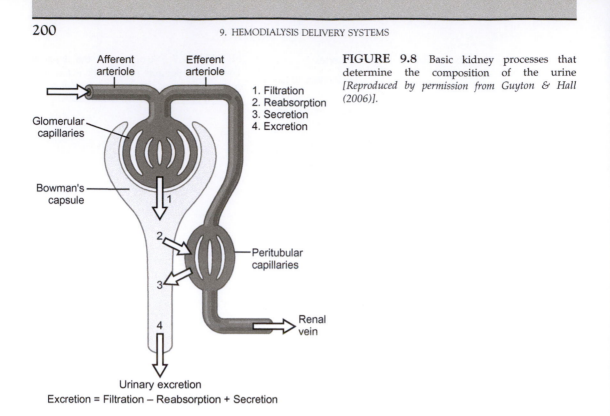

1. Filtration
2. Reabsorption
3. Secretion
4. Excretion

Excretion = Filtration − Reabsorption + Secretion

**FIGURE 9.8** Basic kidney processes that determine the composition of the urine [*Reproduced by permission from Guyton & Hall (2006)*].

II acts on the proximal, distal, and collecting tubule cells to increase $Na^+$ reabsorption and $H^+$ secretion. ADH acts on the collecting tubule and collecting duct cells to increase water and urea reabsorption (Guyton & Hall, 2006).

## RENAL REPLACEMENT THERAPIES

When the kidneys fail, patients have end-stage renal failure (ESRF). The majority of ESRF patients are given renal replacement therapy (RRT), rather than a renal transplant, to replace kidney function. A patient who cannot or does not desire to actively participate in the treatment may choose hemodialysis in a center. HD in a center is typically conducted three times a week for about 4-hr per treatment. Alternatively, a patient may receive HD at home, after being trained for home treatment and given a machine and method to appropriately treat tap water. A patient may also choose peritoneal dialysis, of which there are two forms: more commonly used continuous ambulatory PD and continuous cycling PD.

Although many believe that renal replacement therapy is just diffusion, RRT actually refers to the combination of dialysis, ultrafiltration, and convection. Through dialysis, waste products in the blood are removed as they move across a membrane to dialysate on a concentration gradient. Through ultrafiltration, excess fluid is removed. *Ultrafiltration* is the movement of water molecules across a membrane due to a hydrostatic or osmotic

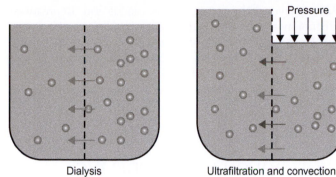

FIGURE 9.9 Dialysis, ultrafiltration, and convection.

Dialysis

Ultrafiltration and convection

pressure gradient. As water molecules move, the dissolved solute molecules are dragged along, which is called *convection* (Figure 9.9).

Dialysate is created by mixing purified water with concentrate from a manufacturer. Water is purified from specific treatment of tap water to remove chemical solutes, bacteria or bacterial products, and particulate matter. The concentrate typically contains sodium bicarbonate, sodium chloride, calcium chloride, potassium chloride, magnesium chloride, acetic acid/citric acid, and dextrose or glucose. The dialysate composition is designed to normalize the electrolyte and pH of the plasma.

We can determine the diffusion and ultrafiltration rates of transport, or flux, $J(t)$, in a blood and dialysate system separated by a semipermeable membrane. First, define membrane properties as $\varepsilon$ = membrane porosity, $l$ = membrane thickness or length of travel, $\tau$ = membrane tortuosity, $r$ = the radius of membrane pores, and $D_{M,i}$ = membrane diffusion coefficient for each solute $i$ (Figure 9.9). Also define the concentration of a solute in blood or dialysate as $c_{B,i}(t)$ or $c_{D,i}(t)$.

The sieving coefficient, $S$, is the fraction that solute blood concentration that reaches the membrane at the side facing blood, $c_{MB,i}(t)$:

$$c_{MB,i}(t) = S \cdot c_{B,i}(t). \tag{9.3}$$

Alternatively, the sieving coefficient is calculated as the ratio of the concentration of solute in dialysate to the concentration of solute in blood.

The isolated diffusion transport rate of each solute, $J_{d,i}(t)$, under the driving force of a concentration gradient, may be derived as

$$J_{d,i}(t) = \frac{\varepsilon D_{M,i} S}{\tau l} \left[ c_{B,i}(t) - c_{D,i}(t) \right]. \tag{9.4}$$

The isolated ultrafiltration transport rate of each solute, $J_{u,i}(t)$, under the driving force of a hydrostatic pressure gradient, may be derived as

$$J_{u,i}(t) = \frac{\varepsilon r^2 S \Delta P}{8 \eta \tau l} c_{B,i}(t), \tag{9.5}$$

where $\Delta P$ is the hydrostatic pressure difference between two phases separated by the membrane, and $\eta$ is the viscosity of fluid in the membrane.

The derivation of Eqs (9.4) and (9.5) is beyond the range of this discussion. Even more complicated is the combination of diffusion and ultrafiltration for simultaneous solutes, when boundary layer effects occur. But from these two simple equations, we see the effects of membrane properties. For high diffusion and ultrafiltration rates, the membrane should be as thin as possible, with as high as possible a porosity, and a sieving coefficient close to 1. Synthetic membranes have been designed to meet these constraints (Strathmann & Gohl, 1990; Ford et al., 2009).

## DIALYSIS ADEQUACY THROUGH UREA MODELING

Hemodialysis and peritoneal dialysis are taxing procedures with potential for harm (including death); so it is important to perform dialysis optimally. The adequacy of the dialysis dose may be measured through urea kinetic modeling or by the removal of other solutes. Ubiquitous urea is used as a surrogate for other waste product toxins because of its ease of measurement and presence in blood.

Urea kinetics can be modeled using one compartment. A compartment refers to a well mixed and kinetically homogenous amount of material and is not necessarily a physiologic space or well delimited physical volume. When we fit data to a compartmental model, each compartment is represented by an exponential. The equations associated with a compartmental model are based on law of conservation of mass.

In this model (Figure 9.10A), our lone compartment represents urea in the total body water that is extracellular and intracellular space, which possesses mass, $m_1$, volume, $v_1$, and concentration, $c_1$. Each flux associated with a compartment is the product of each rate constant times the originating mass. Through dialysis, urea is cleared from this

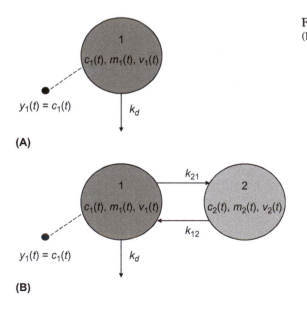

FIGURE 9.10 One- (**A**) and two-compartment (**B**) models of urea kinetics.

(A)

(B)

compartment with rate constant $k_d$, which has units of $min^{-1}$. The change of urea over time may be calculated as

$$\dot{m}_1(t) = -k_d m_1(t). \tag{9.6}$$

Because we measure urea in terms of concentration, we can divide both sides of the equation by volume to obtain

$$\dot{c}_1(t) = -k_d c_1(t). \tag{9.7}$$

For this first-order ordinary differential equation, the result is

$$c_1(t) = c_o e^{-k_d t}, \tag{9.8}$$

where $c_o$ is the initial concentration of urea at $t = 0$. Concentrations can be measured over time from blood samples tested for blood urea nitrogen (BUN), and used to estimate the rate constant, $k_d$. Sampling of this compartment is denoted in Figure 9.10A by the dashed line and output, $y_1(t)$.

If we model urea kinetics with two compartments (Figure 9.10B), then the second compartment represents intracellular urea, which is transported from extracellular to intracellular fluid with rate constant, $k_{21}$, and from intracellular to extracellular fluid with rate constant, $k_{12}$. The concentration equations become

$$\dot{c}_1(t) = -[k_d + k_{21}]c_1(t) + k_{12}c_2(t) \tag{9.9}$$

and

$$\dot{c}_2(t) = k_{21}c_1(t) - k_{21}c_2(t). \tag{9.10}$$

Because concentrations are measured only from the first compartment, not enough information is provided for modeling the three parameters in this system. Therefore, it is difficult to estimate the rate constants without large errors.

In clinical practice, dialysis adequacy is estimated as $Kt/V$. Here, $K$ is the rate of urea clearance, in units of mL/min; $t$ is time, in units of min; and $V$ is urea volume distribution. Since $K$ and $V$ are difficult to determine, $Kt/V$ is estimated from Eq. (9.8). First, we divide both sides by $c_o$ and take the natural log on both sides of this equation:

$$ln\left[\frac{c_1(t)}{c_o}\right] = ln\left[e^{-k_d t}\right] \tag{9.11}$$

$$ln\left[\frac{c_1(t)}{c_o}\right] = -k_d t \tag{9.12}$$

$$-ln\left[\frac{c_1(t)}{c_o}\right] = k_d t. \tag{9.13}$$

The rate constant, $k_d$, is estimated as $K/V$, which has the same units of $min^{-1}$:

$$-ln\left[\frac{c_1(t)}{c_o}\right] \cong \frac{K}{V}t. \tag{9.14}$$

As you see from Eq. (9.14), $Kt/V$ may be estimated from two values of BUN.

# CLINICAL NEED

End-stage renal disease is the final stage of *chronic kidney disease* (*CKD*), which is the persistent and usually progressive reduction in glomerular filtration rate. Chronic kidney disease is caused by diabetes; hypertension; immunological diseases such as lupus nephritis; hereditary factors, such as polycystic kidney disease; or severe allergic reaction or infection (NIDDK, 2010). The stages of CKD are as follows:

1. Stage 1. Evidence of kidney disease such as protein in the urine, with normal or increased GFR (>90 mL/min/1.73 m$^2$)
2. Stage 2. Mild reduction in GFR (60–89 mL/min/1.73 m$^2$)
3. Stage 3. Moderate reduction in GFR (30–59 mL/min/1.73 m$^2$)
4. Stage 4. Severe reduction in GFR (15–29 mL/min/1.73 m$^2$)
5. Stage 5. Kidney failure (GFR < 15 mL/min/1.73 m$^2$).

According to the 2009 United States Renal Data System annual report, about 16% of the American population has chronic kidney disease. This estimate is based on the National Health and Nutrition Examination Survey, which is administered annually to approximately 10,000 individuals. New ESRD incidence is about 350 per million (or 105,000) annually.

Three treatments for ESRD exist: renal transplantation, hemodialysis, and peritoneal dialysis. In terms of life expectancy, quality of life, and long-term cost efficiency, the best treatment for ESRD is renal transplantation. In the United States, the expected remaining lifetimes of the general population against age-matched ESRD transplant and dialysis patients in 2007 were 25.2, 16.4, and 5.9 yr, respectively. However, only 12.3% of patients wait-listed in 2007 received a live-donor transplant within 1 yr of listing. Only 34% of adults wait-listed in 2005 received a deceased-donor transplant within 3 yr of listing. During listing, the probability of death was 4% in the first year of listing, rising to 28% in the fifth year. At 1 yr following transplant, 35% of 2006 recipients had a GFR of 60 to less than 90 mL/min/1.73 m$^2$.

On December 31, 2007, there were 158,739 (30%) ESRD patients with transplants, 341,264 (65%) ESRD patients receiving hemodialysis, and 26,340 patients (5%) receiving peritoneal dialysis. For the dialysis patients, the per-person-per-year total costs varied from just under $60,000 for arteriovenous (AV) fistula access to $72,729 or $79,364 for arteriovenous graft or catheter access, respectively. We discuss these types of vascular access in the next section.

All ESRD costs in 2007 paid by Medicare totaled $23.9 billion, which was 5.8% of the Medicare budget. While Medicare typically provides health insurance for people 65 yr and older, it insures patients of all ages with ESRD. Surprisingly, despite an increasing ESRD population, the ESRD percentage of the Medicare budget has stayed approximately constant over the last 7 yr. This reflects tremendous cost containment, unparalleled in any other healthcare field. In Figure 9.11, the presented ESRD percentages for 2006 and 2007 are lower than in reality because Medicare Part D is not included in available ESRD data. Although total ESRD continues to rise at a constant rate each year, the exponential rise in total Medicare expenditure is responsible for the constant ESRD percentage (U.S. Renal Data System, 2009).

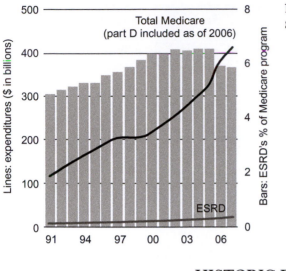

FIGURE 9.11 Costs of Medicare and ESRD programs [*From U.S. Renal Data System (2009)*].

# HISTORIC DEVICES

Without dialysis, urea and other toxins build up in the blood of end-stage renal dialysis patients. This condition of uremia causes nausea, vomiting, intestinal bleeding, itching, convulsions, lethargy, and ultimately coma and death. The suffering of their patients inspired several scientists to develop a method to dialyze the blood to prevent death.

## Early Devices

Building on the work of pharmacologist John Abel, who had experimented with canine dialysis, scientist Georg Haas attempted hemodialysis on several humans with acute renal failure from 1924 to 1928. Abel had passed blood from an animal through a series of porous colloid tubes bathed in saline and returned the blood to the animal. He prevented blood clots with hirudin which he extracted from thousands of leeches. Hass mimicked this procedure, which lasted 30–60 minutes. The procedure's duration was too short to produce a therapeutic effect.

Separately from Hass, nephrologist Willem Kolff read Abel's journal article. He substituted heparin for hirudin and a thin cellophane membrane for the thick colloid tube. Resources were scarce in the Nazi-occupied Netherlands during World War II; so Kolff constructed his first artificial kidney of readily obtainable materials. One hundred thirty feet of cellophane tubing made from sausage casing was wrapped 30 times around a horizontal drum made out of aluminum strips, seated in an enamel tank. As the drum rotated saline in the tank, the patient's blood was exposed to the dialysis bath of what came to be known as the rotating-drum hemodialyzer (Goldstein, 2002) (Figure 9.12). Of 17 patients treated with this device in 1943–1944, two survived (Kolff, 2002).

Kolff continued to develop designs for artificial kidneys. His twin coil design, based on sewing together two loops of cellophane tube and screen, was commercialized by Baxter Laboratories as the U-200 Coil Kidney in 1956. Dialyzer designs by others were based on

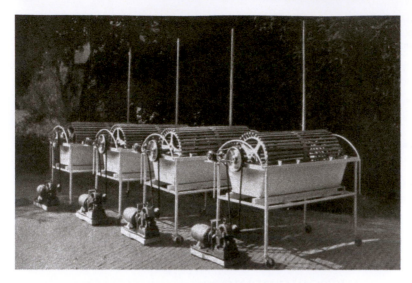

**FIGURE 9.12** Rotating-drum hemodialyzers *[Reproduced by permission from Kolff (2002)].*

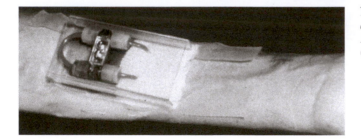

**FIGURE 9.13** Scribner shunt *[Courtesy of University of Washington Archives. Reproduced by permission from Scribner (2002)].*

parallel flow through cellophane tubes, stacked sheets of cellophane, and capillary tubes (Twardowski, 2008).

## Enabling Technology: External Arteriovenous Fistula

Even with improved dialyzers, only acute renal failure patients could be effectively treated, not end-stage renal disease patients. After six or seven dialysis treatments, a physician could no longer connect an ESRD patient to dialysis because patient arteries and veins were damaged. Nephrologist Belding Scribner solved this problem by developing an external arteriovenous shunt.

Scribner connected a U-shaped tube of Teflon to the artery and vein of a dialysis patient. Because this AV shunt was external and removable, it provided a reliable access point for repeated dialysis. Further, between dialysis sessions, the AV shunt provided decreased resistance to flow and increased flow that enlarged the connecting artery and vein (Scribner, 2002) (Figure 9.13).

Scribner's first patient with ESRD, Clyde Shield, was successfully dialyzed in March, 1960, and survived over a dozen years. Suddenly, ESRD patient prognosis switched from 90% fatal to 90% survivable. The institution where Scribner worked, the University of Washington in Seattle, became inundated with requests for dialysis. This first outpatient dialysis center in the world, established in 1962 and currently called the Northwest Kidney Centers, was forced to decide who should be treated, using a two-committee process. The first committee was composed of physicians who determined which patients were suitable for the new treatment. The second committee was composed of seven anonymous community members, who chose the subset of patients to be treated based on age, achievement, and future potential. These two committees constituted the first bioethics committee (Scribner, 2002; Goldstein, 2002). The Seattle group also developed most of the early innovative technology of hemodialysis, including the first small machine that was first used for home HD and that became the prototype of all machines used today.

Willem Kolff and Belding Scribner received the 2002 Lasker Clinical Medical Research Award "for the development of renal hemodialysis, which changed kidney failure from a fatal to a treatable disease, prolonging the useful lives of millions of patients" (Lasker Awards, 2009).

## SYSTEM DESCRIPTION AND DIAGRAM

Scribner shunt vascular access is no longer a standard of care because of the necessity of surgical insertion and high thrombosis and infection rates. Instead, vascular access is most often achieved with an endogenous AV fistula. Now, a side-to-side or end-to-side cephalic vein and radial artery are surgically attached and allowed to mature for 2−6 months (Figure 9.14A). Once mature, these fistulas have long-term patency rates (i.e., 20 yr) and rarely become infected. Less often, an elderly or diabetic patient, whose cephalic veins have already been subjected to phlebotomy and intravenous cannulation, gains vascular access through a synthetic AV graft of ePTFE. The graft surgically connects a brachial or distal radial artery to a basilic vein. Within a week or so, the graft heals within its subdermal tunnel and lasts up to 3 yr before thrombosis or infection occurs (Figure 9.14B).

Least desirable vascular access may be achieved in a patient who requires urgent dialysis and has no fistula or graft (permanent access) through a permanent catheter. The catheter is inserted, under fluoroscopic guidance, through a subcutaneous tunnel into an internal jugular vein, with the catheter tip in the right atrium (Figure 9.14C). This catheter is subject to increased thrombosis or fibrin sheath formation and infection and other complications (Green & Schwab, 2009).

### Hemodialyzers

Since the mid-1980s, the dominant dialyzer design has been the hollow-fiber dialyzer (Figure 9.1). The concurrent use of thousands of synthetic hollow fibers or tubes with about 200-$\mu$m inner diameters, for a total surface area of 0.5−2.0 m$^2$, enable a higher ratio of surface area to capacity and lower wall tension than other designs. These membranes

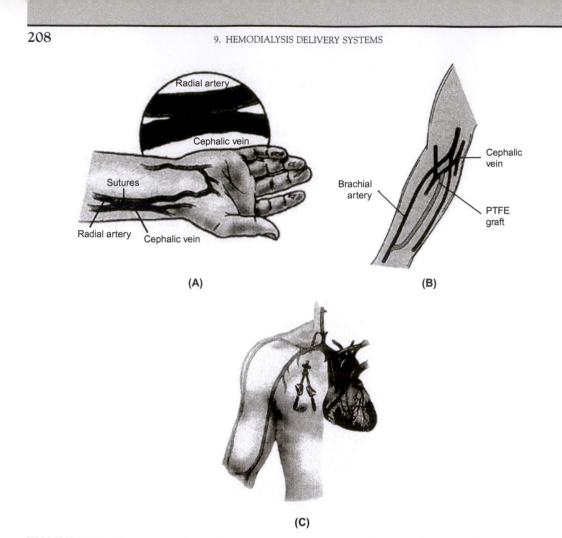

**FIGURE 9.14**    Three types of vascular access. **A**: Exogenous arteriovenous fistula. **B**: Arteriovenous graft. **C**: Permanent catheter. [*A from Ignatavicius (2202). B from Johnson & Fehally (2003). C courtesy of MEDCOMP Corp, Harleysville, Pennsylvania. All figures reproduced from Kallenbach et al. (2005).*]

are made of synthetic engineered thermoplastics, including polyethersulfone, polyacrylonitrile, and polyamides. The membranes are characterized by a spongy matrix, which determines the diffusive permeability, and a thin skin of fine pores on both surfaces, which determines hydraulic permeability and solute retention properties. The fibers are assembled and inserted into a dialyzer shell of acrylate or polycarbonate resin. The fiber bundle is then potted at both ends with silicone rubber before the entire dialyzer is sterilized (Galletti et al., 1995; Ronco & Clark, 2005). A dialyzer may be specified for single use or reuse.

During hemodialysis, blood enters the blood inlet port, moves through the hollow fibers, and exits through the blood outlet port. Meanwhile, dialysate enters the dialysate inlet port, interacts with the hollow fibers, and exits through the dialysate outlet port. The

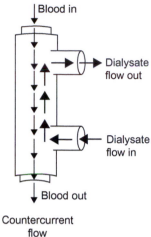

Blood in

Dialysate flow out

Dialysate flow in

Blood out

Countercurrent flow

**FIGURE 9.15** Blood and dialysate flow in the hollow-fiber dialyzer [*Adapted from Kallenbach et al. (2005)*].

blood and dialysate flow in opposite directions to maximize solute movement. Because the hollow fibers are pressurized, both waste products and fluid move from blood to the dialysate and certain solutes such as bicarbonate move from dialysate into blood (Figure 9.15).

## Hemodialysis Delivery Systems

The dialyzer is the key component of a hemodialysis delivery system, which contains two flow circuits (Figure 9.16). In the blood circuit, blood is pumped with peristaltic motion from a patient's artery or vein, mixed with heparin from a syringe pump, and delivered to the dialyzer. Through diffusion, waste products move into the dialysate. The blood then moves from the dialyzer, through an air sensor and a clamping circuit back to a vein. The air sensor uses ultrasound to detect air bubbles; it prevents a large volume of air (60–125 mL) from reaching the general circulation and causing an air embolism. The emergency clamping circuit automatically shuts and the blood pump stops if air is sensed. Pressure sensors before and after the peristaltic pump, as well as after the dialyzer, provide safety inputs to the processing module. If an out-of-range pressure is detected, an alarm sounds and is displayed on the LCD and the pump stops. A single-use blood set made of polyvinyl chloride tubing connects the patient to the blood circuit.

The dialysate circuit may mix dialysate with purified water, concentrated dialysate, and normal electrolytes including sodium and bicarbonate. In Figure 9.16, we assume that the dialysate has already been mixed. First, the electrical conductivity is measured because this is a cost-effective approach for verifying proper composition of the mixed dialysate. Within the conductivity sensor, the magnitude of the current is proportional to the concentration of dissolved ions in solutions. Next, temperature is measured because dialysate temperature outside the physiologic range may cause patient complications. The dialysate flow rate is measured with a flow sensor. Flow may be determined using a number of methods, including the transmission and reception of acoustic pulses, with the transit time

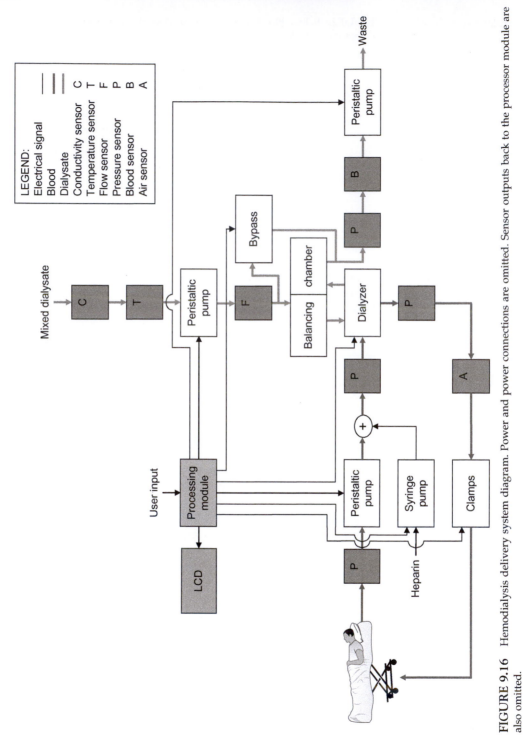

**FIGURE 9.16** Hemodialysis delivery system diagram. Power and power connections are omitted. Sensor outputs back to the procesor module are also omitted.

related to fluid flow. The dialysate then moves through a balancing chamber into the dialyzer, unless a bypass circuit switches dialysate flow past the dialyzer.

When the dialysate exits the dialyzer, it contains waste products and excess water. The dialysate is checked for pressure. It is also checked for blood leakage by transmitting light through the dialysate and detecting it with a photodiode; the hemoglobin in any blood that is present will absorb the light before detection. If a blood leak is detected, an alarm sounds and dialysis is stopped. The dialysate waste is then pumped to the drain. After use, the dialysate circuit is exposed to disinfectant and rinsed (Misra, 2008; AAMI, 2008).

Ultrafiltration is controlled through a balancing chamber and ultrafiltration controller. Dialysate flow in and out of the dialyzer is first regulated by a rigid balancing chamber. Here, the two volumes are separated by a diaphragm. A fixed deflection of the diaphragm by the inflow volume pushes an equal volume out of the chamber. Additionally, the processing module monitors inline blood and dialysate pressures with pressure sensors, and controls ultrafiltration by changing the dialysate peristaltic pump flow rate (Ahmad, 2009).

## KEY FEATURES FROM ENGINEERING STANDARDS

As of 2010, the FDA recognizes several consensus standards for hemodialysis. We discuss requirements from the standards for hemodialyzers and hemodialysis delivery systems: *ANSI/AAMI RD16:2007 Cardiovascular Implants and Artificial Organs—Hemodialyzers, Hemodiafilters, Hemofilters, and Hemoconcentrators* (AAMI, 2007) and *ANSI/AAMI RD5:2003/ (R)2008 Hemodialysis Systems* (AAMI, 2008). FDA does not recognize a specific standard for peritoneal dialysis.

### Hemodialyzer Clearance

Hemodialyzer performance is usually described by solute clearance, even though this measurement is affected by the blood and dialysate flow rates. According to ANSI/AAMI RD16:2007, clearance is determined from the test setup of Figure 9.17. Here, a test solution substituting for blood is moved with a blood pump through the dialyzer, where it interacts with dialysate under ultrafiltration control. The separate test solutions are 15–35 mmol/L urea, 500–1000 μmol/L creatinine, 1–5 mmol/L phosphate (adjusted to pH = 7.4 ± 0.1), and 15–40 mmol/L vitamin B$_{12}$. Clearance of vitamin B$_{12}$, or riboflavin, is used as a surrogate for middle molecule removal, representing uremic toxins. For manufacturer-specified blood and filtrate flow rates, the blood inlet and outlet concentrations are measured during steady state. The clearance, $K$, is then calculated as:

$$K = \left(\frac{c_{BI} - c_{BO}}{c_{BI}}\right)Q_{BI} + \frac{c_{BO}}{c_{BI}}Q_f, \tag{9.15}$$

where $Q$ = flow rate, $c$ = concentration, $BI$ = blood inlet, $BO$ = blood outlet, and $f$ = filtrate. The number of required concentration samples is not specified. The maximum allowable error from published clearance values is not detailed.

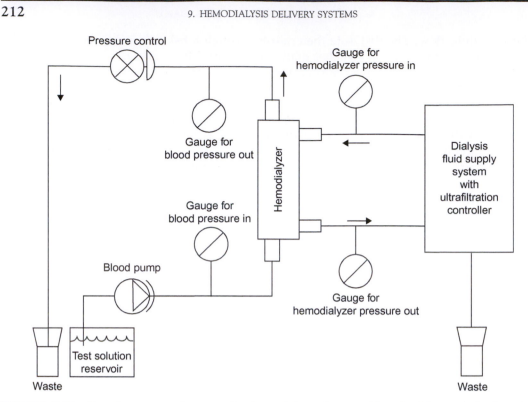

**FIGURE 9.17**   Open-loop system for measuring hemodialyzer clearance *[Reproduced by permission from AAMI (2007)].*

## Blood Circuit Air Protection

ANSI/AAMI RD5:2003/(R)2008 includes requirements to prevent an air embolism. In a hemodialysis delivery system, an air protection sensor must detect an individual bolus of air larger than 1 mL at operating pressure or the presence of a series of microbubbles that occur within a short period of time and total more than 1.5 mL/30 s. When 1.5 mL/30 s is detected, audible and visual alarms must be activated, with the blood pump stopped and venous return clamped to prevent air from reaching the patient.

To test this requirement, a test solution of either saline or whole bovine blood, within the normal operating range of temperature, is used. Graded amounts of air are introduced into the blood circuit *in vitro*, while the device is perfused at the maximum venous pressure capability. It is up to the test engineer to design a test method that "should take into account the principle of detection of air and the location of the air detector." This test method and the sensitivity of the air detector must be disclosed by the manufacturer (AAMI, 2008).

## Temperature Monitoring

The temperature of the dialysate must be kept within the physiologic range. When the temperature has exceeded this range, some patients have complained of weakness and lethargy, with hemolysis (red blood cell rupture) usually present. At 46°C, protein denaturation, hemolysis, and death occur. When the temperature is below this range, only chilliness occurs, but the comatose patient may become hypothermic, and clotting of the hemodialyzer could occur.

Assuming that mixed dialysate normally arrives at the hemodialysis delivery system at a temperature within the physiologic range of 33–40°C, the dialysate must be monitored so that the temperature does not exceed 42°C. Per ANSI/AAMI RD5:2003/(R)2008, if this maximum temperature is exceeded, an audible and visual alarm shall be activated, and blood and dialysate flows shall be stopped. To test this requirement, dialysate at a temperature exceeding the range is fed into the system, with alarm activations verified (AAMI, 2008).

## Ultrafiltration Control System

Hemodialysis delivery systems commonly employ ultrafiltration control. If ultrafiltration control fails, patients may be harmed through excessive fluid removal, failure of fluid removal, or unrecognized infusion of dialysate. Per ANSI/AAMI RD5:2003/(R)2008, ultrafiltration control must function with an overall accuracy of $\pm 5\%$ of the selected ultrafiltration rate or $\pm 100$ mL/h, whichever is greater, over the specified range of operation. The ultrafiltration rate must be displayed.

Testing is conducted with a test solution of dialysate, at a temperature of $37°C \pm 1.5°C$. Flow rates shall be determined volumetrically or with flow meters with a maximum error of $\pm 10\%$ (AAMI, 2008).

## Blood Leak Detection

If the hemodialyzer membrane ruptures, the patient risks blood loss. Per ANSI/AAMI RD5:2003/(R)2008, a blood leak rate of greater than 0.35 mL/min into dialysate must activate an alarm that shuts off the blood pump and minimizes ultrafiltration. To test this requirement, whole bovine blood adjusted to a hematocrit of 25% (percentage of blood consisting of cells) is infused into the dialysate flow path distal to the hemodialyzer connection point, at the minimum and maximum dialysate flow rates. Alarm activation, blood pump shutoff, and ultrafiltration minimization are then verified (AAMI, 2008).

## SUMMARY

A dialyzer is a device that separates substances in solution, or solutes, through unequal diffusion through a semipermeable membrane. When the solution is blood, a dialyzer is called a hemodialyzer. A hemodialysis delivery system, or hemodialysis machine, is an instrument that enables blood to be safely dialyzed of its waste products, with a hemodialyzer and dialysate as its key components. Within the kidney, capillary exchange at the level of the nephron is responsible for many functions, including the regulation of water and electrolyte balances and the excretion of metabolic waste products and foreign chemicals. Each Bowman's capsule in a nephron encases glomerular capillaries, lined by visceral epithelial cells. These capillaries freely filter most substances in plasma, except for proteins. About 20% of renal plasma flowing through the kidney is filtered. The filtered fluid, called the glomerular filtrate, leaves Bowman's capsule and passes through the tubules. There, reabsorption, secretion, and excretion occur, which determine the composition of urine.

When the kidneys fail, patients have end-stage renal failure. The majority of ESRF patients are given renal replacement therapy, rather than a renal transplant, to replace kidney function. Renal replacement therapy is the combination of dialysis, ultrafiltration, and convection. Through dialysis, waste products in blood are removed as they move across a membrane to dialysate on a concentration gradient. Through ultrafiltration, excess fluid is removed. As water molecules move, the dissolved solute molecules are dragged along, which is called convection.

Willem Kolff invented the first successful dialyzer. Belding Scribner developed the external arteriovenous shunt, which enabled end-stage renal disease patients to receive repeated dialysis. Key features of modern dialyzers and hemodialysis systems include hemodialyzer clearance, blood circuit air protection, temperature monitoring, ultrafiltration control, and blood leak detection.

## Acknowledgment

The Author wishes to thank Dr. Suhail Ahmad for his detailed review of this chapter. Dr. Ahmad is a Professor of Medicine at University of Washington and Chief Medical Officer of the Northwest Kidney Centers.

## Exercises

**9.1** Plot the concentration waveform 9_1 that your instructor has provided. Waveform characteristics are documented in the associated header9_1.txt ascii file. Using a compartmental modeling program, fit this urea concentration data set to both a one-compartment and two-compartment model. For the two-compartmental model, assume that $k_{21} = 0.3$ min$^{-1}$. Replot the original data and fitted data for each model. List your final rate constant values, along with the associated coefficient of variation. Which model provides a better fit? Give reasons.

**9.2** Derive the clearance equation given as Eq. (9.15).

**9.3** Ultrafiltration is dependent on the hydrostatic pressure gradient. Describe where in Figure 9.16 this pressure gradient is created and controlled.

**9.4** During 2009, endless debates occurred before the U.S. House of Representatives and Senate passed their versions of a healthcare reform bill. One repeatedly presented false argument was that the federal government would set up so-called death panels to determine which patients would be saved (Baura, 2009). Was Belding Scribner's bioethics committee a death panel? Medicare sets the reimbursement rates for medical services, including those for ESRD patients. Is Medicare a death panel? Should Medicare pay for services to all ESRD patients, regardless of age, or should payments be reserved for those who are at least 65 yr?

Considering the federal graph in Figure 9.11 of annual Medicare costs, do Medicare costs need to be contained? Explain your reasoning.

Read Kolff (1950):

**9.5** Describe three types of artificial kidney designs. Which is the most efficacious? Give reasons.

**9.6** What are three severe complications of hemodialysis? How are these complications prevented?

Read Quinton et al. (1960):

**9.7** How did Scribner and Quinton solve the problem of long-term cannulation of arteries and veins for hemodialysis? Describe the parts of this medical device, which are shown in Figure 9.13.

**9.8** How biocompatible is Teflon in this application? Explain your reasoning. What special steps are taken to minimize the risk of infection?

# References

AAMI (2007). *ANSI/AAMI RD16:2007 cardiovascular implants and artificial organs—Hemodialyzers, hemodialfilters, hemofilters, and hemoconcentrators.* Arlington, VA: AAMI.

AAMI (2008). *ANSI/AAMI RD5:2003/(R)2008.* Arlington, VA: AAMI.

Ahmad, S. (2009). *Manual of clinical dialysis* (2nd ed.). New York: Springer.

Baura, G. D. (2009). Two worlds [Point of View]. *IEEE Engineering in Medicine and Biology Magazine, 28*, 102, 110.

Ford, L. L., Ward, R. A., & Cheung, A. K. (2009). Choice of the hemodialysis membrane. In W. L. Henrich (Ed.), *Principles and practice of dialysis* (4th ed.). Philadelphia: Lippincott Williams & Wilkins.

Galletti, P. M., Colton, C. K., & Lysaght, M. J. (1995). Artificial kidney. In J. D. Bronzino (Ed.), *The biomedical engineering handbook*. Boca Raton: CRC Press.

Goldstein, J. L. (2002). *Albert Lasker clinical medical research award 2002 award presentation*. New York: Lasker Foundation.

Green, C. A., & Schwab, S. J. (2009). Hemodialysis vascular access. In W. L. Henrich (Ed.), *Principles and practice of dialysis* (4th ed.). Philadelphia: Lippincott Williams & Wilkins.

Guyton, A. C., & Hall, J. E. (2006). *Textbook of medical physiology* (11th ed.). Philadelphia: Elsevier Saunders.

Kallenbach, J. Z., Gutch, C. F., Stoner, M. H., & Corea, A. L. (2005). *Review of hemodialysis for nurses and dialysis personnel* (7th ed.). St. Louis, MO: Elsevier Mosby.

Kolff, W. J. (1950). Artificial kidney; treatment of acute and chronic uremia. *Cleveland Clinic Quarterly, 17*, 216–228.

Kolff, W. J. (2002). Lasker Clinical Medical Research Award. The artificial kidney and its effect on the development of other artificial organs. *Nature Medicine, 8*, 1063–1065.

Lasker Awards: Former Winners. (2009). Lasker Foundation.

Levey, A. S., Coresh, J., Green, E. T., Stevens, L. A., Zhang, Y. L., & Hendriksen, S., et al. (2006). Using standardized serum creatinine values in the modification of diet in renal disease study equation for estimating glomerular filtration rate. *Annals of Internal Medicine, 145*, 247–254.

Misra, M. (2008). Basic mechanisms governing solute and fluid transport in hemodialysis. *Hemodialysis International, 12*(Suppl. 2), S25–S28.

NIDDK (2010). *Glomerular Disease Primer: Kidney Disease*. Bethesda, MD: NIH.

NIH (2010). *GFR MDRD calculators for adults*. Bethesda, MD: NIH.

Quinton, W., Dillard, D., & Scribner, B. H. (1960). Cannulation of blood vessels for prolonged hemodialysis. *Transactions American Society for Artificial Internal Organs, 6*, 104–113.

Ronco, C., & Clark, W. R. (2005). Hollow-fiber dialyzers: Technical and clinical considerations. In A. R. Nissenson, & R. N. Fine (Eds.), *Clinical dialysis* (4th ed.). New York: McGraw-Hill.

Scribner, B.H. (2002). Lasker Clinical Medicine Research Award. Medical dilemmas: The old is new. *Nature Medicine, 8*, 1066–1067.

Strathmann, H., & Gohl, H. (1990). Membranes for blood purification: state of the art and new developments. *Contributions Nephrology, 78*, 119–140; discussion, 140–141.

Twardowski, Z. J. (2008). History of hemodialyzers' designs. *Hemodialysis International, 12*, 173–210.

U.S. Renal Data System. (2009). USRDS 2009 Annual data report: Atlas of end-stage renal disease in the United States. In *National institute of diabetes and digestive and kidney diseases* (Ed.). Bethesda, MD: NIH.

CHAPTER

# 10

# Mechanical Ventilators

In this chapter, we discuss the relevant physiology, history, and key features of mechanical ventilators. A *mechanical ventilator* is an instrument that replaces, or assists in, spontaneous breathing. Mechanical ventilators are used in a variety of settings, from the operating room (OR) and intensive care unit (ICU) to the home and transport vehicles. This discussion is limited to ICU ventilators (Figure 10.1) used by adult patients recovering from acute lung injuries.

Upon completion of this chapter, each student shall be able to:

1. Understand the mechanisms underlying pulmonary ventilation and blood flow.
2. Derive the relationship between pressure, flow, and volume in a mechanical ventilator.
3. Describe five key features of mechanical ventilators.

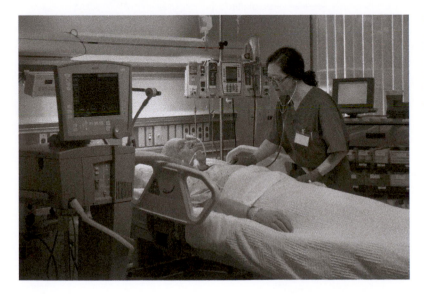

**FIGURE10.1** CareFusion Avea® Mechanical Ventilator. CareFusion is the spinoff of Cardinal Health, which purchased Viasys Healthcare in 2007. (CareFusion, San Diego, California).

# PULMONARY PHYSIOLOGY

In the pulmonary circulation, deoxygenated blood flows from the right atrium and ventricle, through the pulmonary artery, and to the lungs. The blood flow through the lungs is essentially the cardiac output of 5 L/min. Within the lungs, blood flows to the pulmonary capillaries, where gas exchange takes place within the alveoli (Figure 10.2). Newly oxygenated blood then moves from the pulmonary veins to the left heart and into the aorta.

## Pulmonary Blood Flow

Blood flow through the lungs is affected by hydrostatic pressure. In a normal, upright adult, the lung spans a height of approximately 30 cm. This translates into a 23-mmHg vertical pressure gradient, with 15 mmHg above the heart and 8 mmHg below the heart. Due to this pressure difference, the apex of the lung has little flow, while the bottom of the lung has five times as much flow (Figure 10.3).

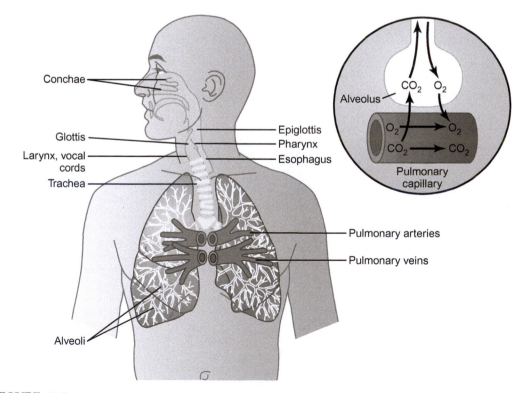

FIGURE 10.2    Respiratory passages. *[Reproduced by permission from Guyton & Hall (2006)].*

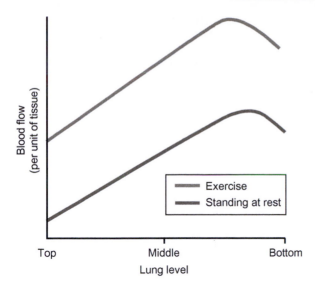

**FIGURE 10.3** Blood flow at different levels in the lung of an upright person at rest and during exercise. *[Reproduced by permission from Guyton & Hall (2006)].*

We can explain these differences by considering three zones of local blood flow within the lungs, as a pulmonary capillary interacts with an alveolus (Figure 10.4). In zone 1, the capillary pressure never increases above the alveolar air pressure during any part of the cardiac cycle. This can have greater impact on overall gas exchange as an abnormal condition, which could occur if a subject's pulmonary systolic arterial pressure is very low due to severe blood loss. As a consequence, no blood flows thorough this lung zone. In zone 2, the capillary pressure is greater than the alveolar air pressure only during systole. Capillary pressure is less than alveolar air pressure during diastole. This occurs at the mid-levels of the lung because the pressure is about 10 mmHg during systole, while the alveolar air pressure is 0 mmHg. So now, blood flows intermittently through this lung zone. Finally, in zone 3, the capillary pressure is always greater than the alveolar air pressure. This occurs in the lower regions of the lung, where blood flows continuously.

## Pulmonary Ventilation

For pulmonary capillaries to receive oxygen at the alveoli, air must be transported to the alveoli. *Pulmonary ventilation* is the inflow and outflow of air between the atmosphere and the lung alveoli. It is the first of four functions that make up respiration. Air moves to and from the lungs through lung expansion and contraction. During normal quiet breathing, inspiration occurs primarily by contraction of the diaphragm. This contraction pulls the lungs downward. During expiration, the diaphragm relaxes, which causes the lungs to compress through elastic recoil (Figure 10.5).

Each lung consists of elastic tissue that floats in the pleural cavity. The pleural cavity is lined by a smooth membrane, the parietal pleura, which is reflected back at its root to become the visceral pleura. The visceral pleura covers the lungs and adjacent structures.

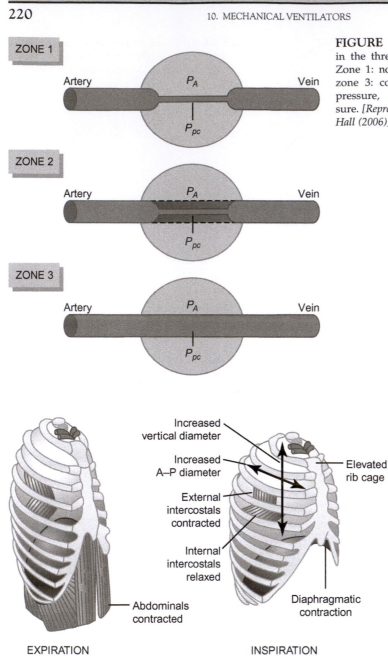

**FIGURE 10.4** Mechanics of blood flow in the three blood flow zones of the lung. Zone 1: no flow, zone 2: intermittent flow, zone 3: continuous flow. $P_A$ = alveolar air pressure, $P_{pc}$ = pulmonary capillary pressure. [*Reproduced by permission from Guyton & Hall (2006)].*

**FIGURE 10.5** Contraction and expansion of the thoracic cage during expiration and inspiration. [*Reproduced by permission from Guyton & Hall (2006)].*

Pleural fluid between the two contacting membranes acts as a lubricant to reduce the friction of the lungs' movement within the pleural space.

At the end of expiration, before inspiration begins, the pleural pressure in the pleural fluid is slightly negative, at about −5 cmH$_2$O, to hold the lungs open at resting level.

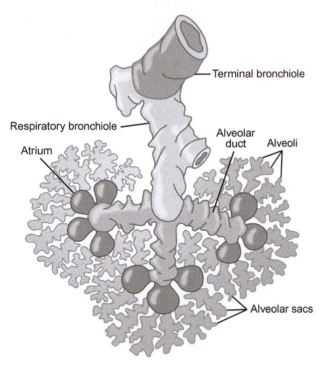

**FIGURE 10.6**  Respiratory unit. *[Redrawn from Miller (1947). Reproduced by permission from Guyton & Hall (2006)].*

Because the glottis is open and no air is flowing, the pressures in all parts of the respiratory tree, including the alveoli, are equalized to atmospheric pressure, which is 0 cmH$_2$O.

During inspiration, contraction of the diaphragm and the expansion of the chest cage by the intercostal muscles pull the lungs downward and outward. This creates more negative pleural pressure, with a fall to about $-7.5$ cmH$_2$O. This fall causes inward flow of about half a liter of air through the trachea, bronchi, and bronchioles to the alveoli in 2 s of normal quiet breathing (Figure 10.6).

During expiration, the pleural pressure reverses itself, and returns to $-5$ cmH$_2$O. The alveolar pressure rises to about $+1$ cmH$_2$O, which forces half a liter of air out of the lungs (Figure 10.7). The difference between the alveolar and pleural pressures is known as the transpulmonary pressure.

We can measure respiratory excursions over time using a spirometer, which is an instrument that measures inspiration and expiration with a flow sensor. As shown in Figure 10.8, a tidal volume, $V_T$, of about 500 mL is normally inspired and expired in each breath. A patient may forcefully inspire about 3000 mL above the tidal volume, which is the inspiratory reserve volume, or forcefully expire about 1100 mL beyond the resting lung volume level at the end of a tidal breath, which is the expiratory reserve volume. Even after the most forceful expiration, a residual volume of about 1200 mL remains in the lung. Together, these four volumes make up the total lung capacity.

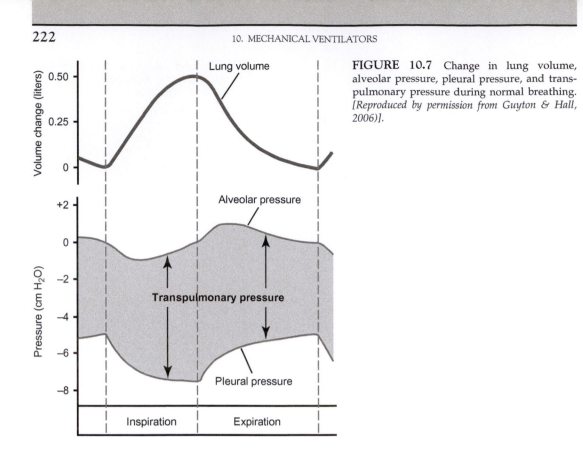

FIGURE 10.7 Change in lung volume, alveolar pressure, pleural pressure, and transpulmonary pressure during normal breathing. [*Reproduced by permission from Guyton & Hall, 2006*].

With pulmonary ventilation, air moves through the alveoli. The rate of alveolar ventilation, $\dot{V}_A$, or alveolar flow, may be calculated as:

$$\dot{V}_A = f \cdot (V_T - V_D), \tag{10.1}$$

where $f$ is the frequency of respiration, in units $\min^{-1}$, and $V_D$ is the physiologic dead space volume. The physiologic dead space volume refers to air that fills the respiratory passages but that does not participate in gas exchange. When poor blood flow is present, the dead space volume may include the alveolar volume because gas exchange is impaired.

## Gas Diffusion

With blood flow to the pulmonary capillaries and air flow to the alveoli, diffusion of gases can occur along a pressure gradient. In the air, the partial pressure of either oxygen ($P_{O_2}$), nitrogen ($P_{N_2}$), or carbon dioxide ($P_{CO_2}$) is equal to its percentage concentration. At sea level, the total pressure of this mixture is 760 mmHg, or 1 atm. Because alveolar air is only partially replaced by atmospheric air with each breath, $P_{O_2}$ decreases from an atmospheric oxygen pressure of 159 mmHg to an alveolar oxygen partial pressure ($P_{A\,O_2}$) of

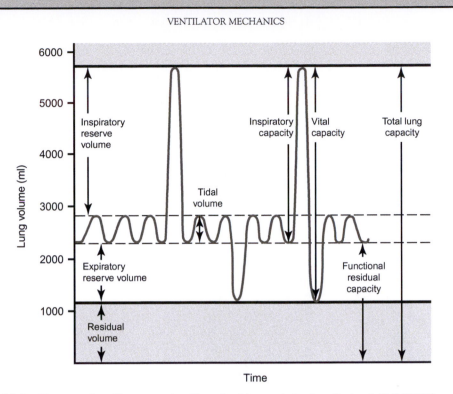

**FIGURE 10.8** Diagram of total lung capacity. *[Reproduced by permission from Guyton & Hall (2006)].*

104 mmHg. Conversely, $CO_2$ increases from an atmospheric $P_{CO_2}$ of 0.3 mmHg to an alveolar $CO_2$ partial pressure ($P_{A\ CO_2}$) of 40 mmHg.

In a gas dissolved in a fluid, the partial pressure of a gas is determined by the ratio of its concentration to the solubility coefficient of the gas. The blood perfusing the pulmonary capillaries is venous blood returning to the lungs from the systemic circulation. Normal venous blood has a $Pv_{O_2}$ of 40 mmHg and a $Pv_{CO_2}$ of 45 mmHg. With these pressure gradients between the alveoli and venous blood in the pulmonary capillaries, oxygen diffuses from the alveoli to the blood, as $CO_2$ diffuses from the blood to the alveoli (Figure 10.9). The alveolar capillaries provide about 98% of blood entering the left atrium. The other 2% passes through the bronchial circulation that supplies the deep tissues of the lung but is not exposed to air. This so-called shunt flow decreases the partial pressure of oxygen that enters the left side of the heart to 95 mmHg, which is the partial pressure of arterial blood ($Pa_{O_2}$) reaching the aorta. The partial pressure of $CO_2$ in the arterial blood ($Pa_{CO_2}$) is 40 mmHg (Guyton & Hall, 2006).

## VENTILATOR MECHANICS

We can model the pressures involved in ventilation by using Newton's third law of motion: For every action, there is an equal and opposite reaction. The transrespiratory pressure, $P_{TR}(t)$, is the pressure at the airway opening minus the pressure at the body

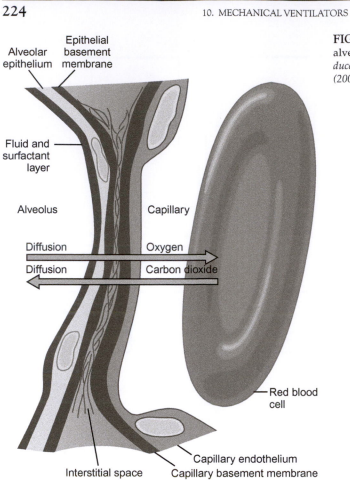

**FIGURE 10.9** Diffusion between the alveolus and pulmonary capillary. *[Reproduced by permission from Guyton & Hall (2006)].*

surface. This pressure is equal to sum of the pressure due to the elastic recoil, $P_E(t)$, and the pressure due to resistance to flow, $P_R(t)$:

$$P_{TR}(t) = P_E(t) + P_R(t). \tag{10.2}$$

During spontaneous breathing, the transrespiratory pressure is equal to the pressure resulting from respiratory muscle contraction, $P_{musc}(t)$. But when a patient requires assistance from a mechanical ventilator to breathe, the transrespiratory pressure includes the mechanical ventilator pressure, $P_{vent}(t)$:

$$P_{TR}(t) = P_{musc}(t) + P_{vent}(t). \tag{10.3}$$

The pressure due to elastic recoil is the product of the elastance, $E$, and lung volume, $V(t)$. Elastance is the change in pressure due to the change in volume, $dP(t)/dV(t)$. For this model, we assume that elastance is constant. As we have already seen in Chapter 6, the pressure due to the resistance to flow is the product of the resistive load, $R$, and flow, $Q(t)$. Because flow is the derivative of volume with

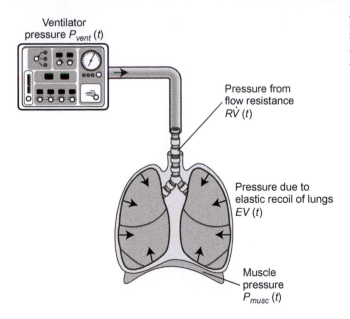

Ventilator pressure $P_{vent}(t)$

Pressure from flow resistance $R\dot{V}(t)$

Pressure due to elastic recoil of lungs $EV(t)$

Muscle pressure $P_{musc}(t)$

**FIGURE 10.10** Model of ventilation representing $P_{musc}(t) + P_{vent}(t) = EV(t) + R\dot{V}(t)$. [Reproduced by permission from Cairo & Pilbeam (2010)].

respect to time, we use $\dot{V}(t)$ to represent flow. Combining these relationships with Eqs. (10.2) and (10.3) results in:

$$P_{musc}(t) + P_{vent}(t) = EV(t) + R\dot{V}(t). \tag{10.4}$$

Eq. (10.4) represents the interaction between the patient and ventilator during inspiration and expiration (Figure 10.10). During spontaneous inspiration, $P_{vent}(t) = 0$; so the muscles generate sufficient pressure to offset elastic recoil and resistive pressures. In contrast, a patient may receive controlled mechanical ventilation, in which she does not trigger breathing. Here, $P_{musc}(t) = 0$; so the ventilator must generate sufficient pressure to offset elastic recoil and resistive pressure.

During passive expiration, pressures from the muscles and ventilator are absent, so $P_{musc}(t) = P_{vent}(t) = 0$. Eq. (10.4) becomes

$$0 = EV(t) + R\dot{V}(t) \tag{10.5}$$

and

$$-R\dot{V}(t) = EV(t). \tag{10.6}$$

The negative sign in Eq. (10.6) indicates that the direction of flow is reversed, which you expect for expiration.

When a patient receives mechanical ventilation, this therapy may be based on pressure, volume, or flow control. Because flow is the derivative of volume and because the three variables are related by Eq. (10.4), only one variable acts as the independent variable for control. After the control variable is chosen, the other two follow.

For example, we can assume that the constants in Eq. (10.4) are unity, with $E = R = 1$. With this assumption, we see that the equation has been modified so that flow plus

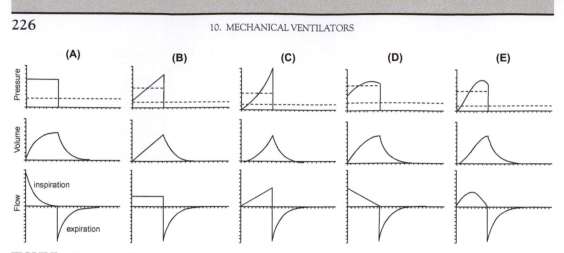

FIGURE 10.11 Idealized ventilator output waveforms. **A**: Constant pressure-controlled inspiration. **B**: Constant flow-controlled inspiration. **C**: Ascending-ramp flow-controlled inspiration. **D**: Descending-ramp flow-controlled inspiration. **E**: sinusoidal flow-controlled inspiration. In all cases, expiration pressure equals zero. *[Reproduced with permission from Chatburn. Figure reproduced with permission from Chatburn (2006)].*

volume equals pressure. This enables us to draw idealized ventilator output waveforms for inspiration and expiration. The waveform sets in Figure 10.11 are based on constant pressure (A), constant flow (B), ascending-ramp flow (C), descending-ramp flow (D), and sinusoidal flow (E) for inspiration. In all cases, the expiration pressure is zero (Chatburn, 2006).

## CLINICAL NEED

In the ICU, mechanical ventilators are indicated for a variety of conditions. Some patients present with apnea due to central nervous system damage, and they are not breathing spontaneously. Others have clinical signs of increased work of breathing, with or without laboratory evidence of impaired gas exchange. These patients in respiratory distress may have a 30–50% greater inspiratory resistance and 100% greater dynamic elastance than normal subjects (Laghi & Tobin, 2006). For many of these patients, measurement of blood gases (described in Chapter 11) reveals either a $Pa_{O_2}$ less than 55 mmHg or a $Pa_{CO_2}$ greater than 50 mmHg, with a corresponding pH less than 7.32 (Amitai et al., 2009).

In the OR, mechanical ventilators are also used because the administration of deep general anesthesia suppresses spontaneous breathing.

## HISTORIC DEVICES

Scientists have always been interested in resuscitating the dead and dying. Legends purport that both Vesalius and Paracelsus attempted to resuscitate humans in the 1500s.

**FIGURE 10.12** Draeger Pulmotor. *[Copyright © the American Society of Anesthesiologist, Inc. This image appears in the Anesthesiology Reflections online collection available at www.anesthesiology.org. Figure reproduced by permission from Bause (2009)].*

## Early Devices

More recently, in 1908, the Draeger Oxygen Apparatus Company division in Pittsburgh began to produce Pulmotors to resuscitate miners. The Pulmotor pumped air or oxygen into the airways of victims of fire or mining accidents (Bause, 2009) (Figure 10.12). Because its air supply was not controlled, excessive pressure could damage internal organs.

As an alternative to forcing air into the lungs, scientists began to develop negative-pressure ventilators. A patient's body was placed in an airtight enclosure, with his head outside the enclosure. The body was subjected to cyclical negative and atmospheric pressure. Negative pressure in the enclosure pulled air into the lungs; atmospheric pressure in the enclosure caused exhalation.

The first clinically successful design of this tank respirator was developed by chemical engineer Philip Drinker and physician Louis Agassiz Shaw. Nicknamed the "iron lung" by the general public, this cylindrical device was powered by an electric motor with two vacuum cleaners, which modified pressure within the enclosure. Inventor John Emerson refined the Drinker-Shaw respirator and reduced the cost to almost half of the original design. In the improved design, the patient lay on a movable bed that slid out of the enclosure as needed. Attendants could access the patient through portal windows (Figure 10.13).

The tank respirator was used to treat patients struck down with polio. Poliomyelitis is caused by the poliovirus, which is transmitted by the oral-fecal route or by ingestion of contaminated water. The first known outbreak in the United States occurred in 1894. This epidemic peaked in 1952, when a record 57,628 cases were reported. During the acute stage, the virus destroys motor neurons that control swallowing, breathing, and limb movement. Severe polio patients were placed in tank respirators until they could breathe spontaneously, usually after 1–2 weeks (National Museum of American History, 2009).

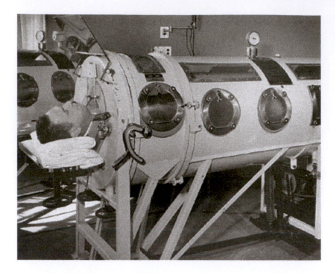

FIGURE 10.13 Man using an Emerson tank respirator equipped with a mirror in the 1950s. Original courtesy of Post-Polio Health International. *[Reproduced by permission from National Museum of American History (2009)].*

Although certainly an improvement in ventilation therapy, the tank respirator's negative pressure was applied to the abdominal wall, as well as to the chest. This created venous pooling in the abdomen and decreased cardiac output.

### Enabling Technology: Positive-Pressure Ventilation

During the polio epidemic of 1952 in Copenhagen, Blegdam Hospital, which was the only hospital serving metropolitan Cophenhagen, did not have sufficient respirators for polio patients. At the recommendation of anesthesiologist Bjorn Ibsen, patients began to be treated with manual positive-pressure ventilation. The patients treated by positive-pressure had decreased mortality rates, compared to those treated with conventional respirators (Lassen, 1953).

As a result, positive-pressure ventilation became more popular in Europe, leading to the construction of mechanical positive-pressure ventilators in Europe and later in the United States. In, 1955, physician Forrest Bird introduced the Bird Mark 7 ventilator, which was the first low-cost, mass-produced mechanical ventilator (Figure 10.14).

## SYSTEM DESCRIPTION AND DIAGRAM

A modern ICU ventilator is electrically powered and microprocessor controlled. The user inputs the desired breathing parameters, such as respiratory rate, tidal volume, set pressure, and inspiratory time, to the processor module. In a generalized system, which is really a combination of deployed architectures, two 50-psi gas sources for air (an internal compressor compresses external air) and for oxygen ($O_2$) provide the pressure to deliver inspiratory gas flow to the patient. Control of the inspiratory flow waveform is governed by the processor module, which controls the air and $O_2$ valves. Two flow sensors provide air and $O_2$ flow data to the processor module. A rigid accumulator serves as the internal

**FIGURE 10.14** Bird Mark 7 Ventilator (Photograph taken by PEMED).

reservoir to supply flow on demand to the patient. The flow from the accumulator is controlled by the processor module through the flow valve, with feedback data from a down-the-line flow sensor.

During inspiration, the exhalation valve is closed to direct all flow to the patient. A pressure sensor in the patient circuit provides feedback to the processor module for maintaining the positive end-expiratory pressure (PEEP). The LCD provides the ventilator user interface (Figure 10.15). After the user inputs desired breathing parameters for a patient, a ventilation mode is delivered. Ventilation mode is defined by the control type used within and between breaths, as well as by the breath sequence within breaths.

## Common Modes of Mechanical Ventilation

As already stated, there are three control variables: pressure, volume, and flow. Because flow is the time derivative of volume, we can simplify control by referring to only pressure and volume control. A third type of control is dual control, in which the control variable switches between pressure and volume, in either order, within a breath.

*Breath sequence* refers to the combinations of mandatory and spontaneous breaths. The clinical intent may be to provide full ventilator support or partial ventilator support to wean a patient off the ventilator. When spontaneous breaths are not allowed between mandatory breaths, the breath sequence is *continuous mandatory ventilation (CMV)*. When spontaneous breaths are allowed between mandatory breaths, the breath sequence becomes *intermittent mandatory ventilation (IMV)*. If no mandatory breaths are allowed, the breath sequence is *continuous spontaneous ventilation (CSV)*.

Combining control type and breath sequence, Chatburn defines eight ventilation modes:

1. Volume control—continuous mandatory ventilation (VC-CMV)
2. Volume control—intermittent mandatory ventilation (VC-IMV)
3. Pressure control—continuous mandatory ventilation (PC-CMV)

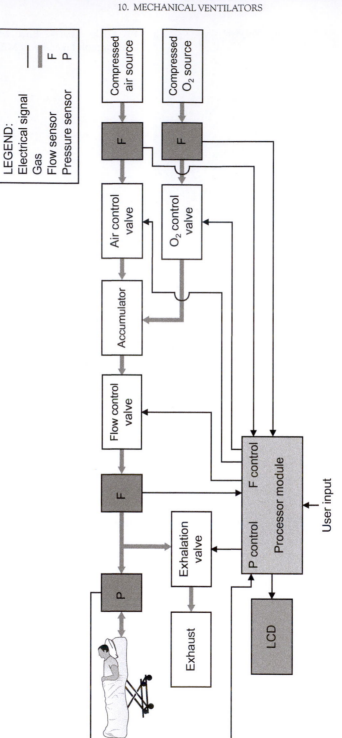

**FIGURE 10.15** Mechanical ventilator system diagram. Power and power connections are omitted.

**4.** Pressure control—intermittent mandatory ventilation (PC-IMV)

**5.** Pressure control—continuous spontaneous ventilation (PC-CSV)

**6.** Dual control—continuous mandatory ventilation (DC-CMV)

**7.** Dual control—intermittent mandatory ventilation (VC-IMV)

**8.** Dual control—continuous spontaneous ventilation (DC-CSV) (Chatburn, 2007)

Please note that VC-CSV is not possible, because the definition of volume control conflicts with the definition of a spontaneous breath (Chatburn, 2007). We describe four of these ventilation modes in greater detail.

A patient who cannot breathe spontaneously may be placed on controlled mandatory ventilation. Such a patient may be suffering from a drug overdose, a neurologic or neuromuscular disorder, or seizure activity requiring sedation and possibly induced paralysis. Preset volume- or pressure-targeted breaths at set intervals are delivered, according to the control variable. The related pressure, volume, and flow waveforms are shown in Figure 10.16.

A patient being weaned from the ventilator may be placed on intermittent mandatory ventilation. As with controlled mandatory ventilation, preset volume- or pressure-targeted breaths at a set interval are delivered. Unlike CMV, between mandatory breaths, the patient may breathe spontaneously from the ventilator circuit. A spontaneous breath starts

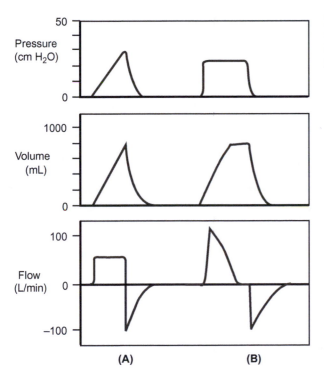

**FIGURE 10.16** Pressure, volume, and flow waveforms for two ventilation modes: **A**: VC-CMV; **B**: PC-CMV. *[Reproduced by permission from Cairo & Pilbeam (2010)].*

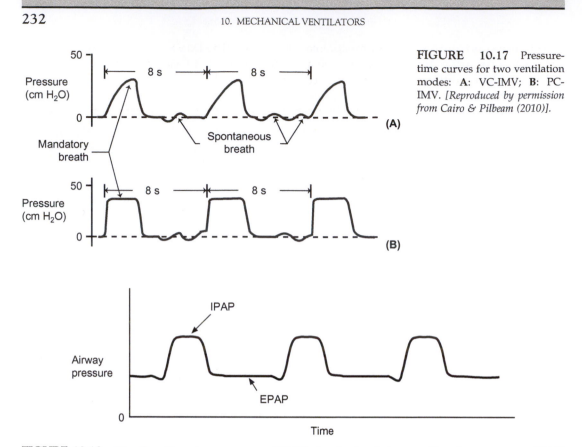

FIGURE 10.17 Pressure-time curves for two ventilation modes: **A**: VC-IMV; **B**: PC-IMV. [*Reproduced by permission from Cairo & Pilbeam (2010)*].

FIGURE 10.18  Bilevel positive airway pressure (BiPAP) showing inspiratory positive airway pressure (IPAP) and expiratory positive airway pressure (EPAP). EPAP is the PEEP. Both pressures are higher than a zero baseline pressure. [*Original figure from Pilbeam (1998). Figure reproduced by permission from Cairo & Pilbeam (2010)*].

from the set baseline, which may be ambient pressure or a positive baseline pressure. For this breath to occur, the patient must assume part of the work of inspiration (Figure 10.17).

A patient receiving noninvasive positive pressure ventilation (NIPPV) through a face mask may have bilevel positive airway pressure administered. This setting, also called BiPAP by one manufacturer, may be used in home ventilators to treat obstructive sleep apnea. With this setting, inspiration and expiration occur at two levels of positive pressure (Cairo and Pilbeam, 2010) (Figure 10.18). Because the patient triggers inspiration, NIPPV is another example of PC-IMV.

## Noninvasive Ventilation

Invasive mechanical ventilation requires endotracheal intubation to complete the patient circuit. *Endotracheal intubation* is the insertion of an artificial airway into the trachea by either the oropharyngeal or nasopharyngeal route. Although invasive ventilation is highly effective and reliable in supporting alveolar ventilation, the intubation procedure is associated with occasional adverse events such as pharyngeal/laryngeal/tracheal tissue damage, self-extubation, and aspiration of gastric contents.

Whenever possible, noninvasive ventilation is a preferable option. The upper airway is intact, airway defense mechanisms are preserved, and patients may eat, drink, verbalize, and expectorate secretions. Because of the convenience in using a face mask, NIPPV is the predominant means of administering noninvasive ventilation (Hill, 2006).

For patients with chronic obstructive pulmonary disease (COPD), NIPPV has been demonstrated to reduce complications, duration of ICU stay, and mortality (Plant et al., 2000). Chronic obstructive pulmonary disease is characterized by airflow limitation that is not fully reversible. We discuss an NIPPV clinical trial for COPD patients in the exercises at the end of the chapter. Noninvasive positive pressure ventilation has not been demonstrated to be more efficacious than invasive ventilation for other types of patients. However, numerous other applications are currently under clinical investigation (Hill, 2006).

## KEY FEATURES FROM ENGINEERING STANDARDS

As of 2010, FDA recommends several consensus standards related to various aspects of mechanical ventilators. One of these standards is *IEC 60601-2-12 Medical Electrical Equipment—Part 2:12: Particular Requirements for the Safety of Lung Ventilators—Critical Care Ventilators* (IEC, 2001).

### Protection from Interruption of Power Supply

If not being weaned, a patient may be completely dependent on a mechanical ventilator for respiratory support. Should the supply power fall below values specified by the manufacturer, the ventilator must sound an alarm for at least 120 s. While the supply is compromised, the patient must be able to breathe spontaneously from the ventilator.

It is not mandatory that the ventilator switch to an internal electrical power source. However, if this switch occurs, the ventilator must indicate the state of this power source, with an alarm sounded before the internal electrical power source is completely depleted.

These requirements are verified by simulating a drop below the supply rates of the manufacturer and checking for the power failure alarm and enablement of spontaneous breathing. If an internal electrical power source is present, the internal power source is depleted to below the minimum value specific by the manufacturer. The internal power failure alarm is then verified (IEC, 2001).

### Maximum Pressure to Patient

The patient must be protected from excess pressure. When the pressure at the patient connection port exceeds 125 cmH$_2$O during normal use, this pressure must not reach the patient. (Pressure is usually set to a lower limit by the clinician.) The respiratory pressure at the patient connection port must be displayed and accurate within ±2% of the full-scale reading +4% of the actual reading. The site of actual pressure measurement may be anywhere at the ventilator breathing system, but the displayed pressure must reference the patient connection port.

These requirements are tested by the administration of various pressures of unspecified magnitude, with "visual inspection and verification of accuracy" (IEC, 2001).

### Active Pressure Limit

The pressure limit in the breathing system for each type of breath must be operator adjustable, or specified by an active breathing algorithm, or some combination of both. When an active limit value is reached, the ventilator must act to reduce the pressure to a level at or below the PEEP value within 200 ms. The method of testing this requirement is unspecified (IEC, 2001).

### Oxygen Monitor

Adequate oxygen must be delivered to the patient. The ventilator is required to be equipped with an oxygen monitor for the measurement of inspiratory oxygen concentration, such as in the inspiratory limb or at the patient connection port. This oxygen monitor must comply with separate standard ISO 7767, which details specific requirements and testing for monitoring patient breathing mixtures. The oxygen monitor must be provided with a high alarm limit (IEC, 2001)

### Protection from Breathing System Leakage

The patient must be adequately ventilated. Leakage from the ventilator breathing system must not exceed 200 mL/min at 50 hPa for ventilators providing tidal volumes greater than 300 ml or 100 mL/min at 40 hPa for tidal volumes between 300 mL and 30 ml. Testing is conducted by sealing all ports, connecting a pressure measuring device, and introducing air until the appropriate pressure is reached. The air flow is then adjusted to stabilize the pressure, with leakage flow recorded (IEC, 2001).

## SUMMARY

A mechanical ventilator is an instrument that replaces or assists in spontaneous breathing. Mechanical ventilators are used in a variety of settings, from the operating room and intensive care unit to the home and transport vehicles. ICU ventilators are used by adult patients recovering from acute lung injuries.

In the pulmonary circulation, deoxygenated blood flows from the right atrium and ventricle, through the pulmonary artery, and to the lungs. The blood flow through the lungs is essentially the cardiac output of 5 L/min. Within the lungs, blood flows to the pulmonary capillaries, where gas exchange takes place with the alveoli. Atmospheric air moves to and from the lung alveoli through lung expansion and contraction. Newly oxygenated blood then moves from the pulmonary veins to the left heart and into the aorta.

We can model the pressures involved in ventilation as $P_{musc}(t) + P_{vent}(t) = EV(t) + R\dot{V}(t)$.

When a patient receives mechanical ventilation ($P_{musc}(t) = 0$), this therapy may be based on pressure, volume, or flow control. Because flow is the derivative of volume and because the three variables are related by this equation, only one variable acts as the independent variable for control. After the control variable is chosen, the other two follow.

Early ventilators were based on negative pressure, but are currently based on positive pressure. Key ventilator features include protection from interruption of power supply, maximum pressure to patient, active pressure limit, oxygen monitoring, and protection from breathing system leakage.

## Exercises

**10.1** Expired respiratory gases are 74% $N_2$, 16% $O_2$, 4% $CO_2$, and 6% $H_2O$. What are the corresponding partial pressures of expired $O_2$ and $CO_2$ at sea level? At the beginning of expiration, what are the pleural, alveolar, and transpulmonary pressures?

**10.2** Mechanical ventilators once used a piston to increase gas pressure. Why is this design no longer needed?

**10.3** Sketch the corresponding flow waveforms for the volume and pressure waveforms given in Figure 10.19.

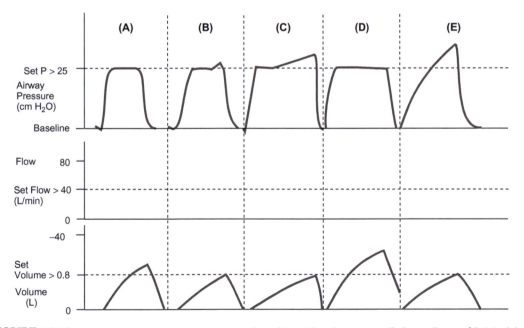

**FIGURE 10.19** Various waveform combinations achievable with volume-controlled ventilation. [*Original from Pilbeam & Cairo (2006). Figure reproduced by permission from Cairo & Pilbeam (2010)*].

**10.4** Why is BiPAP used to treat obstructive sleep apnea?

Read (Lassen, 1953):

**10.5** What was the mortality rate of polio patients during the first month of the epidemic? How was respiratory insufficiency treated in these patients? How did a tank respirator work?

**10.6** Describe the replacement therapy for the tank and cuirass respirators in detail. What was the purpose of using soda lime?

Read Plant et al. (2000):

**10.7** What is a multicenter randomized controlled clinical trial? What was the null hypothesis? What is *power* as the term is used in this study? How was the sample size calculated?

**10.8** What are the normal ranges for arterial pH, partial pressure of oxygen, and partial pressure of carbon dioxide? What were the primary and secondary outcome measures? Is intubation necessary in this type of patient group?

# References

Amitai, A., Sinert, R. H., & Joyce, D. M. (2009). Ventilator management. *eMedicine* (Medscape).

Bause, G. S. (2009). Draeger pulmotor. *Anesthesiology, 110,* 1243.

Cairo, J. M., & Pilbeam, S. P. (2010). *Mosby's respiratory care equipment* (8th ed.). St. Louis, MO: Mosby Elsevier.

Chatburn, R. L. (2006). Classification of mechanical ventilators. In M. J. Tobin (Ed.), *Principles and practice of mechanical ventilation* (2nd ed.). New York: McGraw-Hill.

Chatburn, R. L. (2007). Classification of ventilator modes: Update and proposal for implementation. *Respiratory Care, 52,* 301–323.

Guyton, A. C., & Hall, J. E. (2006). *Textbook of medical physiology* (11th ed.). Philadelphia: Elsevier Saunders.

Hill, N. S. (2006). Noninvasive positive-pressure ventilation. In M. J. Tobin (Ed.), *Principles and practice of mechanical ventilation* (2nd ed.). New York: McGraw-Hill.

IEC (2001). *IEC 60601-2-12 medical electrical equipment—part 2-12: Particular requirements for the safety of lung ventilators—critical care ventilators.* Geneva: IEC.

Laghi, F., & Tobin, M. J. (2006). Indications for mechanical ventilation. In M. J. Tobin (Ed.), *Principles and practice of mechanical ventilation* (2nd ed.). New York: McGraw-Hill.

Lassen, H. C. (1953). A preliminary report on the 1952 epidemic of poliomyelitis in Copenhagen with special reference to the treatment of acute respiratory insufficiency. *Lancet, 1,* 37–41.

Miller, W. S. (1947). *The lung.* Springfield, IL: Charles C. Thomas.

National Museum of American History (2009). *Whatever happened to polio?* Washington, DC: Smithsonian.

Pilbeam, S. P. (1998). *Mechanical ventilation: Physiological and clinical applications* (3rd ed.). St. Louis, MO: Mosby.

Pilbeam, S. P., & Cairo, J. M. (2006). *Mechanical ventilation: Physiological and clinical applications* (4th ed.). St. Louis, MO: Mosby.

Plant, P. K., Owen, J. L., & Elliott, M. W. (2000). Early use of non-invasive ventilation for acute exacerbations of chronic obstructive pulmonary disease on general respiratory wards: A multicentre randomised controlled trial. *Lancet, 355,* 1931–1935.

# Pulse Oximeters

In this chapter, we discuss the relevant physiology, history, and key features of pulse oximeters. A *pulse oximeter* is an instrument that estimates and displays the arterial saturation of oxygen (Figure 11.1). Originally used in the operating rooms of hospitals, pulse oximeters migrated to intensive care units and then to patient clinics.

Upon completion of this chapter, each student shall be able to:

1. Understand the mechanisms underlying the oxyhemoglobin dissociation curve.
2. Explain the relationship between pulse oximeters, the Beer-Lambert law, and the pulse oximetry calibration curve.
3. Describe five key features of pulse oximeters.

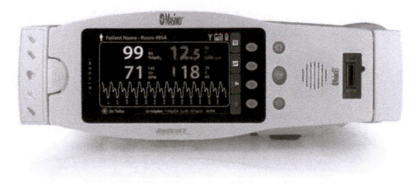

**FIGURE 11.1** Masimo Radical-7 pulse CO-oximeter, which includes pulse oximetry (Masimo Corporation, Irvine, California).

237

# OXYGEN TRANSPORT IN BLOOD

In Chapter 10, we discussed how inspired oxygen is transported to alveoli, where it diffuses to pulmonary capillaries down a pressure gradient. Once in the blood, 3% of oxygen is transported in the dissolved state in the water of plasma and blood cells. Ninety-seven percent is carried in chemical combination with hemoglobin in the erythrocytes, which are the red blood cells.

## Hemoglobin

An *erythrocyte* is a biconcave disc with a mean diameter of about 7.8 microns and volume of 90–95 microns$^3$ (Figure 5.2). Each cell contains hemoglobin, and has the ability to concentrate hemoglobin in cell fluid up to 34 grams in each 100 mL of cells. One hemoglobin molecule is composed of four hemoglobin chains bound loosely together. Each hemoglobin chain is composed of a long polypeptide chain combined with a heme molecule (Figure 11.2).

Within a heme molecule, an oxygen molecule, $O_2$, binds loosely with one of the coordination bonds of the iron atom. This oxyhemoglobin, $HbO_2$, bond is an extremely loose and therefore easily reversible bond. When oxygen reaches the tissues, it is released into tissue fluids.

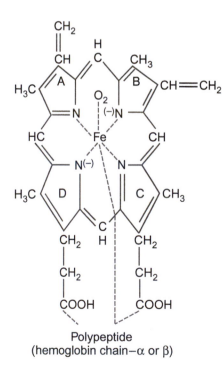

**FIGURE 11.2** Heme molecule structure. *[Reproduced by permission from Guyton & Hall (2006)].*

Polypeptide
(hemoglobin chain–α or β)

## Oxyhemoglobin Dissociation Curve

We can define the arterial saturation of oxygen, $SaO_2$, as the ratio of oxyhemoglobin concentration to the total concentration of arterial hemoglobin available for reversible oxygen binding:

$$SaO_2 = \frac{[HbO_2]}{[Hb] + [HbO_2]}. \tag{11.1}$$

Arterial saturation of oxygen is affected by the partial pressure of arterial oxygen, $P_{O_2}$. At low partial pressures, such as tissue capillary pressure of 40 mmHg, oxygen is released from hemoglobin, leading to a low $SaO_2$ of about 75%. At high partial pressures, such as pulmonary capillary pressure of 95 mmHg, oxygen binds with hemoglobin, leading to high $SaO_2$ of about 95 mmHg. This relationship is illustrated by the oxyhemoglobin dissociation curve (Figure 11.3). The illustrated oxyhemoglobin dissociation curve assumes normal, average blood. But if the pH of blood decreases from its normal value of 7.4−7.6, the dissociation curve shifts to the left about 15%. Similarly, if the pH of blood increases from 7.4−7.2, the dissociation curve shifts to the right about 15% (Figure 11.4).

Other factors that cause the dissociation curve to shift to the right are increased carbon dioxide concentration, increased blood temperature, and increased 2,3-biphosphoglycerate (BPG) concentration. In blood, $CO_2$ reacts with water to form carbonic acid, $H_2CO_3$,

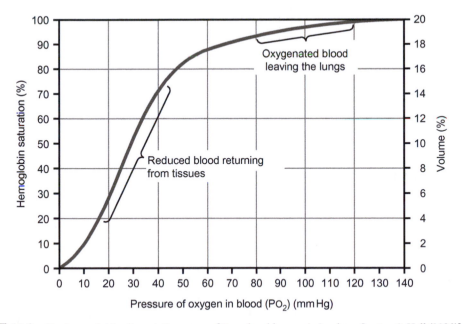

**FIGURE 11.3**  Oxyhemoglobin dissociation curve. *[Reproduced by permission from Guyton & Hall (2006)].*

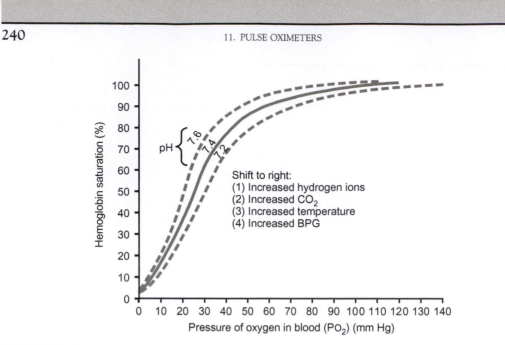

**FIGURE 11.4** Shifts of the oxyhemoglobin dissociation curve. *[Reproduced by permission from Guyton & Hall (2006)].*

which dissociates into hydrogen ions, $H^+$, and bicarbonate ions, $HCO_3^-$. With increased $CO_2$ in the blood, more hydrogen ions are formed, which increases the blood pH and shifts the dissociation curve to the right. When blood temperature increases, oxygen delivery to tissues is increased. This makes more oxygen available for binding to hemoglobin, which shifts the dissociation curve to the right. BPG is a metabolically important phosphate compound that helps keep the dissociation curve shifted slightly to the right at all times. During hypoxic conditions of longer than a few hours, the BPG concentration increases considerably, which shifts the dissociation curve further to the right (Figure 11.4).

## Carbon Monoxide Displacement

Up to this point, we have discussed how oxygen binds to hemoglobin. But carbon monoxide, CO, also binds to hemoglobin at the same location that oxygen binds on the hemoglobin molecule. Because CO binds with about 250 times the tenacity of oxygen, CO easily displaces oxygen. A CO partial pressure of only 0.4 mmHg in alveoli causes CO to compete equally with oxygen for hemoglobin binding (0.4 mmHg is 1/250 of normal alveolar oxygen pressure of 100 mmHg.) Therefore, a CO partial pressure of only 0.6 mmHg can be lethal. The carbon monoxide-hemoglobin dissociation curve is essentially the oxyhemoglobin dissociation curve, except that the *x* axis magnitudes have been divided by a factor of 250 (Figure 11.5). CO-hemoglobin is also known as *carboxyhemoglobin* (Guyton & Hall, 2006).

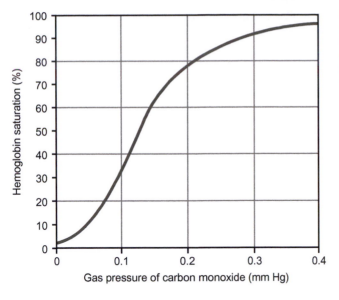

**FIGURE 11.5** Carbon monoxide-hemoglobin dissociation curve. Note the extremely low carbon monoxide pressures at which carbon monoxide combines with hemoglobin. *[Reproduced by permission from Guyton & Hall (2006)].*

## BEER-LAMBERT LAW

As discussed in Chapter 10, the partial pressure of arterial oxygen, $Pa_{O_2}$, is an important parameter for assessing adequate mechanical ventilation. This partial pressure can be inferred through the $SaO_2$ level because these parameters are related through the oxyhemoglobin dissociation curve. Arterial saturation of oxygen can estimated in blood using the Beer-Lambert law, which is also known as Beer's law or the Beer-Lambert-Bouguer law. According to this law, the absorbance of a wavelength of light, $A(\lambda)$, in a homogeneous medium is proportional to the absorption path length, $l$, and concentration, $c$, of the absorbing species:

$$A(\lambda) = \varepsilon(\lambda) \cdot c \cdot l, \tag{11.2}$$

where $\varepsilon(\lambda)$ is the molar absorptivity. The molar absorptivity incorporates a constant for converting natural log to $\log_{10}$. *Absorption* refers to the negated $\log_{10}$ of the ratio of transmitted light intensity, $I(\lambda)$, to incident light intensity (or intensity at $t = 0$), $I_0(\lambda)$:

$$A(\lambda) = -\log_{10} \frac{I_{(\lambda)}}{I_{0(\lambda)}} \tag{11.3}$$

When a number of species, $n$, is present, then the total absorbance, $A_t(\lambda)$, can be calculated as:

$$A_t(\lambda) = \sum_i^n \varepsilon_i(\lambda) \cdot c_i \cdot l_i \tag{11.4}$$

Let us place arterial blood in a glass cuvette of known dimension and transmit two wavelengths of light of known intensity separately through the sample. Assume that the only hemoglobin species present are hemoglobin and oxyhemoglobin. We detect each intensity and calculate the absorbance (Figure 11.6). The molar absorptivities for hemoglobin (or deoxyhemoglobin) and oxyhemoglobin as a function of wavelength are shown in Figure 11.7.

Now we have two equations and two unknowns:

$$A_t(\lambda_1) = \varepsilon_o(\lambda_1) \cdot c_o \cdot l_o + \varepsilon_d(\lambda_1) \cdot c_d \cdot l_d, \tag{11.5}$$

$$A_t(\lambda_2) = \varepsilon_o(\lambda_2) \cdot c_o \cdot l_o + \varepsilon_d(\lambda_2) \cdot c_d \cdot l_d, \tag{11.6}$$

where $o$ = oxyhemoglobin subscript and $d$ = deoxyhemoglobin subscript. We can solve for $c_o$ and $c_d$. Realizing that

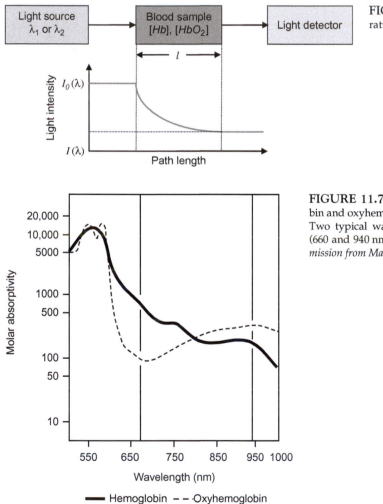

**FIGURE 11.6**  Experimental apparatus for the Beer-Lambert law.

**FIGURE 11.7**  Molar absorptivities of hemoglobin and oxyhemoglobin as functions of wavelength. Two typical wavelengths used in pulse oximetry (660 and 940 nm) are indicated. *[Reproduced by permission from Mackenzie (1985)].*

$$c_o = [HbO_2] \tag{11.7}$$

and

$$c_d = [Hb], \tag{11.8}$$

we substitute these values into the Eq. (11.1), and calculate $SaO_2$. This method, using at least four wavelengths, enables accurate concentration estimation of oxyhemoglobin, deoxyhemoglobin, and other hemoglobin derivatives such as carboxyhemoglobin and methemoglobin (hemoglobin with $Fe^{+3}$ instead of $Fe^{+2}$ ions). It is the basis of CO-oximeter clinical chemistry measurements.

Theoretically, these equations can be applied to arterial blood in the finger, as well as to blood in a glass cuvette, for pulse oximetry.

## ADAPTIVE FILTERING

You may have experienced adaptive filtering when using noise-canceling headphones. An adaptive filter has the ability to adjust its own parameters automatically, and its design requires little or no prior knowledge of signal or noise characteristics (Widrow & Hoff Jr., 1960). Adaptive filtering is applied to pulse oximetry to minimize the effect of patient motion artifact.

In Chapter 1, we briefly discussed how signal and noise can be separated, if they occur in different frequency bands, through frequency-selective filtering (Figure 1.24). Unfortunately, the frequencies of signal and noise in physiologic signals, such as patient motion artifact, often overlap.

One filtering technique that can then be used, under specialized constraints, is adaptive filtering. For an adaptive filter used as an adaptive noise canceler, we assume a signal input, $u(k)$; an original noise associated with the signal, $n_0(k)$; a reference noise source, $n_1(k)$; a filtered output that is an approximation of the original signal, $\hat{y}(k)$; and a linear combination of signal and noise (Figure 11.8). The special constraints for adaptive noise canceling are:

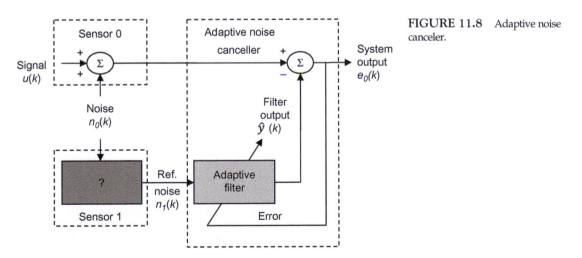

**FIGURE 11.8** Adaptive noise canceler.

1. $u(k)$, $n_0(k)$, $n_1(k)$, and $\hat{y}(k)$ are statistically stationary (not variable with time).
2. $u(k)$, $n_0(k)$, $n_1(k)$, and $\hat{y}(k)$ have zero means.
3. $u(k)$ is uncorrelated with $n_0(k)$ and $n_1(k)$.
4. $n_0(k)$ is correlated with $n_1(k)$.

As shown in Figure 11.8, sensor 0 is used to acquire the signal, which has been contaminated by original noise. Sensor 1 is used to acquire a reference noise signal. The reference noise source is used to model the original noise associated with the signal. The modeled noise is subtracted from the combination of signal and original noise. When the model is accurate, the system output, $e_0(k)$, approximates only the signal. Adaptive noise cancellation was invented by electrical engineer Bernard Widrow in 1960 (Widrow & Hoff Jr., 1960).

Describing this algorithm technology in full detail requires a textbook (Widrow & Stearns, 1985) or at least one book chapter (Baura, 2002). Let us summarize the algorithm by stating that it enables the squared error of the system, $e_0^2(k)$, to be minimized. From Figure 11.8, the error is

$$e_0(k) = u(k) + n_0(k) - \hat{y}(k). \tag{11.9}$$

The squared error, which is the system output power, is

$$e_0^2(k) = u^2(k) + [n_0(k) - \hat{y}(k)]^2 + 2u(k)[n_0(k) - \hat{y}(k)]. \tag{11.10}$$

The expected value, $E$, of a function is its mean. When we take the expected value on both sides of Eq. (11.10), we obtain:

$$E\{e_0^2(k)\} = E\{u^2(k) + \left[n_0(k) - \hat{y}(k)\right]^2 + 2u(k)[n_0(k) - \hat{y}(k)]\} \tag{11.11}$$

$$E\{e_0^2(k)\} = E\{u^2(k)\} + E\{[n_0(k) - \hat{y}(k)]^2\} + E\{2u(k)[n_0(k) - \hat{y}(k)]\}. \tag{11.12}$$

We assumed that $u(k)$ is uncorrelated with $n_0(k)$ and $\hat{y}(k)$. Therefore, the last term in Eq. (11.12) equals zero. So Eq. (11.12) simplifies to

$$E\{e_0^2(k)\} = E\{u^2(k)\} + E\{[n_0(k) - \hat{y}(k)]^2\}. \tag{11.13}$$

When we use an adaptive noise canceler to minimize the system error, the signal input is unaffected. Therefore, minimization of the system output power is

$$E_{min}\{e_0^2(k)\} = E\{u^2(k)\} + E_{min}\{[n_0(k) - \hat{y}(k)]^2\}. \tag{11.14}$$

The adaptive noise canceler is updated with each signal and noise sample. When the minimum system output power is reached, the signal is minimally affected by noise.

Many adaptive noise canceling algorithms have been implemented. Widrow's least mean squares adaptive noise-canceling algorithm iteratively updates an adaptive filter vector, $\mathbf{f}(k)$, for a signal input vector, $\mathbf{u}(k)$, as

$$\mathbf{f}(k + 1) = \mathbf{f}(k) + 2\mu e_0(k)\mathbf{u}(k). \tag{11.15}$$

Here, $\mu$ is a gain constant that regulates the speed and stability of adaptation, which must be set to a value that is both less than the inverse of the maximum system eigenvalue and greater than zero.

# CLINICAL NEED

Pulse oximetry has been a standard of care since 1986. That year, the American Society of Anesthesiologists (ASA) included the following sentence in its Standards for Basic Anesthetic Monitoring: "During all anesthetics, a quantitative method of assessing oxygenation such as pulse oximetry shall be employed" (ASA, 1986). By using pulse oximetry during surgery, anesthesiologists ensure that oxygen is being adequately transported to the tissues during mechanical ventilation. As noted anesthesiologist John Severinghaus observed in 1986, "pulse oximetry is arguably the most significant technological advance ever made in monitoring the well-being and safety of patients during anesthesia, recovery, and critical care" (Severinghaus & Astrup, 1986).

After being successfully deployed in the OR and ICU, pulse oximetry migrated to other hospital units and patient clinics. Medical device engineers and some clinicians consider arterial saturation of oxygen, measured by pulse oximetry, to be the fifth vital sign (Birnbaum, 2009). The normal adult range of this vital sign is 93–100%.

# HISTORIC DEVICES

Oximetry research began to be conducted in the 1930s, based on the Beer-Lambert law. At the time, the Beer-Lambert law would seemingly soon enable continuous, noninvasive estimation of arterial saturation of oxygen.

## Early Devices

In 1931, German physiologist Dr. Ludwig Nicolai began to investigate how light is transmitted through human skin, in an effort to understand the dynamics of tissue oxygen consumption. His associate, Kurt Kramer, confirmed that the Beer-Lambert law could be used to estimate hemoglobin concentration for arterial blood in a glass cuvette with fixed dimensions.

By 1941, physiologist Glenn Millikan developed an oxygen meter for military pilots, which he called an "oximeter." The oximeter transmitted green and infrared wavelengths through the earlobe. Oximeter readings controlled the oxygen supply to the pilot's mask. However, this oximeter and later models could not accurately estimate hemoglobin concentrations (Severinghaus & Astrup, 1986).

In tissue, total absorbance must include two new terms:

$$A_t(\lambda_i) = \varepsilon_o(\lambda_i) \cdot c_o \cdot l_o + \varepsilon_d(\lambda_i) \cdot c_d \cdot l_d + \varepsilon_x(\lambda_i) \cdot c_x \cdot l_x + A_y(\lambda_i). \tag{11.16}$$

The third term in Eq. (11.16) represents variable absorbances not due to arterial blood. The fourth term, $A_y(\lambda_i)$, represents nonspecific sources of optical attenuation such as light scattering and geometric factors.

Biomedical engineer Takuo Aoyagi, who conducted research at Nihon Kohden, is credited with inventing the so-called pulse oximeter by considering these other absorbance terms. Aoyagi recognized that only pulsatile changes in light transmission through living

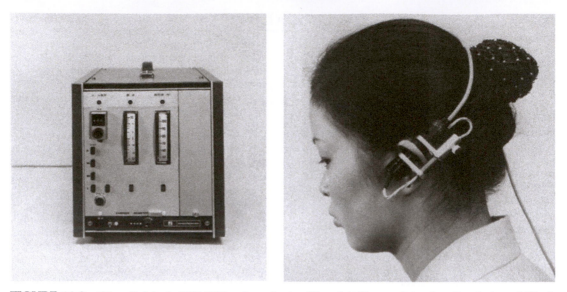

**FIGURE 11.9**  Nihon Kohden's OLV-5100 pulse oximeter. *[Reproduced by permission from Severinghaus & Honda (1987)].*

tissues were necessary for estimating hemoglobin concentrations (Severinghaus & Honda, 1987). The first commercial pulse oximeter was Nihon Kohden's OLV-5100 (Figure 11.9).

Aoyagi's pulsatile observation can be summarized in ratio calculations (Aoyagi, 2003). The ratio, $R$, is calculated as the ratio of the first time derivatives of the two absorbances:

$$R = \frac{\frac{dA_t(\lambda_1)}{dt}}{\frac{dA_t(\lambda_2)}{dt}} = \frac{\varepsilon_o(\lambda_1) \cdot c_o \cdot \frac{dl_o}{dt} + \varepsilon_d(\lambda_1) \cdot c_d \cdot \frac{dl_d}{dt} + \varepsilon_x(\lambda_1) \cdot c_x \cdot \frac{dl_x}{dt} + \frac{d}{dt}\left[A_y(\lambda_1)\right]}{\varepsilon_o(\lambda_2) \cdot c_o \cdot \frac{dl_o}{dt} + \varepsilon_d(\lambda_2) \cdot c_d \cdot \frac{dl_d}{dt} + \varepsilon_x(\lambda_2) \cdot c_x \cdot \frac{dl_x}{dt} + \frac{d}{dt}\left[A_y(\lambda_2)\right]}. \tag{11.17}$$

Because the third and fourth terms in each absorbance are constant, rather than pulsatile, their time derivatives equal zero and can be neglected:

$$R = \frac{\frac{dA_t(\lambda_1)}{dt}}{\frac{dA_t(\lambda_2)}{dt}} = \frac{\varepsilon_o(\lambda_1) \cdot c_o \cdot \frac{dl_o}{dt} + \varepsilon_d(\lambda_1) \cdot c_d \cdot \frac{dl_d}{dt}}{\varepsilon_o(\lambda_2) \cdot c_o \cdot \frac{dl_o}{dt} + \varepsilon_d(\lambda_2) \cdot c_d \cdot \frac{dl_d}{dt}}. \tag{11.18}$$

The pulsatile path lengths for oxyhemoglobin and deoxyhemoglobin are assumed to be equal. Thus, their derivatives cancel out of the numerator and denominator:

$$R = \frac{\varepsilon_o(\lambda_1) \cdot c_o + \varepsilon_d(\lambda_1) \cdot c_d}{\varepsilon_o(\lambda_2) \cdot c_o + \varepsilon_d(\lambda_2) \cdot c_d}. \tag{11.19}$$

Combining Eq. (11.19) with Eqs. (11.1), (11.7), and (11.8) yields

$$SpO_2 \equiv SaO_2 = \frac{\varepsilon_d(\lambda_1) - \varepsilon_d(\lambda_2)R}{\left[\varepsilon_d(\lambda_1) - \varepsilon_o(\lambda_1)\right] - \left[\varepsilon_d(\lambda_2) - \varepsilon_o(\lambda_2)\right]R}. \tag{11.20}$$

The term $SpO_2$ is defined as arterial saturation of oxygen, estimated using pulse oximetry.

In actual implementation, a red light emitting diode (LED) near 660 nm and a near-infrared LED in the range of 890−950 nm are used as light sources. Because these LEDs are not monochromatic light sources, the molar absorptivities cannot be directly used in equations (Flewelling, 1995). Instead, Aoyagi used an approximation of the original ratio:

$$R \approx \frac{I_{AC}(\lambda_1)/I_{DC}(\lambda_1)}{I_{AC}(\lambda_2)/I_{DC}(\lambda_2)}, \tag{11.21}$$

where $AC$ refers to the pulsatile component of transmitted light, and $DC$ refers to the non-pulsatile component of transmitted light (Severinghaus, 2007). The use of Eqs. (11.20) and (11.21) are accurate only for $SpO_2$ values from 90−100% (Kelleher, 1989).

## Enabling Technology: Calibration Curve

In 1981, engineer Scott Wilber, who cofounded Biox Corporation (later acquired by Ohmeda), made pulse oximetry accurate by abandoning the Beer-Lambert law (Wilber, 1983). He measured $R$, as defined in Eq. (11.21), in several subjects who had been exposed to various inspired oxygen fractions. During each $R$ measurement, arterial blood was sampled and later analyzed by a CO-oximeter to determine accurate $SaO_2$ values. The resulting $R$-vs.-$SaO_2$ calibration curve was converted to a software lookup table for later real-time $SpO_2$ estimation within a Biox pulse oximeter. Readings could be obtained from an ear or digit probe. A *probe* is the assembly containing two light sources, two light detectors, and a connector cable. A typical calibration curve, given in Figure 11.10, is unique for each pulse oximeter monitor-probe combination.

Biox marketed its pulse oximeters to pulmonary function laboratories. A Biox competitor, Nellcor, realized the limitations of this lab market and chose to market its pulse oximeter monitors and new disposable probes to anesthesiologists. By 1997, Nellcor had about 90% share of this annual $200 million probe market (IMS Health, 1997; Baura, 2002).

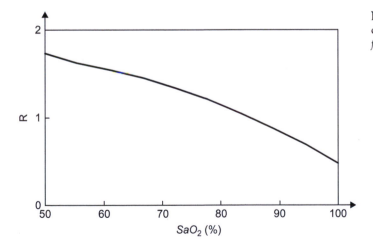

FIGURE 11.10 Sample calibration curve for a pulse oximeter. [*Adapted from ISO (2005)*].

## Enabling Technology: Adaptive Filtering

As pulse oximetry migrated to hospital units other than the OR and ICU, a large limitation emerged. Low signal-to-noise ratio, SNR, greatly affected accurate $SpO_2$ estimation using the calibration lookup table. Low SNR occurs with either high noise or low signal. High noise is present when patients move, which had not occurred when they were anesthetized or sedated. Low noise is present when patients' tissue perfusion is low.

Several pulse oximetry manufacturers attempted to minimize the effect of high noise or low perfusion with empirical signal manipulation. For example, Nellcor's C-LOCK algorithm assumed that a correct ratio could be measured during a short time interval after each QRS complex, if an electrocardiogram were simultaneously acquired with pulse oximetry (Goodman & Corenman, 1990). Because there is no reason for a noiseless ratio to be present during patient motion after each QRS complex, C-LOCK did not minimize the effect of motion artifact (Barker & Shah, 1997).

Electrical engineers Joe Kiani and Mohammed Diab founded Masimo Corporation on the premise that adaptive noise cancellation could minimize the effect of patient motion on pulse oximetry. Because there are two wavelengths, pulse oximetry naturally has two input signals, $u_i(k)$, and two original noise channels, $n_{0i}(k)$. But with only one reference noise source, $n_1(k)$, there is no unique solution for the two original noise sources, $n_{0i}(k)$. The reference noise source is defined as a function of the ratio, $R(k)$:

$$n_1(k) \equiv [u_0(k) + n_{00}(k)] - [u_1(k) + n_{01}(k)]R(k), \tag{11.22}$$

where the summations in Eq. (11.22) are the numerator and denominator of the ratio approximation in Eq. (11.21):

$$[u_0(k) + n_{00}(k)] = \frac{I_{AC}(\lambda_1,k)}{I_{DC}(\lambda_1,k)} \tag{11.23}$$

$$[u_1(k) + n_{01}(k)] = \frac{I_{AC}(\lambda_2,k)}{I_{DC}(\lambda_2,k)}. \tag{11.24}$$

To obtain the unique solution, Masimo's pulse oximeter uses many possible values of the ratio, which correspond to uniformly spaced $SpO_2$ values from about 30% to 100%, to calculate candidate reference noise sources. Each candidate reference noise source is then input to the adaptive noise canceler with the infrared primary input. A corresponding filter output is determined. Masimo assumed that the peak of the output power at the highest saturation corresponds to the true arterial saturation (Diab et al., 1997) (Figure 11.11). This adaptive noise canceler process is repeated once per second, using a 6-s window of $SpO_2$ data (Barker & Shah, 1997).

In a head-to-head clinical comparison of the Nellcor N-200 (which used C-LOCK) and Nellcor N-3000 pulse oximeters, a Masimo prototype system was much more accurate during patient motion. The sensitivity/specificity of alarm detection for the Masimo, Nellcor N-200, and Nellcor N-3000 systems were 100%/100%, 99%/70%, and 84%/64%, respectively. As shown in Figure 11.12, the Masimo prototype could track noiseless reference $SpO_2$ values during tapping and other motion, whereas the Nellcor monitors could not

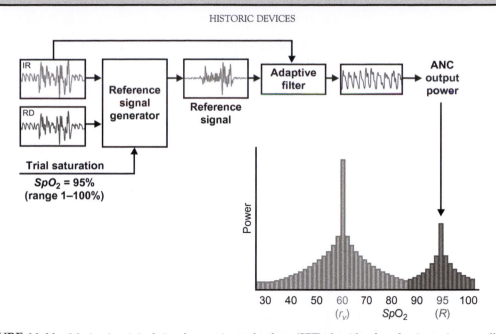

FIGURE 11.11   Masimo's original signal extraction technology (SET) algorithm for adaptive noise cancellation. Ratios, $R$, corresponding to $SpO_2$ values are input to the reference signal generator to output candidate noise sources. IR = infrared wavelength; RD = red wavelength. Each candidate noise source is input to the adaptive filter, with the filter output power then plotted. A peak power corresponding to a lower $SpO_2$ value may correspond to the ratio associated with saturation of venous blood, $r_v$. *[Reproduced with permission from Masimo (2008)].*

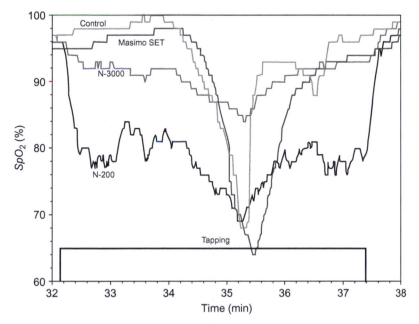

FIGURE 11.12   *$SpO_2$ versus time plot showing a rapid desaturation-resaturation occurring during tapping motion. During this period, a healthy volunteer inspired air from a face mask, in which the fraction of inspired oxygen was adjusted to decrease the control $SpO_2$ values down from 100% to approximately 70% and up to 100% again. N-200 and N-3000 indicate desaturation before it actually occurs. N-200 reads $SpO_2$ near 80% after the control $SpO_2$ value has returned to normal. [Reproduced by permission from Barker & Shah (1997)].*

(Barker & Shah, 1997). Masimo's original adaptive noise cancellation technology is now one of several algorithms that are collectively referred to as *signal extraction technology* (SET).

## SYSTEM DESCRIPTION AND DIAGRAM

A modern pulse oximeter uses Aoyagi's approximation of the ratio (Eq. (11.21)) and Wilber's calibration curve. It may incorporate adaptive noise canceling for minimizing the effects of patient motion and low perfusion.

A disposable or reusable probe is typically positioned on a patient's digit. The clinician powers on the unit, which may run on lead acid batteries or be connected to AC power. An LED driver powers on and off two LEDs at a modulation rate, such as 625 Hz. Typical red and infrared wavelengths used are 660 and 940 nm, respectively. During each modulation cycle, the red LED is powered on, both LEDs are powered off, the infrared LED is powered on, and both LEDs are powered off in sequence. LED light is naturally isolated from the patient.

Light transmitted from an LED and through the finger is received by a photodiode (discussed in detail in Chapter 1). The photodiode current is amplified and undergoes data acquisition. Within the processor module, the digitized data are demodulated to isolate the photoplethysmogram waveform at each wavelength. *Plethysmography* is the measurement of volume. The so-called pleth waveform is misnamed because it is not directly proportional to pulse volume.

The AC and DC components in each waveform are used to calculate the ratio. A lookup table determines the $SpO_2$ value associated with this ratio. Adaptive noise cancellation may also be implemented in the processor module to minimize patient motion and low perfusion effects. The final $SpO_2$ value is transmitted to an LCD for display. Optionally the pleth waveforms may also be displayed (Figure 11.13).

## KEY FEATURES FROM ENGINEERING STANDARDS

As of 2010, the FDA recommends one pulse oximeter consensus standard: *ISO 9919 Medical Electrical Equipment—Particular Requirements for the Basic Safety and Essential Performance of Pulse Oximeter Equipment for Medical Use* (ISO, 2005).

### $SpO_2$ Accuracy

The $SpO_2$ accuracy of a pulse oximeter must be a root-mean-square difference $\leq$ 4.0% $SpO_2$ over the range of 70–100% $SaO_2$. This requirement is verified through clinical study measurements taken over the full range of accuracy claims, complying with clinical study requirements in a separate clinical study standard ISO 14155-2:2003 (ISO, 2003). Functional testers or patient simulators cannot be substituted for validation of accuracy claims. The reference standard for the study is CO-oximeter analysis of

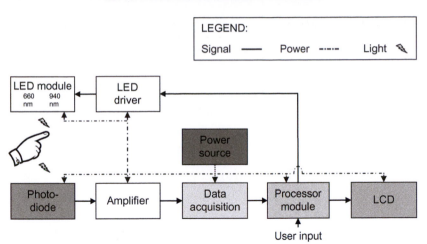

**FIGURE 11.13**   Pulse oximeter system diagram.

simultaneously drawn arterial blood, with at most 1% error. Clinical study protocol parameters such as the number of subjects and exclusion/inclusion criteria are not specified (ISO, 2005).

## Accuracy Under Conditions of Motion

A manufacturer is not required to claim accuracy during patient motion. However, if a manufacturer makes this claim, then accuracy specification during motion must be disclosed in the instructions for use. A summary of test methods used to establish these claims must be disclosed in the technical description (ISO, 2005).

Currently, this requirement applies to only a few manufacturers. In 1995, Nellcor debuted its motion artifact–resistant pulse oximeter, which was based on Kalman filtering. Kalman filtering is a type of adaptive filter (Haykin, 1991). Masimo then sued Nellcor for patent infringement of its signal extraction technology. In their 2006 settlement agreement, Nellcor agreed to pay $265 million to Masimo for sales of infringing products and $65 million in advance royalties for products sold in 2006 (Perkes, 2006).

## Accuracy Under Conditions of Low Perfusion

A manufacturer is not required to claim accuracy during low perfusion. However, if a manufacturer makes this claim, then accuracy specification during low perfusion must be disclosed in the instructions for use. A summary of test methods used to establish these claims must be disclosed in the technical description (ISO, 2005). Again, this requirement applies to only a few manufacturers.

## Signal Inadequacy Indication

When the $SpO_2$ or pulse rate value on a pulse oximeter display is potentially incorrect, an indicator of signal inadequacy must be provided to the operator. Valid indicators include a visual information signal, a low-priority alarm signal, and a non-normalized (noisy) waveform. A normalized waveform does not satisfy this requirement because it likely masks an unreliable signal. A description of the indicator used must be provided in accompanying documents (ISO, 2005).

## Protection From Excessive Temperatures

In normal condition and single-fault condition, the maximum power delivered to the energized pulse oximeter probe must be insufficient to produce a temperature exceeding 41°C at the pulse oximeter probe–tissue interface, when the skin temperature is initially 35°C. Compliance is verified by measuring the maximum surface temperature of the pulse oximeter probe–tissue interface under normal and single fault-conditions. This testing is conducted using the procedure disclosed in the manufacturer's technical description.

The source of this requirement is consideration of infant skin temperature safety. Strong local perfusion can lead to a skin temperature of 35°C. Although a review of literature demonstrated that 42°C may be entirely safe for infants, there were enough conflicting results to lower the maximum temperature requirement to 41°C (ISO, 2005).

## RESPIRATION MONITORS

As already noted, respiration rate is one of the five vital signs. The normal adult range of respiration rate is 12–18 breaths/min. Respiration rate is easily measured during electrocardiogram (ECG) acquisition.

When an ECG is acquired, at least two surface electrodes measure a surface biopotential across the chest. An artifact of this biopotential is a large baseline voltage that changes with the impedance of the lungs during breathing. Normally, this artifact is minimized in the ECG waveform by highpass filtering. But if the acquired ECG is instead sent through a lowpass filter, the respiration waveform is isolated. The respiration waveform is normally much higher in amplitude than the ECG waveform (Figure 11.14).

In patient monitors, such as the Philips MP50 in Figure 7.1, the ECG waveform and respiration waveform appear when the ECG cable is first attached. In the Electrocardiograph Filtering Lab in Chapter 22, you will practice respiration waveform filtering.

## SUMMARY

A pulse oximeter is an instrument that estimates and displays the arterial saturation of oxygen. Originally used in the operating rooms of hospitals, pulse oximeters migrated to intensive care units and then to patient clinics.

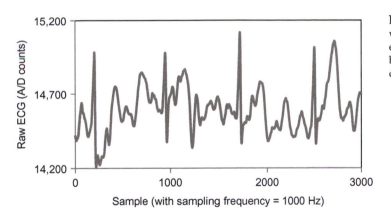

**FIGURE 11.14** A respiration waveform may be filtered from a raw electrocardiogram. Note the varying baseline, which is due to impedance changes in the lung.

Pulse oximetry has been a standard of care since 1986; it allows anesthesiologists to ensure that oxygen is being adequately transported to the tissues during mechanical ventilation. This increase in patient safety is based on the relationship between arterial saturation of oxygen and partial pressure of arterial oxygen, which is known as the oxyhemoglobin dissociation curve.

Originally based on the Beer-Lambert law, pulse oximetry estimation of arterial saturation of oxygen became accurate in the range of 90–100% when Takuo Aoyagi realized that only pulsatile changes in light transmission were necessary for estimating hemoglobin concentrations. Scott Wilber replaced the Beer-Lambert law with a calibration curve, which enabled accurate estimates throughout the range of $SaO_2$. Joe Kiani and Mohammed Diab applied adaptive noise cancellation to pulse oximetry, for accurate estimation in the presence of reduced signal-to-noise ratio.

Key pulse oximetry features include $SpO_2$ accuracy, accuracy under conditions of motion, accuracy under conditions of low perfusion, signal inadequacy indication, and protection from excessive temperatures.

A respiration waveform can be isolated from lowpass filtering of the ECG waveform. It results from lung impedance artifact, due to breathing, that occurs while the ECG is acquired with surface electrodes.

## Exercises

**11.1** Does pulse oximetry estimation of $SaO_2$ under high signal-to-noise ratio conditions involve the Beer-Lambert law? How would the system diagram in Figure 11.13 change if the Beer-Lambert law were used? What components are part of a processor module? Draw a system diagram for a respiration monitor.

**11.2** Examine Bose QuietComfort 15 headphones and its owner's guide, which your instructor has provided. How are the signal input and reference noise source acquired by the circuitry? How is the filtered output transmitted to the user? Design a protocol to test the minimum signal level that you can listen to without hearing a loud, constant-level external noise source. Would this minimum signal level be the same for you and your classmates? Give reasons.

**11.3** Plot the left and right pleth waveforms 11_1 and 11_2 that have been provided by your instructor. Waveform characteristics are documented in the associated header11_1.txt ascii file. The first 30 s of each waveform were acquired without noise; the last 30 s of each waveform were acquired during patient motion. How are the left noise and right noise related? Be specific.

**11.4** What are the advantages of using a disposable, rather than a repeated-use, digit probe? What are the disadvantages? While the majority of pulse oximetry probes are digit probes, some manufacturers have sold probes for forehead use. Do you believe that forehead probes provide accurate readings? Give reasons.

Read Wilber (1983):

**11.5** Describe how the circuitry at the top of Figure 1 works. What specific signals are input into the A/D converter (72)?

**11.6** Assume we modify the circuitry so that the output of the current-to-voltage converter is immediately digitized and sent to an added digital processor. What digital operations would need to be performed to have an output signal that is equivalent to the signal output of the A/D converter (72)? Draw a system diagram that represents these operations.

**11.7** What is the calibration equation given in the patent? Explain how this patent equation is related to Figure 11.10 in this chapter.

**11.8** Claim 1 is an independent claim because other claims (Claims 2–5) depend on it. Is Claim 1 broad or specific in scope? Is a broad or specific claim preferable to a patent assignee? Give reasons.

# References

Aoyagi, T. (2003). Pulse oximetry: Its invention, theory, and future. *Journal of Anesthesiolgy, 17*, 259–266.

ASA (1986). *Standards for basic anesthetic monitoring*. Park Ridge, IL: AHA.

Barker, S. J., & Shah, N. K. (1997). The effects of motion on the performance of pulse oximeters in volunteers (revised publication). *Anesthesiology, 86*, 101–108.

Baura, G. D. (2002). *System theory and practical applications of biomedical signals*. Hoboken, NJ: Wiley-IEEE Press.

Birnbaum, S. (2009). Pulse oximetry: Identifying its applications, coding, and reimbursement. *Chest, 135*, 838–841.

Diab, M. K., Kiani-Azarbayjany, E., Elfadel, I. M., & McCarthy, R. J., et al. (1997). Signal processing apparatus. *US 5,632,272.*

Flewelling, R. (1995). Noninvasive optical monitoring. In J. D. Bronzino (Ed.), *The biomedical engineering handbook*. Boca Raton, FL: CRC Press.

Goodman, D. E., & Corenman, J. E. (1990). Method and apparatus for detecting optical signals. *US 4,928,692.*

Guyton, A. C., & Hall, J. E. (2006). *Textbook of medical physiology* (11th ed.). Philadelphia: Elsevier Saunders.

Haykin, S. S. (1991). *Adaptive filter theory* (2nd ed.). Englewood Cliffs, NJ: Prentice Hall.

IMS Health. (1997). *Hospital supply index*. Plymouth Meeting, IMS America.

ISO (2003). *ISO 14155-2:2003 Clinical investigation of medical devices for human subjects—Part 2: Clinical investigation plans*. Geneva: ISO.

ISO (2005). *ISO 9919:2005 Medical electrical equipment—Particular requirements for the basic safety and essential performance of pulse oximeter equipment for medical use*. Geneva: ISO.

Kelleher, J. F. (1989). Pulse oximetry. *Journal of Clinical Monitoring, 5*, 37–62.

Mackenzie, N. (1985). Comparison of a pulse oximeter with an ear oximeter and an in-vitro oximeter. *Journal of Clinical Monitoring, 1,* 156–160.

Masimo (2008). *Signal extraction technology technical bulletin.* Irvine, CA: Masimo Corporation.

Perkes, C. (2006). Standing up to the giant—How Masimo won its medical-device patent fight with mighty Tyco and collected $330 million. *Orange County Register* (February 9).

Severinghaus, J. W. (2007). Takuo Aoyagi: Discovery of pulse oximetry. *Anesthesia and Analgesia, 105,* S1–4.

Severinghaus, J. W., & Astrup, P. B. (1986). History of blood gas analysis. VI. Oximetry. *Journal of Clinical Monitoring, 2,* 270–288.

Severinghaus, J. W., & Honda, Y. (1987). History of blood gas analysis. VII. Pulse oximetry. *Journal of Clinical Monitoring, 3,* 135–138.

Widrow, B., & Hoff, M., Jr. (1960). Adaptive switching circuits. *IRE WESCON Conv Rec, Part 4.*

Widrow, B., & Stearns, S. D. (1985). *Adaptive signal processing.* Englewood Cliffs: Prentice-Hall.

Wilber, S. (1983). Blood constituent measuring device and method. *US 4,407,290.*

CHAPTER

# 12

# Thermometers

In this chapter, we discuss the relevant physiology, history, and key features of thermometers. A *thermometer* is an instrument that estimates and displays core temperature. Thermometers are ubiquitous and are used in homes, physician clinics, and hospitals (Figure 12.1). All FDA-cleared thermometers are assumed to accurately display core temperature.

Upon completion of this chapter, each student shall be able to:

1. Understand the mechanisms underlying temperature regulation.
2. Critique the accuracy of various types of thermometers.
3. Describe five key features of thermometers.

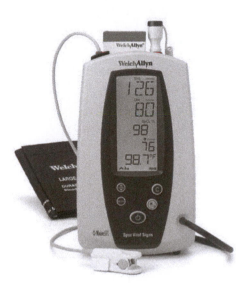

**FIGURE 12.1** Welch Allyn Spots vital signs monitor integrates its SureTemp thermometer, oscillometric blood pressure measurement, and a Masimo pulse oximeter into a single unit (Welch Allyn, Skaneateles, New York).

## THERMOREGULATION PHYSIOLOGY

In Chapter 11, we discussed two vital signs related to pulmonary physiology: arterial saturation of oxygen and respiratory rate. Temperature is covered in this chapter to complete the discussion of the five vital signs. As with body fluid, core temperature is regulated for homeostasis. Under typical conditions, an approximate interthreshold zone of core temperature, 36.0–37.5°C (96.8–99.5°F), is preserved (Figure 12.2).

Core temperature is most accurately measured using a thermistor in an indwelling catheter, such as a pulmonary artery catheter. As mentioned in Chapter 8, pulmonary artery catheter use has declined in recent years, due to increased mortality. Alternatively, core temperature is estimated with reasonable accuracy using a thermistor probe positioned in the rectum (O'Grady et al., 2008).

### Mechanisms of Heat Loss

Body heat is generated in the *core*, that is, the deep organs, such as the liver, brain, and heart, and in the skeletal muscles during exercise. Heat from blood vessels is transferred to the skin, where it radiates to the air and other surroundings.

As shown in Figure 12.3, a nude person sitting inside at normal room temperature loses heat through radiation, conduction, and evaporation. About 60% of total heat is lost through radiation, as infrared (IR) heat rays of wavelengths 5–20 microns radiate from the body. Assuming that body temperature is higher than surrounding wall temperature, more heat radiates from, rather than to, the body. About 3% of total heat is lost through conduction, due to direct contact with surfaces. Approximately 15% of total heat is lost through direct contact with air, with the heat then carried away by convection air currents. Finally, about 22% of total heat is lost through evaporation of water from the skin. Even without sweating, water evaporates from the skin and lungs at about 600–700 mL/day.

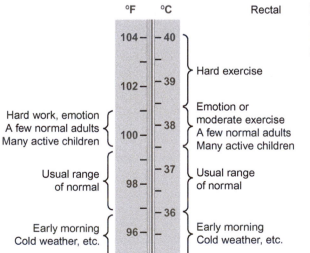

FIGURE 12.2 Estimated interthreshold zone of body core temperature in normal people. [Redrawn from DuBois (1948). *Reproduced by permission from Guyton & Hall (2006)*].

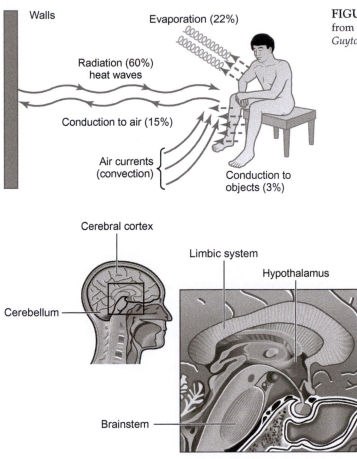

FIGURE 12.3 Mechanisms of heat loss from the body [*Reproduced by permission from Guyton & Hall (2006)*].

FIGURE 12.4 The hypothalamus is part of the diencephalon, which is part of the brain stem. (*Adapted from A.D.A.M., Atlanta, Georgia*).

As a result, skin temperature affects core temperature. The addition of clothing decreases the flow of convection air current and rate of heat loss (Guyton & Hall, 2006).

## Thermoregulation Anatomy

If thermoregulation did not occur, then ambient temperature would affect core temperature. However, the hypothalamus responds to thermosensor data and keeps core temperature within the interthreshold zone.

A *thermosensor* is a neuron that detects cold or warmth. Central, or brain, thermosensors are predominantly warmth-sensitive because the interthreshold zone is positioned very closely to the human upper survival limit of 40°C (104°F). These thermosensors reside in the hypothalamus and other locations. The hypothalamus, along with the thalamus, epithalamus, and subthalamic nucleus, make up the *diencephalon*, which is part of the brain stem (Figure 12.4). Although only a few cubic centimeters in volume, the hypothalamus controls most vegetative and endocrine functions, including body temperature. The

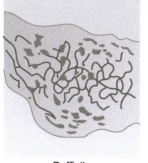

**FIGURE 12.5**  Ruffini's end-organ thermosensor *[Reproduced by permission from Guyton & Hall (2006)].*

Ruffini's
end-organ

increased activity of warmth-sensitive neurons in the preoptic anterior hypothalamus triggers heat-defense responses, whereas decreased activity triggers cold-defense responses.

Peripheral thermosensors reside in the skin and in the oral and urogenital mucous membranes. An example of a skin thermosensor is Ruffini's end organ, which is warmth-sensitive (Figure 12.5). Most skin thermosensors are cold-sensitive, in order to prevent hypothermia. Deep-body thermosensors are located in the esophagus, stomach, and large intraabdominal veins; they respond to core body temperature (Romanovsky, 2007).

Afferent thermosensor activity is transported to the hypothalamus via the sympathetic nervous system (Figure 12.6). The *sympathetic nervous system* is that part of the autonomic nervous system that controls most visceral (internal organ) functions of the body. This system also transports efferent hypothalamic signals to the blood vessels, skin, and skeletal muscles. Neuronal signaling activity is discussed in Chapter 14.

## Thermoregulation

When a change in ambient temperature occurs, the hypothalamus seeks to maintain core temperature in the interthreshold zone through a vasomotor response. In warm temperatures, blood vessels *vasodilate* to increase heat loss. In cold temperatures, they *vasoconstrict* to decrease heat loss (Figure 12.7).

If the vasomotor response is unable to maintain stable core temperature, the hypothalamus induces an appropriate autonomic response of sweating or shivering. As shown in Figure 12.8, different temperatures trigger sweating for evaporative heat loss and shivering for heat production (Mekjavic & Eiken, 2006).

In past decades, scientists believed that there was a specific set point for temperature. However, the independence of anatomically distinct thermoeffectors argues against this engineering notion of a control system with an external reference signal (Romanovsky, 2007).

## SKIN TEMPERATURE VERSUS CORE TEMPERATURE

As we have seen, skin temperature changes with ambient temperature and does not necessarily reflect core temperature. In another condition of hemorrhagic shock, skin

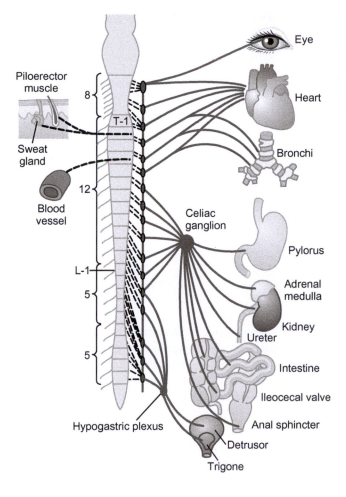

**FIGURE 12.6** Sympathetic nervous system. The black dashed lines represent postganglionic fibers in the gray rami leading from the sympathetic chains into spinal nerves for distribution to blood vessels, sweat glands, and piloerector muscles [*Reproduced by permission from Guyton & Hall (2006)*].

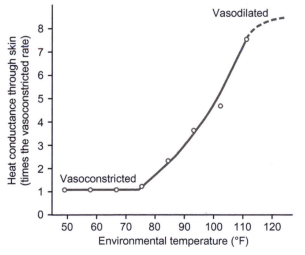

**FIGURE 12.7** Effect of changes in the environmental temperature on heat conductance from the body core to the skin surface. [*Modified from Benzinger (1980.) Reproduced by permission from Guyton & Hall (2006)*].

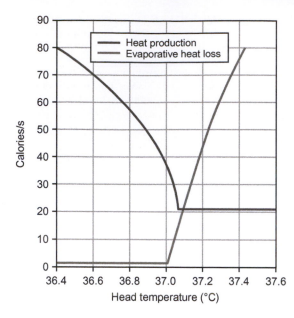

**FIGURE 12.8**   Effect of hypothalamic temperature on evaporative heat loss from the body and on heat production caused primarily by muscle activity and shivering. The temperature level at which increased heat loss begins and heat production reaches a minimum stable level is extremely critical *[Reproduced by permission from Guyton & Hall (2006)].*

temperature also does not correlate with core temperature. Decreased arterial pressure due to decreased blood volume stimulates sympathetic vasoconstriction and therefore cold skin throughout the body. As a result, systemic vascular resistance is increased, raising arterial pressure, and veins constrict to help maintain adequate venous return. Therefore, the image of a mother feeling the temperature of her child's forehead to determine fever should not be the basis of a medical device.

## CLINICAL NEED

Thermometry is used to diagnose high temperature conditions, such as fever and heat stroke, and low temperature exposure.

*Fever* is defined by the Society of Critical Care Medicine as a core temperature ≥ 38.3°C (O'Grady et al., 2008). Fever may be caused by an abnormality in the brain, such as a lesion or tumor; by a bacterial infection, by a viral infection, by the initial 48 hr after surgery, by drug administration, by environmental heat, or by other factors. Upon the diagnosis of fever by a clinician, appropriate therapy can be administered.

*Heat stroke* is defined clinically as a core temperature >40°C. It results from exposure to high environmental temperatures or from strenuous exercise. Due to climate change, heat waves in urban areas in the United States have increased. In a recent epidemiologic study, the incidence of heat stroke varied from 17.6 to 26.5 cases per 100,000 population. The very young or elderly, the poor, the socially isolated, and those without access to air conditioning are most at risk.

During the progression of heat stroke, the hypothalamus first responds with cutaneous vasodilation and splanchnic vasoconstriction. The immunologic and barrier functions of the intestines are altered, which allows the leakage of endotoxins. Multiorgan dysfunction and central nervous system abnormalities, such as delirium, convulsions, or coma, result. After a hospital diagnosis of core temperature >40°C, clinicians work to move core temperature into the normal range and to blunt the physiologic response. Cooling occurs through cold packs, cooling blankets, evaporative cooling with misting and fans, or catheter cooling (Bouchama & Knochel, 2002).

Without intervention, low-temperature exposure may cause core temperature to fall below 29.4°C (84.9°F). When this occurs, the hypothalamus is no longer able to regulate temperature. Now the rate of chemical heat production in each cell is depressed almost twofold for each 10°F decrease in core temperature. Eventually, sleepiness and then coma develop (Guyton & Hall, 2006).

## HISTORIC DEVICES

The clinical measurement of temperature can be traced to the fourth century BCE. Hippocrates used his hand to detect temperature changes of the human body.

### Early Devices

Many centuries later, in 1665, Christian Huygens added a scale to the glass thermometer designed by Sanctorio Sanctorius. This thermometer was based on the expansion of water and was used in the mouth. Soon thereafter, Gabriel Daniel Fahrenheit modified the glass thermometer for use with his new Fahrenheit scale and mercury, which expanded more readily than water or alcohol.

### Enabling Technology: Six-Inch Mercury Thermometer

In 1866, physician Thomas Clifford Allbutt redesigned the mercury thermometer to decrease its length from 12 to 6 in. Correspondingly, the time until the mercury reached a steady-state temperature decreased from 20 to 5 min. Allbutt's design was manufactured by Harvey & Reynolds (Figure 12.9). Using his thermometer, Allbutt was able to conduct clinical studies, mainly of arterial and nervous disorders.

As discussed in Chapter 1, this simple instrument became electronic when engineer Heinz Georgi, a founder of IVAC Corporation, invented the analog electronic thermometer in 1970 (Figure 12.10). Based on readings from a thermistor in its probe and analog circuitry, this thermometer required 30 s to predict and display the steady-state temperature (Georgi, 1972). In their SureTemp digital electronic thermometer in 1994, engineers Tom Gregory and John Stevenson of Diatek decreased the steady state temperature time further from 30 to 4–6 s by preheating the probe (Figure 1.4) (Gregory & Stevenson, 1997). Diatek was later acquired by Welch Allyn.

The success of the electronic thermometer inspired another thermometer modality based on thermal infrared detection. In 1984, engineers Gary O'Hara and David Phillips of

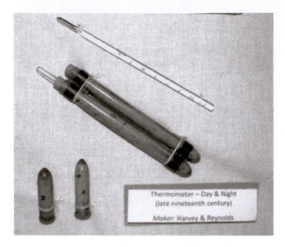

**FIGURE 12.9** Allbutt thermometer, manufactured by Harvey & Reynolds (University of Leeds Museum of the History of Science, Technology, and Medicine, Leeds, United Kingdom).

**FIGURE 12.10** IVAC 811 thermometer. The probe is sitting in its holder. Through several mergers and acquisition, IVAC is part of CareFusion, San Diego, California (Photograph taken by Scientific Equipment Liquidators, Big Lake, Minnesota).

Intelligent Medical Systems used an infrared detector in the ear canal to estimate core temperature (O'Hara & Phillips, 1986).

## SYSTEM DESCRIPTIONS AND DIAGRAMS

Mercury and analog electronic thermometers have been replaced by more modern versions. As discussed in Chapter 1, toxic mercury thermometers have been replaced by thermometers containing nontoxic galinstan, which is a gallium-indium-tin alloy (Figure 1.3B). Analog electronic thermometers have been replaced by faster digital thermometers (Figures 1.4, 12.1).

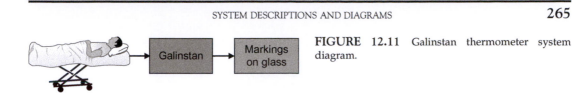

FIGURE 12.11 Galinstan thermometer system diagram.

## Simple Thermometer

A galinstan thermometer is a simple instrument whose measurand is patient temperature. The transducer is the galinstan contained in the glass tube, which expands with temperature. Temperature scale markings on the glass tube act as the display for patient temperature (Figure 12.11).

## Digital Electronic Thermometer

Digital electronic thermometers are an improvement over older analog electronic thermometers. As in analog thermometers, the thermistor in the probe is isolated from the patient by a probe cover. The probe cover prevents cross-contamination between patients. The probe, with probe cover, is inserted into the patient's sublingual pocket in the mouth, rectum, or axilla (armpit).

As discussed in Chapter 1, the thermistor resistance, $R_T(T)$, varies with temperature, $T$:

$$R_T(T) = \exp\left(A_0 + \frac{A_1}{T} + \frac{A_2}{T^2} + \frac{A_3}{T^3}\right),$$ (12.1)

where $T$ = temperature and $A_0$, $A_1$, $A_2$, and $A_3$ are specified by the manufacturer of a hospital-grade thermometer. A less precise temperature-resistance relationship may be used by manufacturers of consumer-grade thermometers. To decrease the time until the steady-state temperature is reached, a heating circuit preheats the probe when the user powers on the thermometer.

Responding to temperature changes, the thermistor circuit output is amplified, digitized, and transmitted to the processor module. To further decrease the time until temperature display, a prediction algorithm is used. Based on initial temperature readings, the processor module predicts the steady-state temperature and sends the prediction to an LCD display. An alkaline battery for the power source ensures that only low voltage is present. The system diagram for a digital electronic thermometer is given in Figure 12.12. With the Welch Allyn SureTemp, this preheated probe and prediction algorithm combination result in oral temperature estimates within 4–6 s.

For galinstan and digital thermometers, care should be taken to obtain an accurate temperature reading. Heat is transferred to the sensor through conduction. For an oral temperature reading, the thermometer or probe should be inserted into the sublingual pocket, beneath the tongue. For proper conduction, the patient should not talk during the measurement. Additionally, the patient should not have eaten, drunk, chewed gum, or smoked within 15 min of the measurement. Similarly, for a rectal temperature reading, the thermometer or probe should be inserted 1 in. into the rectum, using petroleum jelly. The presence of fecal material can affect conduction. In both cases, the sensor should not be removed until the steady-state temperature is reached. For a simple thermometer,

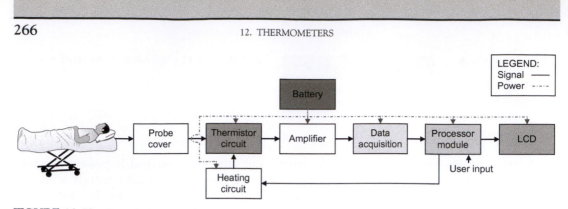

**FIGURE 12.12**   Digital electronic thermometer system diagram.

5 min. should pass. For a digital thermometer, the time to remove the probe is indicated on the display.

Although simple and digital thermometers have been used in the axilla, this location is not recommended. As we have stated, skin temperature does not consistently reflect core temperature.

## Infrared Thermometer

Alternatively, a thermal infrared detector may be used to sense the flux in body radiation, which peaks around the 8-micron wavelength. Typically, a pyroelectric sensor, composed of pyroelectric film and circuitry in a housing filled with dry air or nitrogen, is used for clinical thermometry. Infrared radiation changes the temperature of a thin foil of pyroelectrical material, such as polyvinylidene fluoride. This causes electrical polarization, which generates an electrical charge (Fraden, 1989). A separate thermistor, in thermal equilibrium with the pyroelectric sensor, is also used to determine ambient temperature.

This infrared detector enables us to measure electric flux through an area, that is, the electric field multiplied by the area of the surface projected perpendicularly to the field. Based on the Stefan-Boltzmann equation, the net infrared flux over a broad spectral range, $\Phi_p$, in units of V-m, may be determined as

$$\Phi_p = A\sigma\varepsilon_p\varepsilon_s(T_p^4 - T_s^4), \tag{12.2}$$

where $p$ = patient subscript, $s$ = sensor subscript, $A$ = optical coefficient, $\sigma$ = Stefan-Boltzmann constant, $\varepsilon$ = emissivity, $T$ = surface temperature in degrees Kelvin. *Emissivity* is the relative ability of a material to emit radiation energy. Rearranging the equation yields the patient surface temperature:

$$T_p = \sqrt[4]{T_s^4 + \frac{\Phi_p}{A\sigma\varepsilon_p\varepsilon_s}}. \tag{12.3}$$

Equations (12.2) and (12.3) assume that an absolute blackbody is radiating heat. A blackbody (Figure 12.13) is defined as a reference source of infrared radiation, made in the shape of a cavity and characterized by precisely known temperature of the cavity walls. It

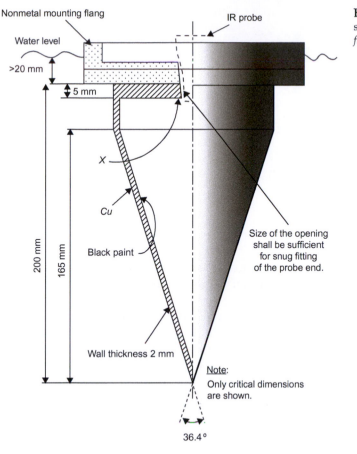

**FIGURE 12.13** A blackbody immersed in water [*Reproduced by permission from ASTM (2009)*].

has an effective emissivity at the cavity opening arbitrarily considered equal to unity. The blackbody cavity is fabricated of a metal with a high thermal conductivity, such as oxygen-free copper. The interior of the cavity is painted with organic enamel paint, with a thickness after drying of 20–50 microns. The cavity is secured to a surface box of low thermal conductivity, such as ABS plastic (ASTM, 2009).

Because a patient's emissivity is not unity but is affected by ambient temperature, the emissivity error in temperature, $e$, may be calculated as

$$e = T_p - \sqrt[4]{\frac{\varepsilon_0}{\varepsilon_p}(T_p^4 - T_a^4) + T_a^4}, \tag{12.4}$$

where $a$ = ambient subscript and $\varepsilon_o$ = original blackbody emissivity (ASTM, 2009). Another source of error is the imperfect transfer of radiation from the anatomic site to the sensor.

Common infrared thermometers sense radiation from the tympanic membrane in the ear or forehead (Figure 12.14). The rationale for using the tympanic membrane is its close proximity to the internal carotid artery, which supplies the hypothalamus. The

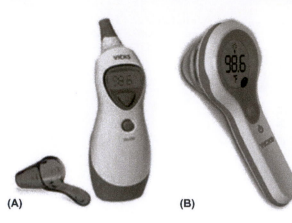

FIGURE 12.14  Infrared thermometers. **A**: Vicks ear thermometer. **B**: Vicks forehead thermometer (Kaz Incorporated, Southborough, Massachusetts).

**(A)**                                    **(B)**

rationale for using the forehead temple is that the superficial temporal artery possesses a high arterial perfusion rate.

A probe is inserted into the ear canal or swept over the temple and other parts of the forehead. A probe cover is typically used to prevent cross-contamination. The probe cover is not necessary for isolation because the pyroelectric sensor at the end of the probe does not make patient contact. The probe acts as a waveguide to direct radiation to the pyroelectric sensor, when the processor module opens the shutter assembly.

The pyroelectric sensor output and ambient temperature thermistor circuit output are amplified, digitized, and transmitted to the processor module. There, initial temperature estimates are calculated using the Stefan-Boltzmann equation. These initial temperature estimates are then compensated for emissivity and other errors. The final temperature estimates are transmitted to an LCD display. An alkaline battery for the power source ensures that only low current is present. The system diagram for an infrared thermometer is given in Figure 12.15.

## KEY FEATURES FROM ENGINEERING STANDARDS

As of 2010, the FDA recommends three consensus standards related to electronic or infrared thermometers. The electronic thermometer standard is *ASTM E1112-00 Standard Specification for Electronic Thermometer for Intermittent Determination of Patient Temperature* (ASTM, 2006). The infrared thermometer standard is *ASTM E1965-98 Standard Specification for Infrared Thermometer for Intermittent Determination of Patient Temperature* (ASTM, 2009). The probe cover standard for either type of thermometer is *ASTM E1104-98 Standard Specification for Clinical Thermometer Probe Covers and Sheaths* (ASTM, 2003).

### Electronic Thermometer Accuracy

Per ASTM E1112-00, any individual reading within the manufacturer's specified temperature range must not deviate by more than a maximum allowable error. In the range of 37.0–39.0°C, the maximum error is ±0.1°C. In the ranges 35.8–36.9°C and 39.1–41.0°C,

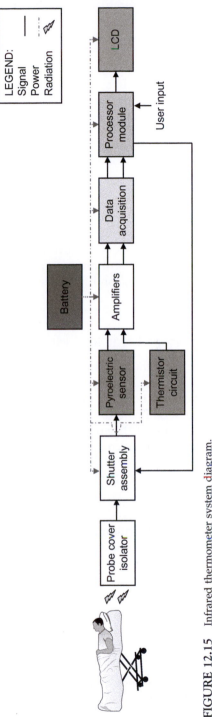

**FIGURE 12.15** Infrared thermometer system diagram.

the maximum error is ±0.2°C. For temperatures <35.8°C or >41.0°C, the maximum error is ±0.3°C.

To test this requirement, a well stirred constant temperature liquid bath, of at least 1 L, is used. The temperature of the bath is accurately calibrated within ±0.03°C on the International Temperature Scale of 1990, as verified by a system whose calibration is traceable to the National Institute of Standards and Technology or other appropriate national standards laboratory. ASTM E1112-00 does not call out specific temperatures for testing, only that the "accuracy requirement be met" (ASTM, 2006).

## Maximum Permissible Laboratory Error for Ear IR Thermometer

Per ASTM E1965-98, a reading within the manufacturer's specified temperature range must not deviate by more than a maximum allowable error. In the range of 36.0–39.0°C, the maximum error is ±0.2°C. For temperatures <36.0°C or >39.0°C, the maximum error is ±0.3°C.

To test this requirement, the blackbody of Figure 12.13 is reliably secured to a water bath, with a volume of at least 2 L. The temperature of the bath is accurately calibrated within ±0.03°C by an immersed contact thermometer, for which the calibration is traceable to a national physical standard. Testing is conducted for the temperatures 35, 37, and 41°C (ASTM, 2009).

## Maximum Permissible Laboratory Error for Skin IR Thermometer

Per ASTM E1965-98, a reading within the manufacturer's specified temperature range must not deviate by more than ±0.3°C.

To test this requirement, the blackbody of Figure 12.13 is reliably secured to a water bath, with a volume of at least 2 L. The temperature of the bath is accurately calibrated within ±0.03°C by an immersed contact thermometer for which the calibration is traceable to a national physical standard. Testing is conducted for the temperatures 23, 30, and 38°C (ASTM, 2009).

## IR Thermometer Clinical Accuracy

The test conditions for electronic thermometers simulate actual conduction conditions during clinical use. In contrast, the test conditions for infrared thermometers do not simulate actual radiation conditions during clinical use. Per ASTM E1965-98, if claims of ear canal (but not skin) clinical accuracy are made, they must be accompanied by the disclosure of methodology and procedures. The manufacturer makes the disclosure available upon request. No specific method for determining clinical accuracy is required (ASTM, 2009).

## Probe Cover Physical Integrity

Per ASTM E1104-98, a clinical thermometer probe cover must be constructed and packaged so that its physical integrity is maintained when applied to, used with, and removed from its thermometer.

To test this requirement, a probe cover, placed and removed from its thermometer according to manufacturer's instructions, is subjected to 8.4 kPa pressure. The pressure is applied for the time interval intended for patient contact, while submerged in water. The water is within the operating temperature environment. No continuous bubble stream must be observed within 5 s (ASTM, 2003).

## FDA CONSENSUS STANDARDS FOR ACCURACY

Many people assume that a medical device that has gone through FDA submission and received either 510(k) notification or premarket approval (PMA) must be accurate and efficacious. In the case of a 510(k)-cleared device, this may or may not be true. A *510(k)* is the section of the Federal Food, Drug and Cosmetic Act that established the FDA's authority to regulate the introduction of medical devices into interstate commerce. By law, a manufacturer requests 510(k) notification, or clearance, for a device that is substantially equivalent in performance to the performance of a legally marketed (predicate) device. As part of a 510(k) submission, a manufacturer generally declares conformity to FDA-recognized consensus standards. The differences between and requirements for 510(k) and PMA submissions are beyond the range of this discussion.

In the case of clinical thermometers, we have observed that the consensus standard accuracy requirements are inconsistent among thermometer types. For an electronic thermometer, a normal temperature of 37.0°C may be measured with an error of ±0.1°C, which equates to an error of ±0.27%. For an ear IR thermometer, a normal temperature of 37.0°C may be measured with an error of ±0.2°C, which equates to an error of ±0.54%. For a skin IR thermometer, a normal temperature of 37.0°C may be measured with an error of ±0.3°C, which equates to an error of ±0.81%.

All these accuracy requirements apply to measurements in a water bath. Although a water bath simulation resembles contact sublingual pocket or rectal measurements, the blackbody in water bath simulation does not resemble ear IR measurement. As shown in Figure 12.16, radiation from the tympanic membrane does not reach the probe and IR sensor directly because the ear canal is curved. This natural physiologic variation degrades ear IR measurement accuracy further. Based on a meta-analysis of 44 studies containing 58 comparisons of 5935 children, the pooled mean temperature difference (ear minus rectal) was −0.29°C, with 95% limits of agreement from −1.32 to +0.74°C (Craig et al., 2002). For a subset of 4098 children with relevant data for further analysis, the pooled sensitivity and pooled specificity for detecting fever, defined as at least 38°C, were 70.8% and 63.7%, respectively (Dodd et al., 2006).

Similarly, the blackbody in water bath simulation does not resemble skin IR measurement. A skin IR thermometer assumes that the radiation from the superficial temporal artery can be accurately measured. Even if this artery possesses a consistently high arterial perfusion rate, the radiation from this artery is affected by skin temperature. As already detailed, skin temperature changes with ambient temperature and does not reflect core temperature during hemorrhagic shock.

In a small study, temporal artery thermometer and pulmonary artery thermistor measurements were simultaneously acquired for 3 hr in 15 adults and 16 children (ages

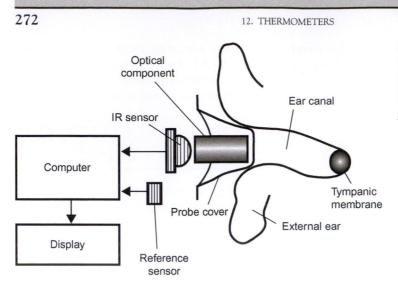

**FIGURE 12.16** The interaction between an infrared (IR) thermometer probe and probe cover and the ear canal and tympanic membrane *[Reproduced by permission from ASTM (2009)].*

9 days—13 yr) recovering from cardiopulmonary bypass. Initially, these patients are hypothermic but often develop fever within hours. The mean difference between 294 adult infrared and core (pulmonary artery catheter) measurements was $1.3 \pm 0.6°C$. The mean difference between 246 pediatric infrared and core measurements was $0.3 \pm 0.5°C$. The increased error in adult measurements may be due to a relatively thicker layer of skin over the artery, compared with pediatric skin (Suleman et al., 2002).

As this chapter was written in 2010, the FDA's Center for Devices and Radiologic Health was beginning to detail long awaited changes to the 510(k) process for 2011. The accuracy of thermometers cleared through this process may, and hopefully will, be improved.

## SUMMARY

Core temperature is regulated for homeostasis. Under typical conditions, an approximate interthreshold zone of core temperature, $36.0-37.5°C$ (or $96.8-99.5°F$), is preserved. When a change in ambient temperature occurs, the hypothalamus seeks to maintain core temperature in the interthreshold zone through a vasomotor response. In warm temperatures, blood vessels vasodilate to increase heat loss. In cold temperatures, blood vessels vasoconstrict to decrease heat loss. If the vasomotor response is unable to maintain stable core temperature, the hypothalamus induces an appropriate autonomic response of sweating or shivering. Different temperatures trigger sweating for evaporative heat loss and shivering for heat production.

Thermometry is used to diagnose high-temperature conditions, such as fever and heat stroke, and low-temperature exposure. A digital thermometer uses thermistor resistance that varies with temperature. An infrared thermometer uses a pyroelectric sensor, composed of pyroelectric film and circuitry within a housing filled with dry air or nitrogen, to generate an electrical charge in response to a change in its temperature. Common infrared thermometers sense radiation from the tympanic membrane in the ear or forehead.

Key thermometer features include electronic thermometer accuracy, ear IR thermometer accuracy, skin IR thermometry accuracy, clinical accuracy, and probe cover physical integrity. FDA consensus standard accuracy requirements vary among thermometer types.

## Acknowledgment

The author wishes to thank Dr. Paul Benkeser for his review of this chapter. Dr. Benkeser is Associate Chair for Undergraduate Studies and Professor, Biomedical Engineering, Georgia Institute of Technology.

### Exercises

**12.1** Develop an equation that describes the resistance changes in an electronic thermometer probe thermistor. The equation should account for the conduction of rectal heat through a plastic probe cover to the probe thermistor, starting at an initial probe temperature. Include a word description of how terms in your equation are related.

**12.2** Plot the thermistor waveform 12_1 that your instructor has provided. Waveform characteristics are documented in the associated header12_1.txt ascii file. The thermistor was not preheated before measurement. What is the steady-state temperature? How much time elapses until the steady state is reached?

**12.3** The thermistor curve appears to be exponential. Fit the thermistor curve to an autoregressive with exogenous input (ARX) model. What is the optimum number of feedforward and feedback parameters? What is your criterion for determining this optimum model? Can the thermistor curve be modeled using ordinary linear differential equations?

**12.4** Rank the accuracy (1 = highest rank) of the following thermometers for determining whether a small child has a fever: axial electronic, oral electronic, rectal electronic, ear infrared, skin infrared.

Read Craig et al. (2002):

**12.5** What is a meta-analysis? What is homogeneity? Why is it important to assess homogeneity across studies used in this meta-analysis?

**12.6** How do mode and offset affect ear thermometer accuracy?

**12.7** What was the hypothesis being investigated? What were the results? Provide statistics. What were sources of heterogeneity?

Read Morley et al. (1992):

**12.8** What was the mean difference between rectal and axillary temperatures at home and at the hospital? Analyze and comment on the Bland-Altman plot in Figure 2. Are axillary thermometery measurements accurate?

# References

ASTM (2003). *ASTM E1104-98 Standard specification for clinical thermometer probe covers and sheaths*. West Conshohocken, PA: ASTM.

ASTM (2006). *ASTM E1112-00 Standard specification for electronic thermometer for intermittent determination of patient temperature*. West Conshohocken, PA: ASTM.

ASTM (2009). *ASTM E1965-98 Standard specification for infrared thermometers for intermittent determination of patient temperature*. West Conshohocken, PA: ASTM.

Benzinger, T. H. (1980). *Heat and temperature fundamentals of medical physiology*. New York: Dowden, Hutchinson & Ross.

Bouchama, A., & Knochel, J. P. (2002). Heat stroke. *New England Journal of Medicine, 346*, 1978–1988.

Craig, J. V., Lancaster, G. A., Taylor, S., Williamson, P. R., & Smyth, R. L. (2002). Infrared ear thermometry compared with rectal thermometry in children: A systematic review. *Lancet, 360*, 603–609.

Dodd, S. R., Lancaster, G. A., Craig, J. V., Smyth, R. L., & Williamson, P. R. (2006). In a systematic review, infrared ear thermometry for fever diagnosis in children finds poor sensitivity. *Journal of Clinical Epidemiology, 59*, 354–357.

DuBois, E. F. (1948). *Fever and the regulation of body temperature*. Springfield, IL: Charles C. Thomas.

Fraden, J. (1989). Infrared electronic thermometer and method for measuring temperature. *US 4,797,840*.

Georgi, H. (1972). Electronic thermometer. *US 3,072,076*.

Gregory, T. K., & Stevenson, J. W. (1997). Medical thermometer. *US 5,632,555*.

Guyton, A. C., & Hall, J. E. (2006). *Textbook of medical physiology* (11th ed.). Philadelphia: Elsevier Saunders.

Mekjavic, I. B., & Eiken, O. (2006). Contribution of thermal and nonthermal factors to the regulation of body temperature in humans. *Journal of Applied Physiology, 100*, 2065–2072.

Morley, C. J., Hewson, P. H., Thornton, A. J., & Cole, T. J. (1992). Axillary and rectal temperature measurements in infants. *Archives of Disease in Childhood, 67*, 122–125.

O'Grady, N. P., Barie, P. S., Bartlett, J. G., Bleck, T., Carroll, K., & Kalil, A. C., et al. (2008). Guidelines for evaluation of new fever in critically ill adult patients: 2008 update from the American college of critical care medicine and the infectious diseases society of america. *Critical Care Medicine, 36*, 1330–1349.

O'Hara, G. J., & Phillips, D. B. (1986). Method and apparatus for measuring internal body temperature utilizing infrared emissions. *US 4,692,642*.

Romanovksy, A. A. (2007). Thermoregulation: Some concepts have changed. Functional architecture of the thermoregulatory system. *American Journal of Physiology Regulatory, Integrative and Comparative Physiology, 292*, R37–R46.

Suleman, M. I., Doufas, A. G., Akca, O., Ducharme, M., & Sessler, D. I. (2002). Insufficiency in a new temporal-artery thermometer for adult and pediatric patients. *Anesthesia and Analgesia, 95*, 67–71.

CHAPTER

# 13

# Electroencephalographs

In this chapter, we discuss the relevant physiology, history, and key features of electro-encephalographs. An *electroencephalograph* is an instrument that enables brain biopotentials called electroencephalograms (EEGs) to be measured, displayed, and analyzed (Figure 13.1). Recordings may be made of spontaneous, or background EEG, as well as an

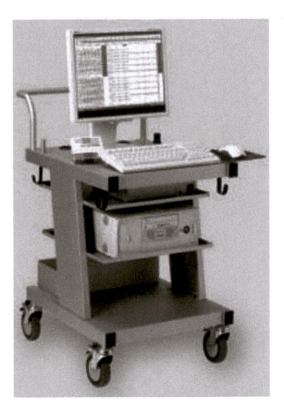

**FIGURE 13.1** CareFusion NicoletOne neurodiagnostic system. The amplifier module is behind the acquisition station keyboard (CareFusion, San Diego, California).

evoked potential that results from a stimulus. In this chapter, we concentrate only on background EEG because evoked potentials have been replaced in clinical practice by magnetic imaging resonance of pertinent neurologic structures.

Upon completion of this chapter, each student shall be able to:

1. Understand the mechanisms underlying the electroencephalogram.
2. Explain the procedures involved in polysomnography, including EEG electrode placement.
3. Describe two key features of electroencephalographs.

## BRAIN PHYSIOLOGY

The brain is composed of three principal components: the cerebrum, brain stem, and cerebellum (Figure 13.2). The *cerebrum*, the largest the part of the brain, receives and processes conscious sensation, generates thought, and controls conscious activity. It is divided into left and right hemispheres. In Chapter 12, we discussed the hypothalamus, which is part of the diencephalon. Other structures making up the diencephalon are the thalamus, epithalamus, and subthalamic nucleus. The diencephalon is part of the *brain stem*, which includes the midbrain, pons, and medulla. The *cerebellum*, in the lower back of the brain, controls motor activities and muscle contractions.

The cerebral hemispheres can be further divided into the frontal, temporal, parietal, and occipital lobes (Figure 13.3). These hemispheres can also be divided into gray matter and white matter. The *gray matter* refers to the cerebral cortex, which is a thin layer of neurons covering the surface of the cerebrum. Although only 2–5 mm thick, it contains about 100 billion neurons. The cortex is subdivided into horizontal layers from the surface, with

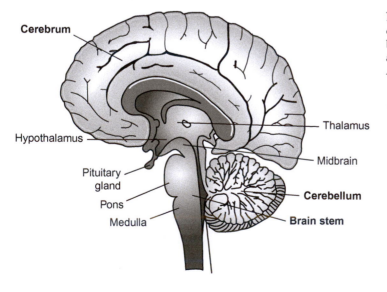

FIGURE 13.2 Three principal components of the brain: cerebrum, brain stem (including the diencephalon), and cerebellum (*Adapted from AMA, Chicago, Illinois*).

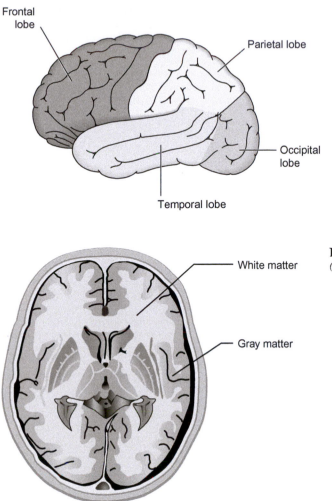

Frontal
lobe

Parietal lobe

Occipital
lobe

Temporal lobe

**FIGURE 13.3** The four lobes of a cerebral hemisphere [*Reproduced by permission from Rosenow (2009)*].

White matter

Gray matter

**FIGURE 13.4** Gray and white matter (*Adapted from A.D.A.M., Atlanta, Georgia*).

layer I being closest to the surface. The *white matter* refers to nerve fibers (Figure 13.4) (Rosenow, 2009; Guyton & Hall, 2006).

## Neural Current Flow

The predominant neuron in the cortex is the pyramidal neuron. These neurons are found in many layers of the cortex and are stacked vertically in columns. Pyramidal neuronal features vary among the different layers, cortical regions, and species. Typically, basal dendrites descend from the base of the soma (cell body), and apical dendrites emanate from the apex. As shown in Figure 13.5, one large apical dendrite connects the soma to tuft and to other dendrites (Figure 13.5) (Spruston, 2008).

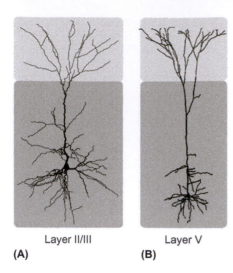

Layer II/III

(A)

Layer V

(B)

**FIGURE 13.5** Typical layer II/III (**A**) and layer V (**B**) pyramidal neurons. The apical tuft is highlighted with a light grey background. [*A: Original figure reproduced by permission from Waters (2003); **B**: Original figure reproduced by permission from Stuart (1998). Figure reproduced by permission from Spruston (2008)*].

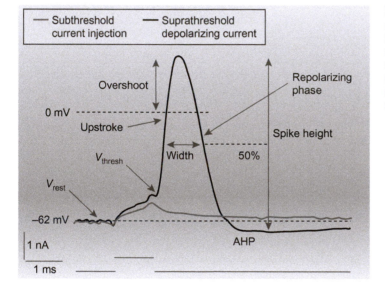

**FIGURE 13.6** Action potential recorded from a rat pyramidal neuron, illustrating commonly measured parameters. AHP = afterhyperpolarization threshold. [*This action potential was modified with permission from Storm (1987). Figure reproduced with permission from Bean (2007)*].

Action potentials, discussed in Chapter 2, arrive from other neurons via dendrites, which terminate in a specialized contact zone, or *synapse*. Each synapse interacts with a dendritic spine on the neuron of interest. The action potential in the afferent neuron causes the release of a neurotransmitter from its nerve terminal that diffuses across the synaptic cleft to the spine. This causes a local change in the postsynaptic potential. The potential difference between the postsynaptic membrane and other parts of the neuronal membrane causes an electrical current to flow along the neuronal membrane and to change the membrane potential of the soma. A new action potential occurs when the neuronal membrane is depolarized beyond a critical level or threshold (Fisch, 1999) (Figure 13.6).

## Generation of Electroencephalograms

When you position two electrodes on the scalp, the biopotential between them can be measured. The electroencephalogram results from the summation of pyramidal neuron postsynaptic potentials. Because only a small fraction of the pyramidal cell current penetrates the cerebral meningeal (membrane) coverings, spinal fluid, and skull to the scalp, the biopotential is usually just $10-100\ \mu V$ in amplitude.

An EEG contains several identifiable rhythms, which are defined by frequency.

- The first identified rhythm, the *alpha rhythm*, ranges between 8 to 13 Hz. The alpha rhythm has its greatest amplitude and persistence in the occipital, posterior, and parietal areas. It reacts to maneuvers such as eye opening, sudden alerting, and mental concentration, which all block the rhythm (Figure 13.7).
- The *beta rhythm* is over 13 Hz.
- The *theta rhythm*, whose frequency range is 4 to <8 Hz, is normally absent or very rare during wakefulness.
- The *delta rhythm*, which is <4 Hz, is present during sleep (Fisch, 1999) (Figure 13.8).

During sleep, EEG changes from a wakefulness stage (stage W), through three non−rapid eye movement (REM) stages (stages N1, N2, N3), to a REM state (stage R). For the majority of patients, the alpha rhythm is present during stage W.

Eyes open     Eyes closed

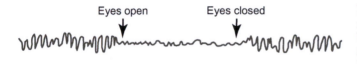

**FIGURE 13.7** Replacement of the alpha rhythm by an asynchronous, low-voltage beta rhythm when the eyes are opened [*Reproduced by permission from Guyton & Hall (2006)*].

Alpha

Beta

Theta

Delta         ] 50 μV

1 s

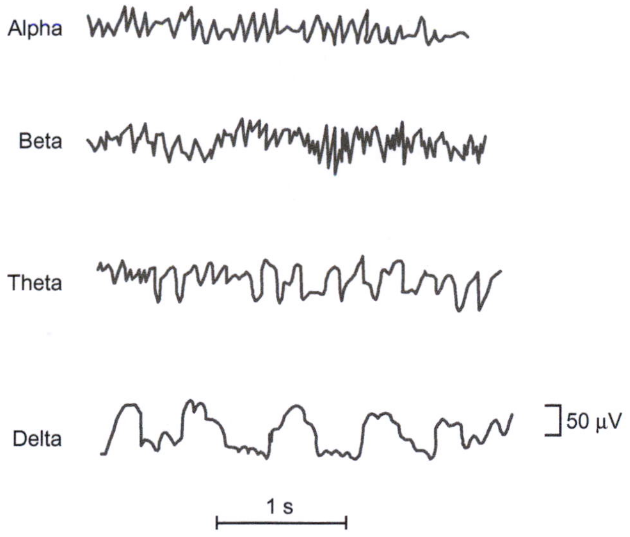

**FIGURE 13.8** Various rhythms present in the normal electroencephalogram [*Reproduced by permission from Guyton & Hall (2006)*].

- During stage N1, the alpha rhythm is attenuated and replaced by low-amplitude, mixed-frequency activity.
- During stage N2, negative sharp waves, called K complexes, and a train of waves with frequency 11–16 Hz, called *sleep spindles*, are present (Figure 13.9).
- During stage N3, waves of frequency 0.5–2 Hz and peak-to-peak amplitude>75 μV are measured in the frontal regions.
- During stage R, rapid eye movement is detected (Iber et al., 2007) (Figure 13.10).

## ORIGINAL 10-20 SYSTEM

As is done for electrocardiograms, EEGs are obtained systematically. The original 10-20 system of electrode placement provides uniform coverage of the scalp. It is called "10-20" because electrodes are placed at intervals of 10% or 20% of distances between bony landmarks (Figure 13.11A). The three bony landmarks used are the *inion*, which is the bony protruberance in the middle of the back of the head; the *nasion*, which is the bridge of the nose directly under the forehead; and the *preauricular point*, which is the depression of bone in front of the ear canal. Electrode locations are denoted by a letter, related to the underlying region, and by a number, related to lateral placement. Odd numbers refer to the left side; even numbers refer to the right size. The subscript z refers to zero or midline sagittal placement. In the modified 10-20 system, four of the electrode locations have been renamed.

As shown in Figure 13.11B, the distance between nasion and inion is measured along the midline. The frontopolar point, Fp, is marked at 10% of the distance from the nasion. The frontal (Fz), central (Cz), and parietal (Pz) points are marked at increments of 20% (30%, 50%, and 70% of the distance from the nasion, respectively). The occipital point (O) is marked at 10% of the distance from the inion. Electrodes are placed only at Fz, Cz, and Pz.

As shown in Figure 13.11C, the distance between the two preauricular points across Cz is measured. Electrodes are placed on the two preauricular points (A1 and A2), as well as at the following distances from the left preauricular point: 10% (T3), 30% (C3), 70% (C4), and 90% (T4).

As shown in Figure 13.11D, the circumference of the head is measured from O through T3, Fp, and T4. Electrodes are placed or checked at the following counterclockwise distances from Fp: 5% (Fp1), 15% (F7), 25% (T3), 35% (T5), 45% (O1), 55% (O2), 65% (T6), 75% (T4), 85% (F8), and 95% (Fp2).

Placement of the last four electrodes requires two measurements each (Figures 13.11E and 13.11F).

- An electrode is placed at F3, which is the midpoint between Fp1 and C3, and between F7 and Fz.
- An electrode is placed at F4, which is the midpoint between Fp2 and C4, and between F8 and Fz.
- An electrode is placed at P3, which is the midpoint between C3 and O1, and between T5 and Pz.

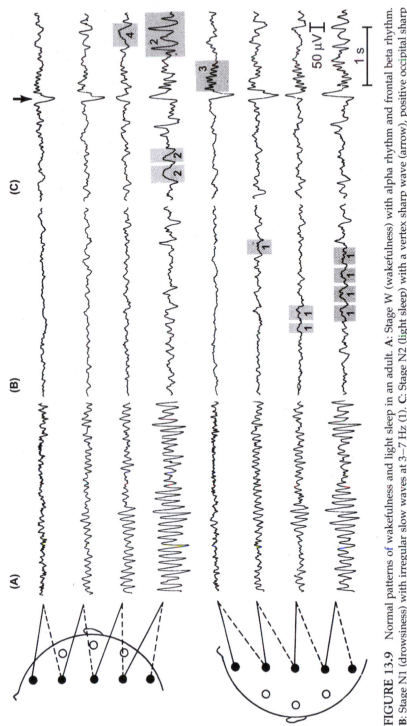

**FIGURE 13.9** Normal patterns of wakefulness and light sleep in an adult. A: Stage W (wakefulness) with alpha rhythm and frontal beta rhythm. B: Stage N1 (drowsiness) with irregular slow waves at 3–7 Hz (1). C: Stage N2 (light sleep) with a vertex sharp wave (arrow), positive occipital sharp transients (2), sleep spindles (3), and slow waves of 2–7 Hz (4) [*Reproduced by permission from Fisch (1999)*].

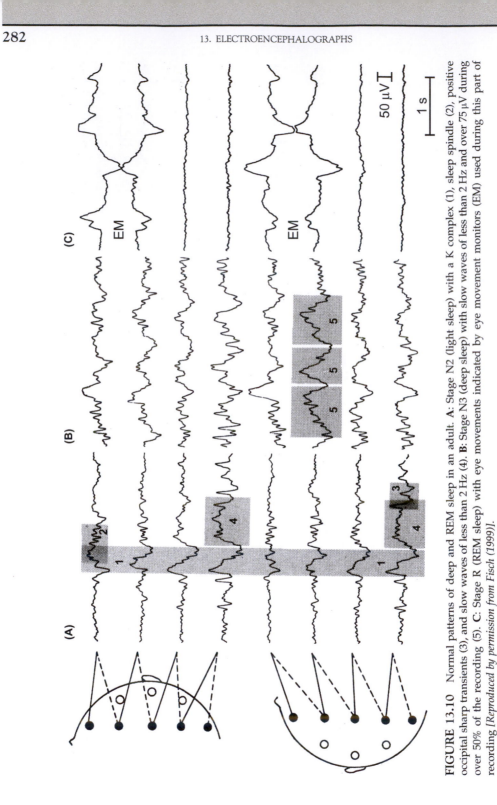

FIGURE 13.10   Normal patterns of deep and REM sleep in an adult. **A**: Stage N2 (light sleep) with a K complex (1), sleep spindle (2), positive occipital sharp transients (3), and slow waves of less than 2 Hz (4). **B**: Stage N3 (deep sleep) with slow waves of less than 2 Hz and over 75 µV during over 50% of the recording (5). **C**: Stage R (REM sleep) with eye movements indicated by eye movement monitors (EM) used during this part of recording [*Reproduced by permission from Fisch (1999)*].

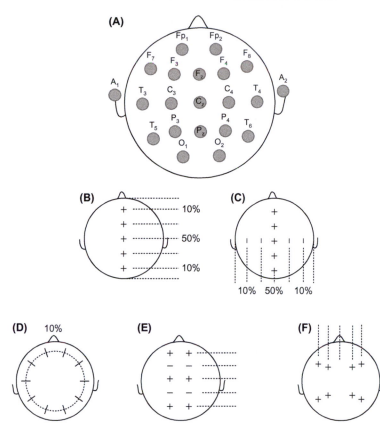

**FIGURE 13.11** Original 10-20 system electrode placement. **A**: Final electrode positions. **B**: Placement of midline electrodes. **C**: Placement of transverse electrodes. **D**: Placement of circumferential electrodes. **E–F**: Placement of last four electrodes [*Reproduced by permission from Fisch (1999)*].

- Finally, an electrode is placed at P4, which is the midpoint between C4 and O2, and between T6 and Pz (Fisch, 1999).

The last electrode, the 22nd one, acts as a reference and is typically positioned on the mastoid bone behind either ear.

## EEG Electrodes

Because our EEG voltage is about 50 times smaller in amplitude than an ECG, we cannot quickly apply the surface electrodes discussed in Chapter 2 to the scalp and expect to acquire a clean EEG waveform. Rather, each electrode location must be first carefully cleansed with an EEG prepping gel. This mildly abrasive gel lowers the impedance at the electrode-skin interface. One Ag/AgCl cup electrode is then glued to the cleansed skin with conductive paste. A typical 10-mm adult cup electrode is shown in Figure 13.12. The paste flows through the cup electrode into its central hole and is covered with gauze so that it does not dry out. To reduce artifact, interelectrode impedances between 100 and 5000 $\Omega$ should be verified before attempting data acquisition.

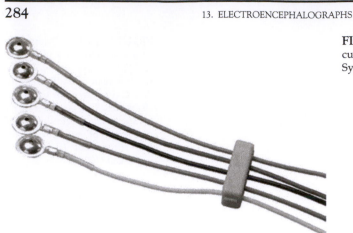

FIGURE 13.12 Philips Medical adult cup EEG electrodes (Philips Medical Systems, Andover, Massachusetts).

## Montages

A voltage may be recorded between any two electrodes, using a biopotential amplifier. Each EEG electrode combination is referred to as a *channel*, rather than as a lead, as is done for ECGs. A combination of simultaneous channels is a *montage*. The standard paper speed is 30 mm/s; EEGs are typically viewed on a computer monitor. For each montage, the viewing sensitivity, in µV/mm, is set by the EEG technologist, so that each channel is viewed with sufficient resolution.

A bipolar montage is formed by placing a series of linked bipolar derivations in straight lines either longitudinally or transversely across the scalp. It is best for analyzing low- to medium-amplitude waveforms that are highly localized. A common electrode reference montage consists of a series of derivations in which the same electrode is used as the reference electrode. With this montage, the location of the maximal potential is recognized by its amplitude, not by phase reversal. An average reference montage consists of derivations in which the reference input is the summation of all electrodes. It allows either localized or more diffusely distributed waveforms to be easily recognized. The best practice is for an EEG technologist to cycle through a minimum set of montages during the course of a study, such as some combination of longitudinal bipolar, transverse bipolar, and referential montages. A typical referential montage is given in Figure 13.13.

## Analysis

A montage is analyzed to determine whether an EEG is abnormal. An abnormal EEG contains epileptiform activity, slow waves, amplitude abnormalities, or certain patterns deviating from normal activity. Epileptiform patterns include spike and sharp wave discharges, and they are usually due to a focal irritative lesion of the cerebral cortex. Local slow waves are often due to circumscribed damage of the white matter, with or without involvement of the cortex. Local amplitude reductions are often due to superficial lesions which reduce the electrical potentials generated in the cortex. Certain patterns deviating from normal activity are linked to specific diseases. Often, an abnormality appears only

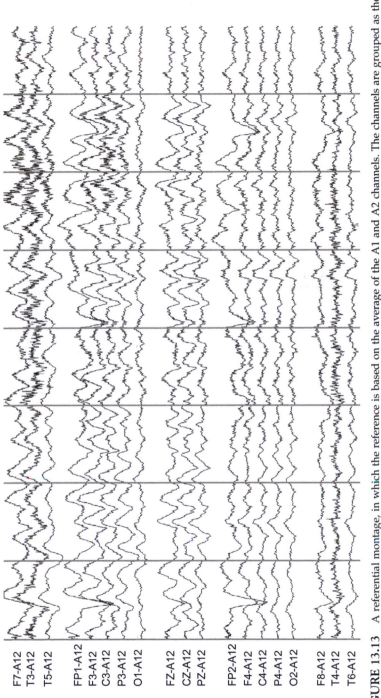
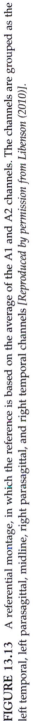

**FIGURE 13.13** A referential montage, in which the reference is based on the average of the A1 and A2 channels. The channels are grouped as the left temporal, left parasagittal, midline, right parasagittal, and right temporal channels [*Reproduced by permission from Libenson (2010)*].

intermittently, in certain head regions, or superimposed on a normal background (Fisch, 1999).

The simultaneous acquisition of video provides correlation of abnormal EEG with clinical manifestations.

# CLINICAL NEED

Electroencephalography is used to localize the source of epilepsy and to diagnose sleep disorders.

## Epilepsy

Epilepsy is not a single condition, but a disorder characterized by the occurrence of at least two unprovoked seizures, taking place within 24 hr. A seizure results from abnormal discharges of cortical neurons. Clinical symptoms of a seizure depend on the location propagation of the epileptic discharge in the cortex. The lifetime likelihood of experiencing at least one epileptic seizure is about 9%; the prevalence of active epilepsy is about 0.8%.

Seizures are classified into partial-onset seizures and generalized-onset seizures. During a *partial-onset seizure*, consciousness is preserved. The six major categories of *generalized-onset seizures* are absence seizure, myoclonic seizure, clonic seizure, tonic seizure, primary generalized tonic-clonic seizure, and atonic seizure. Each type of seizure is associated with particular EEG patterns, which are beyond the range of this discussion.

- An *absence seizure* is a brief episode (usually within 20 s) of impaired consciousness without an aura or postictal confusion. An *aura* is a subjective sensation, such as a bright light or cold breeze, that precedes the seizure.
- A *myoclonic seizure* consists of brief, arrhythmic, jerking, motor movements that last less than 1 s.
- A longer *clonic seizure* evolves from myoclonic seizure clusters.
- A *tonic seizure* consists of sudden-onset tonic extension or flexion of the head, trunk, and/or extremities for several seconds, usually in relation to drowsiness.
- A *tonic-clonic seizure* begins as a tonic seizure for a few seconds and is then followed by clonic rhythmic movements and prolonged postictal confusion (Figure 13.14).
- An *atonic seizure* consists of brief loss of postural tone, often resulting in falls and injuries (Cavazos & Spitz, 2009).

For epilepsy diagnosis, *strobe stimulation*, also known as intermittent photic stimulation, is routinely performed. A strobe light is flashed in the patient's face at varying frequencies of 1−35 Hz for 5−10 s at each frequency, separated by pauses of similar length. The resulting EEG is analyzed to determine whether a normal photic driving response or photoparoxysmal response is present. During a photoparoxysmal response, generalized EEG spike wave discharges are observed (Libenson, 2010).

**FIGURE 13.14** Generalized tonic–clonic seizure in an 18-yr-old girl with a lifelong history of major and minor generalized seizures. **A:** Pattern of slight drowsiness, suddenly interrupted by intermittent spikes and movement artifacts at the onset of a rhythm of generalized repetitive spikes of increasing amplitude and decreasing frequency; it lasts about 10 s. Technician's comments read: "No aura, no cry, but eyes fluttering, head trembling, eyes rolled up in head, both arms stiffened. Tongue blade in mouth." This represents the tonic phase of the tonic–clonic seizure. **B:** Spikes begin to be interrupted by slow waves. This signals the beginning of the clonic phase, which starts with rhythmical, high-amplitude polyspike-and-wave and spike-and-wave complexes at about 1 Hz. Technician's note reads: "Real seizure, seizuring, Eyes fluttering + head shaking" [*Reproduced by permission from Fisch (1999)*].

## Sleep Disorders

Electroencephalograms are also acquired during a sleep study, which is also known as *polysomnography*. While a patient sleeps, EEG and other waveforms and parameters are simultaneously recorded. These include electrooculograms (eyeball rotation biopotentials), chin EMG, leg EMG, airflow, $SpO_2$, and body position. Careful analysis of these waveforms and parameters during sleep enables the appropriate diagnosis of a sleep disorder.

For example, obstructive sleep apnea is diagnosed by the observation of at least five obstructive apneas per hour of sleep, with at least one associated condition. *Obstructive apnea* is the cessation of respiratory airflow for at least 10 s, due to airway blockage (relaxed pharyngeal muscles during sleep). The associated conditions are frequent arousals from sleep, bradytachycardia (alternating bradycardia and tachycardia), and arterial oxygen desaturation.

# HISTORIC DEVICES

Clinical observation of seizures can be traced to the fourth century BCE, when Hippocrates wrote a book about epilepsy. Hippocrates believed that epilepsy was caused by a superfluity of phlegm, which led to subnormal brain consistency.

## Early Devices

Many centuries later, in 1875, physiologist Richard Caton observed localized cortical electrical potentials from the brains of rabbits and monkeys, in response to sensory stimuli. Electrodes positioned on gray matter transmitted a tiny voltage to a galvanometer. Small movements of the galvanometer dial were reflected by a mirror onto a wall 9 ft away. In 1912, the first EEGs were photographed but were recorded without meeting the Nyquist sampling theorem.

## Enabling Technology: Double-Coil Galvanometer

Psychiatrist Hans Berger began to pursue accurate EEG recording in 1924, working with patients with skull defects that exposed the brain. His recording instruments were initially an Edelmann string galvanometer, which was intended for electrocardiography and later a Seimens and Halske double-coil galvanometer, which was the most sensitive instrument of the time. After much experimentation, Berger settled on needle electrodes inserted epidurally in the region of the skull defect and on conditions that minimized electrical interference, such as recording after the main workday ended and other equipment was turned off.

Before his first report in 1929, Berger recorded hundreds of EEGs from his patients and employees at his clinic. He discovered and named the alpha and beta waves. In a 1933 report, Berger included an EEG segment from an 18-yr-old woman during a brief episode of "simple automatic activity with no other movement" (Figure 13.15) (Millett, 2001).

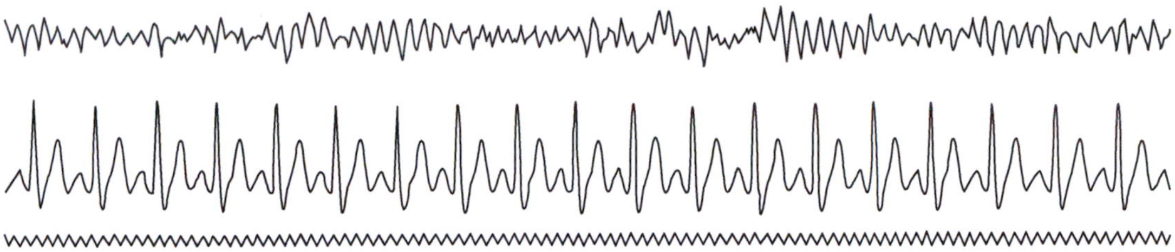

**FIGURE 13.15** A segment from Berger's EEG record during a seizure in an 18-yr-old woman in 1933 [*Reproduced by permission from Goldensohn et al. (1997)*].

Epileptic diagnosis and treatment therapy research quickly followed confirmation of Berger's work. In 1935, neurologists Frederick Gibbs, Hallowell Davis, and William Lennox demonstrated spike-and-wave complexes during clinical absence seizures. The following year, Gibbs et al. demonstrated focal spikes in focal epilepsy. In 1938, EEG activity and clinical seizure behavior began to be correlated when two cameras recorded moving images of an epileptic patient (Goldensohn et al., 1997).

## SYSTEM DESCRIPTION AND DIAGRAM

An electroencephalograph consists of similar parts to those of an electrocardiograph. These parts have been separated into three components: electrodes, amplifier module, and acquisition station (Figure 13.16).

Typically, 22 Ag/AgCl cup electrodes are glued to cleansed skin with conductive paste. For polysonography, other sensors, such as a pulse oximetry sensor, may also be used. The amplifier module isolates and amplifies differential channels (EEG) and DC channels (other sensors). At least 21 EEG electrode voltages are isolated and passed to biopotential amplifiers, with the reference electrode connected to the inverting input pins. Sensor voltages are isolated and passed to amplifiers. All amplified voltages pass through antialiasing filters and a 16-bit analog-to-digital converter. A typical lowpass cutoff frequency is 100 Hz; a typical sampling frequency is 500 Hz. Alkaline batteries as the power source ensure that only low voltage is present. The processor module coordinates data acquisition and sends digital data, via a standard universal serial bus (USB) or Ethernet connection, to the acquisition station.

Although not shown, the amplifier module may be subdivided into a portable head box and base unit. The head box connects to the EEG electrodes and sensors, and it transmits digitized data over a Bluetooth connection or more electrical cabling to the base unit. With the headbox/base unit configuration, the patient is able to move within a short distance of the base unit without cumbersome amplifier module cabling. The photic stimulator (also not shown in the diagram) is often an LED group that the clinician turns on manually.

The acquisition station is a personal computer containing specialized software for processing and displaying EEG data. Besides montages, the software may detect spikes and seizures and display parameter trends on the computer monitor. The acquisition station

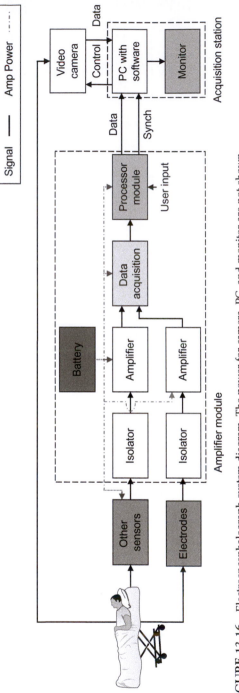

**FIGURE 13.16** Electroencephalograph system diagram. The power for camera, PC, and monitor are not shown.

also controls an optional video camera and synchronizes data acquisition of MPEG4 video and EEG.

## KEY FEATURES FROM ENGINEERING STANDARDS

As of 2010, FDA does not recommend any specific consensus standard for electroencephalography. An applicable standard is *IEC 60601-2-26 Medical Electrical Equipment—Part 2-26: Particular Requirements for the Safety of Electroencephalographs* (IEC, 2002). This standard, like the standard discussed in Chapter 10 for mechanical ventilators, is an addition to main electrical medical device standard *IEC 60601-1:2005 Medical Electrical Equipment—Part 1: General Requirements for Basic Safety and Essential Performance*. IEC 60601-1:2005 contains essential requirements for patient safety and has been slightly modified by AAMI to become ANSI/AAMI ES60601-1:2005 (AAMI, 2005).

IEC 60601-2-26 is not particularly detailed. Because the EEG is so small in amplitude, you would expect it to contain several requirements that ensure the fidelity and accuracy of EEG signals. It does not contain such requirements. Instead, many sections state that "the clauses and subclauses of this section of the General Standard apply." We detail two additions to the General Standard.

### Electrostatic Discharge Prevention

*Electrostatic discharge* (*ESD*) is the transfer of electrical charge between two bodies at different potentials due to direct contact or an induced electrical field. To prevent discharges of several thousand volts, antistatic materials are used in medical device fabrication. According to this requirement, an electroencephalograph may show temporary degradation during discharges but must return to original operation within 10 s without the loss of any stored data. No specific method for testing this requirement is mandated (IEC, 2002).

### Protection from Conducted Disturbances, Induced by Radiofrequency Fields

A *radiofrequency field* is electromagnetic radiation in the frequency range of 3 kHz–300 MHz. In the operating room, an electroencephalograph may be exposed to radio frequency (RF) from an electrosurgical unit used to vaporize tissue. According to this requirement, an electroencephalograph exposed to a conducted radiofrequency voltage via the power supply cord must operate within normal specifications. No specific method for testing this requirement is mandated (IEC, 2002).

## BISPECTRAL INDEX MONITORS

One special type of electroencephalograph is the bispectral index (BIS) monitor, which was used by many anesthesiologists during the early 2000s to minimize anesthesia awareness (Figure 13.17). Anesthesia awareness is the explicit recall of sensory perceptions during general anesthesia, which may lead to anxiety and post-traumatic stress disorder.

The bispectral index parameter was invented by Covidien Aspect Medical Systems. BIS was inspired by the bispectrum, which is a statistical calculation for stochastic (random) signals. We can assume that the EEG, $u(k)$, is random, rather than deterministic (periodic) like the ECG. The bispectrum, $B(f_1, f_2)$, looks at the relationship between the Fourier transforms, $U(f)$, at two frequencies and at a combination frequency, such that

$$B(f_1, f_2) = \left| \sum_{i=1}^{L} U_i(f_1) U_i(f_2) U_i^*(f_1 + f_2) \right|. \tag{13.1}$$

A high bispectrum value indicates a phase coupling among the triplet of frequencies, $f_1$, $f_2$, and $(f_1 + f_2)$ (Rampil, 1998).

In a multicenter clinical trial, the BIS monitor was shown to reduce the incidence of anesthesia awareness (Myles et al., 2004). A special multi-electrode is applied to a patient's forehead, and used to capture EEGs. These EEGs are processed to calculate the bispectral index parameter, which is *not* equivalent to the bispectrum. BIS was derived from analysis of 5000 hr of EEG recordings during 1500 anesthesia administrations; it is the linear combination of two time domain calculations and two frequency domain calculations. Ranging from 0 to 100, a BIS value below 60 has been shown to prevent anesthesia awareness (Baura, 2008).

More recently, a study of 2000 patients at Washington University demonstrated that monitoring of end-tidal anesthesia gas was as effective for reducing anesthesia awareness as BIS monitoring in patients receiving volatile anesthetic gas during surgery (Avidan et al., 2008).

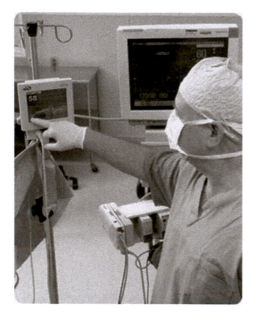

FIGURE 13.17  BIS VISTA monitoring system being used by a clinician (Covidien Aspect Medical Systems, Norwood, Massachusetts).

# SUMMARY

An electroencephalograph is an instrument that enables brain biopotentials called electroencephalograms to be measured, displayed, and analyzed. Recordings may be made of spontaneous, or background, EEG, as well as of an evoked potential that results from a particular stimulus. We concentrated only on background EEG because evoked potentials have been replaced in clinical practice by magnetic imaging resonance of pertinent neurologic structures.

When you position two electrodes on the scalp, the biopotential between them can be measured. This electroencephalogram results from the summation of pyramidal neuron postsynaptic potentials. Because only a small fraction of the pyramidal cell current penetrates the cerebral meningeal coverings, spinal fluid, and skull to the scalp, the biopotential is usually just $10-100\,\mu V$ in amplitude.

An EEG contains several identifiable rhythms, which are defined by frequency. The first identified rhythm, the alpha rhythm, ranges between 8 and 13 Hz. The alpha rhythm has its greatest amplitude and persistence in the occipital, posterior, and parietal areas. It reacts to maneuvers such as eye opening, sudden alerting, and mental concentration, which all block the rhythm. The beta rhythm is over 13 Hz. The theta rhythm, whose frequency range is 4 to <8 Hz, is normally absent or very rare during wakefulness. The delta rhythm, which is <4 Hz, is present during sleep.

During sleep, EEG changes from a wakefulness stage (stage W), through three non-REM stages to an REM state. For the majority of patients, the alpha rhythm is present during stage W. During stage N1, the alpha rhythm is attenuated and replaced by low-amplitude, mixed-frequency activity. During stage N2, negative sharp waves, called K complexes, and a train of waves with frequency $11-16$ Hz, called sleep spindles, are present. During stage N3, waves of frequency $0.5-2$ Hz and peak-to-peak amplitude $>75\,\mu V$ are measured in the frontal regions. During stage R, rapid eye movement is detected.

A voltage may be recorded between any two electrodes, using a biopotential amplifier. Each EEG electrode combination is referred to as a channel, rather than a lead, as with ECGs. A combination of simultaneous channels is referred to as a montage. For each montage, the viewing sensitivity, in $\mu V/mm$, is set by the EEG technologist, so that each channel is viewed with sufficient resolution.

Electroencephalography is used to localize the source of epilepsy and to diagnose sleep disorders. Key electroencephalograph features include electrostatic discharge prevention and protection from conducted disturbances, induced by radiofrequency fields.

## Exercises

**13.1** In 2009, the *Las Vegas Sun* profiled a patient whose diagnosis of epilepsy and subsequent brain surgery changed his life. Watch the short film about Chris Stones: http://www.lasvegassun.com/videos/2009/oct/04/2959/. Note the part of this video that documents Stones' brain surgeries. What was the purpose of each of the three hospital procedures he received? Name all the medical devices that were videotaped during these procedures.

**13.2** Why is Ag/AgCl, rather than KCl, used for the electrode-electrolyte interface? How does interelectrode impedance greater than $5000\,\Omega$ increase artifact? How does the electroencephalograph measure this impedance?

**13.3** What type of montage is shown in Figure 13.10? Identify each channel. For example, the channel (F8-T4) is not shown.

**13.4** Using only the eight electrodes in Figure 13.10, design a switching network that enables the user to switch between data acquisition of a bipolar, common electrode reference, or average reference montage.

**13.5** Write a frequency response requirement for electroencephalographs. Provide justification for your requirement.

An exciting new application of electroencephalography and thermometry is the treatment of perinatal hypoxia-ischemia. Read (Hellstrom-Westas et al., 2006):

**13.6** What is amplitude-integrated electroencephalography (a-EEG)? Which electrodes are used? What patterns in a full-term infant are associated with seizures?

Read (Gluckman et al., 2005):

**13.7** Why is aEEG used in the protocol? What specific aEEG inclusion criteria are used? Is aEEG important in analyzing study results? Explain your reasoning.

**13.8** What was the hypothesis being investigated? What were the results? Describe all the sensors that were part of the study protocol.

# References

AAMI (2005). *ANSI/AAMI ES60601-2-12 Medical electrical equipment—Part 1: General requirements for basic safety and performance* (3rd ed.). Arlington, VA: AAMI.

Avidan, M. S., Zhang, L., Burnside, B. A., Finkel, K. J., Searleman, A. C., & Selvidge, J. A., et al. (2008). Anesthesia awareness and the bispectral index. *New England Journal of Medicine, 358,* 1097–1108.

Baura, G. D. (2008). *A biosystems approach to industrial patient monitoring and diagnostic devices.* San Rafael, CA: Morgan Claypool.

Bean, B. P. (2007). The action potential in mammalian central neurons. *Nature Reviews Neuroscience, 8,* 451–465.

Cavazos, J. E., & Spitz, M. (2009). *Seizures and epilepsy, overview and classification.* emedicine.medscape.com.

Fisch, B. J. (1999). *Fisch and Spehlmann's EEG primer: Basic principles of digital and analog EEG* (3rd ed.). Amsterdam: Elsevier.

Gluckmann, P. D., Wyatt, J. S., Azzopardi, D., Ballard, R., Edwards, A. D., & Ferriero, D. M., et al. (2005). Selective head cooling with mild systemic hypothermia after neonatal encephalopathy: Multicentre randomised trial. *Lancet, 365,* 663–670.

Goldenshon, E. S., Porter, R. J., & Schwartzkroin, P. A. (1997). The American Epilepsy Society: An historic perspective on 50 years of advances in research. *Epilepsia, 38,* 124–150.

Guyton, A. C., & Hall, J. E. (2006). *Textbook of medical physiology* (11th ed.). Philadelphia: Elsevier Saunders.

Hellstrom-Westas, L., Rosen, I., De Vries, L. S., & Greisen, G. (2006). Amplitude-integrated EEG classification and interpretation in preterm and term infants. *NeoReviews, 7,* e76–e87.

Iber, C., Ancoli-Israel, S., Chesson, A. L., & Quan, S. F. (2007). *The AASM manual for the scoring of sleep and associated events.* Westchester, IL: American Academy of Sleep Medicine.

IEC (2002). *IEC 60601-2-26 Medical electrical equipment—Part 2-26: Particular requirements for the safety of electroencephalographs.* Geneva: IEC.

Libenson, M. H. (2010). *Practical approach to electroencephalography.* Philadelphia: Elsevier Saunders.

Millett, D. (2001). Hans Berger: From psychic energy to the EEG. *Perspectives in Biology and Medicine, 44,* 522–542.

Myles, P. S., Leslie, K., McNeil, J., Forbes, A., & Chan, M. T. (2004). Bispectral index monitoring to prevent awareness during anaesthesia: The B-Aware randomised controlled trial. *Lancet*, *363*, 1757–1763.

Rampil, I. J. (1998). A primer for EEG signal processing in anesthesia. *Anesthesiology*, *89*, 980–1002.

Rosenow, J. M. (2009). Anatomy of the nervous system. In E. S. Krames, P. H. Peckham, & A. R. Rezai (Eds.), *Neuromodulation*. London: Elsevier.

Spruston, N. (2008). Pyramidal neurons: dendritic structure and synaptic integration. *Nature Reviews Neuroscience*, *9*, 206–221.

Storm, J. F. (1987). Action potential repolarization and a fast after-hyperpolarization in rat hippocampal pyramidal cells. *Journal of Physiology*, *385*, 733–759.

Stuart, G., et al. (1998). Determinants of voltage attenuation in neocortical pyramidal neuron dendrites. *Journal of Neuroscience*, *18*, 3501–3510.

Waters, J., et al. (2003). Supralinear Ca2+ influx into dendritic tufts of layer 2/3 neocortical pyramidal neurons in vitro and in vivo. *Journal of Neuroscience*, *23*, 8558–8567.

CHAPTER

# 14

# Deep Brain Stimulators

With this chapter, we begin to move from diagnosing brain disorders using electroencephalographs to treating brain disorders using neurostimulators. *Neurostimulation* is the "electrical process of inhibition, stimulation, modification, regulation, or therapeutic alteration of activity in the central, peripheral, or autonomic nervous systems. It is inherently nondestructive, reversible, and adjustable." Neurostimulation is a subset of neuromodulation, which is "the science of how electrical, chemical and mechanical interventions can modulate nervous system function" (Krames et al., 2009).

Various neurostimulators are used as therapies. Spinal cord stimulators treat chronic pain. Vagus nerve stimulators treat epilepsy. Cochlear implants provide alternative stimulation for the partial restoration of hearing. Functional electrical stimulators activate paralyzed muscles.

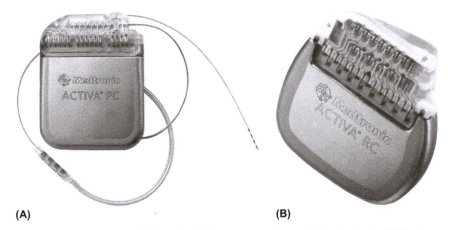

**(A)**                                             **(B)**

FIGURE 14.1   Medtronic Activa PC (**A**) and RC (**B**) neurostimulators. Only the Activa PC is shown with leads and lead extensions (Medtronic, Minneapolis, Minnesota).

In this chapter, we discuss the relevant physiology, history, and key features of deep brain stimulators. A deep brain stimulator is an electrical stimulator that discharges electrical current within the globus pallidus of the basal ganglia or subthalamic nucleus of the diencephalon, as a treatment for tremor associated with *Parkinson's disease* (PD) (Figure 14.1). This disorder was first formally described by British physician James Parkinson in 1817 as a "shaking palsy." The treatment is reversible, with PD symptoms recurring after the stimulation current stops.

Upon completion of this chapter, each student shall be able to:

1. Understand the mechanisms underlying Parkinson's disease.
2. Explain the procedures involved in deep brain stimulator implantation, including microelectrode recordings.
3. Describe five key features of deep brain stimulators.

## BASAL GANGLIA

*Deep brain stimulation (DBS)* is therapy administered below the cerebrum. Often, stimulation occurs in the basal ganglia. These groups of neurons lie deep beneath the cerebral hemispheres, and are comprised of the caudate nucleus, the putamen, and the globus pallidus (Figure 14.2).

The basal ganglia are involved in the control of both motor and cognitive functions. Neurotransmission of dopamine in the putamen and caudate is the initial step in signaling pathways that pass through the globus pallidus, the subthalamic nucleus, and the thalamus sequentially to the motor cortex. The putamen and caudate are collectively called the *striatum*.

### Parkinson's Disease

Normally, pigmented midbrain neurons in the substantia nigra pars compacta project to the putamen and caudate, where they release dopamine. When more than half of the dopaminergic nerve terminals in the striatum are affected, the motor impairments of Parkinson's disease arise (Figure 14.3). The mechanisms underlying these dopaminergic nerve terminal effects vary. PD patients have one or more characteristic findings, including resting tremor, rigidity, slowed movement, decreased dexterity, small handwriting, flexed posture, gait disorder, and imbalance. As the movement disorder progresses, dementia can develop over several years.

### Levodopa Therapy

When a patient becomes sufficiently bothered by Parkinsonian symptoms, levodopa (L-dopa) may be prescribed. Dopamine administration is ineffective because dopamine cannot pass between endothelial tight junctions of the blood-brain barrier (BBB) to enter the brain. L-dopa, a naturally occurring amino acid that crosses the BBB, is an intermediate in the dopamine synthesis pathway. Within 30 min after oral L-dopa administration, a patient may recover from previous impairments in speech, dexterity, and gait.

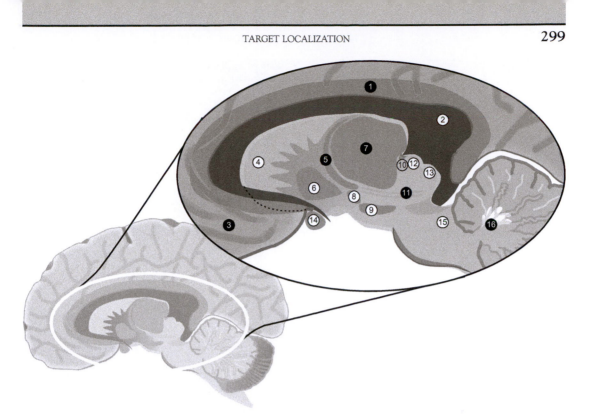

**FIGURE 14.2** Deep brain structures adjacent to frontal lobe (3): 1 = cingulated cortex, 2 = corpus callosum, 4 = caudate-putamen complex, 5 = zona incerta, 6 = globus pallidus, 7 = thalamus, 8 = subthalamic nucleus, 9 = substantia nigra, 10 = posterior commissure, 11 = H fields of Forel, 12 = superior colliculus, 13 = inferior colliculus, 14 = optic chiasm, 15 = superior cerebellar peduncle, 16 = dentate nucleus *[Reproduced by permission from Visser-Vandewalle et al. (2009)].*

However, L-dopa does not relieve all Parkinsonian symptoms. Tremor may not improve, and retropulsive (involuntary backward) imbalance almost never improves. After 2 yr of L-dopa treatment, the patterns of response may change. Within 5 yr, up to half of L-dopa patients have motor fluctuations, dyskinesias (inability to perform voluntary muscle movements), or both (Lewitt, 2008).

## TARGET LOCALIZATION

Before a DBS electrode can be implanted to administer therapy, the subthalamic nucleus (STN) or global pallidus internus (GPi) target must be localized. Typical active contact locations are shown in Figure 14.4.

Localization occurs in three steps. First, the patient's head is fit preoperatively with a standard stereotactic frame. A stereotactic frame is a rigid frame that provides an immobile three-dimensional coordinate system for target localization The patient's brain is imaged using magnetic resonance imaging (MRI). MRI enables soft tissue structures to be

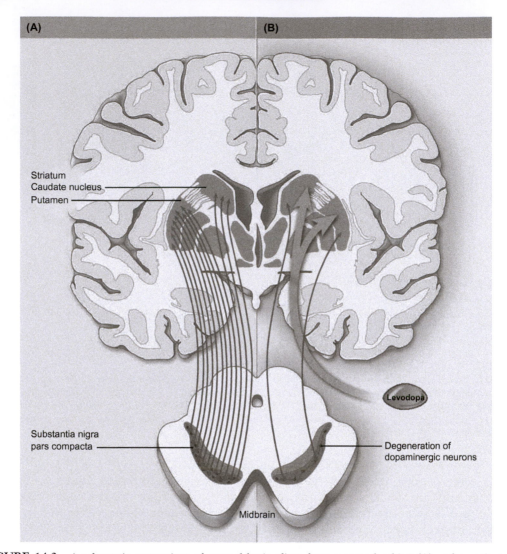

**FIGURE 14.3** A schematic comparison of coronal brain slices from a control subject (**A**) and a patient with Parkinson's disease (**B**) illustrates the major neurodegenerative loss of dopamine-synthesizing neurons in the substantia nigra pars compacta in the brain stem, projecting to striatal nuclei (caudate and putamen) in the cerebrum. Exogenous levodopa administered for treatment of Parkinson's disease is transported to the brain, where it enhances striatal dopaminergic neurotransmission [*Reproduced by permission from Lewitt (2008)*].

visualized, with respect to the frame landmarks (Figure 14.5). The acquired images are subject to inherent nonlinearities, imperfect image fusion, mechanical errors in the stereotactic system or micropositioner hardware, and dynamic brain shift during the procedure.

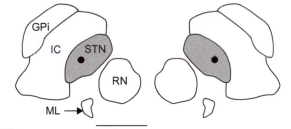

**(A)** STN

1cm

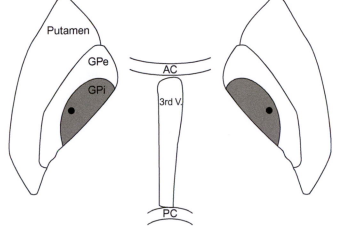

**(B)** GPi

**FIGURE 14.4** Typical active contact locations (shown as *block dots*) with respect to axial plane anatomy drawn from the Schaltenbrand and Wahren human brain atlas. The axial planes selected are those from the atlas that are closest to the vertical position of the average active contact. **A**: STN active contact, shown on the plane 4.5 mm inferior to the intercommissural line. **B**: GPi active contact, shown on the plane passing through the intercommissural line. IC = internal capsule, RN = red nucleus, ML = medial lemniscus, GPe = globus pallidus externis, AC = anterior commissure, PC = posterior commissure, 3rd V. = third ventricle [*Adapted from Schaltenbrand & Wahren (1977), with permission. Figure reproduced by permission from Sillay & Starr (2009)*].

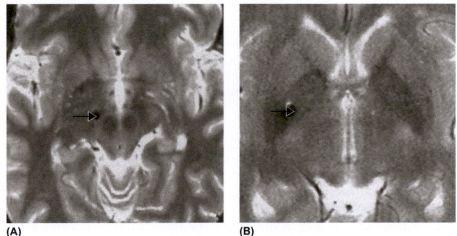

**(A)**　　　　　　　　　　　**(B)**

**FIGURE 14.5** Electrode locations on postoperative MRI, at the axial level of the active contacts following programming for optimal clinical benefit. Axial FSE images. **A**: GPi, at the level of the AC and PC. **B**: STN, at an axial plane 4 mm inferior to the AC-PC plane. The artifact generated by the lead is indicated with an arrow [*From Starr (2002), with permission from S. Karger AG, Basel. Figure reproduced by permission from Sillay & Starr (2009)*].

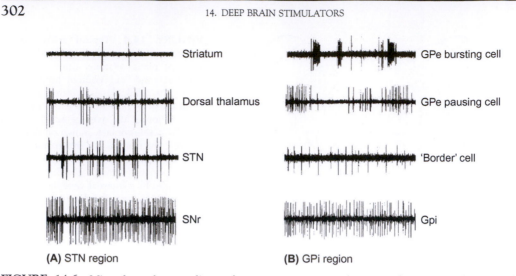

**FIGURE 14.6** Microelectrode recordings of spontaneous neuronal activity from the thalamus and basal ganglia in the region of STN (**A**) and GPi (**B**). Each trace represents a 1-s recording in patients with Parkinson's disease [*Modified from Starr (2003), with permission from Dekker Publishing. Figure reproduced by permission from Sillay & Starr, 2009)*].

Second, after the brain is exposed, microelectrode recording (MER) is used to further localize in real time the target in reference to stereotactic space. A typical bipolar electrode is fabricated from a platinum-iridium alloy, insulated with glass, and has an outer diameter of about 1 mm. The electrode is advanced from the cortical surface to the target. As the electrode tip passes through nuclear groups and white matter tracts, single-action potentials may be recorded. Voltage characteristics enable structures to be identified. For example, STN approach is typically characterized by an increase in background signal and STN firing rates of 20–50 spikes/s. Similarly, passage from the external to internal GP is typically characterized by an increase in firing rate to 80–100 spikes/s (Figure 14.6).

Finally, somatotopy, or associated movement, is used to fine-tune the microelectrode location. The clinician listens for the modulation of action potential discharge in relation to passive movements of opposite side limbs. Findings of only leg-related activity along an electrode trajectory indicate a relatively medial localization within the motor territory (Sillay & Starr, 2009).

# CLINICAL NEED

According to the National Institute of Neurologic Disorders and Stroke, at least 500,000 individuals in the United States suffer from Parkinson's disease. The annual incidence of new cases is estimated as 11.2–13.4 per 100,000, or about 40,000 new cases annually. This incidence rate is affected by race and ethnicity. Non-Hispanic whites and Hispanics are more likely to be diagnosed than African-Americans and Asians (Van Den Eeden et al., 2003; Dahodwala et al., 2009). The likelihood of developing PD increases with age.

Environmental factors and genetic makeup increase the risk of developing PD. Rare genetic mutations cause PD in 10% of the PD population. Neurotoxins that directly target the substantia nigra and/or basal ganglia also induce PD. For example, intravenous exposure to 1-methyl-4-phenyl-1,2,3,6-tetrahydropyridine (MPTP) causes PD in humans (Bronstein et al., 2009).

In patients with advanced PD, bilateral stimulation of STN or GPi over 3 months was associated with significant improvement in motor function, compared to levodopa therapy (Deep-Brain Stimulation for Parkinson's Disease Study Group, 2001). More recently, a larger randomized controlled trial of 255 patients over 6 months of bilateral stimulation reaffirmed the improvement in motor function. However, this larger study also demonstrated that DBS is associated with increased risks of falls (14 vs. 5, $p = 0.03$) and dystonia events (8 vs. 1, $p = 0.02$) (Weaver et al., 2009).

# HISTORIC DEVICES

Electrical stimulation of the living human brain can be traced to 1874. Patient Mary Rafferty was being treated for an aggressive basal cell carcinoma of the scalp at Good Samaritan Hospital in Cincinnati. A 2-in.-diameter portion of her skull had eroded, exposing her brain. After 1 month of treatment, physician Robert Bartholow asked for permission to electrically stimulate her brain.

During the first experiment, Bartholow discovered Rafferty could not feel mechanical stimulation of her brain. During a second experiment, low-amplitude stimulation across two electrodes inserted into the dura mater caused muscular contractions. During a third experiment, higher-amplitude stimulation induced convulsions, coma, and eventually death. After surviving international criticism of his experimental protocol, Bartholow accepted a medical chair at Jefferson Medical College and cofounded the American Neurological Association (Morgan, 1982).

## Early Devices

The first surgical intervention for PD was ablation. Neurosurgeon R. Meyers ablated the basal ganglia to control parkinsonian tremor of his patient in 1942. Other lesion sites included parts of the thalamus, globus pallidus, or internal capsule. The introduction of human stereotaxis in 1947 enabled reproducible navigation to an intended surgical target.

Introduction of levodopa therapy in 1968 temporarily suspended the apparent need for movement disorders surgery. However, lesional stereotactic surgery reemerged in the 1990s for patients experiencing levodopa therapy complications. Stereotactic targets included the thalamic nucleus ventralis intermedius (VIM), STN, and GPi (Sillay & Starr, 2009).

## Enabling Technology: Replacement of Ablation by Stimulation

French neurosurgeon Alim-Louis Benabid first recognized that high-frequency VIM stimulation could replace thalamotomy in the treatment of tremor (Benabid et al., 1987).

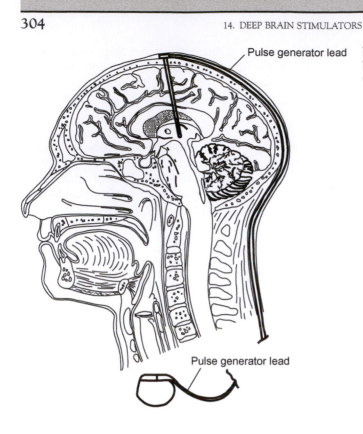

Pulse generator lead

**FIGURE 14.7** An illustration of deep brain stimulation, taken from US 5,716,377 (*Rise & King, 1998*).

Pulse generator lead

He worked with Medtronic to develop an implantable stimulator to treat Parkinson's disease (Figure 14.7). Benabid also conducted animal studies to optimize the stimulation target, which led him to stimulate the subthalamic nucleus to treat advanced PD symptoms. A group in Switzerland demonstrated that the globus pallidus was also an effective target (Breit et al., 2004).

Deep brain stimulation for suppression of essential or Parkinsonian tremor was approved by the FDA in 1997 (CDRH, 1997). Essential tremor typically occurs in the hands first, during their use. Parkinsonian tremor of the hands is most prominent when the hands are at rest. Originally, it was hypothesized that DBS acted like a functional lesion. It is now hypothesized that DBS increases output from the stimulated structure, in addition to suppressing local neuronal activity. The exact mechanism of action remains a matter of debate (Hiner et al., 2009).

## SYSTEM DESCRIPTION AND DIAGRAM

Deep brain stimulators are closely related to pacemakers. One or two DBS leads are implanted, with each electrode at the local target, through a 14-mm burr hole drilled in the skull. The leads are anchored at the skull and connect to percutaneous extension wires.

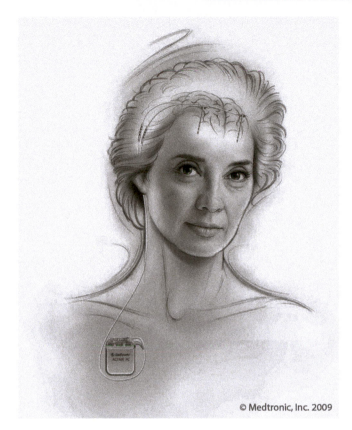

**FIGURE 14.8** Connections between DBS leads, percutaneous extensions, and implantable pulse generator (Medtronic, Minneapolis, Minnesota).

© Medtronic, Inc. 2009

These extension wires are inserted through a percutaneous tunnel in the head and neck, allowing connection to the implantable pulse generation (IPG).

The IPG resembles a cardiac pacemaker pulse generator, except that the parallel EGM sensing circuit is replaced with a series impedance (Z) measurement circuit. The impedance is calculated as the ratio of simultaneously measured stimulation voltage and current. The IPG is implanted in the same subcutaneous pocket (in the pectoralis fascia below the clavicle) as a pacemaker pulse generator (Figure 14.8). A programmer is used to adjust IPG settings. A system diagram for a deep brain stimulator system is given in Figure 14.9.

## Leads

In a Medtronic lead, each lead is made of platinum/iridium, with a polyurethane outer jacket. The lead body diameter is 1.27 mm; the lead length is 10–50 cm. Four electrodes at the end of a lead are spaced either 0.5 or 1.5 mm apart (Medtronic, 2008) (Figure 14.10). After the initial lead placement, an external test generator is used to verify that proper stimulation is possible. The leads are then secured in a lead anchoring device on the skull, before the pulse generator is implanted several weeks later.

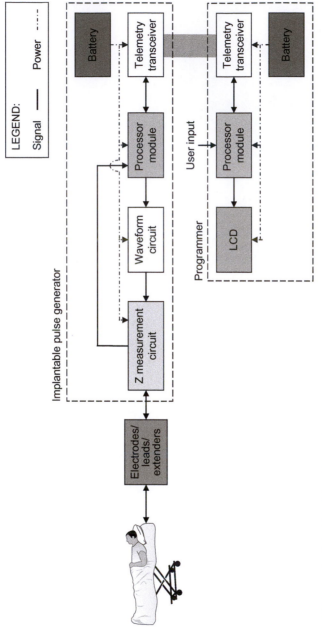

FIGURE 14.9　Deep brain stimulator system diagram.

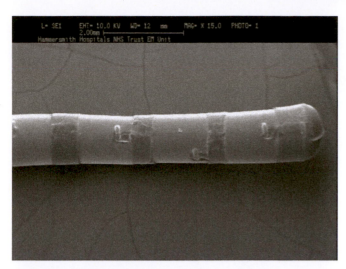

**FIGURE 14.10** Scanning electron micrograph of the surface and tip of the control Medtronic 3387 electrode, showing the four potentially active electrical contacts. Bar = 2 mm *[Reproduced by permission from Moss et al. (2004)].*

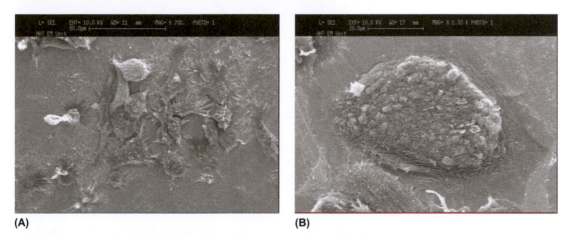

(A)  (B)

**FIGURE 14.11** Scanning electron micrographs of an explanted electrode at two magnifications. **A:** Cells on surface, bar = 50 μm. **B:** Foreign body giant cell, 20 μm *[Reproduced by permission from Moss et al. (2004)].*

After implantation, host reactions, such as the foreign body reaction, occur. In 21 explanted leads, recovered after 3–31 months from PD or dystonia patients, foreign body giant cells were always observed (Figure 14.11) (Moss et al., 2004). Increased tissue impedance over time may contribute to the required increase in stimulation current from the device during chronic patient therapy.

## Implantable Pulse Generator

During a second surgery, the pulse generator is implanted and connected by neck-tunneled extensions to the leads. Each extension is fabricated from silver MP35N wire,

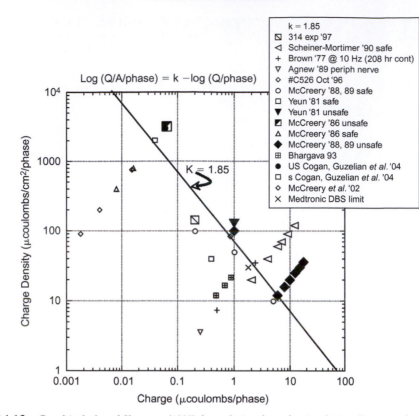

FIGURE 14.12    Graphical plot of Shannon (1992) formulation for safe stimulation. Data are derived from published and unpublished experiments. Filled-in symbols were deemed unsafe by the investigators. The open arrowhead symbols are from studies using imbalanced biphasic pulse applied to muscle through intramuscular stimulating electrodes [*Reproduced by permission from Mortimer & Bhadra (2009)*].

with an outer jacket of polyurethane. The pulse generator has a hermetically sealed titanium case.

Biphasic stimulation in the safe stimulation range predicted by Shannon is used. In 1992, Shannon constructed a mathematical model, based on data reported by several investigators, that predicts safe levels of stimulation (Shannon, 1992). In Figure 14.12, Medtronic's DBS limit of 30 $\mu C/cm^2$/phase is shown in the plot of charge density vs. charge (Mortimer & Bhadra, 2009). Typical pulse generator settings are 2.5–3.5 V amplitude, 60–120 $\mu$sec pulse width, and 130–185 pulses/sec (Medtronic, 2008).

The original IPG battery, used with bilateral stimulation, lasted 3–5 yr. It was based on lithium thionyl chloride (Li/SOCl$_2$) chemistry. These batteries have much higher energy densities than do Li/I$_2$ pacemaker batteries. Single-channel DBS requires 0.08–1-mA average current, while cardiac pacing requires 8–30-$\mu$A average current (Peckham et al., 2001).

To further increase battery longevity, an IPG with a rechargeable Li ion battery was recently developed and approved by the FDA (CDRH, 2009). These secondary Li ion

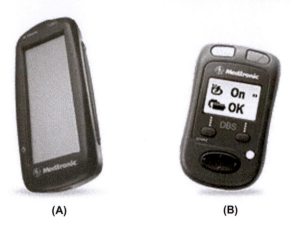

**(A)**                                    **(B)**

FIGURE 14.13  Medtronic N'Vision Model 8840 clinician programmer (**A**) and DBS patient programmer (**B**) (Medtronic, Minneapolis, Minnesota).

batteries differ from primary Li/I$_2$ batteries in that the anodes are not lithium metal, allowing for safe recharging under controlled conditions (Takeuchi et al., 2004).

The battery life of the associated Activa RC neurostimulator (Figure 14.1) is estimated to be 9 yr. If a patient chooses to recharge her battery with inductive coupling, a recharger is plugged into a wall outlet and positioned with a holster over the IPG. According to user instructions, "Charging sessions can take from a few minutes to more than 12 hours, depending on the charging efficiency, how often you charge, and your therapy needs" (Medtronic, 2007). Due to the electromagnetic coupling required, rechargeable pulse generators can be implanted only at a maximum tissue depth of 1 cm, compared to the typical maximum depth of 4 cm for nonrechargable devices.

## Stimulator Programmers

The clinician programmer is a battery-operated computer that enables four IPG therapy programs to be saved and DBS data to be downloaded and displayed on an LCD. DBS data include percentage of time the stimulator is on, recharge periods, low-battery periods, battery voltage, and electrode impedance. These data are transmitted and received over radiofrequency by the programmer head and IPG (Figure 14.13). A patient programmer, which has limited functionality, allows a patient to turn therapy on of off, change therapy programs, and check battery status.

## KEY FEATURES FROM ENGINEERING STANDARDS

As of 2010, FDA recommends a specific consensus standard for neurostimulators: *ANSI/AAMI/ISO 14708-3:2008 Implants for Surgery—Active Implantable Medical Devices—Part 3: Implantable Neurostimulators* (AAMI, 2009).

## Stimulation Pulse Characteristics

When sensory nerves are stimulated, the stimulation pulse must be stable to prevent unintentional changes in stimulation intensity that the patient might perceive as shocking, intermittent stimulation, or loss of therapy.

The pulse characteristics of amplitude, width, and rate are required to be determined for each channel and output (monopolar, bipolar) mode. Any observed variations in pulse shape are to be described. The measurements are to be made at minimum, typical, and maximum load impedances, as well as at a nominal impedance of $499 \, \Omega \pm 1\%$. Acceptable performance is not specified (AAMI, 2009).

## Battery Indication

For an implantable neurostimulator dependent on an implanted source of electrical energy such as a battery, an indication must be provided that gives an advanced notice of energy source depletion. The manufacturer defines the expected remaining service life following this notice.

Compliance to this requirement is confirmed by inspection of a design analysis provided by the manufacturer, supported by the manufacturer's calculations and data from test studies, as appropriate. No minimum battery life requirement is specified (AAMI, 2009).

## Biologic Effects

After implantation, a neurostimulator, lead, or lead extension must not release significant particulate matter within the body. During device development, this performance is tested by removing the implantable part aseptically from its sterilized package and immersing it in a saline solution bath of approximately 9 g/L in a neutral glass container, with at least five times the volume of the implantable part surface area. The container is covered with a glass lid and maintained in an agitated state at $37 \pm 2°C$ for 8–18 hr. A reference saline bath sample of similar volume is also prepared. After 8–18 hr have elapsed, the excess average count of particles in the test bath, compared to the reference bath, must not exceed 100/mL for particles greater than 5.0 μm and must not exceed 5/mL greater than 25 μm (AAMI, 2009).

## Effect of Miscellaneous Medical Treatments

Patients with implanted neurostimulators may require diagnosis or therapy from procedures such as MRI, positron emission tomography (PET) scans, therapeutic ultrasound, and lithotripsy. Implanting and follow-up physicians are advised to consider the potential effects of these procedures on the implanted neurostimulator. During the regulatory approval cycle, the approving agency considers the device's compatibility with these procedures by reviewing the manufacturer's documentation, including results from modeling, a design or risk assessment, test studies, or other appropriate means (AAMI, 2009).

## Immunity from Electromagnetic Interference

The implantable parts of the neurostimulator must not be susceptible to electrical influences due to external electromagnetic fields in the range of 10 Hz–30 MHz. These fields are generated by power frequency equipment and appliances.

Test coils must be capable of producing a uniform field in a plane parallel to the coils over an area of minimum radius 7.5 cm. The neurostimulator is placed into a saline bath of 0.27 S/m conductivity at the center of the central plane. The conductive lead is wrapped in a spiral around the neurostimulator so that the entire device fits inside the uniform area. The total area enclosed by a lead of length, $l$, must equal $0.09\,l^2$. Frequencies are then either swept or stepped in the following order for each decade: 1, 2, 3, 4, 5, 6, 7, 8, 9. The starting frequency of each decade is 0.1, 1, 10, 100, 1,000, 10,000 kHz.

To pass this requirement, the neurostimulator, during and after the test, must operate as intended, without loss of function and without unintentional responses. Additionally, after the test, the neurostimulator must still conform to device specifications.

Three other tests in ANSI/AAMI/ISO 14708-3:2008 assess protection from static magnetic fields, electromagnetic fields in the range of 30–450 MHz, and electromagnetic fields in the range of 450 MHz–3 GHz (AAMI, 2009).

## SUMMARY

Neurostimulation is the electrical process of inhibition, stimulation, modification, regulation, or therapeutic alteration of activity in the central, peripheral, or autonomic nervous systems. It is inherently nondestructive, reversible, and adjustable. Various neurostimulators are used as therapies. A deep brain stimulator is an electrical stimulator that discharges electrical current within the globus pallidus of the basal ganglia or subthalamic nucleus of the diencephalon, as a treatment for Parkinson's disease.

Normally, pigmented midbrain neurons in the substantia nigra pars compacta project to the putamen and caudate, where they release dopamine. When more than half of the dopaminergic nerve terminals in the striatum are gone, the motor impairments of Parkinson's disease arise. The mechanisms underlying these dopaminergic nerve terminals vary. PD patients have one or more characteristic findings, including resting tremor, rigidity, slowed movement, decreased dexterity, small handwriting, flexed posture, gait disorder, and imbalance. As the movement disorder progresses, dementia can develop over several years.

Deep brain stimulators are closely related to cardiac pacemakers. One or two DBS leads are implanted, with each electrode at the local target, through a 14-mm burr hole drilled in the skull. The leads are anchored at the skull, and connect to percutaneous extension wires. These extension wires are inserted through a percutaneous tunnel in the head and neck, allowing connection to the implantable pulse generator. The IPG is implanted in the same subcutaneous pocket, in the pectoralis fascia below the clavicle, as a pacemaker pulse generator. An external programmer is used to adjust IPG settings.

Key deep brain stimulator device features include stimulation pulse characteristics, battery depletion indicator, prevention of unintended biologic effects, effects of miscellaneous medical treatments, and immunity from electromagnetic interference.

## Acknowledgment

The author thanks Dennis Hepp for his detailed review of this chapter. Mr. Hepp is Managing Director at Rivertek Medical Systems.

## Exercises

**14.1** In 2009, Maurice Porter received a deep brain stimulator to treat essential tremor. Watch the short film about his surgery: http://www.pbs.org/kcet/wiredscience/video/255-deep_brain_stimulation.html. (Porter's story is contained within a parallel story of patient Steven Gulies.) What were the baseline tests that were conducted on Porter? Name all the medical devices that were videotaped during Porter's procedure.

**14.2** Draw a system diagram for a microelectrode recording system that measures single-action potentials. In consideration of these low-amplitude currents, describe the necessary requirements for electrode impedance, analog filtering, amplification, and analog-to-digital conversion.

**14.3** Why is more current required for deep brain stimulation than for cardiac pacing? Give specific reasons. Download the Activa recharging user manual: http://professional.medtronic.com/wcm/groups/mdtcom_sg/@mdt/@neuro/documents/documents/dbs-rc-37751-ptmanl.pdf. Name three specific advantages and three specific disadvantages for a PD patient using a rechargeable battery.

**14.4** Compare the rigor of neurostimulator requirements presented in this chapter with the rigor of pacemaker requirements from Chapter 3. Be specific.

**14.5** Is a deep brain stimulator a medical instrument? What developments are necessary before a deep brain stimulator can provide demand stimulation? Be specific.

Read Benabid et al. (1987):

**14.6** What was the new experimental therapy that was compared to standard of care VIM thalamotomy? What is a paresthesia? What type of device is a Medtronic Pisces? Describe the stimulation frequency in detail.

**14.7** For the 18 patients who received thalamotomies, how was tremor affected? For the 6 patients who were implanted with deep brain electrodes, how was tremor affected? How were the tremors affected in the patient who received both treatments? What are some of the stimulation factors, besides frequency, that could be modified to improve treatment results?

Read Deep-Brain Stimulation for Parkinson's Disease Study Group (2001):

**14.8** Describe the study design in detail. How did the reversible effect of stimulation complicate the trial? What is UPDRS? Describe some of the specific metrics in UPDRS.

**14.9** What is the Wilcoxon rank-sum test? Why was this statistic used? What were some of the other improvements during stimulation in the off-medication state besides an improved UPDRS score?

# References

AAMI (2009). *ANSI/AAMI/ISO 14708-3:2008 implants for surgery—active implantable medical devices—Part 3: Implantable neurostimulators*. Arlington, VA: AAMI.

Benabid, A. L., Pollak, P., Louveau, A., Henry, S., & De Rougemont, J. (1987). Combined (thalamotomy and stimulation) stereotactic surgery of the VIM thalamic nucleus for bilateral Parkinson disease. *Applied Neurophysiology, 50,* 344–346.

Breit, S., Schulz, J. B., & Benabid, A. L. (2004). Deep brain stimulation. *Cell Tissue Research, 318,* 275–288.

Bronstein, J., Carvey, P., Chen, H., Cory-Slechta, D., Dimonte, D., & Duda, J., et al. (2009). Meeting report: Consensus statement-Parkinson's disease and the environment: Collaborative on health and the environment and Parkinson's Action Network (CHE PAN) conference 26–28 June 2007. *Environmental Health Perspectives, 117,* 117–121.

CDRH (1997). *Approval order P960009*. Rockville, MD: FDA.

CDRH (2009). *Approval order P960009 S051*. Rockville, MD: FDA.

Dahodwala, N., Siderowf, A., Xie, M., Noll, E., Stern, M., & Mandell, D. S. (2009). Racial differences in the diagnosis of Parkinson's disease. *Movement Disorders, 24,* 1200–1205.

Deep-Brain Stimulation for Parkinson's Disease Study Group (2001). Deep-brain stimulation of the subthalamic nucleus or the pars interna of the globus pallidus in Parkinson's disease. *New England Journal of Medicine, 345,* 956–963.

Hiner, B. C., Molnar, G. F., & Kopell, B. H. (2009). Movement disorders: Anatomy and physiology relevant to deep brain stimulation. In E. S. Krames, P. H. Peckham, & A. R. Rezai (Eds.), *Neuromodulation*. London: Elsevier.

Krames, E. S., Peckham, P. H., Rezai, A. R., & Aboelsaad, F. (2009). What is neuromodulation? In E. S. Krames, P. H. Peckham, & A. R. Rezai (Eds.), *Neuromodulation*. London: Elsevier.

Lewitt, P. A. (2008). Levodopa for the treatment of Parkinson's disease. *New England Journal of Medicine, 359,* 2468–2476.

Medtornic (2007). *Recharger 37751 charging system user manual*. Minneapolis: Medtronic.

Medtronic (2008). *DBS lead kit for deep brain stimulation implant manual*. Minneapolis: Medtronic.

Morgan, J. P. (1982). The first reported case of electrical stimulation of the human brain. *Journal of the History of Medicine and Allied Sciences, 37,* 51–64.

Mortimer, J. T., & Bhadra, N. (2009). Fundamentals of electrical stimulation. In E. S. Krames, P. H. Peckham, & A. R. Rezai (Eds.), *Neuromodulation*. London: Elsevier.

Moss, J., Ryder, T., Aziz, T. Z., Graeber, M. B., & Bain, P. G. (2004). Electron microscopy of tissue adherent to explanted electrodes in dystonia and Parkinson's disease. *Brain, 127,* 2755–2763.

Peckham, P. H., Keith, M. W., Kilgore, K. L., Grill, J. H., Wuolle, K. S., & Thrope, G. B., et al. (2001). Efficacy of an implanted neuroprosthesis for restoring hand grasp in tetraplegia: A multicenter study. *Archives of Physical Medicine and Rehabilitation, 82,* 1380–1388.

Rise, M. T., & King, G. W. (1998). Method of treating movement disorders by brain stimulation. US 5,716,377.

Schaltenbrand, G., & Wahren, W. (1977). *Introduction to stereotaxis with an Atlas of the human brain*. Stuttgart: Thieme.

Shannon, R. V. (1992). A model of safe levels for electrical stimulation. *IEEE Transactions on Biomedical Engineering, 39,* 424–426.

Sillay, K. A., & Starr, P. A. (2009). Deep brain stimulation in Parkinson's disease. In E. S. Krames, P. H. Peckham, & A. R. Rezai (Eds.), *Neuromodulation*. London: Elsevier.

Starr, P. A. (2002). Placement of deep brain stimulators into the subthalamic nucleus or globus pallidus internus: Technical approach. *Stereotactic and Functional Neurosurgery, 79,* 118–145.

Starr, P. A. (2003). Technical considerations in movement disorders surgery. In M. Schulder (Ed.), *The handbook of stereotactic and functional neurosurgery*. New York: Dekker.

Takeuchi, E. S., Leising, R. A., Spillman, D. M., Rubino, R., Gan, H., & Takeuchi, K. J., et al. (2004). Lithium batteries for medical applications. In G. Nazri, & G. Pistoia (Eds.), *Lithium batteries: Science and technology*. Norwell, MA: Kluwer.

Van den Eeden, S. K., Tanner, C. M., Bernstein, A. L., Fross, R. D., Leimpeter, A., & Bloch, D. A., et al. (2003). Incidence of Parkinson's disease: Variation by age, gender, and race/ethnicity. *American Journal of Epidemiology*, *157*, 1015—1022.

Visser-Vandewalle, V., Temel, Y., & Ackermans, L. (2009). Deep brain stimulation in Tourette's syndrome. In E. S. Krames, P. H. Peckham, & A. R. Rezai (Eds.), *Neuromodulation*. London: Elsevier.

Weaver, F. M., Follett, K., Stern, M., Hur, K., Harris, C., & Marks, W. J., Jr., et al. (2009). Bilateral deep brain stimulation vs. best medical therapy for patients with advanced Parkinson disease: A randomized controlled trial. *JAMA*, *301*, 63—73.

# Cochlear Implants

In this chapter, we continue our discussion of neurostimulators and delve into the relevant physiology, history, and key features of cochlear implant (CI) systems. A cochlear implant system is an electrical stimulator that discharges electrical current to spiral ganglia, giving rise to action potentials in the auditory nerve fibers. It is also a medical instrument that can measure intracochlear evoked potentials, electrical field potentials generated by the electrodes, and electrode impedance. It consists of three main components: an external sound processor, an implanted stimulator, and a programmer (Figure 15.1). The programmer consists of an interface box and software running on a personal computer (PC).

**FIGURE 15.1** MED-EL MAESTRO CI System, consisting of the OPUS 2 with the D Coil, the CONCERTO Cochlear Implant, and the FineTuner remote control (MED-EL, Innsbruck, Austria).

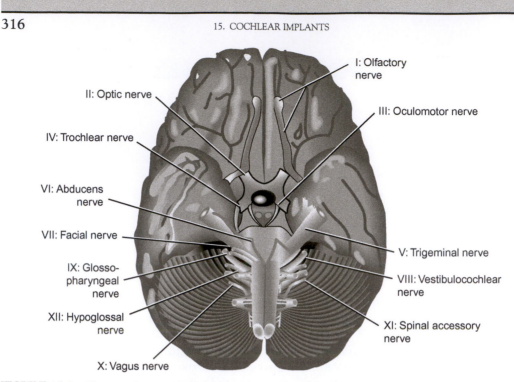

**FIGURE 15.2**    The cranial nerves *(Adapted from A.D.A.M., Atlanta, Georgia).*

Upon completion of this chapter, each student shall be able to:

1. Identify the mechanisms of hearing and sensorineural hearing loss.
2. Understand how speech processing strategy and the time of cochlear implantation affect speech recognition skills acquisition.
3. Describe five key features of cochlear implants.

## AUDITORY PHYSIOLOGY

An organ, such as the ear or eye, may act as a sensor to measure external stimuli. The output of this sensor is transmitted through a sensory nerve to the brain for further processing. For example, 12 paired cranial nerves innervate each side of the head, neck, and trunk. The nerves are named and numbered from anterior to posterior locations. Cranial nerve VIII is the vestibulocochlear, or auditory, nerve (Figure 15.2).

We begin this chapter by discussing how the ear (Figure 15.3) translates sound into signals that are transmitted by the auditory nerve to the brain for hearing perception.

### Hearing

*Sound* consists of the mechanical vibrations that are transmitted as longitudinal pressure waves in a medium such as air. Sound is measured in *decibels*, often in sound pressure level

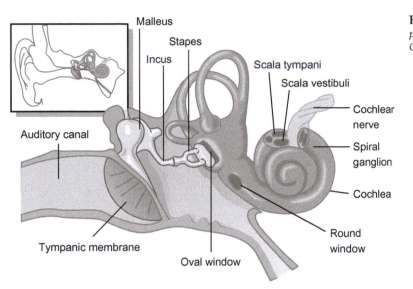

**FIGURE 15.3** The ear [*Reproduced by permission from Guyton & Hall (2006)*].

Malleus

Stapes

Incus

Scala tympani

Scala vestibuli

Cochlear nerve

Auditory canal

Spiral ganglion

Cochlea

Tympanic membrane

Round window

Oval window

(dB SPL), comparing a measured pressure level to the minimum pressure fluctuation detected by the ear. Assuming a threshold to audibility, $P_0$, to be $2 \times 10^{-5}$ Pa at 1000 Hz, sound is measured as 10 times the log of the squared ratio of the measured pressure, $P$, to threshold to audibility pressure. This calculation simplifies to 20 times the log of the pressure ratio:

$$dB \ SPL = 10 \log\left(\frac{P}{P_0}\right)^2 = 20 \log\left(\frac{P}{P_0}\right). \tag{15.1}$$

Using this equation, the level of the threshold of audibility is about 0–20 dB SPL for young adults with normal hearing, and the level of a hair dryer is about 80–90 dB SPL.

Sound pressure waves enter the ear through the cartilage of the *auricle* and are directed through the auditory canal to the tympanic membrane. The ridges and hollows of the auricle create additional cavities with resonances that are capable of influencing sounds with much higher frequencies. These two outer ear structures act as a natural filter to amplify frequencies critical to human speech by as much as 20 dB (May & Niparko, 2009).

In the middle ear, the tympanic membrane receives the pressure waves and transmits them to the bones that are collectively known as the ossicles. The *ossicles* are the malleus, incus, and stapes. The lateral end of the handle of the malleus is attached to the center of the tympanic membrane. As the pressure waves move through the ossicles, they decrease in amplitude but increase in force. The mechanical movement of the ossicles, as well as the real difference between the tympanic membrane and the stapes footplate, provides impedance matching. These combined actions overcome the impedance mismatch between air and the fluid-filled cochlea. Simultaneously, pressure waves travel through the air to the oval window, but the sensitivity for hearing is about 60 dB less than for ossicular transmission.

Next, the pressure waves begin to move through the incompressible fluid of the *cochlea*, or inner ear. The cochlea consists of three adjacent tubes: the scala vestibuli, the scala media, and the scala tympani (Figure 15.4). The cochlea is coiled upon itself for about

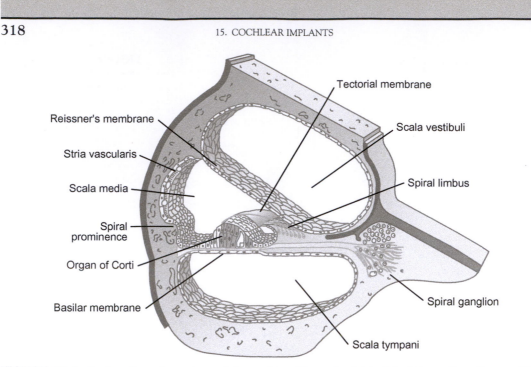

**FIGURE 15.4**   Section through one turn of the cochlea. *[Drawn by Sylvia Colard Keene. From Fawcett (1986). Figure reproduced by permission from Guyton & Hall (2006)].*

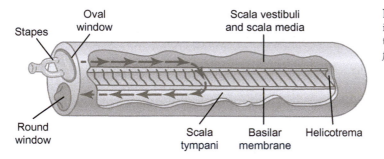

**FIGURE 15.5**   Movement of fluid in the cochlea after forward thrust of the stapes *[Reproduced by permission from Guyton & Hall (2006)].*

2.5 turns. The scala tympani and scala media are separated from each other by the osseous spiral ligament, on top of which lies the basilar membrane. The basilar membrane supports the organ of Corti, which contains sensory hair cells and supporting cells.

When pressure waves move the footplate of the stapes inward against the fluid of the cochlea, the basilar membrane is set in motion (Figure 15.5). Different pressure wave frequencies have different patterns of transmission (Figure 15.6). When a pressure wave reaches the portion of the basilar membrane that has a natural resonant frequency equal to the input sound frequency, the basilar membrane vibrates at its maximum, such that the wave energy dissipates and dies. Thus, sound frequencies from 20 Hz to 20 kHz become mapped by their physical distance of travel, or distance from the stapes (Figure 15.7).

As the basilar membrane vibrates, hair cells within the organ of Corti also move. There are about 15,000 inner hair cells at birth, which decline with age (May & Niparko, 2009).

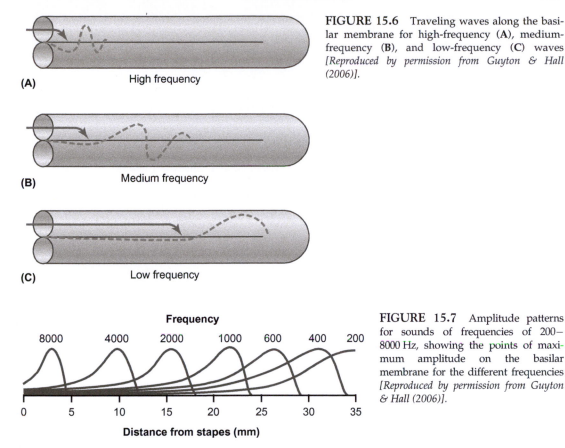

**FIGURE 15.6** Traveling waves along the basilar membrane for high-frequency (**A**), medium-frequency (**B**), and low-frequency (**C**) waves [*Reproduced by permission from Guyton & Hall (2006)*].

**FIGURE 15.7** Amplitude patterns for sounds of frequencies of 200–8000 Hz, showing the points of maximum amplitude on the basilar membrane for the different frequencies [*Reproduced by permission from Guyton & Hall (2006)*].

Hair cells, more properly called hair bundles, are composed of tens to hundreds of actin-based stereocilia. Hair bundle deflections toward the tallest stereocilia open transduction channels (Figure 15.8). Potassium ions pour through the channels, depolarizing the hair bundles and causing the release of the neurotransmitter glutamate. Spiral ganglion neurons innervating the hair bundles respond by firing action potentials. The spiral ganglion cells give rise to the auditory nerve fibers that constitute the auditory nerve, which transmits these action potentials to the brain (Vollrath et al., 2007; May & Niparko, 2009).

## Sensorineural Hearing Loss

When a problem with the outer or middle ear prevents sound from being properly conducted to the inner ear, the hearing loss is a *conductive loss*. In contrast, damage to the inner ear results in a *sensorineural hearing loss*. Cochlear implants are indicated for patients with sensorineural hearing loss whose hearing thresholds are 70–90 dB or poorer at 1000 Hz (CDRH, 2001).

Congenital sensorineural hearing loss, or hearing loss present at birth, can be traced to many factors, with about 50% due to genetic mutations, 25% due to environmental factors,

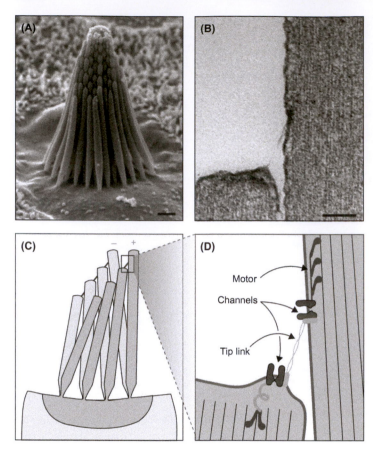

**FIGURE 15.8** Hair cell anatomy and physiology. **A:** Hair bundle in a bullfrog saccule, comprising ~60 stereocilia (scale bar = 1 micron). **B:** Two stereocilia and the tip link extending between them (scale bar = 0.1 micron). **C:** Deflection of hair bundle. Stereocilia tips remain in contact but shear with deflection. **D:** Schematic model of transduction. Shearing with positive deflection increases tension in tip links, which pull open a transduction channel at each end. Myosin motors slip or climb to restore resting tension. An elastic gating spring likely exists between a channel and the actin cytoskeleton [*Reproduced by permission from Vollrath et al. (2007)*].

and 25% of unknown etiology. Environmental factors that lead to congenital hearing loss include cytomegalovirus, ototoxic medications, rubella, and other viruses. These factors may cause cochlear swelling and hair cell loss.

Acquired sensorineural hearing loss is due to many factors, including aging, noise, ototoxicity, cardiovascular disease, hypoxia, bacterial and viral agents, and infectious disease. Very often the cause is unknown. Aging generally causes hair cell loss in the organ of Corti and degeneration of some auditory nerve fibers. Sustained noise or a single exposure acoustic trauma may cause permanent hair cell loss. Ototoxic medications, such as some antibiotics, also cause hair cell loss. Children who contract mumps or measles might lose hair cells and possibly spiral ganglia and nerve fibers (Almond & Brown, 2009).

## SPEECH PROCESSING

If sufficient spiral ganglia and the auditory nerve are intact, hearing may be augmented by providing electrical stimulation to the cochlea. Exactly reproducing the original auditory processing of speech is impossible. However, brain plasticity enables certain patients to decode new electrical stimulation patterns over time as speech when frequency-place

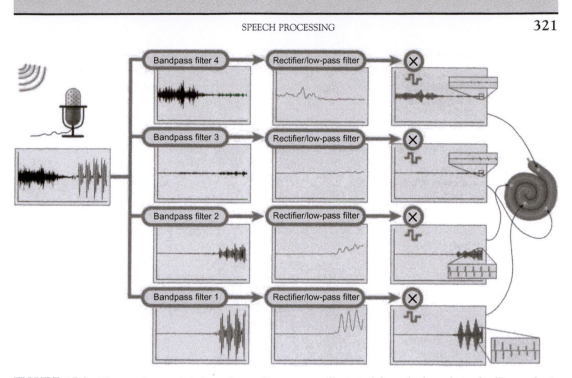

**FIGURE 15.9**   The continuous interleaved sampling strategy illustrated for only four channels. (Preemphasis filtering and envelope compression are not illustrated.) *[Reproduced by permission from Dorman & Wilson (2004)].*

coding (the location of basilar membrane vibration) and temporal coding (synchronization or phase-locking of ganglia to the period of a pressure wave) are at least grossly preserved. Adult patients who had language skills before deafness can associate new stimulation patterns with their memories of sounded speech. Pediatric patients younger than the age of 2 yr can learn spoken language at a normal or near-normal rate because their brains quickly adapt to electrical stimulation. But older children who did not have aural stimulation during their early years have a much more difficult time acquiring speech and oral language skills. What would normally be auditory areas of their brains have been reallocated for different tasks before cochlear implantation.

In many systems, cochlear implant processing is based on the *continuous interleaved sampling (CIS)* processing strategy. For this processing strategy, acoustic sound is first filtered to preemphasize speech frequencies less than 1.2 kHz. The prefiltered speech is bandpass filtered into 4–22 channels. Each frequency range is then rectified and lowpass filtered (at a relatively high cutoff frequency, such as 200–400 Hz), to obtain a speech envelope. Each envelope signal is compressed with a nonlinear mapping function in order to map the wide dynamic range of sound into the narrow dynamic range of electrically evoked hearing. The extracted envelope is used to modulate a biphasic pulse train. Each biphasic pulse channel output is directed to a single intracochlear electrode, with low- to high-frequency channels assigned to apical-to-basal electrodes. This mimics the order of frequency mapping in the normal cochlea. The pulse trains are interleaved in time, so that pulses across channels are not simultaneous and channel interaction is reduced (Figure 15.9).

Figure 15.9 simplifies how each envelope signal was originally mapped to an electrode channel. A single pulse train was sent from the transmitter coil to the receiver coil. Four channels of envelope data were interleaved onto this pulse train, with each channel using *pulse amplitude modulation* (*PAM*). In PAM, the amplitudes of regularly spaced pulses are proportional to the corresponding sample values of a continuous envelope signal, $e(t)$. For each channel, the PAM signal, $s(t)$, is mathematically equivalent to the convolution of $e_\delta(t)$, the instantaneously sampled version of $e(t)$, and the pulse, $h(t)$:

$$s(t) = e_\delta(t) * h(t) \qquad (15.2)$$

At the receiver end, the pulse train was demultiplexed to obtain separate channels of envelope data.

Various processing strategies preserve *formants*, which are groups of overtone pitches or vocal tract resonances, in spoken words. For any word, there is no unique signature involving specific frequencies. Rather, a formant pattern encompassing the relationship between two dominant magnitude peaks allows speech to be flexible. This flexibility enables normal speech variations, such as speech patterns from people in northern versus southern U.S. states, to be accurately recognized. When normal speech is bandpass filtered and transformed into four amplitude-modulated sine waves, normal-hearing listeners can understand about 90% of words in simple sentences (Figure 15.10) (Dorman & Wilson, 2004; Wilson & Dorman, 2009).

# CLINICAL NEED

According to the U.S. Census Bureau's Survey of Income and Program Participation, about 8 million children (aged 6–17 yr) and adults in the United States in 2002 were deaf or hard of hearing. About 1 million of these individuals are functionally deaf. *Functionally deaf* is defined as deaf or unable to hear normal conversation at all, even with a hearing aid. Fifty percent of those who are functionally deaf are at least 65 yr (Mitchell, 2006).

Each year, about one in every 1000 newborns has congenital hearing loss. Because universal newborn hearing screening is mandatory in 37 states and the District of Columbia, many of these cases are identified in hospitals and birthing centers. Newborn hearing screening is typically conducted using otoacoustic emissions and/or auditory brainstem response testing before the infant leaves the hospital. Neither test requires a behavioral response from the newborn. Both tests can be quickly and easily administered by trained personnel. When the infant does not pass the screening in the hospital, then follow-up testing is done later on using additional diagnostic tests.

After cochlear implantation, adult and pediatric patients are tested for recognition of spoken speech, often using single words and sentences. Adult CI users score, on average, about 50–60% on the most difficult monosyllabic word recognition test; however, variability is high with patients performing both better and worse. On average, adult patients with short durations of deafness prior to implantation fare better than patients with long periods of deafness. Other factors, such as the cause of deafness and central auditory processing, can also contribute to performance.

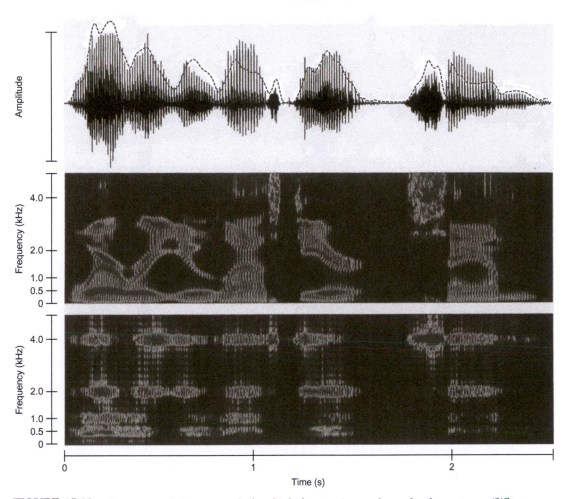

**FIGURE 15.10** Human speech is composed of multiple frequencies, as shown for the sentence, "Where were you last year, Sam?" The waveform for the sentence is shown at top (black trace), along with the speech envelope (dashed line). In the middle panel, the same sentence is plotted according to its component frequencies. The energy at each point in the frequency spectrum is indicated on a scale from low (light grey) to high (dark grey). For any one word, a formant is the relationship between the dark grey areas. The bottom panel shows the same audio signal after being processed to remove all information except the envelopes of four contiguous bands from 300 Hz to 5 kHz, with center frequencies of 0.5, 1, 2, and 4 kHz. Remarkably, this simplified signal is almost as intelligible as actual speech *[Reproduced by permission from Dorman & Wilson, 2004)]*.

For pediatric cochlear implant users, the effect of early auditory deprivation can be demonstrated in research *cortical evoked potentials* (*CEP*). The CEP can be recorded as an electrocephalogram that is analyzed in response to a spoken syllable, such as "ba." In congenitally deaf children implanted (*n* = 12) before the age of 4 yr, the CEP latency moved to within 5 months into the normal latency range. The children had heard nothing for up to 3.5 yr but showed age-appropriate time of cortical activity in response to speech.

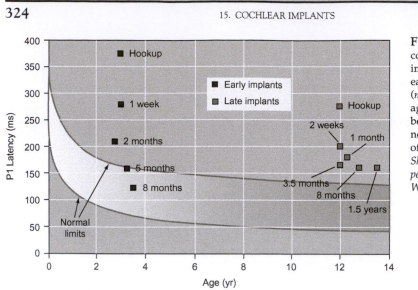

FIGURE 15.11 CEPs in congenitally deaf children implanted before ($n = 12$ early implants) and after ($n = 8$ late implants) 4 yr of age over time. The area between the 2 curves is the normal range as a function of age [*Data courtesy of Anu Sharma. Figure reproduced by permission from Dorman & Wilson (2004)*].

However, in congenitally deaf children implanted after ($n = 8$) the age of 4 yr, the CEP latency did not move into the normal latency range even after 1.5 yr of CI use (Figure 15.11) (Dorman & Wilson, 2004; Wilson & Dorman, 2009).

# HISTORIC DEVICES

Clinical stimulation of the auditory system can be traced to Alessandro Volta's work in the late 1700s. Volta, who invented the voltaic battery from dissimilar metals, believed that the physiologic effects of nerve stimulation are affected by the type of nerve stimulated, but not by the type of stimulus used. He stimulated his own tongue to elicit taste, his eye to elicit light, and his ears to elicit sound. Specifically, Volta produced noise by applying two probes connected to a battery of 30—40 silver-zinc elements into both of his ears (Piccolino, 2000). He described this sound as "a boom within the head," followed by a sound resembling thick soup boiling (Volta, 1800).

## Early Implantable Devices

In 1957, otolaryngologist André Djorno and scientist Charles Eyries collaborated to implant the first auditory prosthesis. The patient who received the induction coil was able to discriminate between lower- and higher-frequency stimuli but could not understand speech.

Djorno and Eyries's work was communicated to otolaryngologist William House by one of House's patients. In 1962, the House group implanted an induction coil, embedded in PMMA, into a patient's cochlea. The induction coil received square wave stimulation, which enabled the patient to perceive the rhythm of music and speech and to discriminate between male and female voices (Doyle et al., 1963). This first device evolved into the House 3M single-channel cochlear implant, which obtained FDA approval in 1984

**FIGURE 15.12** The House 3M cochlear implant consisted of a single-channel speech processor, transmitter (oval piece), microphone (with tie pin—type fastener), and implant (not shown). The speech processor was worn on the body. The transmitter coil was placed over the implant behind the ear. The microphone was attached to clothing (*Reproduced by permission from the Hearing Aid Museum, Stewartstown, Pennsylvania*).

(Figure 15.12). More than 1000 House 3M devices were implanted between 1972 and the mid-1980s.

In 1975, NIH sponsored a thorough evaluation of patients who had received single-channel cochlear implants. After 13 patients received psychoacoustic, audiologic, and vestibular testing, it was concluded that single-channel devices could not support speech understanding. However, the patients' speech production, lip reading, and quality of life were all enhanced with the device. Later NIH funding spurred research at several institutions. Work funded by a variety of sources at University of California San Francisco, University of Melbourne, Australia, and Technical University of Vienna, Austria, became the foundation of cochlear implants manufactured by Advanced Bionics, Cochlear Corporation, and MED-EL Corporation, respectively (Eisen, 2009).

## Enabling Technology: Continuous Interleaved Sampling Speech Processing Strategy

By the late 1980s, some investigational multichannel cochlear implants used the *compressed analog* (*CA*) processing strategy. After a signal was compressed, it was bandpass filtered into four frequency bands. Each band was transmitted simultaneously to four electrodes. In 1991, bioengineer Blake Wilson demonstrated that CIS processing (Figure 15.9, described in a previous section) significantly improved the speech recognition of seven patients. These patients, who had used a 4-channel cochlear implant with CA processing from 2 to 5 yr, received CIS processing with the same implant for a few days before testing. Factors contributing to improved performance included nonsimultaneous stimuli between electrodes, 5—6 CIS channels versus 4 CA channels, representation of rapid envelope variations through the use of high pulse rates, and preservation of amplitude cues with channel-by-channel compression (Wilson et al., 1991). This speech processing strategy is still used in FDA-approved cochlear implants today, although it is no longer the default processing strategy in these implants.

# SYSTEM DESCRIPTION AND DIAGRAM

Current cochlear implants use a processor that is worn behind the ear. Sound is sensed by the microphone of the processor and is processed using the CIS or another processing strategy. Encoded data parameters are transmitted through the skin from the transmitter to receiver over radio frequency. Magnets hold the processor on the head so that it can communicate with the implant. The received data are decoded into stimulation inputs for 8 or more channels, which are sent to the electrode array. The electrode array resides in the cochlea, where it stimulates remaining spiral ganglia and the auditory nerve (Figure 15.13).

Cochlear implant systems from three manufacturers have been approved by the FDA for market release: Advanced Bionics, Cochlear Corporation, and MED-EL. To simplify the discussion, we focus specifically on the MED-EL architecture. A system diagram for the MED-EL cochlear implant is given in Figure 15.14.

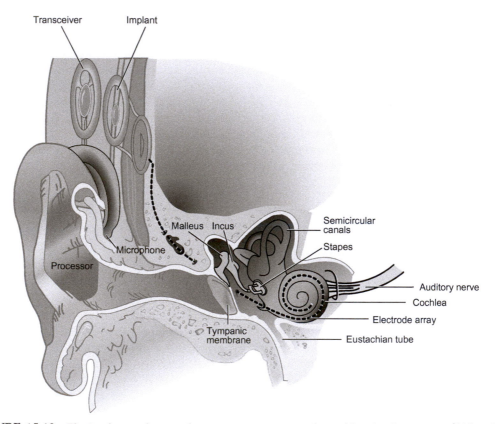

**FIGURE 15.13**   The implant and external processor components of a cochlear implant system [(*Adapted from Papsin & Gordon (2007)*].

## Processor

As shown in Figure 15.14, sound (in the range of 70–8500 Hz) is converted to an electrical signal by a microphone and processed by the digital signal processor (DSP) module, according to a preselected processing strategy, such as CIS. The extracted signal features are encoded for transmission, amplified, and transmitted over radiofrequency (12 MHz, 0.6 MB/s) by the transceiver coil to the implant. The transceiver also receives electrode voltage data from the implant, which are sent to the processor module.

When the user switches on the processor, the processor module provides feedback by causing a red LED to flash on and off. The number of flashes corresponds to the listening environment program that is currently selected. Four programs are stored in the processor module. Including four different programs allows users to select a program for a particular listening situation, such as listening to speech in a noisy background. Adults who wish to select another program, change microphone sensitivity, or change program volume have access to a remote control (not shown) that communicates with the processor by radiofrequency. A 3.5-mm stereo jack input allows sounds from an MP3 player or teacher's microphone to enter the processor module directly. A rechargeable 3.7-V lithium ion battery provides a full day of about 16 hr operation.

## Implant

The transceiver coil in the implant receives the encoded signal, as well as about 20–40 mW of power. The power is required by an ASIC that provides the decoder,

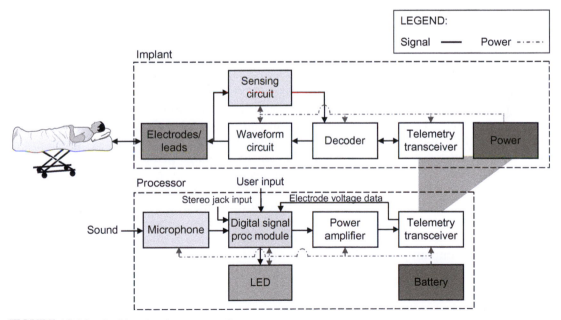

FIGURE 15.14  Cochlear implant system diagram.

waveform circuit, and sensing circuit functions. The received data parameters are decoded and used to construct the proper biphasic stimulation waveforms for selected channels. This circuit includes 24 current sources for 12 pairs of electrodes. Each channel provides about 0–1.2 mA of current, with a total stimulation rate of 51 kHz. A channel cannot be stimulated by DC current because a capacitor is serially connected to block its delivery.

By providing about 10 mA current between any two electrodes, the impedance can be determined from the measured voltage. Evoked compound action potentials (ECAPs), which are discussed in the Electrode Array subsection, can also be measured.

The Concerto implant is hermetically sealed in titanium housing. An alternate Pulsar package for the same electronics is made from $Al_2O_3$. For the Concerto package, the reference electrode is part of the implant. For the Pulsar package, the more typical reference electrode is a three-fold, multistranded platinum iridium wire, encased in silicone rubber, which is implanted outside the cochlea between the temporalis muscle and the cranium. The transceiver magnet is unaffected by up to 0.2 T of magnetic resonance imaging (MRI) scanning.

## Electrode Array

Stimulation current is transmitted to the active electrode array. The electrode array is composed of single-strand platinum iridium wire and 24 platinum electrodes, encased in silicone rubber (Figure 15.15). The electrodes are arranged as paired interconnected surfaces, which result in 12 monopolar stimulation channels. The interelectrode spacing is 2.4 mm, across an active distance of 26.4 mm. The electrode array may be inserted up to 31 mm into the cochlea.

Typically, monopolar, rather than bipolar, stimulation is used because monopolar stimulation requires less current for producing auditory percepts, without a degradation in performance. Further, differences in the currents required to produce equally loud percepts across individual electrodes are substantially lower with monopolar than with bipolar stimulation, which can simplify speech processor fitting for patients (Wilson & Dorman, 2009; Zeng et al., 2008; MED-EL, 2010).

The electrodes perform sensing, as well as electrical stimulation. The electrically evoked compound action potential (ECAP) may be measured, which reflects synchronous auditory nerve activity elicited by electrical stimulation. The typical neural response waveform for a biphasic stimulus is characterized by a negative voltage peak, $N1$, with a latency of 0.2–0.4 ms, followed by a positive voltage peak, $P2$, with a latency of 0.5–0.8 ms (Figure 15.16). ECAP amplitude is calculated as the difference between the $N1$ and $P2$ amplitudes, with amplitudes increasing monotonically with increasing stimulation current (Alvarez et al., 2010). (An animation of the cochlear implant procedure can be viewed at http://www.fda.gov/medicaldevices/productsandmedicalprocedures/implantsandprosthetics/cochlearimplants/ucm133345.htm.)

FIGURE 15.15    MED-EL standard electrode array (MED-EL, Innsbruck, Austria).

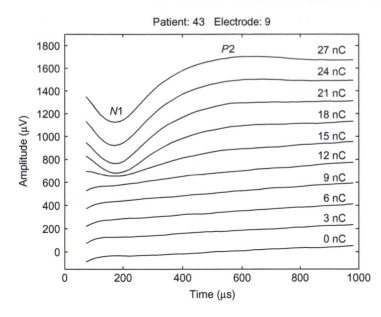

Patient: 43   Electrode: 9

**FIGURE 15.16** ECAP response acquired with increasing stimulation level, for electrode 9 of patient 43 in a specific study. The unit nC is nanoCoulomb [*Reproduced by permission from Alvarez et al. (2010)*].

## Programmer

Approximately 4 weeks after successful implant surgery, a patient receives the external processor. An audiologist uses fitting software and an interface box to connect a PC to the processor. Each channel is tested to ensure that it is electrically functional (neither a short or open circuit) and does not cause unpleasant nonauditory sensation such as dizziness or other facial nerve stimulation. For each viable channel, the threshold, T, and most comfortable loudness, C, levels are determined. The difference between these levels is the *dynamic range*, which is a function of processing strategy and electrode configuration. The C levels are also balanced across the electrode array. For pediatric patients without previous sound exposure, ECAP levels provide additional feedback for the fitting procedure, although their correlation with thresholds and comfort levels is not very strong. The final fitting map, or program, is downloaded to the external processor (Zeng et al., 2008). Up to four programs may be stored in the external processor, and accessed with the remote control. The patient may later receive a second implant for bilateral cochlear implantation.

## KEY FEATURES FROM ENGINEERING STANDARDS

Although the FDA recommends seven consensus standards related to acoustics or speech intelligibility as of 2010, it does not recommend a consensus standard specific to cochlear implant hardware. The first cochlear implant standard was recently approved in Europe. We discuss requirements from this standard: EN 45502-2-3 *Active Implantable Medical*

*Devices—Part 2-3: Particular Requirements for Cochlear and Auditory Brainstem Implant Systems* (BSI, 2010).

## Output Amplitude

When each channel is programmed to its maximum amplitude and pulse width, the amplitude is required to be accurate to within ±5%, "taking all errors into consideration." Testing is conducted with a 70-dB-SPL input applied to the microphone. Each output is connected to a 1000-±1-$\Omega$ load resistor. Measurements of peak outputs are made using an oscilloscope.

The mean of amplitudes and pulse widths and their ranges are recorded. No specific number of test devices is specified (BSI, 2010).

## Immunity to MRI Scanning

Immunity during MRI scanning is desirable because some implanted patients require repeated MRI for monitoring of central nervous system disease. If an implant meets this requirement, the manufacturer declares the safe MRI field strengths.

To test MRI safety, an implant sample is completely immersed in a plastic container filled with 9 g/L saline, and placed in the center of an MRI machine. Two identical covered plastic containers are selected with a volume that is sufficient to submerge the implant and that can accommodate the volume saline that is 3 ± 0.3 times the volume of the implant. Before being placed in one container, the implant is stored for 24 hr at the temperature of the scanning location in the MRI department. Both containers, filled with 9 g/L saline, are stored for 24 hr before the test in the same location. The containers are then placed in a position in the MRI machine judged to receive the highest amount of RF power. An MRI test sequence representing the worst-case clinical scan typically performed, with the highest absorption rate, is initiated and run for at least 15 min.

It must be verified that, after the scan, the implant still conforms to manufacturer's specifications. A reduction in strength of the internal magnet is acceptable if the manufacturer makes available an alternative fixation method and appropriate information in the labeling (BSI, 2010).

## Effect of Tensile Forces

Between birth and adulthood, pediatric patients are subject to skull growth of 12 ± 5 mm from the round window to the sino-dural angle (most lateral aspect of the mastoid bone). Leads need to be designed to withstand elongation, which could be experienced during skull growth.

Tensile force is tested on two specimens. Specimen A is the implant and its lead, in the condition shipped to the customer. Specimen B is the implantable lead without the stimulator. Both specimens are preconditioned for at least 10 days in a bath of approximately 9 g/L saline at 37 ± 5°C. Immediately prior to testing, the leads are rinsed in distilled or deionized water and then wiped free of surface water.

For Specimen A, one clamp is attached to the stimulator, and the other is attached to the most distal part of the lead subject to elongation. The initial distance between clamping points is measured. The lead is then subjected to elongation of at least 15 mm or a tensile force of minimum 1 N, whichever is reached first. The tensile stress is to be sustained for at least 1 min and then relieved. The test is repeated for the other lead. Specimen A is then returned to the saline bath for at least 1 hr of immersion.

For Specimen B, both sides of the lead are clamped. The initial distance between clamping points is measured. The lead is then subjected to elongation of at least 15 mm or a tensile force of minimum 1 N, whichever is reached first. The tensile stress is to be sustained for at least 1 min and then relieved. The lead is then subjected to a 1-kHz square- wave voltage, with a peak-to-peak voltage of twice the maximum voltage of the implant system. The voltage is applied for a minimum of 15 s between each combination of conducting pairs inside the lead. The impedance between points in each pair is measured.

For Specimen A, the electrical continuity of each conduction path and the insulation between the wires in each pair inside each lead must be verified. For Specimen B, it must be verified that the impedance between each pair of conducting wires exceeds 100 kΩ. For both specimens, no permanent functional damage must be exhibited (BSI, 2010).

## Effect of Direct Impact

In consideration of pediatric patients, who are more likely than adult patients to fall and damage their implants, the implantable part of the device is required to be constructed so that it is not damaged by impacts during normal use.

To test this requirement, two implant samples are each subjected to an applied force. Each sample is clamped into a testing apparatus and then subjected to 1.5-±5%-J impact energy by a pendulum hammer or vertical hammer. The sample is affixed to a rigid and flat supporting surface so that the side facing the cranial bone during normal use lies evenly on the surface. A piece of 3-mm, 40–60° silicone is placed evenly over the implant. One sample is struck perpendicularly. The second sample is struck off-centrically at what is considered to be the weakest exposed point of the stimulator.

It must be verified that, after impact, both stimulator samples still meet manufacturer amplitude and pulse width, impedance measurement accuracy, level of MRI safety, and hermeticity requirements (BSI, 2010).

## Effect of Atmospheric Pressure

For patients who want to scuba dive, manufacturers provided an atmospheric pressure specification. To test this requirement, an implant is placed in a suitable water pressure chamber and cycled 20 times from ambient pressure to a maximum pressure. This maximum must be 1.5 times the pressure specified in the manufacturer's documentation. The rate of pressure change must be at least 100 kPa/min; the maximum pressure must be held for at least 1 min.

No specific results are mandated. Compliance is confirmed by inspection of test procedures and results (BSI, 2010).

# SUMMARY

A cochlear implant system is an electrical stimulator that discharges electrical current to spiral ganglia, giving rise to action potentials in the auditory nerve fibers. It is also a medical instrument that can measure intracochlear evoked potentials, electrical field potentials generated by the electrodes, and electrode impedance. It consists of three main components: an external sound processor, an implanted stimulator, and a programmer.

When pressure waves move the footplate of the stapes inward against the fluid of the cochlea, the basilar membrane is set in motion. Different pressure wave frequencies have different patterns of transmission. When a pressure wave reaches the portion of the basilar membrane that has a natural resonant frequency equal to the input sound frequency, the basilar membrane vibrates at its maximum such that the wave energy dissipates and dies. Thus, sound frequencies from 20 Hz to 20 kHz become mapped by their physical distance of travel, or the distance from the stapes.

As the basilar membrane vibrates, hair cells within the organ of Corti also move. There are about 15,000 inner hair cells at birth, which decline with age. Hair cells, more properly called hair bundles, are composed of tens to hundreds of actin-based stereocilia. Hair cell deflections toward the tallest stereocilia open transduction channels. Potassium ions pour through the channels, depolarizing the hair bundles and causing the release of the neurotransmitter glutamate. Spiral ganglion neurons innervating the hair bundles respond by firing action potentials. The spiral ganglion cells give rise to the auditory nerve fibers that constitute the auditory nerve, which transmits these action potentials to the brain.

Sensorineural hearing loss results from damage to the inner ear. If sufficient spiral ganglia and the auditory nerve fibers are intact, hearing may be augmented by providing electrical stimulation to the cochlea. Exactly reproducing the original cochlear processing scheme of speech is impossible because these firing patterns are too complex. However, brain plasticity enables certain patients to decode new electrical stimulation patterns over time as speech when frequency-place coding and temporal coding are at least grossly preserved. In many systems, cochlear implant processing is based on the continuous interleaved sampling processing strategy.

The current architecture of cochlear implants includes a processor, implant with electrode array, and programmer. Key cochlear implant features include output amplitude, immunity to MRI scanning, the effect of tensile forces, the effect of direct impact, and the effect of atmospheric pressure.

## Acknowledgment

The author thanks Dr. Claude Jolly for his detailed review of this chapter. Dr. Jolly is Director of Electrode Development for MED-EL.

### Exercises

**15.1** Using an audio acquisition system provided by your instructor, speak "Where were you last year, Sam?" into a microphone, and create a spectrogram of these words. Compare your spectrogram to those of two classmates. Plot all three spectrograms. How many formants are present? How do the formants differ among speakers?

**15.2** Draw the system diagram for a Siemens 700 SP hearing aid, which is worn behind the ear. Include access to TEK, but not the other optional accessories. Under what physiologic conditions in the outer, middle, and inner ears is this hearing aid effective? Be specific.

**15.3** What electrode biocompatibility issues occur during and after cochlear implantation? Be specific. Can a patient hear music after implantation? Explain your reasoning.

**15.4** What are the issues complicating the design of a completely implantable cochlear implant (i.e., no external processor)? Be specific.

Read Doyle et al. (1963):

**15.5** Where did electrode stimulation occur in these patients? What was the refractory limitation in frequency? How was this overcome? How is the audio added to the pulse train stimulation?

**15.6** How were two electrodes permanently implanted? Describe how sound is transmitted to this system.

Read Alvarez et al. (2010):

**15.7** Describe the ECAP acquisition protocol in detail. Why was the T level not considered?

**15.8** What is the specific hypothesis being tested? Describe the model. Was the hypothesis proven or disproven? Give reasons.

# References

Almond, M., & Brown, D. J. (2009). The pathology and etiology of sensorineural hearing loss and implications for cochlear implantation. In J. K. Niparko (Ed.), *Cochlear implants: Principles and Practices* (2nd ed.). Philadelphia: Lippincott Williams & Wilkins.

Alvarez, I., De La Torre, A., Sainz, M., Roldan, C., Schoesser, H., & Spitzer, P. (2010). Using evoked compound action potentials to assess activation of electrodes and predict C-levels in the Tempo + cochlear implant speech processor. *Ear Hearing, 31*, 134–145.

BSI (2010). *EN 45502-2-3:2010 Active implantable medical devices—Part 2–3: Particular requirements for cochlear and auditory brainstem implant systems.* Brussels: BSI.

CDRH (2001). *Approval order P000025.* Rockville, MD: FDA.

Dorman, M. F., & Wilson, B. S. (2004). The design and function of cochlear implants. *American Scientist, 92*, 436–445.

Doyle, J. B., Jr., Doyle, J. H., Turnbull, F. M., Abbey, J., & House, L. (1963). Electrical stimulation in eighth nerve deafness. A preliminary report. *Bulletin of the Los Angeles neurological society, 28*, 148–150.

Eisen, M. D. (2009). The history of cochlear implants. In J. K. Niparko (Ed.), *Cochlear implants: Principles and Practices* (2nd ed.). Philadelphia: Lippincott Williams & Wilkins.

Fawcett, D. W. (1986). *Bloom and Fawcett: A textbook of histology* (11th ed.). Philadelphia: WB Saunders.

Guyton, A. C., & Hall, J. E. (2006). *Textbook of medical physiology* (11th ed.). Philadelphia: Elsevier Saunders.

May, B. J., & Niparko, J. K. (2009). Auditory physiology and perception. In J. K. Niparko (Ed.), *Cochlear implants: Principles and Practices* (2nd ed.). Philadelphia: Lippincott Williams & Wilkins.

Med-El (2010). *Hear. There and everywhere. MAESTRO cochlear implant system.* Durham, NC: MED-EL.

Mitchell, R. E. (2006). How many deaf people are there in the United States? Estimates from the survey of income and program participation. *Journal of Deaf Studies and Deaf Education, 11*, 112–119.

Papsin, B. C., & Gordon, K. A. (2007). Cochlear implants for children with severe-to-profound hearing loss. *New England Journal of Medicine, 357*, 2380–2387.

Piccolino, M. (2000). The bicentennial of the Voltaic battery (1800–2000): The artificial electric organ. *Trends in Neurosciences, 23*, 147–151.

Vollrath, M. A., Kwan, K. Y., & Corey, D. P. (2007). The micromachinery of mechanotransduction in hair cells. *Annual Review of Neuroscience, 30*, 339–365.

Volta, A. (1800). On the electricity excited by the mere contact of conducting substances of different species. *Philosophical Transactions of the Royal Society of London, 90*, 403–431.

Wilson, B. S., & Dorman, M. F. (2009). The design of cochlear implants. In J. K. Niparko (Ed.), *Cochlear implants: Principles and Practices* (2nd ed.). Philadelphia: Lippincott Williams & Wilkins.

Wilson, B. S., Finley, C. C., Lawson, D. T., Wolford, R. D., Eddington, D. K., & Rabinowitz, W. M. (1991). Better speech recognition with cochlear implants. *Nature, 352*, 236–238.

Zeng, F. G., Rebscher, S., Harrison, W. V., Sun, X., & Feng, H. (2008). Cochlear implants: System design, integration and evaluation. *IEEE Transactions on Biomedical Engineering, 1*, 115–142.

# 16

# Functional Electrical Stimulators

In this final chapter on neurostimulators, we move from the brain and head to the spinal nerves. The spinal nerves and cranial nerves (discussed in Chapter 15) make up the peripheral nerves. The spinal nerves connect the brain and spinal cord with skin sensory receptors and consciously controlled muscles. When a spinal cord injury occurs, a patient faces significant neurologic dysfunction and disability.

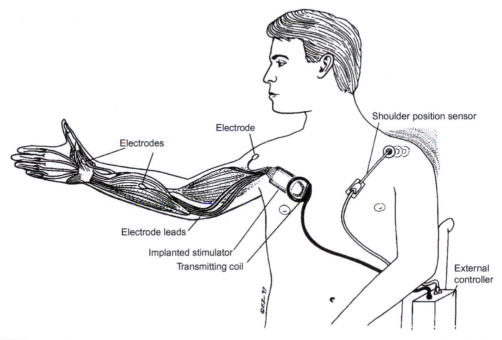

**FIGURE 16.1** The most recent implantable FES system approved by the FDA: the NeuroControl Freehand. *[Original figure reprinted by permission of NeuroControl Corp. This figure reproduced by permission from Peckham et al. (2001)].*

**(A)**                          **(B)**

**FIGURE 16.2** The NESS H200 (A) and L300 (B) stimulate hand and leg muscles with surface electrodes (Bioness, Inc., Valencia, California).

*Functional electrical stimulation (FES)* is "the application of electrical current to excitable tissue to supplement or replace function that is lost in neurologically impaired individuals" (Peckam & Knutson, 2005). We have already discussed one functional electrical stimulator: the cochlear implant. In this chapter, we specifically consider implantable functional electrical stimulators for restoration of limb function in spinal cord injury (SCI) patients (Figure 16.1). Surface functional electrical stimulators for spinal cord injury provide some utility (Figure 16.2), but are beyond the range of this discussion. The holy grail in FES is to enable paraplegics to walk (Loeb & Davoodi, 2005).

Upon completion of this chapter, each student shall be able to:

1. Understand the mechanism underlying spinal cord injury.
2. Explain how functional electrical stimulation can enable muscle contraction in SCI patients.
3. Describe how the bion and direct neural interface systems can be incorporated into functional electrical stimulation.

## SPINAL NERVES

The spinal cord continues from the brain stem. As it passes through the vertebrae, the spinal cord branches into 31 pairs of spinal nerves: 8 cervical, 12 thoracic, 5 lumbar, 5 sacral, and 1 cocygeal nerves. As shown in Figure 16.3, some of these nerves emerge below the pedicle of the vertebra for which they are named.

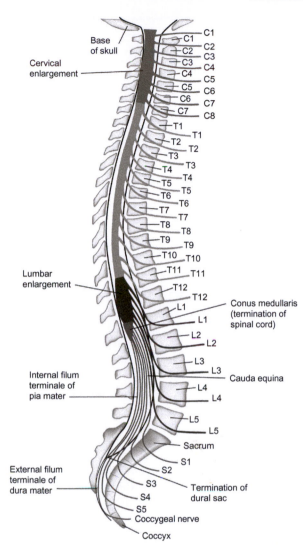

**FIGURE 16.3** The organization and anatomy of the spinal cord and spinal roots. C = cervical, T = thoracic, L = lumbar, S = sacral [*Reproduced by permission from Rosenow (2009)*].

At each spinal cord segment, a pair of spinal nerves is formed. The *dorsal horn* is the location of cell bodies on which the afferent sensory nerves from the periphery synapse. It gives rise to several dorsal rootlets, which converge to form a dorsal root. Similarly, the anterior, or ventral, horn is the location of cell bodies upon which the efferent motor nerves to the periphery synapse. It gives rise to several ventral rootlets, which converge to form a ventral root. The dorsal and ventral roots fuse to form a spinal nerve, which is a mixed nerve carrying both sensory and motor fibers. Each nerve travels only 1−2 mm before branching into the smaller nerve fibers that supply muscles (University of Michigan, 2010) (Figure 16.4).

A peripheral nerve fiber consists of bundles of *axons*. The axon of one motor neuron extends from the anterior horn of the spinal cord to the muscle. A sensory axon extends from the point of sensation, such as the skin, to the dorsal root ganglion, which is just

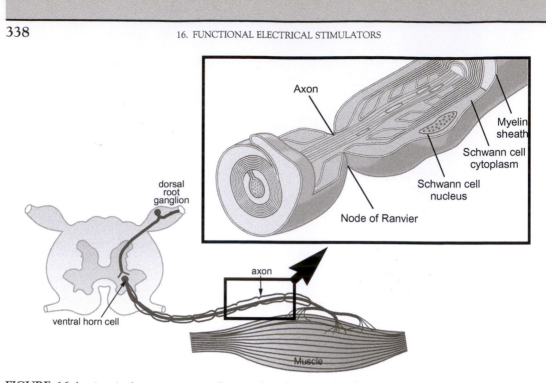

**FIGURE 16.4**   A spinal nerve connects the spinal cord to muscles. *[Based on figures reproduced by permission from Guyton & Hall (2006)].*

medial to where the dorsal and ventral roots merge, and into the dorsal horn of the spinal cord. Either axon may be as long as 1 m.

An action potential from the spinal cord propagates down a myelinated fiber. The *myelin sheath*, deposited by Schwann cells, acts an insulator, such that depolarization occurs only at the unmyelinated nodes of Ranvier (Figure 16.4). Because depolarization jumps long intervals across the axis of the nerve fiber, the conduction velocity is much faster ($\leq 100$ m/s) than if the fiber were unmyelinated ($\geq 0.25$ m/s).

## Neuromuscular Junction

From the spinal cord, an action potential propagates along a nerve fiber until it enters the belly of a muscle. There, it propagates through branches of the nerve fiber, which stimulate from three to several hundred skeletal muscle fibers. Each nerve terminates at a neuromuscular junction, with a muscle fiber, or *myofibril*, near its midpoint. When the action potential reaches the neuromuscular junction, about 125 vesicles of acetylcholine are released into the synaptic space (Figure 16.5). Acetylcholine receptors in the myofibril membrane receive acetylcholine, open ion channels, and create an end plate potential (Figure 16.6). This end plate potential initiates an action potential that spreads along the muscle membrane and causes muscle contraction (Guyton & Hall, 2006).

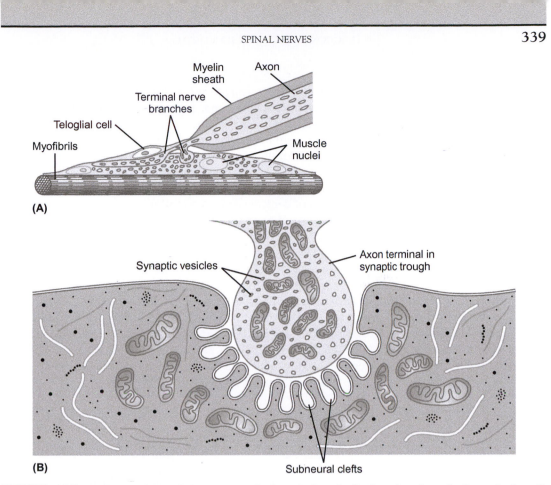

(A)

(B)

**FIGURE 16.5** Different views of the motor end plate. **A**: Longitudinal section through the end plate. **B**: Electron micrographic appearance of the contact point between a single axon terminal and the muscle fiber membrane. *[Redrawn from Fawcett, as modified from Couteaux, in Fawcett (1986). Figure reproduced by permission from Guyton & Hall (2006)].*

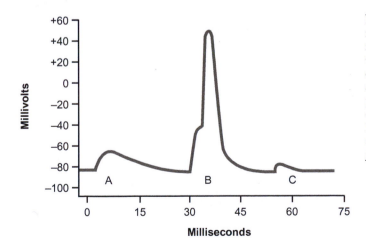

**FIGURE 16.6** End plate potentials, in mV. **A**: Weakened end plate potential recorded in a curarized muscle, too weak to elicit an action potential. **B**: Normal end plate potential eliciting a muscle action potential. **C**: Weakened end plate potential caused by botulinum toxin that decreases end plate release of acetylcholine, again too weak to elicit a muscle action potential *[Reproduced by permission from Guyton & Hall (2006)].*

## Spinal Cord Injury

The muscle signaling network is disrupted by *spinal cord injury*, which is defined as any injury to the spinal cord via blunt or penetrating trauma. Injury results from primary and second processes. The *primary process* results from the initial injury and includes energy transfer to the spinal cord, spinal cord deformation, and persistent postinjury cord compression. These mechanisms occur within seconds to minutes after injury, and they lead to immediate cell death, axonal disruption, and vascular and metabolic changes. The *secondary injury* process, which begins within minutes of injury and lasts for weeks to months, is not well understood. Following the necrosis of central cord gray matter and astrocyte degeneration, scar tissue develops, which may prevent axonal regeneration (Gupta et al., 2010).

SCI is classified according to sensory and motor levels for the right and left sides, using dermatomes. A *dermatome* is an area of skin that is mainly supplied by a single spinal nerve (Figure 16.7). By testing each dermatone, the lowest segment where motor and sensory function is normal on both sides is determined. This level is defined as the *neurologic level of injury*. For example, if a patient cannot sense or move his knees, the classification is L2. To be classified as a *complete* injury, the patient must have no voluntary anal contraction, no anal sensation, and total paralysis in the S4−5 dermatones. Otherwise, the injury is considered *incomplete* (ASIA, 2006).

*Tetraplegia* is injury in the cervical region, with associated loss of muscle strength in all four extremities. *Paraplegia* is injury in the thoracic, lumbar, or sacral regions.

## ELECTRICAL STIMULATION

After spinal cord injury, action potentials no longer propagate down a spinal nerve to its muscles. FES may provide the electrical stimulation for muscle contraction. Generally, FES activates nerve instead of muscle because the threshold charge for directly eliciting muscle fiber action potentials is much greater than the threshold for producing nerve action potentials. For example, the strength-duration curves for feline tibialis anterior nerve and muscle tissue are shown in Figure 16.8. For the same 500-μs pulse, 3 mA and 38 mA are required to stimulate the nerve and muscle, respectively.

For effective FES, the lower motor neurons must be intact from the anterior horns of the spinal cord to the neuromuscular junctions in the target muscles. Further, the neuromuscular junction and muscle tissue must be healthy, as is the case for spinal cord injury. Unlike deep brain stimulation, pinpoint target accuracy is not required. Direct nerve stimulation depolarizes the cell membranes of nearby neurons. When a critical depolarization threshold is reached, an influx of sodium ions from the extracellular space to the intracellular spaces produces an action potential that propagates in both directions away from stimulation target. Action potentials that propagate proximally in the peripheral nerve axons ultimately are annihilated at the cell body. Action potentials that propagate distally are transmitted to a neuromuscular junction, eventually leading to muscle contraction.

Muscle contraction strength is controlled by biphasic pulse frequency, amplitude, and duration. Beneath a critical pulse frequency of 12−15 Hz, the muscle responds with a series of twitches. Above this fusion frequency range, assuming the muscles have been

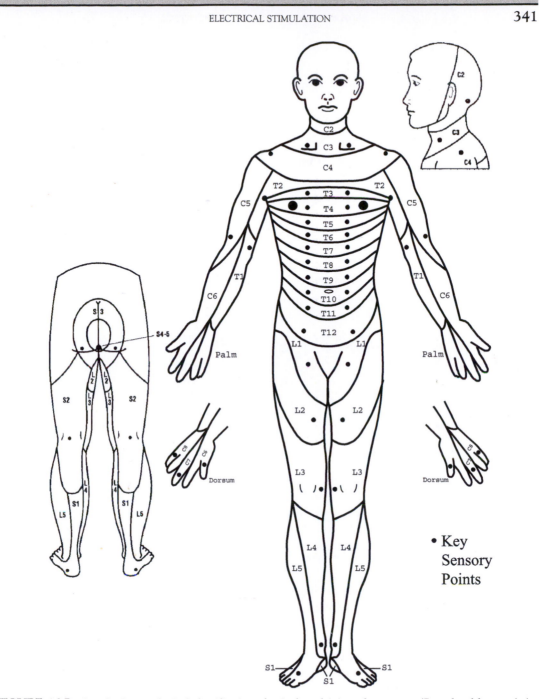

**FIGURE 16.7** Standard neurological classification of spinal cord injury dermatones (*Reproduced by permission from the American Spinal Injury Association, Atlanta, Georgia*).

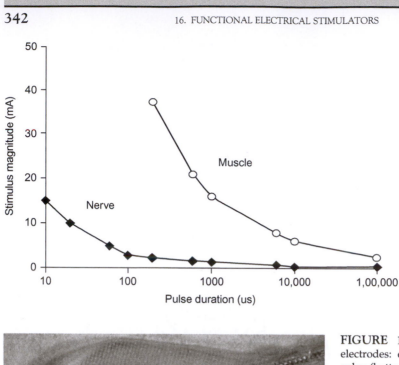

FIGURE 16.8 Strength-duration curves for nerve and muscle tissue. Stimulus magnitude required to produce a constant muscle response in normal and pharmacologically denervated feline tibialis anterior. [*Modified after Figure 20 in Mortimer (1981). Figure reproduced by permission from Peckham & Knutson (2005)*].

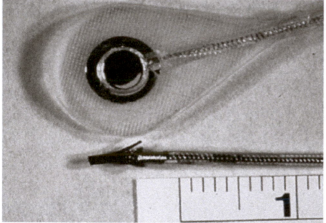

FIGURE 16.9 Common muscle-based electrodes: epimysial (top) and intramuscular (bottom) electrodes [*Reproduced from Akers et al. (1997). Figure reproduced by permission from Tyler & Polasek (2009)*].

conditioned to have relatively long-duration twitches, the response becomes a smooth contraction. This is known as *temporal summation*. Increasing the pulse amplitude and/or pulse duration increases the number of motor neurons activated, which is known as *spatial summation*.

As with other neurostimulators, the stimulator may be voltage- or current-controlled. Voltage-controlled stimulation results in more variable motor responses because the delivered current is affected by changes in tissue impedance. Current-controlled stimulation is typically used in FES to increase the likelihood of repeated muscle responses and maintain the stimulus charge within safe levels (Peckham & Knutson, 2005).

Stimulation current is delivered with different electrodes. Epimysial or intramuscular electrodes are commonly used for monopolar stimulation (Figure 16.9). An epimysial

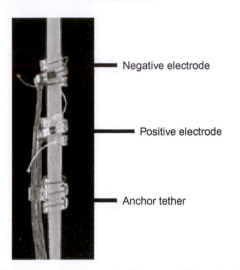

- Negative electrode
- Positive electrode
- Anchor tether

**FIGURE 16.10**   Huntington Helix self-sizing electrode *[Courtesy of Cyberonics, Inc. Figure reproduced by permission from Tyler & Polasek (2009)].*

electrode is sewn to the muscle. An intramuscular electrode may be implanted via a needle. Self-sizing extraneural electrodes provide bipolar stimulation. They optimize the perimeter area and interface with the nerve by applying pressure below critical levels to stop blood flow (Figure 16.10).

## CLINICAL NEED

According to the National Spinal Cord Injury Statistical Center, approximately 262,000 individuals in the United States suffered from spinal cord injuries in 2009. About 12,000 new cases occur each year, with about 40% due to motor vehicle crashes. The average age at injury is 40.2 years; about 80% of injuries occur to males. The most frequent neurologic category at discharge is incomplete tetraplegia (NSCISC, 2009).

Since the military conflicts in Afghanistan and Iraq began in 2001, 39,992 soldiers have been reported as wounded in action, as of September 22, 2010 (DoD, 2010). Returning soldiers often experienced multiple traumas, including traumatic brain injury, amputation, hearing and vision-related injuries, and SCI. Between May 2002 and September 2008, 113 soldiers received Veteran's Administration treatment and/or rehabilitation after sustaining an SCI while on active duty (Weaver et al., 2009).

Despite numerous laboratory studies and clinical trials, no reliable effective treatment of SCI has been developed. Although clinical management advances in the past 15 years have led to improvement in SCI patient survival and long-term outcomes, no clinically relevant therapeutic intervention exists. Current areas of investigation include early acute management, including early surgical intervention; pharmacologic interventions, such as erythropoietin; and cellular transplantation, such as peripheral nerve micrografts (Gupta et al., 2010).

## HISTORIC DEVICES

Functional electrical stimulation began as a therapy for restoring function to paralyzed limbs. Simple hand motions, such as grasp and release, can be recovered, but human gait is harder to achieve. The gain cycle consists of two phases. While one foot is in the stance phase (heel strike, foot flat, toe off), the other foot is in the swing phase (acceleration, deceleration).

### Early Devices

In 1960, neurophysiologist Wladimir Liberson invented a stimulator that assisted hemiplegic patients, who had lost control of one side of their bodies, in walking. A pressure switch under the heel of one foot activated a stimulator when the heel was raised. The stimulator transmitted current via surface electrodes to the peroneus longus muscle, which could raise the foot (Figure 16.11).

This FES application led to the investigation of footdrop systems by other investigators and to the investigation of systems that enabled seated paraplegics to stand and walk. By 1997, sequential percutaneous and surface stimulation of the quadriceps and other muscles, using 16 channels, enabled three paraplegic patients to walk 34 meters (Kobetic et al., 1997).

Footdrop systems also inspired the investigation of tetraplegic hand motion. Bioengineer P. Hunter Peckham's group at the Cleveland FES Center used percutaneous stimulation of the finger flexor and extensor muscles to enable key grip (Peckham et al., 1980).

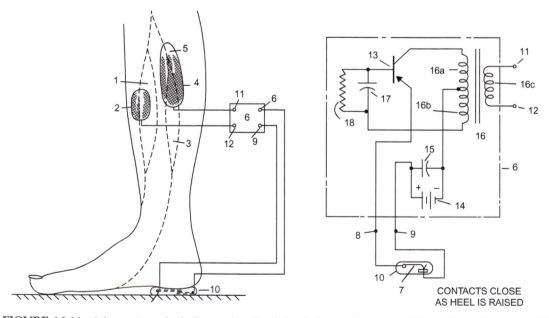

**FIGURE 16.11** Liberson's method of muscular stimulation in human beings to aid in walking. *[Taken from US 3,344,792; Offner & Liberson (1967)].*

Physiologist Giles Brindley's group at the Institute of Psychiatry in London used epimysial electrode stimulation of the forearm to enable power and key grip (Perkins et al., 1994; Kilgore et al., 2009).

## The Freehand Neurostimulator

The Freehand neurostimulator, developed by Peckham's group, is the only implanted FES system to receive FDA approval. Similar in architecture to cochlear implants, this system consisted of an 8-channel implant, epimysial electrodes surgically positioned in the forearm and hand, an external control unit (ECU), a transmitter coil, and an external shoulder sensor (Figure 16.1). Forty-six patients initially trialed the FES system for hand grasping and release. Their results were the basis of the PMA submission that received FDA approval in 1997. The Freehand system was commercialized by NeuroControl Corporation (CDRH, 1997).

## SYSTEM DESCRIPTION AND DIAGRAM

A system diagram for the Freehand is shown in Figure 16.12. The patient wears a shoulder sensor, which consists of a joystick mounted on the chest and a logic switch. By moving his shoulder, the patient alerts the processor module to turn the system on for a grasp movement. The ECU then powers an LED, and preprogrammed, amplified control signals are sent via the external transmitter coil to the implant. The ECU can also be powered on with an external switch; it is powered by a rechargeable battery.

The transmitter coil is positioned on the skin, directly above the implant. The implant consists of CMOS circuitry in a hermetically sealed titanium case, implanted in a subcutaneous

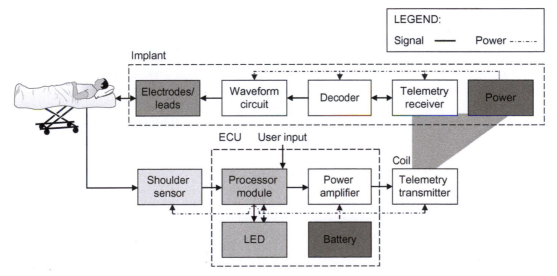

**FIGURE 16.12** NeuroControl Freehand system diagram.

pocket below the clavicle. The receiver in the implant receives the control signals and power. Once decoded, the control signals describe which of 8 electrodes should provide biphasic stimulation, as well as their frequencies (12 or 16 Hz) and pulse widths (0–200 ms). Stimulation currents are sent through the leads to 7 epimysial electrodes that have been sutured onto hand and arm muscles. Stimulation current is also transmitted to an 8th sensory epimysial electrode implanted in the shoulder, so that the patient directly senses the stimulation. Each epimysial electrode is fabricated from platinum/iridium, coated with silicone elastomer.

A small quick shoulder motion locks the grasp, and turns off the stimulation current. Another small quick shoulder motion unlocks the grasp. The control is proportional, such that the shoulder positional change determines the amount of force produced by the hand (with a maximum current of 20 mA). Another shoulder motion changes the setting from lateral grasp to palmer grasp.

The electrode mapping function is not shown in the system diagram. After electrode and implant surgery, muscle conditioning is initiated by programming the stimulator to cycle automatically through the stimulated grasp patterns for 8 hr while the patient is sleeping. Through trial and error, the clinician determines a coordinated stimulation pattern, so that muscles are activated in a sequence that produces the desired function. The clinician then programs the control parameters using an external personal computer. Rehabilitation and training with the neurostimulator require 3–6 months.

For the first 50 Freehand implants, the median patient age was 32 yr; 82% were male. The median time between injury and study enrollment was 4.6 yr. One patient reported constipation associated with the use of the device and did not complete rehabilitation training. In the other 49 patients, pinch force in lateral pinch increased significantly from 0.3–12 N ($p < 0.001$). In 48/49 patients, pinch force in palmer pinch increased significantly from 0–6.6 N ($p < 0.001$) (Smith et al., 1987; CDRH, 1997; Peckham et al., 2001).

## Marketing Limitations

The Freehand system was discontinued by NeuroControl in 2002, after 225 patients had been implanted. Even though patient satisfaction was high, economics did not justify continued manufacture. Of 12,000 new SCI cases each year, only 1000 cases are C5/C6. The other neurostimulator applications discussed, parkinsonian and essential tremors and deafness, have many more potential patients. NeuroControl's investors persuaded NeuroControl to target the larger stroke market. The cost of the device and implantation surgery was about $50,000, which was not completely covered by Medicare. In 2002, insurers were beginning to cover these costs when the Freehand was discontinued.

The Freehand was not freely recommended by physiatrists, who are the physicians who treat spinal cord injuries, to their patients. Even though orthopedic surgeons, who implanted the Freehand, were generally pleased with device performance, physiatrists inherently distrusted implanted devices. Some physiatrists feared that their patients would be used to test potentially dangerous devices. NeuroControl also admitted that it did not promote the Freehand well to physiatrists during the early stages of clinical trials (Cavuoto, 2004).

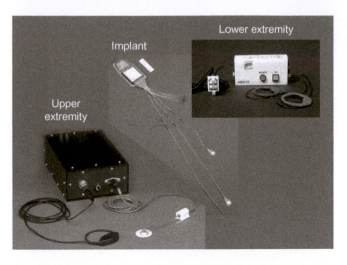

**FIGURE 16.13** Cleveland FES upper- and lower-extremity systems [*Reproduced by permission from Kilgore et al. (2009)*].

# EMERGING TECHNOLOGIES

Peckham's group and others at the Cleveland FES Center continue to investigate FES systems. The upper extremity system controller has moved from external shoulder positioning and implanted wrist sensor positioning to implanted electromyogram positioning. A similar system, implanted in the lower extremities, enables paraplegic patients to stand (Kilgore et al., 2009). Both upper- and lower-extremity Cleveland FES systems continue to be large in size, with capabilities split between an external control unit and implant (Figure 16.13).

The only implanted FES system to receive FDA approval, the Freehand neurostimulator, is no longer marketed. Miniaturization, with all functionality moved to the implant, and sensory feedback would change the architecture to that of a standard pacemaker pulse generator and could improve physiatrist acceptance. However, a higher-capacity battery would be required.

In this section, we describe other emerging architectural alternatives that could improve acceptance.

## Bions

Eliminating electrode leads would significantly simplify implant surgery and decrease the risk of potential infections. One potential technology that replaces an electrode and lead is the bion. The bion is a microstimulator that contains a transceiver for data and battery recharging, a goniometry sensor to detect relative motion of a nearby permanent magnet, an electrode anode, and an electrode cathode. The circuitry is hermetically sealed in a ceramic case (Figure 16.14).

For three days of continuous operation without charge, the total current consumption is limited to $40\,\mu A$. The stimulator communicates with an external master control unit

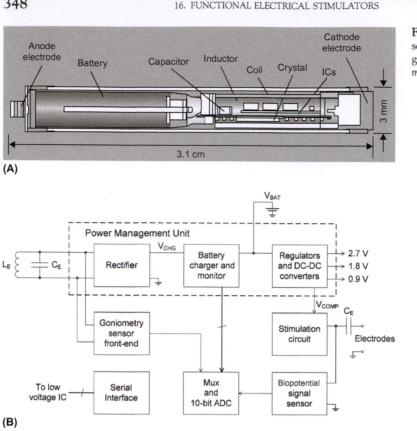

**FIGURE 16.14** Bion cross section (**A**) and block diagram (**B**) [*Reproduced by permission from Lee et al. (2009)*].

through a 400-MHz wireless link that can support a 2.5-MB/s bit rate. The communication system theoretically supports 852 stimulators (Lee et al., 2009).

An earlier bion prototype was recently implanted to stimulate the occipital nerve. In this clinical trial, stimulation was hypothesized to provide relief from refractory primary headache disorders. In each patient, one bion was inserted through a small 1-cm incision using bion placement tools (Figures 16.15 and 16.16).

At 6-month follow-up, 8/9 patients were still using the device. One patient stopped using the bion at 4 months because she stated "the battery recharging schedule was too demanding". Specifically, she recharged the battery 1.5 hr to use the bion 1.5 hr (Trentman et al., 2009).

## Neural Interface Systems

Direct brain control of electrical stimulation would significantly simplify the FES architecture. Rather than being used to manipulate an external or implanted sensor, the brain would be reconnected to the muscles requiring stimulation. This strategy is referred to as

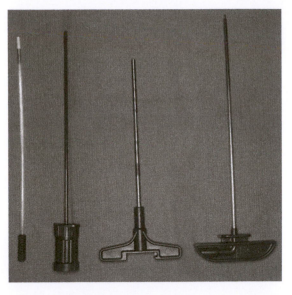

**FIGURE 16.15** The bion placement tools and holder *[Reproduced by permission from Trentman et al. (2009)].*

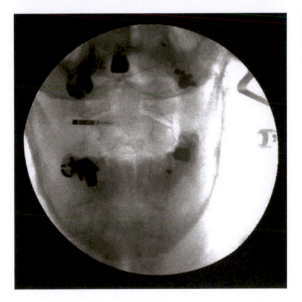

**FIGURE 16.16** AP skull film showing bion after insertion. The device is subcutaneous in the occipital region *[Reproduced by permission from Trentman et al. (2009)].*

a *direct neural interface system (NIS).* In a surrogate NIS, a cognitive, perceptual or sensory neural signal substitutes for a desired action (Figure 16.17).

Both direct and surrogate neural interface systems consist of three parts: the sensor, decoder, and controller (Figure 16.18). A brain signal is recorded, interpreted as a useful command, and used to effect some desired action. Direct neural interface systems research currently emphasizes the sensing and decoding functions.

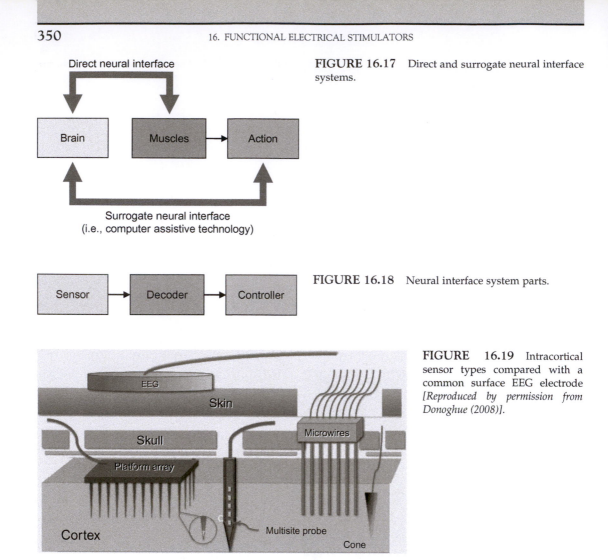

**FIGURE 16.17**   Direct and surrogate neural interface systems.

**FIGURE 16.18**   Neural interface system parts.

**FIGURE   16.19** Intracortical sensor types compared with a common surface EEG electrode *[Reproduced by permission from Donoghue (2008)].*

In direct neural interface systems, action potentials, or so-called spikes, and field potentials (FPs) are sensed. Field potentials are more complex than spikes because they represent transmembrane current flow, summed across neuron groups of varying size, frequency, and spatial distribution. FPs recorded by scalp electrodes are EEGs. FPs recorded inside the skull, close to the cortical surface, are electrocorticograms (ECoG); FPs recorded intraparechymally are the local field potential (LFP). Because EEG sensing has limited bandwidth and significant noise and muscle artifact, simultaneous ECoG and LFP sensing is preferred. The commonly used electrodes for intracortical sensing are platform arrays, multisite probes, and cone electrodes (Figure 16.19).

Sensing occurs in the primary motor cortex (MI), which is located in the posterior part of the precentral gyrus in humans and macaque monkeys. Based on nonhuman primate studies, the MI is a well-known source of neuronal output to the spinal cord for arm

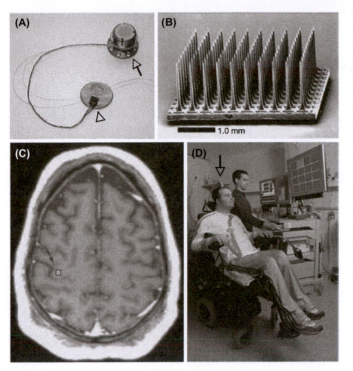

**FIGURE 16.20** Intracortical sensor and placement, in one subject. **A:** BrainGate sensor, resting on a U.S. penny, connected by a 13-cm ribbon cable to percutaneous Ti pedestal (arrow). **B:** Scanning electron micrograph of sensor. **C:** Preoperative axial T1-weighted MRI of the brain. The arm/hand "knob" of the right precentral gyrus (arrow) corresponds to approximate location of sensor implant site. **D:** Subject is sitting in a wheelchair, mechanically ventilated through a tracheostomy. The gray box (arrow) connected to the percutaneous pedestal contains amplifier and signal-conditioning hardware; cabling brings the amplified neural signals to computers sitting beside the participant, who is looking at the monitor [*Reproduced by permission from Hochberg et al. (2006).*

movements. In a restricted early study of humans with tetraplegia, spiking and FPs remained in the motor cortex years after SCI. Anatomical landmarks of the human arm/hand area of the motor cortex, which is defined by a "knob" in the precentral gyrus, appear to be reliable functional landmarks as well. There, movement intention is sufficient to activate MI neurons (Donoghue, 2008; Donoghue & Hochberg, 2009).

A BrainGate platform array was recently used to record MI activity in a 25-yr-old male with complete tetraplegia (Figure 16.20). The array consisted of 100 tapered 1- or 1.5-mm long microelectrodes in a $10 \times 10$ grid, with electrodes spaced by 400 microns. The assembly, which formed a $4 \times 4$-mm base, was carved from a block of boron-doped silicon, with each electrode isolated by a glass layer. The electrodes were connected by internal microcabling to a titanium pedestal that was attached on the skull and passed through the skin for connection to external electronics (Donoghue et al., 2007).

During each research session, the subject was asked to imagine movements such as opening/closing of his hand. Single and multiunit data were recorded (Figure 16.21) and used for decoding. In one set of trials, the subject imagined controlling a computer mouse, while a technician moved a cursor through a succession of randomly positioned targets. A response matrix, $\underline{R}(t)$, contained the fire rate over a 1-s history for each neuron. A linear filter, $\mathbf{f}(t)$, was constructed from the response matrix as

$$\mathbf{u}(t) = \underline{R}(t) \cdot \mathbf{f}(t) = \underline{R}(t)[\underline{R}(t)^T \underline{R}(t)]^{-1} \underline{R}(t)^T \mathbf{k}(t), \qquad (16.1)$$

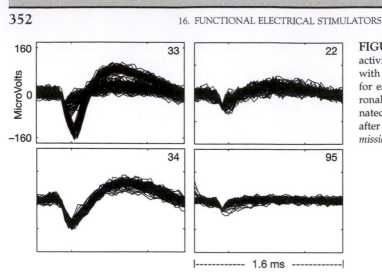

**FIGURE 16.21** Discriminated neural activity at electrodes 33, 34, 22, and 95, with 80 superimposed action potentials for each unit. At electrode 33, two neuronal units could be reliably discriminated. Recordings took place 90 days after array placement [*Reproduced by permission from Hochberg et al. (2006)*].

where $t$ = trial, $\mathbf{k}(t)$ contains the kinematic $x$-$y$ positions, and $\mathbf{u}(t)$ is the kinematic reconstruction of where to move. The resulting controlled movements are shown at the following sites (Hochberg et al., 2006):

- http://www.nature.com/nature/journal/v442/n7099/extref/nature04970-s5.mov
- http://www.nature.com/nature/journal/v442/n7099/extref/nature04970-s3.mov

## Robotic Exoskeleton

Perhaps in the short term, it is best to follow NeuroControl's lead and concentrate on other FES applications. Other technologies may solve the clinical need of enabling SCI patients to walk. At the Cleveland FES Center, a hybrid system of exoskeletal bracing and multichannel FES is under investigation (Kobetic et al., 2009). Alternatively, if you ignore FES, a robotic exoskeleton may facilitate gait.

The ReWalk robotic exoskeleton, developed by Argo Medical Technologies in Israel, is described as a light wearable brace support suit. By integrating DC motors, rechargeable batteries, and an array of sensors with computer-based control, upper-body attempts to walk are detected and used to initiate walking processes. Crutches are still required during walking (NIH, 2008) (Figure 16.22).

The ReWalk is not an elegant solution, but it may be functional for paraplegics. The intent of this noninvasive solution is to replace the wheelchair. The system is currently undergoing clinical trials in Israel and at MossRehab, which is part of Albert Einstein Medical Center in Philadelphia (MossRehab, 2009). Clinical trial use can be viewed at http://www.argomedtec.com/ (select "ReWalk on ABC: Good Morning America").

## KEY FEATURES FROM ENGINEERING STANDARDS

The neurostimulator consensus standard discussed in Chapter 14 applies to implantable functional electrical stimulators: *ANSI/AAMI/ISO 14708-3:2008 Implants for Surgery—Active Implantable Medical Devices—Part 3: Implantable Neurostimulators* (AAMI, 2009).

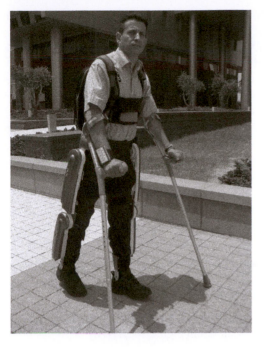

**FIGURE 16.22** The ReWalk system is undergoing clinical trials in Israel and Philadelphia (Argo Medical Technologies, Israel).

# SUMMARY

Functional electrical stimulation is the application of electrical current to excitable tissue to supplement or replace function that is lost in neurologically impaired individuals. We have already discussed one functional electrical stimulator: the cochlear implant. In this chapter, we specifically considered implantable functional electrical stimulators for restoration of limb function in spinal cord injury patients. The holy grail in FES is to enable paraplegics to walk.

From the spinal cord, an action potential propagates along a nerve fiber until it enters the belly of a muscle. There, it propagates through branches of the nerve fiber, which stimulate from three to several hundred skeletal muscle fibers. Each nerve terminates at a neuromuscular junction, with a muscle fiber, or myofibril, near its midpoint. When the action potential reaches the neuromuscular junction, about 125 vesicles of acetylcholine are released into the synaptic space. Acetylcholine receptors in the myofibril membrane receive acetylcholine, open ion channels, and create an end plate potential. This end plate potential initiates an action potential that spreads along the muscle membrane and causes muscle contraction. The muscle signaling network is disrupted by spinal cord injury.

Generally, FES activates nerve instead of muscle because the threshold charge for directly eliciting muscle fiber action potentials is much greater than the threshold for producing nerve action potentials. Muscle contraction strength is controlled by biphasic pulse frequency, amplitude, and duration. Beneath a critical pulse frequency of 12–15 Hz, the muscle responds with a series of twitches. Above this fusion frequency range, the response becomes a smooth contraction. Current-controlled stimulation is typically used in FES to increase the likelihood of repeated muscle responses and to maintain the stimulus charge at safe levels. The electrodes used are epimysial or intramuscular electrodes for monopolar stimulation and self-sizing extraneural electrodes for bipolar stimulation.

The only implanted FES system to receive FDA approval, the Freehand neurostimulator, is no longer marketed. This system was not fully implantable. Miniaturization, with all functionality moved to the implant, and sensory feedback would change the architecture to that of a standard pacemaker pulse generator and could improve physiatrist acceptance. However, a higher capacity battery would be required.

An implantable functional electrical stimulator would be subject to the same consensus standard requirements as other neurostimulators.

## Acknowledgment

The author thanks Dennis Hepp for his detailed review of this chapter. Mr. Hepp is Managing Director at Rivertek Medical Systems.

## Exercises

**16.1** Which is more important for successful commercialization of a medical device: great technology or great marketing? Give reasons, including examples from the marketplace.

**16.2** Draw a system diagram of the BrainGate system, as shown in Figure 16.20. Why is the system response (i.e., a cursor drawing a circle) so slow? Be specific.

**16.3** Redraw the Freehand system diagram using a neural interface system controller and pacemaker pulse generator architecture.

**16.4** What biocompatibility issues occur during and after implantation of direct neural interface electrodes? Be specific.

**16.5** Prepare a table that compares the energy requirements for an implantable pacemaker, ICD, deep brain stimulator, cochlear implant, and fully implantable FE stimulator. The categories for comparison should include: stimulation frequency, current (*not* voltage) amplitude, pulse width, battery life, battery chemistry. When do you believe a fully implantable FE stimulator will be market released? Be specific.

Read Peckham et al. (2001):

**16.6** What tests were used to assess Freehand patient outcomes? What adverse events occurred? Did the risks justify the benefits?

Read Hochberg et al. (2006):

**16.7** How influential are shoulder movements in cursor motions? Be specific.

**16.8** How easy or difficult will it be to develop a generalized decoder for spinal cord injury patients for the same functionality (i.e., playing Neural Pong)? Provide an explanation.

**16.9** Name three issues of the current system that need to be resolved before it can be clinically trialed in U.S. patients, with the intention of submitting clinical data to the FDA for premarket approval. Be specific.

# References

AAMI (2009). *ANSI/AAMI/ISO 14708-3:2008 Implants for surgery—Active implantable medical devices—Part 3: Implantable neurostimulators.* Arlington, VA: AAMI.

Akers, J. M., & Peckham, P. H., et al. (1997). Tissue response to chronically stimulated implanted epimysial and intramuscular electrodes. *IEEE Transactions on Rehabilitation Engineering, 5*(2), 207–220.

ASIA (2006). *Standard neurological classification of spinal cord injury.* Atlanta, GA: American Spinal Injury Association.

Bloom, W., & Fawcett, D. W. (1986). *A textbook of histology* (11th ed.). Philadelphia: W. B. Saunders.

Cavuoto, J. (2004). Neural engineering's image problem. *IEEE Spectrum, 41,* 32–37.

CDRH (1997). *Approval order P950035.* Rockville, MD: FDA.

DOD. (2010). U.S. casualty status.

Donoghue, J. P. (2008). Bridging the brain to the world: A perspective on neural interface systems. *Neuron, 60,* 511–521.

Donoghue, J. P., & Hochberg, L. R. (2009). Designing a neural interface system to restore mobility. In E. S. Krames, P. H. Peckham, & A. R. Rezai (Eds.), *Neuromodulation.* London: Elsevier.

Donoghue, J. P., Nurmikko, A., Black, M., & Hochberg, L. R. (2007). Assistive technology and robotic control using motor cortex ensemble-based neural interface systems in humans with tetraplegia. *Journal of Physiology, 579,* 603–611.

Gupta, R., Bathen, M. E., Smith, J. S., Levi, A. D., Bhatia, N. N., & Steward, O. (2010). Advances in the management of spinal cord injury. *Journal of the American Academy of Orthopedic Surgeons, 18,* 210–222.

Guyton, A. C., & Hall, J. E. (2006). *Textbook of medical physiology* (11th ed.). Philadelphia: Elsevier Saunders.

Hochberg, L. R., Serruya, M. D., Friehs, G. M., Mukand, J. A., Saleh, M., & Caplan, A. H., et al. (2006). Neuronal ensemble control of prosthetic devices by a human with tetraplegia. *Nature, 442,* 164–171.

Kilgore, K. L., Keith, M. W., & Peckham, P. H. (2009). Stimulation for return of upper and lower extremity function. In E. S. Krames, P. H. Peckham, & A. R. Rezai (Eds.), *Neuromodulation.* London: Elsevier.

Kobetic, R., To, C. S., Schnellenberger, J. R., Audu, M. L., Bulea, T. C., & Gaudio, R., et al. (2009). Development of hybrid orthosis for standing, walking, and stair climbing after spinal cord injury. *Journal of Rehabilitation Research and Development, 46,* 447–462.

Kobetic, R., Triolo, R. J., & Marsolais, E. B. (1997). Muscle selection and walking performance of multichannel FES systems for ambulation in paraplegia. *IEEE Transactions on Rehabilitation Engineering, 5,* 23–29.

Lee, E., Matei, E., Gord, J., Hess, P., Nercessian, P., & Stover, H., et al. (2009). A biomedical implantable FES battery-powered micro-stimulator. *IEEE Transactions on Circuits and Systems, Part I: Regular Papers, 56,* 2583–2596.

Loeb, G. E., & Davoodi, R. (2005). The functional reanimation of paralyzed limbs. *IEEE Engineering in Medicine and Biology Magazine, 24,* 45–51.

Mortimer, J. T. (1981). Motor prostheses. In J. M Brookshart, & V. B. Mountcastle (Eds.), *Handbook of physiology—The nervous system II* (pp. 155–187). Bethesda, MD: American Physiological Society.

Mossrehab (2009). *MossRehab conducting exclusive U.S. trial of ReWalk.* Philadelphia: Albert Einstein Medical Center.

NIH (2008). *A study testing safety and tolerance of the ReWalk exoskeleton suit.* Bethesda, MD: NIH. http://clinicaltrials.gov

NSCISC (2009). *Spinal cord injury facts and figures at a glance.* Birmingham: National Spinal Cord Injury Statistical Center.

Offner, F. F., & Liberson, W. T. (1967). Method of muscular stimulation in human beings to aid in walking. US 3,344,792.

Peckham, P. H., Keith, M. W., Kilgore, K. L., Grill, J. H., Wuolle, K. S., & Thrope, G. B., et al. (2001). Efficacy of an implanted neuroprosthesis for restoring hand grasp in tetraplegia: a multicenter study. *Archives of Physical Medicine and Rehabilitation, 82,* 1380–1388.

Peckham, P. H., & Knutson, J. S. (2005). Functional electrical stimulation for neuromuscular applications. *Annual Review of Biomedical Engineering, 7,* 327–360.

Peckham, P. H., Mortimer, J. T., & Marsolais, E. B. (1980). Controlled prehension and release in the C5 quadriplegic elicited by functional electrical stimulation of the paralyzed forearm musculature. *Annals of Biomedical Engineering, 8,* 369–388.

Perkins, T. A., Brindley, G. S., Donaldson, N. D., Polkey, C. E., & Rushton, D. N. (1994). Implant provision of key, pinch and power grips in a C6 tetraplegic. *Medical & Biological Engineering & Computing, 32*, 367–372.

Rosenow, J. M. (2009). Anatomy of the nervous system. In E. S. Krames, P. H. Peckham, & A. R. Rezai (Eds.), *Neuromodulation*. London: Elsevier.

Smith, B., Peckham, P. H., Keith, M. W., & ROSCOE, D. D. (1987). An externally powered, multichannel, implantable stimulator for versatile control of paralyzed muscle. *IEEE Transactions on Biomedical Engineering, 34*, 499–508.

Trentman, T. L., Rosenfeld, D. M., Vargas, B. B., Schwedt, T. J., Zimmerman, R. S., & Dodick, D. W. (2009). Greater occipital nerve stimulation via the Bion microstimulator: implantation technique and stimulation parameters. Clinical trial: NCT00205894. *Pain Physician, 12*, 621–628.

Tyler, D. J., & Polasek, K. H. (2009). Electrodes for the neural interface. In E. S. Krames, P. H. Peckham, & A. R. Rezai (Eds.), *Neuromodulation*. London: Elsevier.

University of Michigan. (2010). Medical gross anatomy: Spinal cord and spinal nerve.

Weaver, F. M., Burns, S. P., Evans, C. T., Rapacki, L. M., Goldstein, B., & Hammond, M. C. (2009). Provider perspectives on soldiers with new spinal cord injuries returning from Iraq and Afghanistan. *Archives of Physical Medicine and Rehabilitation, 90*, 517–521.

CHAPTER

# 17

# Intraocular Lens Implants

In this chapter, we discuss the relevant physiology, history, and key features of intraocular lens implants. An *intraocular lens* (*IOL*) is a medical device that was originally developed to replace an opacified native lens, which is commonly referred to as a cataract. The IOL restores a patient's vision after cataract surgery (Figure 17.1). We restrict this discussion to IOLs for cataract patients, rather than for other patients who may have an IOL implanted.

Upon completion of this chapter, each student shall be able to:

1. Understand the mechanisms underlying cataract and posterior capsule opacification formation.
2. Explain the ultrasound procedures utilized in cataract removal.
3. Describe five key features of IOLs.

FIGURE 17.1   Abbott Medical Optics Tecnis multifocal intraocular lens (AMO, Santa Ana, California).

## OCULAR PHYSIOLOGY

As discussed in Chapter 15, an organ such as the ear may act as a sensor to measure external stimuli. The output of this sensor is transmitted through a sensory nerve, such as the auditory nerve, to the brain for further processing. We begin this chapter by discussing how the eye translates light into signals transmitted by the optic nerve to the brain for vision processing.

### Vision

When light enters the eye, it first passes through the cornea (Figure 17.2). The *cornea* is the transparent external surface of the eyeball and is continuous with the white *sclera*, which is the supporting wall of the eyeball. The cornea acts as a first lens to focus light as it moves through the aqueous humor and pupil to the *lens*. The aqueous humor, derived from plasma, provides nutrition to ocular structures and maintains intraocular pressure. The *pupil* is the opening in the iris, the colored muscle that controls the amount of entering light. The *anterior chamber* is the space between the cornea and iris; the *posterior chamber* is the space between the iris and lens.

The lens focuses light through the *vitreous humor*, a jelly-like fluid, to the retina in the back of the eye. The suspended lens curvature changes from moderately convex to very

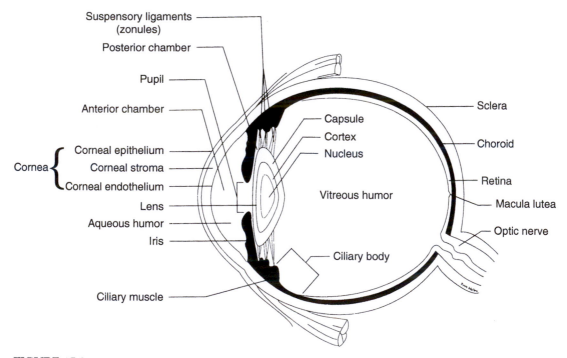

**FIGURE 17.2**    Eye structures *[Reproduced by permission from Refojo (2004)]*.

convex, depending on the pull and relaxation of suspensory ligaments (zonules). This so-called *accommodation* focus, which is controlled by the parasympathetic nerves, enables a young person to see clearly at near and far distances.

Light passes through layers of the retina to the layer of *rods* and *cones* (Figure 17.3). Rhodopsin in the rods and cone pigments in the cones react to the light, causing direct current to be transmitted through the optic nerve.

The refractive power of a lens system to bend light rays is measured in diopters. One *diopter* is equal to 1 m divided by its focal length. A normal eye has an axial length of about 24 mm. In a normal eye, the total refractive power is about 59 diopters when a lens focuses distant vision.

The bending of light within the eye as it moves between tissue boundaries can also be described. The refractive index, $n$, is defined as the ratio of the sine of the incidence angle, normal to the boundary, $\theta_i$, to the sine of the refraction angle, normal to the boundary, $\theta_f$:

$$n = \frac{\sin \theta_i}{\sin \theta_f}. \tag{17.1}$$

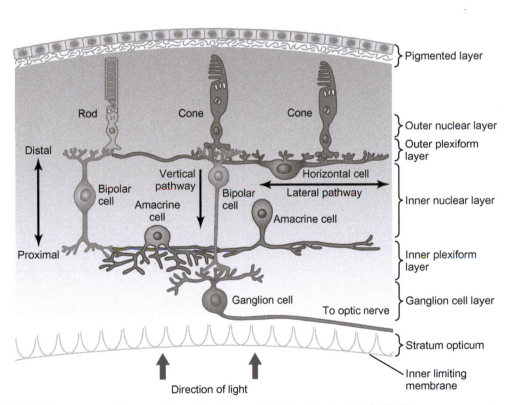

**FIGURE 17.3** Layers of the retina [*Reproduced by permission from Guyton & Hall (2006)*].

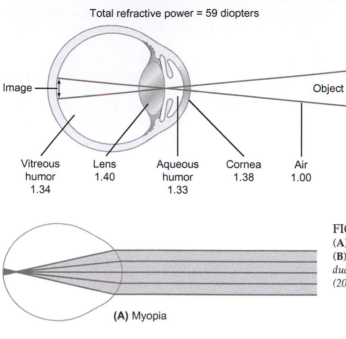

Total refractive power = 59 diopters

Image —

Object

Vitreous
humor
1.34

Lens
1.40

Aqueous
humor
1.33

Cornea
1.38

Air
1.00

**FIGURE 17.4** Eye total refractive power and component refractive indices [*Reproduced by permission from Guyton & Hall (2006)*].

(A) Myopia

(B) Hyperopia

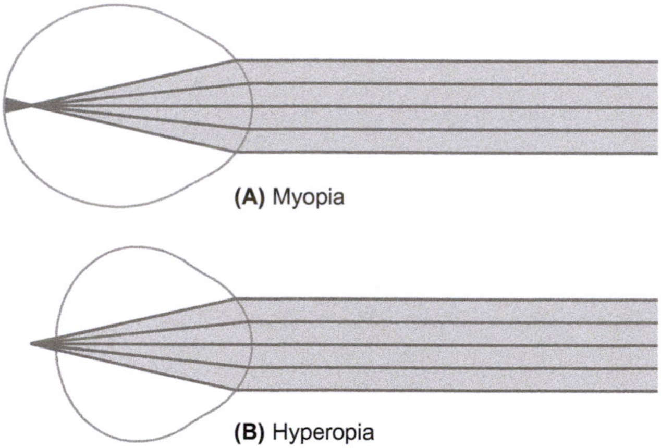

**FIGURE 17.5** Parallel light rays focus (A) in front of the retina in myopia and (B) behind the retina in hyperopia [*Reproduced by permission from Guyton & Hall (2006)*].

The refractive index of air is 1; the refractive index of water is 1.33. The refractive index of a lens, on average, is 1.40 (Guyton and Hall, 2006) (Figure 17.4).

An eye may be unable to bend parallel light sufficiently for focusing on the retina. This may occur when the eyeball is too long or too short or when lens accommodation fails. When the focused light is in front of the retina, myopia (nearsightedness) occurs. When the focused light is behind the retina, hyperopia (farsightedness) occurs (Figure 17.5). An external concave lens (contact lens or eyeglass lens) corrects myopia; an external convex lens corrects hyperopia (Figure 17.6)

## Cataracts

A lens consists of several components, most of which are shown in Figure 17.2. The central nucleus and surrounding cortex are each formed from lens fiber cells. The *lens capsule* is a smooth transparent basement membrane surrounding the cortex, which consists mainly of type IV collagen. It is manipulated during accommodation. Between the anterior lens capsule and outermost layer of lens fiber cells is a single layer of simple epithelial

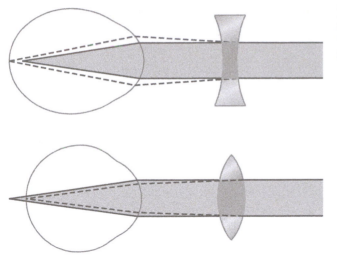

**FIGURE 17.6** Correction of myopia with a concave lens and correction of hyperopia with a convex lens [*Reproduced by permission from Guyton & Hall (2006)*].

cells (not shown). These lens epithelial cells (LECs) differentiate into lens fiber cells (Song et al., 2009).

As a lens ages, lens fibers are compressed centrally and the lens becomes harder. With time, the lens loses its transparency and may form a cataract. A nuclear schlerotic cataract is formed when the lens center (nucleus) opacifies. This is associated with yellowing of the lens. A cortical cataract is formed when discrete opacities (cortical spokes) develop in the cortex. A posterior subcapsular cataract is formed when granular opacities develop in the central posterior cortex, just under the posterior capsule (Figure 17.7). Cataract formation causes vision to painlessly blur, with progressive vision loss that eventually interferes with daily tasks. Some patients also complain of glare, loss of contrast sensitivity, change in refraction, or monocular diplopia (double vision) (Johns et al., 2003; Asbell et al., 2005).

## ULTRASOUND

As we prepare to discuss cataract treatment, we introduce several ultrasound concepts. Ultrasound is used to measure lens parameters and to remove the lens from its capsular bag before an IOL can be implanted.

### A-Scan Biometry

*Ultrasound* consists of mechanical vibrations above the frequencies of human hearing, which is about 20 kHz. Ultrasound imaging uses reflected mechanical energy, in the range of 1–50 MHz, to determine the boundaries between tissues and produce images. *Amplitude-mode (A-mode) scanning* is the simplest one-dimensional type of ultrasound imaging. Ophthalmologists refer to A-mode scanning as A-scan biometry. A-mode results in a plot of the amplitude of backscattered echo vs. the time after transmission of an ultrasound pulse around 10 MHz.

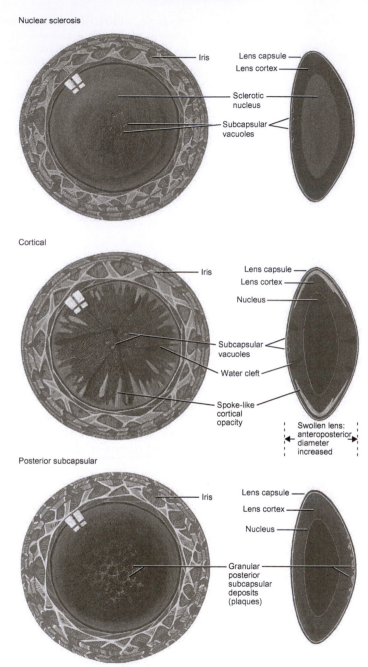

**FIGURE 17.7** Types of cataracts *[Reproduced from Johns and colleagues (2003) with permission from CIBAVision, Inc. Figure reproduced by permission from Asbell et al. (2005)].*

To understand A-mode, assume that a propagating wave from a transducer has a planar wave front, that the tissue is perfectly homogeneous, and that no energy is lost as the wave passes through tissue. Then the longitudinal particle displacement, $W(z, t)$, can be described by the one-dimensional, linearized, lossless wave equation for propagation of sound in fluids:

$$\frac{\partial^2 W(z,t)}{\partial z^2} = \frac{1}{c^2} \cdot \frac{\partial W(z,t)}{\partial t}, \tag{17.2}$$

where $c$ is the sound propagation velocity in tissue, $z$ is longitudinal distance, and $t$ is time. The value of $W$ is generally a few 10ths of a nanometer. The particle velocity in the $z$ direction, $v(z, t)$, is given by the time derivative of the particle displacement:

$$v(z,t) = \frac{\partial W(z,t)}{\partial t}. \tag{17.3}$$

The value of $v(z, t)$ is generally a few cm/s. The pressure, $P(z, t)$, of the ultrasound wave at a particular point in the $z$ direction is given by:

$$P(z,t) = \rho \cdot c \cdot v(z,t), \tag{17.4}$$

where $\rho$ is the tissue density. Pressure is measured in Pascals.

Because the transducer undergoes sinusoidal motion, pressure and particle velocity can be written as

$$P(z,t) = P_o\, e^{j\omega t} \tag{17.5}$$

$$v(z,t) = v_o\, e^{j\omega t}, \tag{17.6}$$

where $P_o$ and $v_o$ are the peak pressure and particle velocity, respectively. The propagating wave can be represented by a series of compression and rarefraction waves (Figure 17.8).

We can characterize wave propagation through the acoustic impedance of tissue, $Z$, which is defined as the ratio of pressure to propagation velocity:

$$Z = \frac{P(z,t)}{v(z,t)}. \tag{17.7}$$

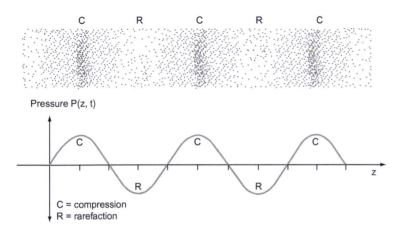

FIGURE 17.8 Propagating longitudinal ultrasound wave.

The acoustic impedance can be calculated as the product of tissue density and sound velocity in tissue:

$$Z = \rho \cdot c. \tag{17.8}$$

These values for various tissues and water are given in Table 17.1. Now let the longitudinal wave be generated in a transducer and transmitted into tissue with an incidence angle of 90° (Figure 17.9). As the wave strikes each tissue boundary, a fraction of the wave energy is reflected, while the remaining fraction is transmitted.

The pressure reflection coefficient, $R_P$, is the ratio of the pressures of the reflected, $P_r(z, t)$, and incident, $P_i(z, t)$, waves:

$$R_P = \frac{P_r(z,t)}{P_i(z,t)} = \frac{Z_2 - Z_1}{Z_2 + Z_1}. \tag{17.9}$$

TABLE 17.1    Sound propagation velocity and acoustic impedance for various materials

|  | Sound Propagation Velocity (c) (m/s) | Impedance (Z) (MegaRayls) |
| --- | --- | --- |
| Air | 330 | 0.0004 |
| Blood | 1550 | 1.61 |
| Bone | 3500 | 7.8 |
| Brain | 1540 | 1.58 |
| Fat | 1450 | 1.38 |
| Muscle | 1580 | 1.7 |
| Liver | 1570 | 1.65 |
| Vitreous humor | 1520 | 1.52 |
| Water @ 20°C | 1482.3 | 1.482 |

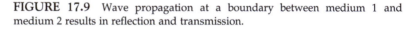

FIGURE 17.9    Wave propagation at a boundary between medium 1 and medium 2 results in reflection and transmission.

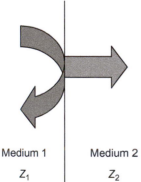

Medium 1 | Medium 2

$Z_1$ | $Z_2$

According to boundary conditions, the acoustic pressures on both sides of the boundary are equal, and the particle velocities normal to the boundary are equal. These conditions give us the relationship

$$T_P = R_P + 1, \tag{17.10}$$

where $T_P$ is the pressure transmission coefficient. Combining Eqs. (17.9) and (17.10) yields

$$T_P = \frac{Z_2 - Z_1}{Z_2 + Z_1} + 1 \tag{17.11}$$

$$T_P = \frac{Z_2 - Z_1 + Z_2 + Z_1}{Z_2 + Z_1} \tag{17.12}$$

$$T_P = \frac{2Z_2}{Z_2 + Z_1}. \tag{17.13}$$

The ultrasound transducer is also a receiver, which receives reflected wave echoes. By keeping track of the time and amplitude of received waves, the tissue boundaries can be determined (Webb, 2003). For A-mode scanning through the eye, echoes are received after the wave strikes the cornea, anterior lens, cataract, posterior lens, retina, and sclera (Figure 17.10). Greater impedance differences at a tissue boundary result in higher amplitudes. A-mode is one of the techniques used to predict the optimal IOL refractive power to be implanted.

## Emulsification

At lower frequencies, such as 35—45 kHz, ultrasound may be used to fragment a cataractous lens. An ultrasound probe tip, inserted into a small incision in the lens capsule,

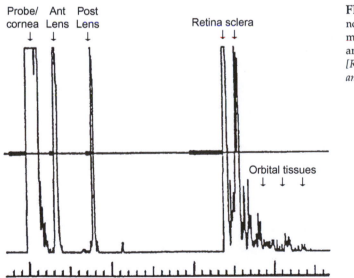

FIGURE 17.10 A-mode scan for a normal eye. If a cataract were present, multiple spikes would occur between the anterior and posterior lens spikes *[Reproduced by permission from Waldron and Aaberg Jr. (2008)].*

emulsifies the lens. For many years, the mechanisms behind this emulsification were attributed to the jackhammer effect and cavitation. The *jackhammer effect* is the result of cycles of negative and positive pressure that mechanically cut the cataract directly. *Cavitation* is the growth, oscillation, and collapse of micron-sized bubbles in liquids due to an acoustic field (Packer et al., 2005).

Recently, eight extracted human cataracts were subjected to emulsification under controlled pressure conditions. The volume of lens material entering the probe tip after one 100 ms burst of ultrasound energy was estimated as the linear movement of material toward the tip, using video images. The presence of cavitation was controlled through ultrasound power level and hyperbaric chamber pressure.

In this study, cavitation around the probe tip primarily occurred for longitudinal ultrasonic power levels of at least 30%. Cavitation bubble formation was observed during the backstroke or as the tip moved away from the lens material; cavitation bubble collapse was observed during the forward displacement of the tip. Cavitation at any power level was successfully suppressed when the pressure in the hyperbaric chamber increased at least 2.0 bar above atmospheric pressure (1 bar = 100,000 Pa) (Figure 17.11). At 0 and 2 bar, the relative lens travel into the phaco tip after a single burst did not change. This study strongly suggests that cavitation plays no role in lens fragmentation and that the jackhammer effect is the dominant emulsification mechanism (Zacharias, 2008).

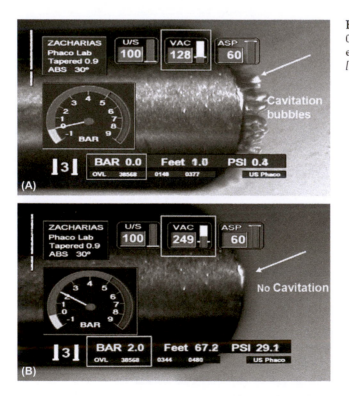

**FIGURE 17.11** Cavitation was present at 0 bar (**A**) but not at 2 bar (**B**) during cataract emulsification experiments with 100% power *[Reproduced by permission from Zacharias (2008)].*

# CLINICAL NEED

Cataract is the leading cause of blindness worldwide, and the leading cause of vision degeneration in the United States. It is responsible for about 60% of all Medicare vision costs.

According to a study funded by the National Eye Institute, approximately 20.5 million (17.2%) Americans older than 40 years have cataract in either eye. Women have a significantly higher age-adjusted prevalence of cataract than men (odds ratio = 1.37). These estimates are based on U.S. Census data from 2000. The total number of persons with cataract is predicted to be 30.1 million by 2020 (Congdon et al., 2004).

The majority of cataracts occur with age. Other underlying causes are congenital, traumatic, or metabolic. Diabetes is a known risk factor for cataracts.

Cataract extraction is the only effective treatment for cataract. Extraction restores sight but leaves a patient significantly disabled. The spectacles required to correct vision are extremely thick, leading to intrinsic visual distortion. The extracted cataract is currently replaced with an intraocular lens.

# HISTORIC DEVICES

The clinical treatment of cataract can be traced to before 29 CE, when Roman physician Aurelius Cornelius Celsus described couching in a medical encyclopedia, which compiled earlier Greek texts. *Couching* is the laying of cataractous opaque material on a couch or bed at the bottom of the eye. During couching, a needle penetrated the eye, moved through the pupil, and separated upper ligaments from the lens. After separation, the lens lay in the vitreous cavity. Though vision was improved, the patient still saw blurry images without a lens.

## Early Devices

In 1748, French surgeon Jacques Daviel introduced a surgical technique in which various tools were used to expose the lens. A needle and spatula were used to enter the lens capsule and remove the cataract (Figure 17.12). This is a form of extracapsular surgery. Unfortunately, because general anesthesia was not introduced until the late 1840s, this procedure required the services of a strong assistant to restrain the patient's head during surgery (Rucker, 1965).

## Enabling Technology: Poly(Methyl Methacrylate) Lens

By the 1940s, cataract extraction through extracapsular surgery was standard practice. After observing that World War II aviators could tolerate shards of poly(methyl methacrylate) (PMMA) aircraft canopies in their eyes, British surgeon Harold Ridley began to experiment with PMMA as an implantable lens to restore vision in these aphakic (natural-lens-removed) patients. PMMA, also known as plexiglass or acrylic, has a refractive index

17. INTRAOCULAR LENS IMPLANTS

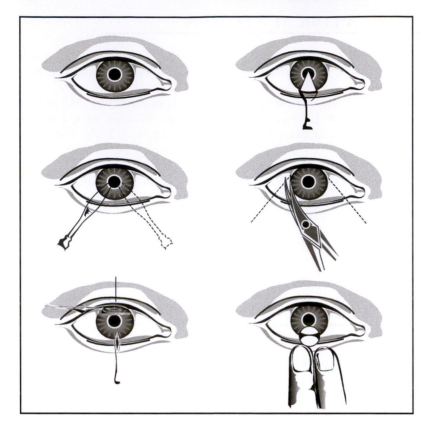

**FIGURE 17.12** Daviel's illustration of his technique of cataract extraction [Daviel, 1753. *Adapted from Rucker* (1965)].

of 1.49, which is close to the refractive index of a natural lens. Ridley implanted his first IOL in his patient's posterior chamber in 1949 and demonstrated that it could be tolerated in the eyes of some patients without inflammation for at least 2 yr. This IOL was 8.35 mm in diameter and 2.40 mm in thickness (Figure 17.13) (Ridley, 1952). Ridley's then radical idea that a foreign body could be placed into the delicate tissues of the eye marked the beginning of the modern era of medical implants.

## Enabling Technology: Phacoemulsification

Although Ridley demonstrated that IOLs were plausible, ophthalmologists did not adopt this procedure for decades. Ophthalmologist Charles Kelman's new phacoemulsification procedure, which simplified cataract extraction and IOL implant, created the demand for a better lens (Goldstein, 2004).

In 1967, Kelman invented an ultrasound device whose probe tip could be inserted in a small incision in the lens capsule and used to fragment the cataract. The tip also provided irrigation fluid for cooling and suction for aspiration of the emulsified lens. Kelman named this procedure *phacoemulsification* (Figure 17.14) (Kelman, 1967). He received the 2004 Lasker Clinical Medical Research Award "for revolutionizing the surgical removal of

**FIGURE 17.13** Insertion of PMMA lens beneath lower part of iris *[Reproduced by permission from Ridley (1952)].*

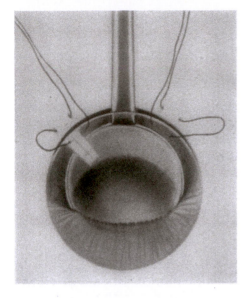

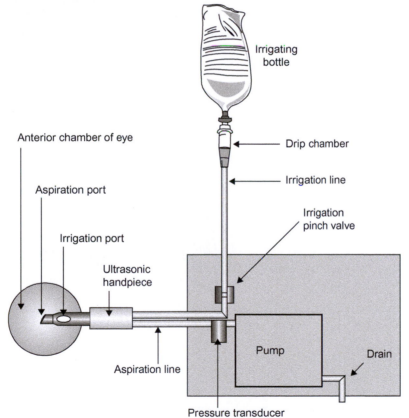

**FIGURE 17.14** Phacoemulsification system *[Adapted from Seibel (2005).*

cataracts, turning a 10-day hospital stay into an outpatient procedure, and dramatically reducing complications" (Lasker Awards, 2009).

## SYSTEM DESCRIPTION AND DIAGRAM

In parallel with phacoemulsification progressing toward medical community acceptance, ophthalmologists experimented with the location of IOL implantation. IOLs were first implanted in the anterior chamber, followed by implantation in the posterior chamber ciliary sulcus (immediately behind the iris) or lens capsular bag. Implantation in the capsular bag after extracapsular surgery eliminated corneal damage and iris chafing, reduced inflammation, and improved central fixation.

### Intraocular Lens Designs

IOL design evolved to incorporate two fixation haptics with the central optic. The haptics allow an IOL to be centered within the capsular bag. Newer designs are manufactured as one piece, with the central optic and haptics fabricated from the same material (Figure 17.1).

IOL design also evolved from providing monofocal to multifocal correction. With original monofocal IOLs, a patient had corrected distance vision but required spectacles for near vision. Multifocal IOLs enable many patients to become spectacle independent. One of the newer multifocal designs incorporates an anterior modified prolate surface and a fully diffractive posterior surface (Figures 17.1 and 17.15). We discuss this design further in Exercise 17.3.

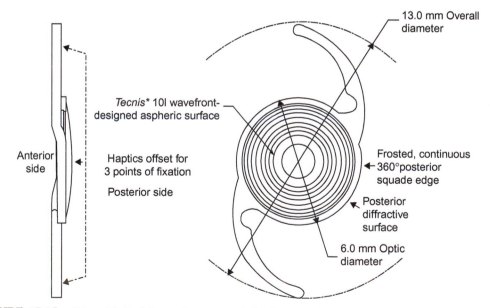

**FIGURE 17.15**   Abbott Medical Optics Tecnis's multifocal intraocular lens characteristics (AMO, Santa Ana, California).

IOL materials evolved from Ridley's original PMMA to silicones, hydrophobic acrylics, and hydrophilic acrylics. Although phacoemulsification requires at most a 3-mm incision to enter the eye, a PMMA IOL requires a 6-mm incision. To eliminate surgically induced astigmatism (unfocused vision) and the need for sutures, IOLs began to be fabricated from softer materials that could be folded before lens capsule insertion. Many silicone IOLs are manufactured from poly(dimethyl siloxane). The surface energy and hardness of PMMA can be modified by surface treatment. Coating PMMA with Teflon AF renders its surface hydrophobic. Modifying the surface of PMMA with heparin renders its surface hydrophilic. Additionally, a chromophore such as benzotriazole may be added during manufacture to absorb ultraviolet radiation (Werner, 2008). The Tecnis IOL in Figure 17.15 is fabricated from a UV-blocking hydrophobic acrylic.

A biocompatibility system diagram for an intraocular lens is given in Figure 17.16. We assume that the lens is implanted in the capsular bag. If the capsular bag is ruptured during cataract surgery, then the lens may be implanted in the ciliary sulcus. IOL efficacy is affected by design, material, and surface modification. As stated in the next section, efficacy is also affected by lens epithelial cells and aqueous humor (AH) proteins.

A routine phacoemulsification and IOL implant procedure can be viewed at http://www.youtube.com/watch?v=HDIoVedWWUQ&feature=related.

## Intraocular Lens Complications

Immediate postoperative complications from cataract removal and IOL implantation are wound leak, inflammation, corneal edema, and endophthalmitis. *Endophthalmitis* is the extensive intraocular inflammation associated with infection. It is the most severe of these immediate complications and occurs in 0.08–0.10% of cases (Asbell et al., 2005).

Long term, the most common complication from cataract surgery is *posterior capsular opacification* (*PCO*) (Figure 17.17). Decreased visual acuity induced by PCO is reported to occur in 20–40% of patients, 2–5 yr after surgery. In children, PCO rates range from 40% to 100%. This second opacification (often called a second cataract) is induced by a wound-healing response after cataract surgery. Leftover LECs proliferate and migrate across the posterior capsule, then undergo lens fiber regeneration and epithelial-to-mesenchymal transition. These myofibroblasts deposit collagen, which leads to PCO development. Cytokines and

FIGURE 17.16 Intraocular lens biocompatibility system diagram.

IOL (design, material, surface mod, leftover LEC, AH proteins)

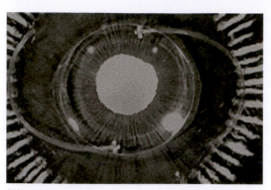

**(A)**

FIGURE 17.17 Photographs from human eyes obtained postmortem showing different degrees of opacification of the anterior capsule with **A**: a hydrophobic acrylic IOL, **B**: another hydrophobic acrylic IOL. **C**: PMMA IOL. *[Reproduced by permission from Steinert (2010).]*

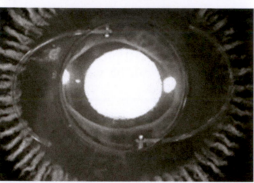

**(B)**

**(C)**

growth factors in the aqueous humor influence the behavior of these LECs after surgery (Awasthi et al., 2009). Pearl-type (Elschnig pearl) PCO morphology has been observed to significantly change during a 24-hr period (Neumayer et al., 2006) (Figure 17.18).

PCO is treated by creating an opening in the posterior capsule using a neodymium yttrium aluminum garnet (Nd-YAG) laser. A pressure wave, with a wavelength of 1.064 nm, disrupts the anterior vitreal side of the capsule, leaving a central opening.

To prevent PCO, surgeons attempt to remove all residual LECs during cataract surgery. Further, manufacturers are designing new IOLs with sharp edges because the created

(A) (B) (C) (D)

Day 1

Day 2

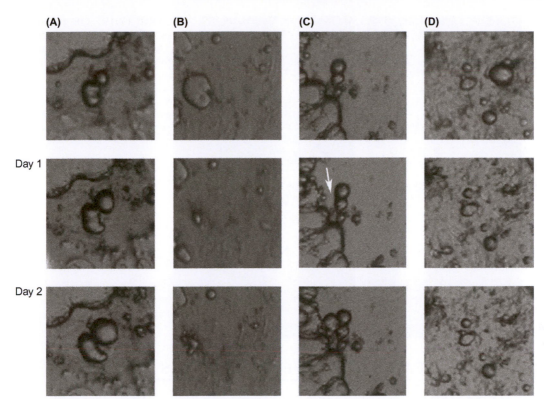

**FIGURE 17.18** Examples of daily changes in Elschnig pearl morphology. **A**: Major increase in pearl size; the upper pearl shows a major increase in diameter between days 0 and 1 and between days 1 and 2. **B**: Major decrease in pearl size; the large pearl on day 0 showed a major decrease in diameter between day 0 and 1 but no further decrease between day 1 and 2. **C**: Appearance of a pearl; at day 2 a newly formed pearl filled a gap (arrows) visible at day 0 and 1. **D**: Disappearance of a pearl; the large pearl apparent at day 0 disappeared between day 0 and 1. Side length of shown images is 770 μm [*Reproduced by permission from Neumayer et al. (2006)*].

sharp capsular bend inhibits LEC migration onto the posterior capsule (Awasthi et al., 2009) (Figure 7.19). Recently, 34 cataract patients were randomly implanted with a rounded-edge (IOL) in one eye and a sharp posterior optic edge in the other eye. The two silicone IOL designs were identical except for the edge profile. After 3 yr, the Automated Quantification of After-cataract (AQUA) PCO scores in these patients were significantly higher for the round edge (2.04 vs. 0.64, $p < 0.001$). The range of the AQUA PCO score is 0–10 (Buehl et al., 2007) (Figure 17.19).

## Other Intraocular Lenses

IOLs continue to be improved for aphakic patients. The Staar Surgical Toric IOL was approved by the FDA in 1991 to correct astigmatism (nonspherical eye shape) (CDRH, 1991). The Bausch & Lomb Crystalens was approved by FDA in 2003 to restore accommodation (CDRH, 2003). IOLs have also been developed for phakic patients to treat myopia.

**(A)**

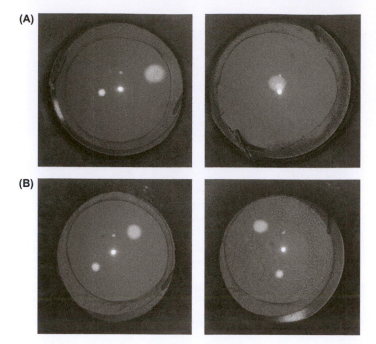

**(B)**

FIGURE 17.19 Effect of optic edge design in silicone intraocular lenses on posterior capsule opacification. Retroillumination images from two representative patients (**A** and **B**) 3 yr after surgery. Sharp-edge (Clariflex) eyes are shown on left, and round edge (SI40) eyes on right [*Reproduced by permission from Buehl et al. (2007)*].

## KEY FEATURES FROM ENGINEERING STANDARDS

As of 2010, FDA recommends 12 specific consensus standards for intraocular lenses. We discuss requirements from three important IOL standards that are part of *ISO 11979 Ophthalmic Implants—Intraocular Lenses*:

- *ISO 11979-3:2006 Part 3: Mechanical Properties and Test Methods* (ISO, 2006a).
- *ISO 11979-5:2006 Part 5: Biocompatibility* (ISO, 2006b).
- *ISO 11979-7: 2006 Part 7: Clinical Investigations* (ISO, 2006c).

### Dynamic Fatigue Durability

Each haptic loop must be durable. All loops are required to be capable of withstanding, without breaking, 250,000 cycles of near sinusoidal deformation of ±0.25 mm around a compressed distance. The compression frequency must be within 1–10 Hz.

This requirement is tested in IOLs intended for capsular bag placement by clamping an IOL center that is 5.0 mm away, compressed, from a testing plate. The flat-surfaced testing plate must be produced from a low-friction material to minimize haptic frictional constraint. The number of test samples is not specified. To pass, no tested loop must break (ISO, 2006a).

### Recovery of Properties Following Simulated Surgical Manipulation

IOLs that are intended to be folded or compressed during implantation must still meet manufacturing specifications after deformation. This requirement is tested in 10 lenses of each of the dioptric powers with the smallest and largest cross-sectional dimensions. Folding instructions supplied by the manufacturer are followed, with the lens deformed 3–20 min. After release from the deformed state, the lenses must relax at *in situ* conditions for up to 24 ±2 hr. They are then tested for dioptric power, image quality, overall diameter and sagitta, and surface and bulk homogeneity. To pass, all lenses must remain within manufacturer specifications (ISO, 2006a).

## Hydrolytic Stability

Implanted IOLs must remain stable over time. To test this requirement, an IOL is exposed to simulated conditions that approximate five years in an aqueous environment of 35 ±2°C. The actual study time is multiplied by the following factor, $F$, assuming an inner eye temperature of 35°C:

$$F = 2.0^{\frac{(T-35)}{10}}, \tag{17.14}$$

where $T$ is the accelerated temperature used for simulation. After the test period, the exposure medium is qualitatively and quantitatively analyzed for any chemical entities. The IOL is then examined by light microscopy, at $10\times$ or higher, and by scanning electron microscopy, at $500\times$ or higher. Optical transmittance spectra in the ultraviolet and visible spectral regions, dioptric power, and refractive index are also assessed.

To pass, significant differences in the exposure medium and IOL must not be detected. Specific descriptions of what constitutes a significant difference in exposure medium, spectral transmittance, and refractive index are not given. A significant difference in surface appearance is the observation of bubbles, dendrites, breaks, or fissures. A significant difference in dioptric power would be a change $\geq \pm 0.25$ diopters for a 20-diopter lens. The number of required samples for testing is not given (ISO, 2006b).

## Effect of Nd-YAG Laser Exposure

When posterior capsule opacification occurs, a Nd-YAG laser may be used to restore vision by creating a hole or an optical path through the lens capsule. Because an IOL may be nicked by the laser, no cytotoxic substances may be released after Nd-YAG laser exposure.

To test this requirement, each of five sterile, finished IOLs is immersed in an optical cuvette containing 2 mL physiological saline. Each is then exposed to 50 single laser pulses, set at an energy level of 5 mJ. The laser is focused on the posterior surface of the IOL and refocused for each pulse. Spots should be distributed evenly over the central 3 mm of the IOL optic. The exposure media for all five lenses is pooled and tested for cytotoxicity, according to *ISO 11979*, as described in Chapter 6 (ISO, 2006b).

## Clinical Study

New IOL designs must be tested in at least 300 human subjects to assess safety and performance. The study must be conducted in two phases to minimize potential risks. During

**TABLE 17.2**    Posterior chamber IOL adverse event rates *[Adapted from ISO (2006c)]*

| Adverse event | SPE rate[c] % | Threshold rate[d] % | Number of subjects = 100 — Max. number of cases allowed before SPE rate exceeded[e] | Threshold rate[d] % | Number of subjects = 300 — Max. number of cases allowed before SPE rate exceeded[e] |
|---|---|---|---|---|---|
| *CUMULATIVE:* | | | | | |
| Cystoid macular oedema | 3.0 | 8.9 | 6 | 6.0 | 14 |
| Hypopyon | 0.3 | 3.0 | 1 | 1.8 | 3 |
| Endophthaimitis[a] | 0.1 | 3.0 | 1 | 1.0 | 1 |
| Lens disiocated from posterior chamber | 0.1 | 3.0 | 1 | 1.0 | 1 |
| Pupillary block | 0.1 | 3.0 | 1 | 1.0 | 1 |
| Retinal detachment | 0.3 | 3.0 | 1 | 1.8 | 3 |
| Secondary surgical intervention [b] | 0.8 | 4.2 | 2 | 2.6 | 5 |
| *PERSISTENT:* | | | | | |
| Comeal stroma oedema | 0.3 | 3.0 | 1 | 1.8 | 3 |
| Cystoid macular oedema | 0.5 | 4.2 | 2 | 2.2 | 4 |
| Iritis | 0.3 | 3.0 | 1 | 1.8 | 3 |
| Raised IOP requiring treatment | 0.4 | 4.2 | 2 | 1.8 | 3 |

[a]*Endophthalmitis is defined as inflammatory reaction (sterile or infectious) involving the vitreous body.*
[b]*Excludes posterior capsulotomies.*
[c]*The SPE rate is the safety and performance endpoint.*
[d]*The threshold rate is the minimum rate detectable as statistically significantly different from the SPE rate (greater than the SPE rate in the case of adverse events: less than the SPE rate in the case of BCVA).*
[e]*The maximum number of cases allowed before SPE rate exceeded are the maximum number of subjects with that adverse event that can occur in a clinical investigation before the rate in that investigation becomes statistically significantly greater than the SPE rate.*

Phase 1, an initial 100 subjects are studied. If the results of Phase 1 are acceptable, then the remaining subjects are studied.

In an uncontrolled study of a posterior chamber IOL, the adverse events and visual acuity rates are compared to safety and performance endpoints in Tables 17.2 and 17.3, respectively. The passing rates are detailed in these tables (ISO, 2006c).

**TABLE 17.3** Overall postoperative best spectacle-corrected visual acuity (BCVA) 0.5 (6/12, 20/40) or better *[Adapted from ISO (2006c)]*

| Lens type | SPE rate[a] % | Threshold rate[b] % | Number of subjects = 100 Min. number of cases allowed before less than SPE rate [c] | Number of subjects = 300 Threshold rate[b] % | Min. number of cases allowed before less than SPE rate [c] |
|---|---|---|---|---|---|
| Anterior chamber IOL | 80.4 | 69.6 | 74 | 74.3 | 230 |
| Posterior chamber IOL | 92.5 | 84.4 | 88 | 88.3 | 270 |

[a]*The SPE rate is the safety and performance endpoint.*
[b]*The threshold rate is the minimum rate detectable as statistically significantly different from the SPE rate (greater than the SPE rate in the case of adverse events; less than the SPE rate in the case of BCVA).*
[c]*The minimum number of cases allowed before less than SPE rate are the minimum number of subjects with BCVA 0.5 or better that can occur in a clinical investigation before the rate in that investigation becomes statistically significantly less than the SPE rate.*

# SUMMARY

When light enters the eye, it first passes through the cornea. The cornea is the transparent external surface of the eyeball and is continuous with the white sclera, which is the supporting wall of the eyeball. The cornea acts as a first lens to focus light as it moves through the aqueous humor and pupil to the lens.

The lens focuses light through the vitreous humor, a jelly-like fluid, to the retina in the back of the eye. The suspended lens curvature changes from moderately convex to very convex, depending on the pull and relaxation of suspensory ligaments. Light passes through layers of the retina to the layer of rods and cones. Rhodopsin in the rods and cone pigments in the cones react to the light, causing direct current to be transmitted through the optic nerve.

As a lens ages, lens fibers are compressed centrally, and the lens becomes harder. With time the lens loses its transparency and may form a cataract. To restore sight, the cataract is fragmented and removed by phacoemulsification. An intraocular lens is then implanted to replace the cataractous lens.

Current IOL designs incorporate two fixation haptics with the central optic, allowing the IOL to be centered within the capsular bag. The IOLs can provide multifocal correction. They are fabricated from silicones, hydrophobic acrylics, and hydrophilic acrylics. To prevent posterior capsular opacification, manufacturers are designing new IOLs with sharp edges because the created sharp capsular bend inhibits LEC migration onto the posterior capsule.

Key IOL features include dynamic fatigue durability, recovery of properties following simulated surgical manipulation, hydrolytic stability, effect of Nd-YAG laser exposure, and clinical study.

## Acknowledgment

The author thanks Dr. Arlene Gwon for her detailed review. Dr. Gwon is a Clinical Professor of Ophthalmogy at University of California, Irvine.

## Exercises

**17.1** An accurate A-scan is critical for determining the refractive power needed for an IOL that will be implanted. The method described in the text incorporates an error due to corneal compression by the A-scan probe. To avoid corneal compression, a small sclera shell may be placed between a patient's upper and lower eyelids. After the shell is filled with fluid, the probe is inserted into the shell without making contact with the cornea. Redraw the A-scan in Figure 17.10 to reflect avoidance of corneal compression when this second method is used. Label all spikes with detected structures.

**17.2** Do other wound-healing responses occur in the eye after cataract surgery? Give specific evidence.

**17.3** What are the reported advantages of an anterior modified prolate surface and a fully diffractive posterior surface for a multifocal IOL? How would you test these reported advantages in a bench study? What is your key measurement? Be specific.

**17.4** In Chapters 2–17, we studied requirements from the following standards bodies: AAMI, ISO, ASTM, BSI, IEC. Please rank the relative rigor of consensus standards from each standards body, with 1 being the most lenient and 10 being the strictest. Name three types of requirements that contribute to a rigorous standard. Be specific.

Read Ridley (1952):

**17.5** Describe Ridley's acrylic lens design. How is this lens inserted and retained in place?

**17.6** What were the results of the 27 operations? What was the criterion for success? Can this criterion be used to determine IOL success today?

Read Kelman (1967):

**17.7** With what other techniques did Kelman experiment before settling on low-frequency emulsification for cataract removal? Describe the experimental system.

**17.8** Draw a system diagram for the phacoemulsification system.

## References

Asbell, P. A., Dualan, I., Mindel, J., Brocks, D., Ahmad, M., & Epstein, S. (2005). Age-related cataract. *Lancet, 365,* 599–609.

Awasthi, N., Guo, S., & Wagner, B. J. (2009). Posterior capsular opacification: A problem reduced but not yet eradicated. *Archives of Ophthalmology, 127,* 555–562.

Buehl, W., Menapace, R., Findl, O., Neumayer, T., Bolz, M., & Prinz, A. (2007). Long-term effect of optic edge design in a silicone intraocular lens on posterior capsule opacification. *American Journal of Ophthalmology, 143,* 913–919.

CDRH (1991). *Approval order P880091.* Rockville, MD: FDA.

CDRH (2003). *Approval order P030002.* Rockville, MD: FDA.

Congdon, N., Vingerling, J. R., Klein, B. E., West, S., Friedman, D. S., & Kempen, J., et al. (2004). Prevalence of cataract and pseudophakia/aphakia among adults in the United States. *Archives of Ophthalmology, 122,* 487–494.

Daviel, J. (1753). Sur une nouvelle methode de guerir la cataracte par l'extraction du cristalin. *Mémoire de l'Académie de Chirurgie, 2,* 337.

Goldstein, J. L. (2004). How a jolt and a bolt in a dentist's chair revolutionized cataract surgery. *Nature Medicine, 10,* 1032–1033.

Guyton, A. C., & Hall, J. E. (2006). *Textbook of medical physiology* (11th ed.). Philadelphia: Elsevier Saunders.

ISO (2006a). *ISO 11979-3:2006 Ophthalmic implants—Intraocular lenses—Part 3: Mechanical properties and test methods.* Geneva: ISO.

ISO (2006b). *ISO 11979-5:2006 Ophthalmic implants—Intraocular lenses—Part 5: Biocompatibility.* Geneva: ISO.

ISO (2006c). *ISO 11979-7:2006 Ophthalmic implants—Intraocular lenses—Part 7: Clinical investigations.* Geneva: ISO.

Johns, K. J., Feder, R. S., Hammill, B. M., Miller-Meeks, M. J., Rosenfeld, S. I., & Perry, P. E. (2003). *Basic and clinical science course, Section 11: Lens and cataract.* San Francisco: American Academy of Ophthalmology.

Kelman, C. D. (1967). Phaco-emulsification and aspiration. A new technique of cataract removal. A preliminary report. *American Journal of Ophthalmology, 64,* 23–35.

Lasker Awards: Former Winners. (2009). Lasker Foundation.

Neumayer, T., Findl, O., Buehl, W., & Georgopoulos, M. (2006). Daily changes in the morphology of Elschnig pearls. *American Journal of Ophthalmology, 141,* 517–523.

Packer, M., Fishkind, W. J., Fine, I. H., Seibel, B. S., & Hoffman, R. S. (2005). The physics of phaco: A review. *Journal of Cataract and Refractive Surgery, 31,* 424–431.

Refojo, M. F. (2004). Ophthalmological Applications. In B. D. Ratner, A. S. Hoffman, F. J. Schoen, & J. E. Lemons (Eds.), *Biomaterials science* (2nd ed.). San Diego, CA: Elsevier.

Ridley, H. (1952). Intra-ocular acrylic lenses; a recent development in the surgery of cataract. *British Journal of Ophthalmology, 36,* 113–122.

Rucker, C. W. (1965). Cataract: A historical perspective. *Investigative Ophthalmology, 4,* 377–383.

Seibel, B. S. (2005). *Phacodynamics: Mastering the tools and techniques of phacoemulsification surgery* (4th ed.). Thorofare, NJ: Slack.

Song, S., Landsbury, A., Dahm, R., Liu, Y., Zhang, Q. & Quinlan, R. A. Functions of the intermediate filament cytoskeleton in the eye lens. *Journal of Clinical Investigation, 119,* 1837–1848.

Steinert, R. F. (Ed.), (2010). *Cataract surgery* (3rd ed.). Philadelphia: Elsevier Saunders.

Waldron, R. G., & Aaberg Jr., T. M. (2008). A-scan biometry. emedicine.medscape.com.

Webb, A. (2003). *Introduction to biomedical imaging.* Hoboken, NJ: John Wiley/IEEE Press.

Werner, L. (2008). Biocompatibility of intraocular lens materials. *Current Opinion in Ophthalmology, 19,* 41–49.

Zacharias, J. (2008). Role of cavitation in the phacoemulsification process. *Journal of Cataract and Refractive Surgery, 34,* 846–852.

# Total Hip Prostheses

In this chapter, we discuss the relevant physiology, history, and key features of total hip prostheses. A *hip prosthesis* is a medical device that replaces a damaged hip joint (Figure 18.1). The study of orthopedic devices, such as the total hip prosthesis, is important because orthopedics is a high-growth area in the medical device industry.

**FIGURE 18.1** Johnson & Johnson DuPuy total hip prosthesis consisting of Duraloc 100 acetabular shell (shown in cross section, middle, and bottom), Marathon liner, and Co-Cr alloy femoral component *[Reproduced by permission from Engh et al. (2006)].*

Upon completion of this chapter, each student shall be able to:

1. Understand the mechanisms underlying bone's ability to endure joint contact forces.
2. Explain the relationship between prosthesis wear and osteolysis.
3. Describe four key features of hip prostheses.

## HIP PHYSIOLOGY

The hip consists of two bones, cushioned by a synovial joint capsule. Within this ball-and-socket joint, a convex femoral head is inserted into a concave acetabulum within the pelvis (Figure 18.2).

### Bone

Bone tissue consists of approximately 30% collagen fibrils, 60% ceramic crystalline mineral, and 10% water, by weight. The collagen fibrils, which are mostly type I collagen, are arranged in stacked sheets, and contain several noncollagenous proteins. They are mineralized by an impure form of naturally occurring hydroxyapatite [$Ca_{10}(PO_4)_6(OH)_2$], organized as nanometer-sized crystals (Bartel et al., 2006). The mineralized fibrils may be held together by a nonfibrillar organic matrix that acts as a glue (Figure 18.3). When a force is applied to the bone, the glue may resist the separation of mineralized collagen fibers, which counteracts the formation of cracks (Fantner et al., 2005). The identity of this glue may be the bone phosphoprotein osteopontin (Zappone et al., 2008).

Bone can be classified as cortical or trabecular. Dense *cortical*, or *compact*, *bone* contains tightly packed sheets. Spongy *trabecular*, or *cancellous*, *bone* is organized into a series of rods and plates, interspersed with large marrow spaces. Cortical bone is found in the outer layers of most bones. Trabecular bone is found within flat and irregular bones, such as the

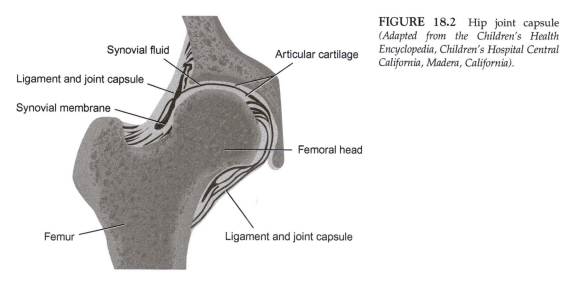

**FIGURE 18.2** Hip joint capsule (*Adapted from the Children's Health Encyclopedia, Children's Hospital Central California, Madera, California*).

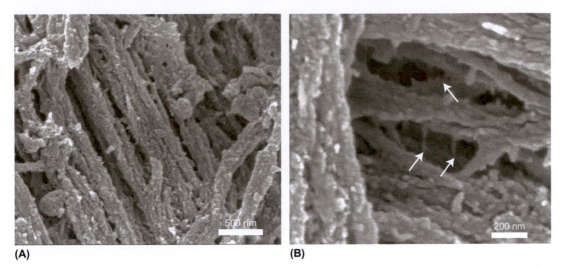

**(A)**                                                                         **(B)**

**FIGURE 18.3**   Scanning electron microscope images of mineralized collagen fibrils on a fracture surface of human trabecular bone. **(A)** Fracture surface. **(B)** Individual collagen fibrils are held together with glue filaments (arrows), which might resist separation of the filaments *[Reproduced by permission from Fantner et al. (2005)].*

sternum, and at the ends of all long bones. The femoral head and pelvis are constructed of trabecular bone, with a thin outer layer of cortical bone.

Because of its structure, trabecular bone is anisotrophic. *Anisotropy* is the exhibition of properties (such as mechanical strength) with different magnitudes, dependent on the direction of measurement (Keaveny et al., 2001). The predominant alignment of the trabecular rods and plates determines the preferred orientation. Mechanical property testing should be conducted along this orientation (Bartel et al., 2006; Keaveny et al., 2001).

Common mechanical properties are the elastic modulus and ultimate strength (Table 18.1). The elastic modulus and ultimate strength are determined from a plot of measured stress (applied force per unit area) vs. strain (the fractional change in dimension) when a material undergoes deformation. As shown in Figure 18.4, the elastic modulus is the initial slope of this nonlinear curve; the ultimate strength is the stress at which the material breaks. Even though it has a relatively low elastic modulus and ultimate strength, trabecular bone resists fracture during movement.

Bone is constantly resorbed, grown, and reinforced. These processes enable bone to adapt mechanically to changes in habitual loading or implantation of a prosthesis. *Osteoclasts* are formed from precursor monocytes, and are giant multinucleated cells that secrete an acid over exposed bone tissue. They ingest the debris formed from acid-dissolved bone, and expel it to the bloodstream. *Osteocytes* trapped with the bone matrix signal surface cells and attract them toward damaged areas for repair. These surface *bone lining cells* are dormant osteoblasts, which cover 90% of trabecular bone surfaces. *Osteoblasts* secrete *osteoid*, the collagenous matrix that forms the foundation of bone (Bartel et al., 2006).

TABLE 18.1    Mechanical properties of dominant orthopedic biomaterials

| Material | Elastic Modulus (GPa) | Ultimate Strength (MPa) | Hardness HVN |
|---|---|---|---|
| Human vertebra, 15–87 yr | 0.067 | 2.4 | — |
| Human proximal femur, 58–85 yr | 0.441 | 6.8 | — |
| UHMWPE | 0.5–1.3 | 30–40 $t$ | 60–90 MPa |
| PMMA | 1.8–3.3 | 38–80 $t$ | 100–200 MPa |
| $Al_2O_3$ | 366 | 3790$c$/310 $t$ | 20–30 GPa |
| $ZrO_2$ | 201 | 7500$c$/420 $t$ | 12 GPa |
| Co-Cr alloy ASTM F75 | 210–253 | 655–1277 $t$ | 300–400 |
| Ti alloy Ti-6AL-4V ASTM 136 | 116 | 965–1103 $t$ | 310 |

$t$ = tension; $c$ = compression.
Source: Keaveny et al. (2001); Hallab et al. (2004b).

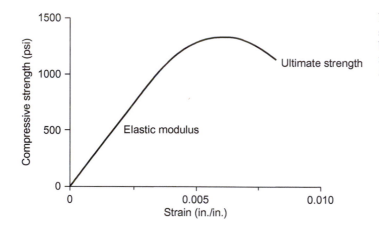

FIGURE 18.4 An early stress/strain diagram for human trabecular bone in the femoral head. SI units are not used [Reproduced by permission from Schoenfeld et al. (1974)].

## Synovial Joint Capsule

The acetabulum and head of the femur are covered with articulating cartilage to reduce friction and absorb shock. They are sealed in a synovial joint capsule. This ball-and-socket joint capsule consists of dense fibrous connective tissue lined with synovial membrane, which forms a sleeve around the articulating bones to which it is attached. The synovial membrane secretes synovial fluid for joint lubrication. Accessory ligaments provide bone-to-bone connection, in order to restrict motion and to protect the capsule from loading.

## Contact Forces

The hip joint experiences substantial contact forces during activity. These forces are difficult to model, but can be measured *in vivo* using instrumented hip prostheses (Figure 18.5).

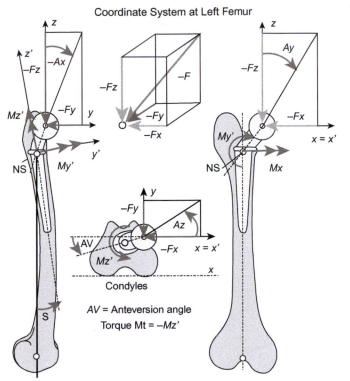

Coordinate System at Left Femur

AV = Anteversion angle
Torque Mt = –Mz'

**FIGURE 18.5** Coordinate system for measured hip contact forces [*Reproduced by permission from Bergmann et al. (2001)*].

At the prosthetic femoral head, the hip contact force vector, $F$, can be resolved into components $F_x$, $F_y$, and $F_z$. The z-axis is parallel to the idealized femoral long axis; the x-axis is parallel to the dorsal contour of the femoral condyles in the transverse plane. The y-axis is the medial axis. The negative direction points toward the femoral head. The contact force causes a torsional moment, $M_t$, around the intersection point, NS, of shaft and neck axes of the prosthesis. A positive torsional moment rotates the femoral head inwards. It is equal to the moment in the negative z' direction, where the z'-axis is the prosthesis' stem axis. The moment, $M$, can be resolved into components $M_x$, $M_{y'}$, and $M_{z'}$.

Contact forces were measured in four patients. Each instrumented hip prosthesis was a modification of a cemented standard titanium femoral component, with a polyethylene liner. Three semiconductor strain gauges were placed within the hollow neck. The neck was sealed by a laser-welded plate that contained a transmission antenna. Force sensor data were digitized, using a sampling frequency of 200 Hz, and transmitted by radio frequency to a receiver. The error of each sensor was ±1%. Data were recorded during slow walking, normal walking, fast walking, moving up stairs, moving down stairs, standing up, sitting down, standing on 1-2-1 legs, and knee bend.

The average forces and moments, as a total and in the x-, y-, and z-directions, are illustrated in Figures 18.6 and 18.7. The peak hip contact force of 260% body weight occurred while moving down stairs. The peak torsional prosthesis moment of 2.24% body weight-meters occurred while moving up stairs (Bergmann et al., 2001). These results are consistent with other peak force study results in the range of 250 to 350% body weight (Brand, 1994).

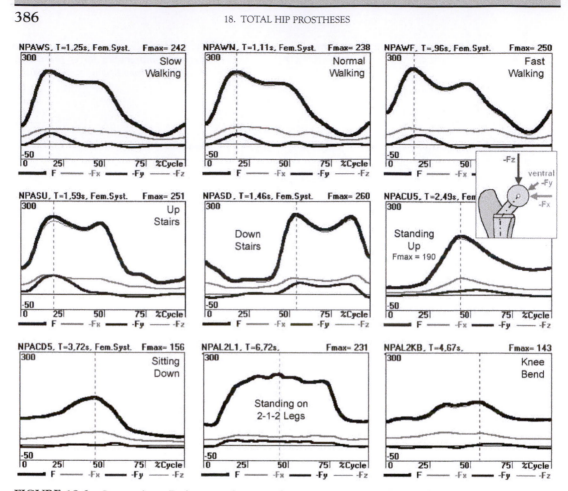

**FIGURE 18.6**  Contact force $F$ of averaged patient during nine activities. Contact force $F$ and its components $-F_x$, $-F_y$, $-F_z$, $F$, and $-F_z$ are nearly identical. The scale range is $-50$ to $+300\%$ body weight. Cycle duration and peak force $F_p = F_{max}$ are indicated in diagrams [*Reproduced by permission from Bergmann et al. (2001)*].

## WEAR-MEDIATED OSTEOLYSIS

We begin this total hip prosthesis discussion by introducing the concept of wear. The hip prosthesis is the first implanted device we have discussed with two articulating surfaces subjected to dynamic loading. Over time, the relative motion of these surfaces produces particle debris, which is referred to as *wear*. Wear is primarily caused by three processes (Hallab et al., 2004b). *Abrasion* occurs when a harder surface or surface with harder asperities (particles) plows grooves into the softer material. *Adhesion* occurs when touching asperities adhere together, causing the tips of the asperities to break off. Microscopic *fatigue* occurs when subsurface cracks are formed from alternating loading/unloading, causing surface delamination (Figure 18.8).

Hip prosthetic wear is affected by prosthetic, patient, and surgical factors that subdivide into numerous design and environmental variables. Some design variables include

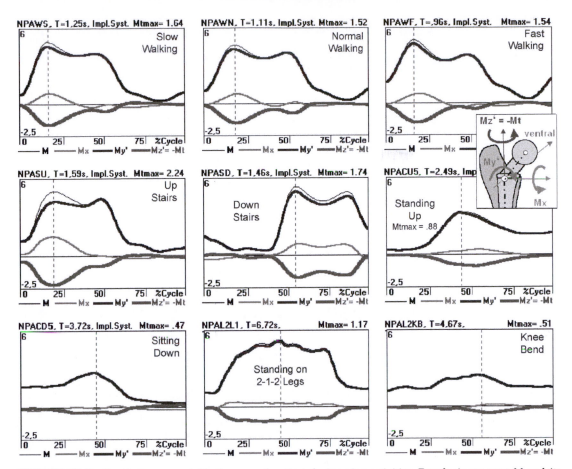

**FIGURE 18.7** Prosthetic moment $M$ of averaged patient during nine activities. Prosthetic moment $M$ and its components $-M_x$, $-M_{y'}$, $-M_{z'} = -M_t$. $M$ and $-M_{z'}$. $M$ and $M_{z'}$ are nearly identical. The scale range is $-2.5$ to $+6\%$ body weight-meters. Negative $M_z$ cause inward torsion around the femoral stem. Cycle duration $T$ and peak torsional moment $M_{tp} = M_{tmax} = -M_{zp}$ are indicated in diagrams *[Reproduced by permission from Bergmann et al. (2001)].*

bearing materials, surface finishes, and manufacturing processes (prosthesis), as well as desired component position and orientation with respect to bone (surgical). Some environmental variables include bone structure and weight (patient), as well as variation about a desired position or orientation (surgical). These variables interact in a complex manner that cannot be reduced to a simple equation (Brown & Bartels, 2008). For a cobalt-chromium (Co-Cr) head bearing on an ultrahigh-molecular-weight polyethylene (UHMWPE) liner, the linear wear rate is generally on the order of 0.1 mm/year (Hallab et al., 2004b), with particulate generation as high as $1 \times 10^6$ particles per step or per cycle.

Particle debris generated by wear induces the formation of an inflammatory reaction, which promotes a foreign-body granulation tissue response that may invade the bone-implant interface (Hallab et al., 2004b). In fibrous membranes adjacent to retrieved prostheses,

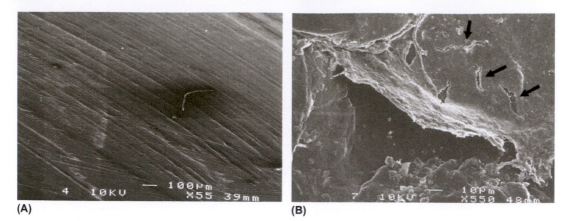

**FIGURE 18.8** Highly cross-linked UHMWPE liners, explanted from patients at the time of revision total hip arthroplasty, exhibit wear, as seen under the scanning electron microscope. **(A)** Adhesion. **(B)** Fatigue *[Reproduced by permission from Bradford et al. (2004)].*

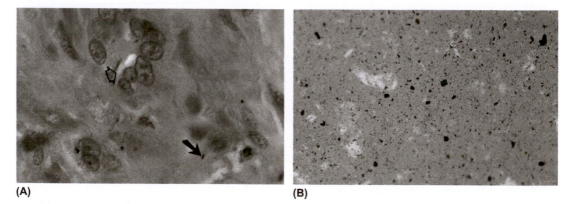

**FIGURE 18.9** Photomicrograph of the same tissue before and after tissue digestion. **(A)** Before digestion, a foreign-body giant cell is present that contains polarizable material, presumably ultra-high–molecular-weight polyethylene (open arrow). Small particles of metal are also present (solid arrow) (×1040). **(B)** After digestion, a photomicrograph was made under partially polarized light. There are many particles of dark metal and polarizable polyethylene (×420) *[Reproduced by permission from Margevicius et al. (1994)].*

foreign-body giant cells are associated with wear particles (Figure 18.9). Biologically active polyethylene wear particles appear to have an average particle size in the 0.5 micron diameter (Margevicius et al., 1994, Campbell et al., 1995, Schmalzried & Callaghan, 1999).

Foreign-body giant cell phagocytosis of wear particles initiates the release of cytokines and other inflammatory mediators. Increased cytokine release stimulates osteoclastic activity, aseptic prosthesis loosening, and focal bone resorption. The resulting *osteolysis*, or bone loss, compromises osseointegration (Charnley, 1975) and stability of the prosthesis (Harris, et al., 1976), which leads to clinical failure (Wang et al., 2004).

# CLINICAL NEED

A patient primarily receives a total hip prosthesis as treatment for arthritic pain. According to the Centers for Disease Control and Prevention, in 2005, 46.4 million U.S. persons had doctor-diagnosed arthritis and 17.4 million had arthritis-attributable activity limitations. As the population ages, an estimated 67 million adults may have arthritis by 2030 (CDC, 2006).

The most common form of arthritis is *osteoarthritis* (*OA*), in which the whole joint experiences focal and progressive articular cartilage loss. Concurrently, the bone underneath the cartilage changes, which may result in marginal outgrowths and increased thickness of the bony envelope. The synovium may show modest inflammatory infiltrates, and the ligaments may be lax.

Systemic factors that predispose people to OA include age, gender (women after 50 yr), ethnicity (African-Americans), bone density estrogen replacement therapy in postmenopausal women (lowered risk), and genetics. A local biomechanical factor such as obesity, a joint injury, a joint deformity, sports participation, or muscle weakness may then contribute to the severity of OA (Felson et al., 2000a).

Osteoarthritis often involves either the hip or knee joint. In 2007, there were 260,000 total hip arthroplasties (THAs) in the United States (Feder, 2008). *Arthroplasty* is the reconstruction of a normal joint. According to an NIH Scientific Conference summary, "total joint arthroplasty represents the most significant advancement in the treatment of OA in the past century," and "is the mainstay of surgical treatment of the OA hip, knee, and glenohumeral [shoulder] joint... By all measures, total joint replacement is among the most effective of all medical interventions; the pain and disability of end-stage OA can be eliminated, restoring patients to near-normal function" (Felson et al., 2000b).

# HISTORIC DEVICES

Clinical treatment of painful and deformed hip joints dates back to the 1820s, when the affected femoral and acetabular bones were removed. To restore mobility, surgeons attempted to use materials such as wooden blocks and porcine tissue as hip joint substitutes.

## Early Devices

In 1890, Themistocles Gluck developed the first prosthetic hip, which was a femoral head carved from ivory, secured in place with pumice and plaster of paris. Later, in 1923, Marius Smith-Petersen improved the design by using a molded glass cup, which interfaced between the femoral head and acetabular cup. Smith-Petersen iterated the design, eventually producing a molded cup from a Co-Cr alloy named Vitallium in 1938 (Figure 18.10).

The molded cup evolved to include a short stem, which was eventually elongated. In 1956, orthopedic surgeons G. K. McKee and John Watson-Farrar divided the prosthesis

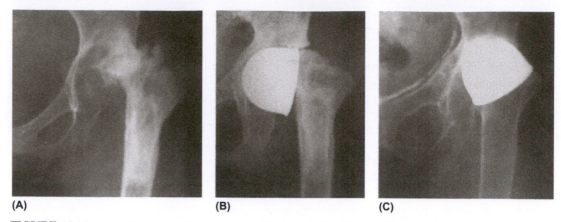

**(A)**                              **(B)**                              **(C)**

**FIGURE 18.10**    Forty-five year follow-up of a Smith-Petersen Vitallium molded cup. **(A)** Patient at age 25 yr in 1949, with severe degenerative hip due to congenital dislocation. **(B)** After cup implantation in 1949. **(C)** Follow-up in 1995. Patient walked with a limp but was free from pain. Notice resorption of the neck *[Reproduced by permission from Mahaligam et al. (1996)].*

**FIGURE 18.11**    Charnley's stainless steel femoral head prosthesis mated to PTFE *[Reproduced by permission from Charnley (1960)].*

into two pieces: a Co-Cr spherical femoral head with a long stem and a Co-Cr acetabular cup (Hallab et al., 2004b).

## Enabling Technologies: Low Friction and PMMA Cement

Orthopedic surgeon Sir John Charnley began to investigate elements of total hip arthroplasty (THA) in 1949. He first focused on quantifying the friction and lubrication of animal and artificial joints. These experiments led him to create, in 1956, an acetabulum of polytetrafluorethylene (PTFE) as a substitute for articular cartilage to reduce friction (Figure 18.11). After observing significant PTFE wear rates in his patients, Charnley changed the acetabular material to UHMWPE, which has relatively greater wear resistance, low friction, and high-impact strength.

Charnley also reexamined the process of gluing a prosthesis into bone with resin, replacing this process with grouting of poly(methyl methacrylate) (PMMA) cement in 1960. By forcing bone cement into all available interstices, the weight of the prosthesis body was dispersed over a large area of bone, improving fixation by a factor of 200 times (Gomez & Morcuende, 2005). John Charnley received the 1974 Lasker Clinical Medical Research Award "for his conceptual and technical contributions to total hip joint replacement, which have opened new horizons of research and treatment in arthritis and crippling joint diseases" (Lasker Awards, 2009).

## SYSTEM DESCRIPTIONS AND DIAGRAMS

Charnley's low-friction arthroplasty of the 1960s survives today as a metal femoral component, with stem and ball, articulating against an acetabular component, with a metal shell and a liner that is usually fabricated from polyethylene (Figure 18.1). The titanium (Ti) or Co-Cr alloy femoral stem may be cemented or press-fit into place. If cement is used, the following steps are included: femoral preparation to diminish interface bleeding, reduced cement porosity by vacuum mixing, preheating of the stem and polymer, pressurization with a cement gun, stem geometries that increase the intramedullary pressure, and cement intrusion into the cancellous structure of the bone. Typically, older individuals receive cemented stems because the chance for revision after 20 yr is minimal. With press-fit fixation, the drilled hole has a slightly smaller cross-sectional area than the stem and requires force to insert. Press-fit femoral stems are often proximally coated to encourage proximal osseointegration.

The Ti or Co-Cr alloy acetabular component is also press-fitted into place, with optional screws used to achieve initial stabilization. The shell has a grit-blasted or plasma-sprayed rough surface that can be either porous- or hydroxyapitite-coated for additional bone growth. The original polyethylene acetabular component evolved to a cross-linked polyethylene liner, which is fabricated from UHMWPE exposed to gamma irradiation and then further treated to reduce residual free radicals (Parvizi et al., 2009). Residual free radicals such as $H_2O_2$ make polyethylene susceptible to oxidative degradation (McKellop, et al., 1999, Engh, et al., 2006).

We already noted contact forces in the free body diagram of Figure 18.5. A biocompatibility system diagram for the hip prosthesis is given in Figure 18.12. Hip prosthesis efficacy is predominantly affected by osteolysis, which can be traced to wear debris. Wear is a function of prosthetic, patient, and surgical factors that subdivide into design and environmental variables.

An animation of the THA procedure can be viewed at http://www.nhs.uk/Livewell/arthritis/Pages/Hipoperationanimation.aspx. A patient radiograph before and after THA is shown in Figure 18.13.

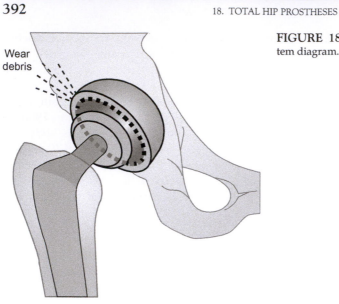

Wear
debris

Hip prosthesis (prosthetic, patient, and
surgical factors)

**FIGURE 18.12**   Hip prosthesis biocompatibility system diagram. *(Adapted from New York Times, New York).*

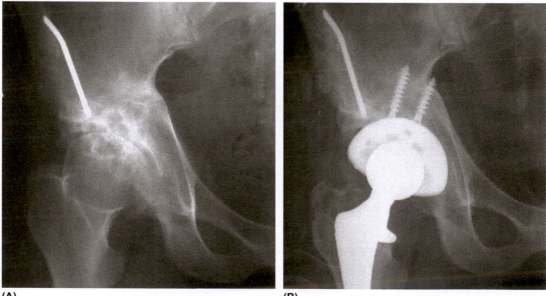

**(A)**                                                      **(B)**

**FIGURE 18.13**   Radiographs before (**A**) and after (**B**) THA. This patient had five previous surgeries for developmental dysplasia of the hip *[Reproduced by permission from Berger (2009)].*

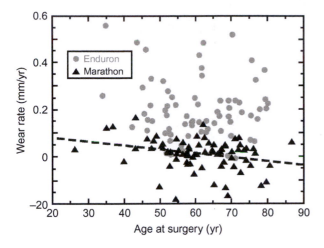

**FIGURE 18.14** The linear wear rate data show that the Marathon liners are wearing at a substantially lower rate as compared with the non-cross-linked Enduron liners. The dashed black line represents the relationship between Marathon wear rates and age at surgery ($r = 0.29$, $p = 0.01$). There was no relationship between Enduron wear rates and age at surgery ($p = 0.47$) [Reproduced by permission from Engh et al. (2006)].

## Bearings and Wear

Minimization of wear debris from hip prosthetic materials has been an industry goal for decades. After promising *in vitro* hip simulator studies, several manufacturers began to highly cross-link their polyethylene liners. A study of 230 THA patients who received the same DuPuy Co-Cr alloy femoral component and shell (Figure 18.1), with either a conventional or cross-linked UHMWPE liner, provided evidence of wear minimization. The patients were matched for gender, diagnosis, age at surgery, weight, and body mass index. Half of the patients ($n = 116$) received the non-cross-linked Enduron UHMWPE liner; half ($n = 114$) received the cross-linked Marathon UHMWPE liner. Linear wear rate was calculated as the rate of femoral head penetration per year, as observed from three pelvic x-rays over a minimum 4-yr follow-up. One orthopedic surgeon reviewed all x-rays to assess osteolysis, while blinded to liner type. As shown in Figure 18.14, the linear wear rate was significantly less in the cross-linked liners: $0.01 \pm 0.07$ vs. $0.20 \pm 0.13$ mm/yr ($p < 0.001$). The incidence of $>1$ cm$^2$ osteolyis was also significantly less: 6.3 vs. 22.2% ($p = 0.002$) (Engh et al., 2006).

Historically, a small femoral head was used with a polyethylene liner to minimize wear. Larger diameter femoral heads, which reduce dislocation, are now used with higher modulus (Table 18.1) metal-on-metal or ceramic-on-ceramic bearing surfaces. For these bearing surfaces, linear wear rate has been estimated by the Implant Wear Symposium 2007 Engineering Work Group as 5 μm/yr or 0–3 μm/yr, respectively (Clarke & Manley, 2008). However, clinical wear results vary considerably (Huber et al., 2010, Bolland et al., 2011).

Issues emerged after the bearing surfaces were changed. Metal-on-metal bearings are associated with *metal reactivity*, which is a lymphocyte-dominated reaction to metal ions (Hallab et al., 2004a). Metal reactivity may cause osteolysis (Schmalzried, 2009), which is being observed earlier and at a higher rate (Park et al., 2005, Korovessis et al., 2006) than with other bearing materials. Soft tissue reactions, which have been classified as pseudotumors, are increasingly observed around metal-on-metal bearings. Systemic effects, such as

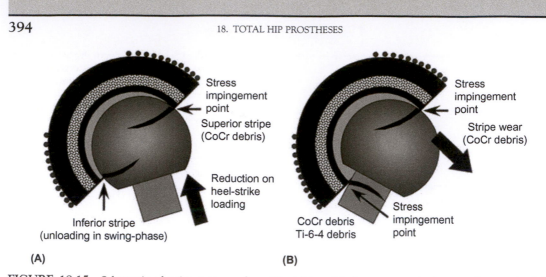

**(A)**                                                    **(B)**

**FIGURE 18.15**   Schematic of stripe wear modes with rigid acetabular component rims wearing on the ball counterface and resulting in a wear stripe on the acetabular component bevel. **(A)** With slight subluxation in the swing phase of gait, the acetabular component edge will create an inferior and somewhat equatorial ball stripe. During reduction at heel-strike, a superior but equatorial stripe will be created. Both stripes will produce ceramic-on-ceramic debris. **(B)** At any impingement event, the destabilizing forces will produce a more polar stripe on the ball, and the neck-cup impingement will produce debris from both the metallic femoral neck and the ceramic ball. Co-Cr = cobalt chromium, Ti-6-4 = grade 5 titanium *[Reproduced by permission from Clarke & Manley (2008)].*

decreased total lymphocyte count in patients and DNA changes induced by synovial fluid, are also present (Crawford et al., 2010). Although the potential long-term toxicity of metal-on-metal is unknown, the increasing negative reports prompted the editors of the *Journal of Arthroplasty* to issue the following warning in 2010:

> The lack of clinical advantage with metal bearings and the significant downsides to the use of metal-on-metal mean that, in the authors' opinion, this bearing option should be used with great caution, if at all (Crawford et al., 2010).

More recently, the FDA issued 145 orders for postmarket surveillance studies to 21 metal-on-metal hip prosthesis manufacturers on May 6, 2011. These manufacturers were required to submit a research protocol that addresses metal-on-metal adverse events (CDRH, 2011).

Another commonly implanted bearing material is ceramic. Harder ceramic-on-ceramic bearings made of second-generation alumina or zirconia materials have a small risk of brittle fracture. The more serious complication is squeaking, which is estimated to occur in 0.5 to 20% of implants (Walter et al., 2007, Restrepo et al., 2008, Feder, 2008, Baek & Kim, 2008). In the laboratory, it is possible to generate squeaking after formation of a wear stripe, which results from dragging a ceramic head across the rim of a ceramic liner under load (Taylor et al., 2007) (Figure 18.15). The wide variation of *in vivo* squeaking reports and the consistency of *in vitro* simulations suggest that variation in surgical technique and specific biomechanical conditions are both required for squeaking to occur (Clarke & Manley, 2008).

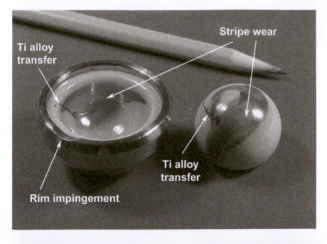

FIGURE 18.16 Photograph of the retrieved acetabular component from a patient with squeaking hip with evidence of impingement. Note the indentation of the metal rim generated by the femoral neck [*Reproduced by permission from Restrepo et al. (2008)*].

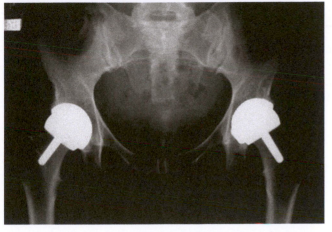

FIGURE 18.17 Bilateral hip resurfacing prostheses [*Reproduced by permission from Muirhead-Allwood et al. (2009)*].

In 999 patients receiving 1056 THAs with alumina ceramic bearings from January 2002 to December 2004, 28 patients (with 30 hips), or 2.7%, complained of squeaking. The squeaking developed between 0.6 and 2.7 yr after THA; it was not associated with other symptoms such as pain or instability. All acetabular and femoral components in these patients were found to be well fixed, with no evidence of loosening. Six squeaky hips underwent revision surgery to replace the bearings with metal-on-polyethylene bearings. In four of these hips, substantial posteroinferior neck-rim impingement was observed (Figure 18.16) (Restrepo et al., 2008).

In 2006, the proportions of bearing type sales in the United States were 71% metal-on-polyethylene, 20% metal-on-metal, 5% ceramic-on-ceramic, and 4% ceramic-on-polyethylene (Griffin & Bradbury, 2009). In the next few years, until metal-on-metal bearing issues were recognized, the use of metal-on-metal bearings increased dramatically because of the popularity of hip resurfacing procedures (Crawford et al., 2010).

Typically, a high risk (Class III) medical device must receive premarket approval from the FDA before it can be sold in the United States. As of May, 2011, a high risk metal-on-metal acetabular component requires only 510(k) clearance (without mandatory clinical testing) before market release because it is a preamendment Class III device. For a more thorough discussion of FDA regulation, see (Baura, 2013).

## Hip Resurfacing

Hip resurfacing is an option for younger patients receiving THA. During this procedure, the bone in the proximal femur is conserved by capping of the femoral head, in anticipation of a later revision of traditional THA. The cap is a Co-Cr alloy femoral resurfacing component, which articulates against a Co-Cr alloy acetabular component (Figure 18.17). The first resurfacing system received premarket approval from FDA in 2006 (CDRH, 2006). As stated above, increased and significant risks have been observed with the use of metal-on-metal bearings.

# KEY FEATURES FROM ENGINEERING STANDARDS

As of May, 2011, the FDA recommends six specific consensus standards for hip prostheses. We discuss requirements from three of these standards. Surprisingly, the two standards related to wear measurement are not recommended. Because wear is such an important issue, these standards are included:

- *ASTM F1440-92:2008 Standard Practice for Cyclic Fatigue Testing of Metallic Stemmed Hip Arthroplasty Femoral Components Without Torsion* (ASTM, 2008).
- *ASTM F1612-95:2005 Standard Practice for Cyclic Fatigue Testing of Metallic Stemmed Hip Arthroplasty Femoral Components Without torsion* (ASTM, 2005).
- *ASTM F1875-98:2009 Standard Practice for Fretting Corrosion Testing of Modular Implant Interfaces: Hip Femoral Head-Bore and Cone Taper Interface* (ASTM, 2009).
- *ISO 14242-1:2000 Implants for Surgery—Year of Total Hip-Joint Prostheses—Part 1: Loading and Displacement Parameters for Wear-Testing Machines and Corresponding Environmental Conditions for Test* (ISO, 2000a).
- *ISO 14242-2:2002 Implants for Surgery—Wear of Total Hip-Joint Prostheses—Part 2: Methods of Measurement* (ISO, 2000b).

## Femoral Component Fatigue without Torsion

As we saw when contact forces were measured during activities such as walking, the femoral component must support substantial forces without failing. One recognizable mode of failure occurs with the distal portion of the stem firmly anchored, while medial proximal support is lost because of proximal cement breakdown. As body loads are applied through the head of the prosthesis, significant stem stresses can result at the area where the cement is still firmly anchored. The test described in ASTM F1440:2008 simulates this condition.

A test stem and ball must be constrained by PMMA, with a grouting thickness of at least 1 cm, within a test fixture. Load application must pass through the ball center and be within a 10 ±1° angle of the distal stem axis. An unspecified force is applied, at a frequency of at most 30 Hz. The force cycles between an unspecified minimum value, $F$, and a maximum of $(10 \cdot F)$. The amount of horizontal deflection of the head during testing is recorded. The test continues until the specimen fails or until a predetermined number of cycles, such as 10 million cycles, has been applied.

Test results are to be used for relative comparison with other devices only (ASTM, 2008).

## Femoral Component Fatigue with Torsion

As an additional test for femoral component fatigue, the femoral component fatigue test without tortion, which was described in the previous section, is repeated with one added constraint. As before, a femoral component is positioned in its test fixture, with a load applied that passes through the ball center, with a $10 \pm 1°$ angle of the distal stem axis. Additionally, the angle between the distal stem axis and the line of load application in the anterioposterior projection must be $9 \pm 1°$.

During testing, the amount of horizontal deflection of the head during testing is recorded. The test continues until the specimen fails or until a predetermined number of cycles, such as 10 million cycles, has been applied. Test results are to be used for relative comparison with other devices only (ASTM, 2005).

## Femoral Component Corrosion

Femoral component fatigue testing, without torsion, may be conducted concurrently with corrosion testing. Corrosion products and particulate debris may be released from the ball/stem interface, which is subjected to micromotion during normal activity. These products and debris could stimulate adverse biological reactions and lead to accelerated wear of the articulation.

To conduct corrosion testing, the original fatigue testing apparatus is placed within a sleeve. A test solution of 5–100 mL is added to the sleeve. The distilled water solution may either contain 0.9% sodium chloride or 0.9% sodium chloride and 10% calf serum. Care is taken to ensure that the contact area between the ball and bearing is not exposed to the solution. A cyclic load of 3 kN, with a minimum load of 300 N and a maximum load of 3.3 kN, is applied, at a frequency of 5 Hz. The test is terminated after 10 million cycles.

After the test is terminated, the fluid is collected and analyzed for total metal content and particle characterization. Specific analysis techniques are not mandated. Test results are to be used for relative comparison with other devices only (ASTM, 2009).

## Wear: Test Apparatus

Wear is measured *in vitro* using a standard procedure. According to ISO 14242-1:2002, the femoral and acetabular components are tested in a fluid medium of $25 \pm 2\%$ calf serum, diluted with deionized water. The fluid test medium is filtered through a 2-micron filter and has a protein mass concentration of not less than 17 g/L.

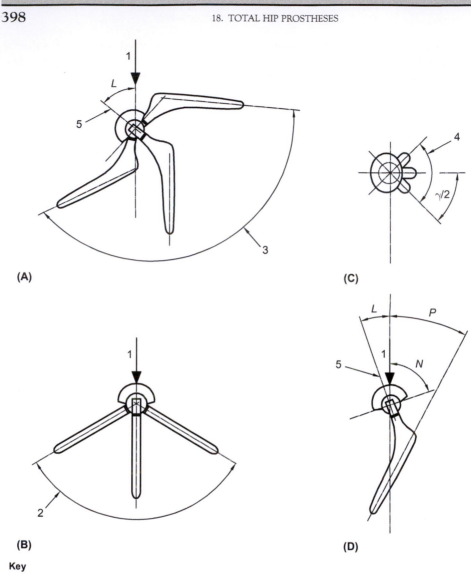

**Key**

| | |
|---|---|
| 1 | Load axis |
| 2 | Flexion/extension angle |
| 3 | Abduction/adduction angle |
| 4 | Inward/outward rotation angle |
| 5 | Polar axis of acetabular component |

*L*   Inclination of the polar axis of the acetabular component to the load line

*N*   Inclination of face of acetabular component equal to 60° ± 3° or as specified by the manufacturer

*P*   Inclination of stem axis to load line in mid-position of abduction/adduction range

**FIGURE 18.18**   Angular movement of femoral component and orientation of components relative to the load line. **(A)** Abduction/adduction. **(B)** Flexion/extension. **(C)** Inward/outward rotation. **(D)** Orientation of acetabular component and femoral component in mid-position to the load line *[Reproduced by permission from ISO (2000a)].*

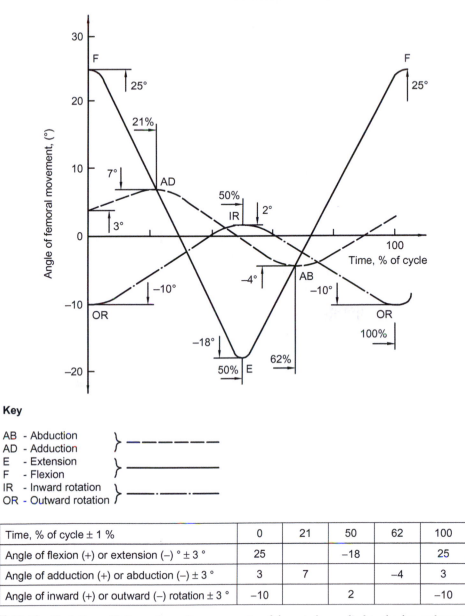

| Time, % of cycle ± 1 % | 0 | 21 | 50 | 62 | 100 |
|---|---|---|---|---|---|
| Angle of flexion (+) or extension (−) ° ± 3 ° | 25 | | −18 | | 25 |
| Angle of adduction (+) or abduction (−) ± 3 ° | 3 | 7 | | −4 | 3 |
| Angle of inward (+) or outward (−) rotation ± 3 ° | −10 | | 2 | | −10 |

**FIGURE 18.19**    Variation with time of angular movement and force to be applied to the femoral test specimen [*Reproduced by permission from ISO (2000a)*].

Before adding the fluid test medium, the femoral component of the testing specimen is mounted in a test machine with its stem in the abduction/adduction position of 10 ±3°, as shown in Figure 18.18D, and in the inward/outward rotation position $\gamma/2°$, as shown in Figure 18.18C. The value of $\gamma$ is unspecified. The acetabular component is mounted with the polar axis vertical, as shown in Figure 18.18B, and inclined at an angle of 30 ±3°, as shown in Figure 18.18A.

The fluid test medium is then introduced to completely immerse the contact surfaces of the test specimen. It is maintained at a temperature of 37 ±2°C. The test machine is started, so that cycles of flexion/adduction/rotation (described in Figure 18.19) are executed. The operating frequency is 1 ±0.1 Hz.

The test is executed for $5 \times 10^6$ cycles, or until the articulating surfaces are broken up or delaminated. At $5 \times 10^5$ cycles and multiples of $1 \times 10^6$ cycles, the test is stopped to take wear measurements. The number of test specimens is unspecified (ISO, 2000a).

## Wear: Test Measurements

Wear testing involves a test specimen and possibly a nonarticulating control specimen. The control specimen is subjected to the loading regime of Figures 18.18 and 18.19, without the angular displacements. A control specimen is required if polymers are the object of investigation. The specimens are weighed before and after testing, with wear calculated from mass changes.

For the initial weighing, the specimens are soaked in the fluid test medium for 48 ±4 hr. After removal from the fluid, the specimens are cleaned in an ultrasonic cleaner, with care taken to avoid abrasion. They are dried with a jet of filtered inert gas, and soaked in propan-2-ol for 5 min ±15 s. They are then redried with a jet of filtered inert gas and a vacuum of better than 13.3 ±0.13 Pa for at least 30 min. The weights are recorded using a balance with an accuracy of ±0.1 mg. This procedure, starting with the specimen soaking, is repeated until the incremental mass change of the specimen over 24 hr is less than 10% of the cumulative mass change.

The test and control specimens are mounted in the test machine. The test specimen undergoes wear testing; the control specimen is subjected to the test loads. During each scheduled stop, the test and control specimens are weighed. For several test and control specimens, gravimetric wear, $W(n)$, is calculated as:

$$W(n) = m_l(n) + m_G(n), \tag{18.1}$$

where $n$ is the cycles of loading, $m_l(n)$ is the average uncorrected mass loss of the test specimens, and $m_G(n)$ is the average mass gain of the control specimens over the same period. No maximum acceptable wear mass is specified (ISO, 2000b).

## SUMMARY

A hip prosthesis is a medical device that replaces a damaged hip joint. The hip consists of a convex femoral head inserted into a concave acetabulum within the pelvis, cushioned by a synovial joint capsule.

The hip joint experiences substantial contact forces during activity. The peak hip contact force of approximately 260% body weight occurs while moving down stairs. The peak

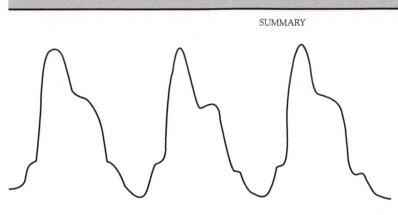

**FIGURE 18.20** Idealized contact force in the left hip during walking.

torsional prosthesis moment of approximately 2% body weight-meters occurs while moving up stairs. These contact forces, in parallel with a predisposition to osteoarthritis, may cause severe arthritic pain. A patient primarily receives a total hip prosthesis for this pain.

Charnley's low-friction arthroplasty of the 1960s survives today as a metal femoral component, with stem and ball, articulating against an acetabular component, with a metal shell and a liner that is usually fabricated from cross-linked polyethylene. The Ti or Co-Cr alloy femoral stem may be cemented or press-fit into place. The Ti or Co-Cr alloy acetabular component is also press-fitted into place, with optional screws used to achieve initial stabilization.

Hip prosthesis efficacy is predominantly affected by osteolysis, which can be traced to wear debris. Wear is a function of prosthetic, patient, and surgical factors that subdivide into design and environmental variables. Harder metal-on-metal and ceramic-on-ceramic bearing surfaces are associated with other effects. Metal-on-metal bearings are associated with metal reactivity, which is a lymphocyte-dominated reaction to metal ions that may cause osteolysis. Ceramic-on-ceramic (alumina or zirconia) bearings made of second-generation materials have a small risk of brittle fracture, but they may squeak.

Key hip prosthesis features include assessments of femoral component fatigue without torsion, femoral component fatigue with torsion, femoral component corrosion, and wear.

## Exercises

**18.1** THA patients find their squeaky ceramic hips to be annoying and embarrassing. Watch the short film from the *New York Times*: http://www.nytimes.com/interactive/2008/05/09/business/20080511_SQUEAK_GRAPHIC.html#step1. How would you describe the squeaking sound? What hip prosthesis bearing type would you recommend to your grandfather? Give reasons.

**18.2** The elements of the gait cycle can be observed in hip contact force data over time. Identify the stance phase, swing phase, heel strike, and toe off in the walking activity force plot shown in Figure 18.20. Does more force occur during the stance or swing phase?

**18.3** Draw a free-body diagram of the condition tested during the femoral component fatigue testing without torsion requirement, before fracture. The free-body diagram should contain two devices. Consider the femoral stem and ball to be one femoral component device, and the acetabular cup and liner to be one acetabular component device. Label the location where the femoral component could fracture.

**18.4** Does cement fixation of the femoral stem *currently* cause osteolysis? Provide specific evidence for your position.

Read Charnley(1960):

**18.5** What is the coefficient of friction? What was the significance of this measurement to Charnley? How did he measure the coefficient of friction with a pendulum? How did ankle joint measurements compare to metal-oil measurements?

**18.6** What was Charnley's hypothesis for improving arthroplasty results? With what materials did Charnley experiment? Describe reasons for his material choice.

Read Mont et al. (2009):

**18.7** How do the total hip arthroplasty (THA) and hip resurfacing procedures differ? Why would a patient choose hip resurfacing as a first option? Are the study results of hip resurfacing comparable to those of THA? Give reasons.

**18.8** Using the *NEJM* deep brain stimulation study protocol from Chapter 14 as a reference standard, what are some of the limitations of this study protocol? Do the hip resurfacing results in this study generalize for the hip-patient population?

# References

ASTM. (2005). *ASTM F1612-95:2005 Standard practice for cyclic fatigue testing of metallic stemmed hip arthroplasty femoral components without torsion.* West Conshohocken, PA: ASTM.

ASTM (2008). *ASTM F1440-92:2008 Standard practice for cyclic fatigue testing of metallic stemmed hip arthroplasty femoral components without torsion.* West Conshohocken, PA: ASTM.

ASTM (2009). *ASTM F1875-98:2009 Standard practice for fretting corrosion testing of modular implant interfaces: Hip femoral head-bore and cone taper interface.* West Conshohocken, PA: ASTM.

Baek, S. H., & Kim, S. Y. (2008). Cementless total hip arthroplasty with alumina bearings in patients younger than fifty with femoral head osteonecrosis. *Journal of Bone and Joint Surgery—American, 90,* 1314–1320.

Bartel, D. L., Davy, D. T., & Keaveny, T. M. (2006). *Orthopaedic biomechanics: mechanics and design in musculoskeletal systems.* Upper Saddle, River, NJ: Pearson Prentice Hall.

Baura, G. D. (2013). U. S. medical device regulation. An introduction to biomedical product development, market release and postmarket surveillance. Manuscript in preparation.

Berger, R. A. (2009). The minimally invasive total hip arthoropalsty with two incisions. In T. E. Brown, Q. Cui, W. M. Mihalko, & K. J. Saleh (Eds.), *Arthritis and Arthroplasty: The hip.* Philadelphia: Saunders Elsevier.

Bergmann, G., Deuretzbacher, G., Heller, M., Graichen, F., & Rohlmann, A., et al. (2001). Hip contact forces and gait patterns from routine activities. *Journal of Biomechanics, 34,* 859–871.

Bolland, B. J., Culliford, D. J., Langton, D. J., Millington, J. P., & Arden, N. K., et al. (2011). High failure rates with a large-diameter hybrid metal-on-metal total hip replacement: clinical, radiological and retrieval analysis. *Journal of Bone and Joint Surgery—British, 93,* 608–615.

Bradford, L., Baker, D. A., Graham, J., Chawan, A., & Ries, M. D., et al. (2004). Wear and surface cracking in early retrieved highly cross-linked polyethylene acetabular liners. *Journal of Bone Joint Surgery—American, 86-A,* 1271–1282.

Brand, R. A., Pedersen, D. R., Davy, D. T., Kotzar, G. M., & Heiple, K. G., et al. (1994). Comparison of hip force calculations and measurements in the same patient. *Journal of Arthroplasty, 9,* 45–51.

Brown, T. D., & Bartel, D. L. (2008). What design factors influence wear behavior at the bearing surfaces in total joint replacements? *Journal of the American Academy of Orthopedic Surgeons, 16*(Suppl 1), S101–S106.

Campbell, P., Ma, S., Yeom, B., McKellop, H, & Schmalzried, T. P., et al. (1995). Isolation of predominantly submicron-sized UHMWPE wear particles from periprosthetic tissues. *Journal of Biomedical Material Research, 29,* 127–131.

CDC (2006). Prevalence of doctor-diagnosed arthritis and arthritis-attributable activity limitation—United States, 2003–2005. *Morbidity and Mortality Weekly Report, 55,* 1089–1092.

CDRH (2006). *Approval order P040033.* Rockville, MD: FDA.

CDRH (2011). *Metal-on-metal hip implants: FDA's role and activities.* Silver Spring, MD: FDA.

Charnley, J. (1960). The lubrication of animal joints in relation to surgical reconstruction by arthroplasty. *Annals of the Rheumatic Diseases, 19,* 10–19.

Charnley, J. (1975). Fracture of femoral prostheses in total hip replacement. A clinical study. *Clinical Orthopaedics and Related Research, 111,* 105–120.

Clarke, I. C., & Manley, M. T. (2008). How do alternative bearing surfaces influence wear behavior? *Journal of the American Academy of Orthopedic Surgery, 16*(Suppl 1), S86–S93.

Crawford, R., Ranawat, C. S., & Rothman, R. H. (2010). Metal on metal: Is it worth the risk? *Journal of Arthroplasty, 25,* 1–2.

Engh, C. A., JR, Stepniewski, A. S., Ginn, S. D., Beykirch, S. E., & Sychterz-Terefenko, C. J., et al. (2006). A randomized prospective evaluation of outcomes after total hip arthroplasty using cross-linked marathon and non-cross-linked Enduron polyethylene liners. *Journal of Arthroplasty, 21,* 17–25.

Fantner, G. E., Hassenkam, T., Kindt, J. H., Weaver, J. C., & Birkedal, H., et al. (2005). Sacrificial bonds and hidden length dissipate energy as mineralized fibrils separate during bone fracture. *Nature Materials, 4,* 612–616.

Feder, B. J. (2008). *That must be Bob. I hear his new hip squeaking.* New York Times (May 11).

Felson, D. T., Lawrence, R. C., Dieppe, P. A., Hirsch, R., & Helmick, C. G., et al. (2000a). Osteoarthritis: New insights. Part 1: The disease and its risk factors. *Annals of Internal Medicine, 133,* 635–646.

Felson, D. T., Lawrence, R. C., Hochberg, M. C., McAlindon, T., & Dieppe, P. A., et al. (2000b). Osteoarthritis: New insights. Part 2: Treatment approaches. *Annals of Internal Medicine, 133,* 726–737.

Gomez, P. F., & Morcuende, J. A. (2005). A historical and economic perspective on Sir John Charnley, Chas F. Thackray Limited, and the early arthoplasty industry. *Iowa Orthopaedic Journal, 25,* 30–37.

Griffin, W., & Bradbury, T. L. (2009). Acetabular component options in total hip arthroplasty. In T. E. Brown, Q. Cui, W. M. Mihalko, & K. J. Saleh (Eds.), *Arthritis and Arthroplasty: The hip.* Philadelphia: Saunders Elsevier.

Hallab, N. J., Anderson, S., Caicedo, M., Skipor, A., & Campbell, P., et al. (2004a). Immune responses correlate with serum-metal in metal-on-metal hip arthroplasty. *Journal of Arthroplasty, 19*(Suppl. 3), 88–93.

Hallab, N. J., Jacobs, J. J., & Katz, J. L. (2004b). Orthopedic applications. In B. D. Ratner, A. S. Hoffman, F. J. Schoen, & J. E. Lemons (Eds.), *Biomaterials Science* (2nd ed.). San Diego, CA: Elsevier.

Harris, W. H., Schiller, A. L., Scholler, J., Freiberg, R. A., & Scott, R., et al. (1976). Extensive localized bone resorption in the femur following total hip replacement. *Journal of Bone and Joint Surgery, 58,* 612–618.

Huber, M., Reinisch, G., Zenz, P., Zweymuller, K., & Lintner, F. Postmortem study of femoral osteolysis associated with metal-on-metal articulation in total hip replacement: an analysis of nine cases. *Journal of Bone and Joint Surgery—American, 92,* 1720–1731.

ISO (2000a). *ISO 14242-1 Implants for surgery—Wear of total hip-joint prostheses—Part 1: Loading and displacement parameters for wear-testing machines and corresponding environmental conditions for test.* Geneva: ISO.

ISO (2000b). *ISO 14242-2 Implants for surgery—Wear of total hip-joint prostheses—Part 2: Methods of measurement.* Geneva: ISO.

Keaveny, T. M., Morgan, E. F., Niebur, G. L., & Yeh, O. C. (2001). Biomechanics of trabecular bone. *Annual Review of Biomedical Engineering, 3,* 307–333.

Lasker Awards: Former Winners. (2009). Lasker Foundation.

Korovessis, P., Petsinis, G., Repanti, M., & Repantis, T. (2006). Metallosis after contemporary metal-on-metal total hip arthroplasty. Five to nine-year follow-up. *Journal of Bone and Joint Surgery—American, 88,* 1183–1191.

Mahalingam, K., & Reidy, D. (1996). Smith-Petersen vitallium mould arthroplasty: a 45-year follow-up. *Journal of Bone Joint Surgery—British, 78,* 496–497.

Margevicius, K. J., Bauer, T. W., McMahon, J. T., Brown, S. A., & Merritt, K. (1994). Isolation and characterization of debris in membranes around total joint prostheses. *Journal of Bone Joint Surgery—American, 76,* 1664–1675.

McKellop, H., Shen, F., Lu, B., Campbell, P., & Salovey, R. (1999). Development of an extremely wear-resistant ultra high molecular weight polyethylene for total hip replacements. *Journal of Orthopaedic Research, 17,* 157–167.

Mont, M. A., Marker, D. R., Smith, J. M., Ulrich, S. D., & McGrath, M. S. (2009). Resurfacing is comparable to total hip arthroplasty at short-term follow-up. *Clinical Orthopaedics and Related Research, 467,* 66–71.

I. MEDICAL DEVICES

Muirhead-Allwood, S. K., Kabir, C., & Sandiford, N. (2009). Hip resurfacing arthroplasty. In T. E. Brown, Q. Cui, W. M. Mihalko, & K. J. Saleh (Eds.), *Arthritis and Arthroplasty: The Hip*. Philadelphia: Saunders Elsevier.

Park, Y. S., Moon, Y. W., Lim, S. J., Yang, J. M., & Ahn, G., et al. (2005). Early osteolysis following second-generation metal-on-metal hip replacement. *Journal of Bone and Joint Surgery– American, 87*, 1515–1521.

Parvizi, J., Ghanem, E., & Rothman, R. H. (2009). Current designs of uncemented total hip arthroplasty. In T. E. Brown, Q. Cui, W. M. Mihalko, & K. J. Saleh (Eds.), *Arthritis and Arthroplasty: The Hip*. Philadelphia: Saunders Elsevier.

Restrepo, C., Parvisi, J., Kurtz, S. M., Sharkey, P. F., & Hozack, W. J., et al. (2008). The noisy ceramic hip: Is component malpositioning the cause? *Journal of Arthroplasty, 23*, 643–649.

Schmalzried, T. P. (2009). Metal-metal bearing surfaces in hip arthroplasty. *Orthopedics, 32*.

Schmalzried, T. P., & Callaghan, J. J. (1999). Wear in total hip and knee replacements. *Journal of Bone Joint Surgery—American, 81*, 115–136.

Schoenfeld, C. M., Lautenschlager, E. P., & Meyer, P. R., JR. (1974). Mechanical properties of human cancellous bone in the femoral head. *Medical and Biological Engineering & Computing, 12*, 313–317.

Taylor, S., Manley, M. T., & Sutton, K. (2007). The role of stripe wear in causing acoustic emissions from alumina ceramic-on-ceramic bearings. *Journal of Arthroplasty, 22*(Suppl 3), 47–51.

Walter, W. L., O'Toole, G. C., Walter, W. K., Ellis, A., & Zicat, B. A. (2007). Squeaking in ceramic-on-ceramic hips. *Journal of Arthroplasty, 22*, 496–503.

Wang, M. L., Sharkey, P. F., & Tuan, R. S. (2004). Particle bioreactivity and wear-mediated osteolysis. *Journal of Arthroplasty, 19*, 1028–1038.

Zappone, B., Thurner, P. J., Adams, J., Fantner, G. E., & Hansma, P. K. (2008). Effect of Ca2 + ions on the adhesion and mechanical properties of adsorbed layers of human osteopontin. *Biophysical Journal, 95*, 2939–2950.

CHAPTER

# 19

# Drug-Eluting Stents

In this chapter, we move from discussing medical devices to combination products. A *combination product* is comprised of two or more regulated medical products: medical device, drug, biologic. We began a discussion of stents in Chapter 8 by introducing bare metal stents. Here, we continue by introducing drug-eluting stents. The first-generation drug-eluting stent was a bare metal stent, which was coated with a drug-releasing polymer (Figure 19.1).

Upon completion of this chapter, each student shall be able to:

1. Define a combination product.
2. Understand how design iterations occur during the system development process of a medical device.
3. Describe how a coronary artery responds to various stent component design choices.

FIGURE 19.1  Taxus Express coronary stent (Boston Scientific, Natick, Massachusetts).

## COMBINATION PRODUCTS

The Code of Federal Regulations (CFR) is the codification of the general and permanent rules published in the Federal Register by the executive departments and agencies of the Federal Government. According to Title 21, Section 3.2(e) of the CFR [21 CFR 3.2(e)],

A combination product includes:

**(1)** A product comprised of two or more regulated components, i.e., drug/device, biologic/device, drug/biologic, or drug/device/biologic, that are physically, chemically, or otherwise combined or mixed and produced as a single entity;
**(2)** Two or more separate products packaged together in a single package or as a unit and comprised of drug and device products, device and biological products, or biological and drug products;
**(3)** A drug, device, or biological product packaged separately that according to its investigational plan or proposed labeling is intended for use only with an approved individually specified drug, device or biological product where both are required to achieve the intended use, indication, or effect an where upon approval of the proposed product the labeling of the approved product would need to be changed, for example, to reflect a change in intended use, dosage form, strength, route of administration, or significant change in dose; or
**(4)** Any investigational drug, device, or biological product packaged separately that according to its proposed labeling is for use only with another individually specified investigational drug, device, or biological product where both are required to achieve the intended use, indication, or effect (U.S. Code of Federal Regulations, 2010).

The first combination product we discussed was the steroid-eluting pacemaker lead, which is a combination medical device and drug. As shown in Figure 3.13, the terminating electrode in the medical device sits adjacent to a silicone rubber plug compounded with the drug dexamethasone sodium phosphate. The drug is eluted over time to minimize the formation of scar tissue and, more importantly, to minimize the increase in impedance.

A *drug-eluting stent* is a bare metal stent medical device, coated with a polymer-drug formulation that controls the release rate of the drug (Figure 19.1). Similar to the steroid-eluting pacemaker lead, an antiproliferative drug is eluted from the stent over time to minimize cell proliferation and consequently restenosis of the treated arterial segment. In Chapter 20, we discuss the development of the artificial pancreas, which is a medical device/biologic combination product.

Several decades ago, a combination product may have been separately approved by all appropriate FDA Centers, which required substantial time and resources. Medical devices are approved by the Center for Devices and Radiologic Health (CDRH). Drugs are approved by the Center for Drug Evaluation and Research (CDER); biologics are approved by the Center for Biologics Evaluation and Research (CBER). In previous chapters, when we referred to the "FDA," we were referring to the FDA CDRH.

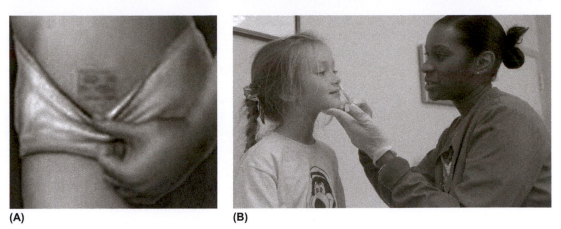

**(A)**                                                    **(B)**

**FIGURE 19.2**  Combination products in which the medical device is not the primary mode of action. **A:** Shire Daytrana ADHD patch (Shire USA, Inc., Wayne, Pennsylvania). **B:** MedImmune Vaccines FluMist (MedImmune, Gaithersburg, Maryland).

To streamline this process, the FDA created the Office of Combination Products (OCP) in 2002. OCP assigns an FDA center primary jurisdiction for review of a combination product when the jurisdiction is unclear or in dispute. Assignment occurs through determination of the primary mode of action. *Primary mode of action* is the single mode of action of a combination product that provides the most important therapeutic action of the combination product.

If we look at the steroid-eluting pacemaker lead and drug-eluting stent, the primary mode of action is the medical device. With or without a drug, a pacemaker lead transmits electrical stimulation to heart muscle, and a bare metal stent initially improves patency of a coronary artery after percutaneous coronary intervention (PCI) by providing mechanical support of the treated segment.

With other medical device/drug or medical device/biologic combinations, the drug or biologic is the primary mode of action. For example, a Shire Daytrana patch transmits the drug methylphenidate to children with attention deficit hyperactivity disorder (ADHD). MedImmune Vaccine FluMist is a live nasal influenza vaccine that is administered by syringe (Figure 19.2).

For a more thorough discussion of the FDA regulation of combination products, please see (Baura, 2013).

## DRUG DELIVERY USING COATINGS

In a drug-eluting stent, the drug is delivered by a drug carrier coating. This biocompatible coating must provide consistent dosing, controlled release kinetics, and structural integrity (Ranade et al., 2004). *Structural integrity* is the ability of the coating to withstand crimping onto a balloon dilatation catheter, sterilization, and deployment without compromising the physical integrity of the coating.

For drug-eluting stents, a pharmacological agent is commonly embedded in a polymer matrix coated on the surface of a stent, thereby controlling the time release of the drug after implantation. A *polymer* is comprised of one or more specific chemical subunits (monomers) covalently bonded to one another from hundreds to millions of times. Polymers are often referred to as *macromolecules*.

Drug release is typically achieved via a physical mechanism, specifically diffusion of a drug from the polymer matrix or by swelling of that matrix and subsequent dissolution of the drug. However, drug release may also be achieved by a chemical mechanism if the coating polymer is degradable. Factors influencing drug release rate are:

- The formulation of drug and polymer.
- The detailed polymer chemistry.
- The thermodynamics of mixing between polymer and drug.
- The molecular weight of the polymer.
- The distribution of drug within the polymer matrix.

Examples of polymers used in drug-eluting stents are poly(styrene-*b*-isobutylene-*b*-styrene) (SIBS), poly(ethylene-*co*-vinyl acetate) (PEVA), poly(*n*-butyl methacrylate) (PBMA), poly(vinylidene fluoride-*co*-hexafluoropropylene) (PVDF-HFP), and poly(D,L-lactide) (PDLLA). The synthesis of these polymers is beyond the range of this discussion. The structures of some polymers are shown in Figure 19.3.

PDLLA has the additional advantage of being degradable and, ultimately, bioresorbable (Kohn et al., 2004). The term *bioresorbable* is often used synonymously with *bioabsorbable*, although there is a subtle distinction between the two terms. A *bioresorbable polymer* is chemically degraded in the body into nontoxic products, which are eliminated from the body or metabolized; conversely, the final degradation products of a *bioabsorbable polymer* are only eliminated by the body.

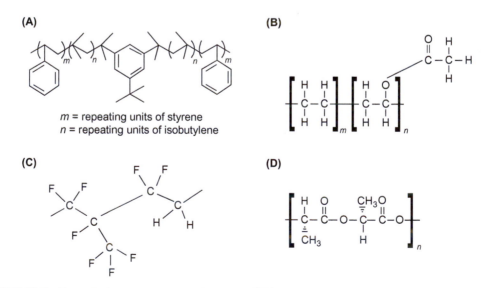

FIGURE 19.3   Drug-eluting stent coating polymers. **A:** SIBS. **B:** PEVA. **C:** PVDF-HPF. **D:** PDLLA.

With respect to PDLLA, each monomer subunit contains an ester bond, which is vulnerable to hydrolysis (decomposition through reaction with water). In an aqueous environment, PDLLA is progressively reduced to smaller molecular weights. PDLLA domains less than 2 microns in characteristic dimension may be phagocytosed by macrophages. Otherwise, PDLLA degrades to lactic acid (Figure 19.4), which readily deprotonates to form lactate. Lactate is then processed via the Krebs cycle (Ormiston et al., 2008) or other metabolic pathways (Philp et al., 2005).

Drug dosing is optimized so that polymer-drug elution moves to steady state within a short period, such as 30 days. For the Taxus Express stent, SIBS is mixed with 1 mcg/mm$^2$ paclitaxel (PTx). Paclitaxel inhibits antimitotic microtubules, which suppresses cell division. This eventually results in disruption of smooth cell migration and proliferation, which occurs before neointimal hyperplasia.

Drug release is dependent on formulation and is correlated with paclitaxel loading. After implantation, there is a burst release of PTx, followed by a longer lasting, slower sustained release (Ranade et al., 2004) (Figure 19.5). The chosen paclitaxel dose determines its effect, which ranges from cytostatic to cytotoxic. The safe use of paclitaxel in this application is enabled by metering the dose through a polymer carrier.

For the Taxus Express stent, the drug is dispersed as particulates in the polymer, rather than dissolved in the polymer. In other words, the initial loading concentration, $c_o$, is greater than the saturation concentration, $c_s$. For dispersed systems ($c_o > c_s$), precipitated regions are considered nondiffusion and disappear as a function of drug release to create a moving boundary problem. If we assume the polymer matrix covers a planar slab of surface area, $A$, and treat the problem as a pseudorelease state, then the change in drug mass over time, $dm(t)/t$, can be approximated according to the Higuchi equation:

$$\frac{dm(t)}{t} = \frac{A}{2}\left[\frac{Dc_sc_o}{t}\right]^{\frac{1}{2}},$$
(19.1)

where $D$ = diffusion coefficient (Liechty et al., 2010; Heller & Hoffman, 2004).

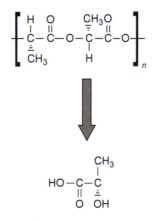

FIGURE 19.4    PDLLA degrades to lactic acid.

The SIBS matrix on the manufactured Taxus Express stent maintains its structural integrity. The smooth conformal SIBS coating on the stent does not crack after ethylene oxide sterilization and expansion (Ranade et al., 2004) (Figure 19.6).

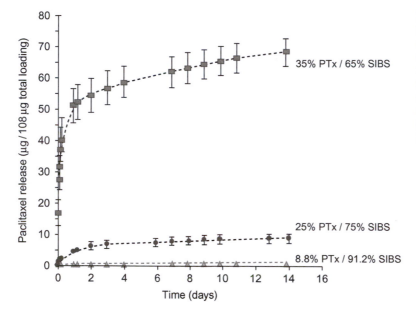

FIGURE 19.5 *In vitro kinetic drug-release profiles of PTx from SIBS polymer stent coatings containing 8.8, 25, and 35% PTx over 14 days in PBS-Tween20 at 37°C. Data are an average from three stents at each time point [Reproduced by permission from Ranade et al. (2004)].*

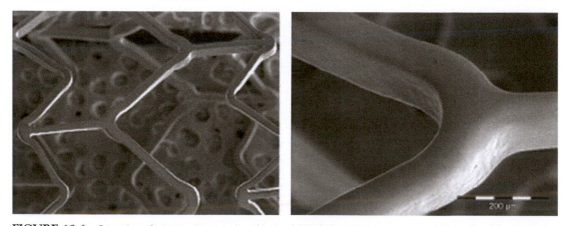

FIGURE 19.6   Scanning electron micrographs of typical TAXUS stents postexpansion *[Reproduced by permission from Ranade et al. (2004)].*

## CLINICAL NEED

A drug-eluting stent treats coronary artery disease (CAD). CAD affects about 17 million Americans that are age 20 yr and older. Each year, 610,000 new and 325,000 recurrent myocardial infarctions occur (AHA, 2009).

As discussed in Chapter 8, CAD treatment evolved from deployment of balloon angioplasty in 1977 to the FDA approval of the Palmaz-Schatz coronary stent in 1994. Cardiologist Andreas Gruentzig designed and deployed balloon catheters to restore blood flow in stenosed coronary arteries. Unfortunately, endothelial injury from balloon inflation initiates a wound-healing response that leads to restenosis. Long-term failure due to restenosis occurs in 30–60% of patients treated by balloon angioplasty, most frequently within 6 months (Bauters & Isner, 1998; Rajagopal & Rockson, 2003). A bare metal stent may be deployed immediately after balloon angioplasty to minimize thrombus formation and to provide a scaffold to maintain vessel patency. However, fibrous cap rupture and stent strut penetration into the plaque lipid core may increase neointimal growth, which results in restenosis (Farb et al., 2002). This long-term failure due to restenosis occurs in 20–30% of patients treated with bare metal stents (Bauters & Isner, 1998; Rajagopal & Rockson, 2003).

Because neointimal hyperplasia poses the most significant drawback for bare metal stents, the addition of an antiproliferative agent is a logical progression in medical devices for the treatment of CAD.

## ENGINEERING DESIGN

The medical device iterations for treating coronary artery stenosis are examples of engineering design. According ABET, which accredits engineering programs,

> Engineering design is the process of devising a system, component, or process to meet desired needs. It is a decision-making process (often iterative), in which the basic sciences, mathematics, and the engineering sciences are applied to convert resources optimally to meet these stated needs (ABET, 2009).

In an established company with sufficient resources for product development, a product is generally developed, marketed, and retired according to the system development process described in Chapter 1 (Figure 1.32). As illustrated in the design control part of this process, system requirements become design requirements and prototypes, which are then tested. Related research conducted before product development begins is not shown.

For a medical device company, the FDA CDRH mandates that design control be conducted as part of Good Manufacturing Processes. The design control process in the shaded area in Figure 1.32 is translated into the waterfall design process of Figure 19.7. Based on user needs, requirements specifications and design specifications are written, which constitute design inputs. Prototypes are built during the design process, which are then tested. During verification, testing is conducted to verify that each requirement in the requirements specifications has been met. During validation, testing is conducted to validate that the medical device meets user needs. Although not specifically shown, risk analysis is also

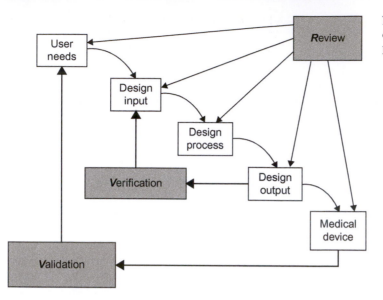

FIGURE 19.7 Application of design controls to waterfall design process (CDRH, 1997).

conducted for the medical device being developed. At each point of the process, a design review is conducted.

Let us look at how design iterations occur in medical device product development. First, we assume that an established medical device company has sufficient resources for product development, well-defined essential requirements, and well executed design verification and validation. Design iterations primarily occur during the design control process, although some product improvements may be made after initial market release. Well-defined essential requirements steer product developers toward component and system design iterations for prototypes. Prototypes are tested. Tests that fail are reported to the product developers, who further iterate and improve their designs. The final designs meet requirements and user needs when tested.

Masimo pulse oximeters (discussed in Chapter 11) are designed according to this design process. Evidence for Masimo's well-controlled design process is the low number of adverse events reported by clinical users in the the FDA Manufacturers and User Facility Design Experience (MAUDE) database. From January 1, 1997 to June 30, 2010, 52 adverse events for *all types* of Masimo equipment were reported and documented (CDRH, 2010).

In contrast, 1717 adverse events were reported and documented in MAUDE for the Guidant Ancure system from January 1, 1997 to June 30, 2010 (CDRH, 2010). As discussed in Chapter 8, the Ancure system was marketed only from 1999 to 2001. The Ancure system was developed by a then recently acquired division of Guidant called Endovascular Technologies.

In general, a higher probability of adverse events exists for implanted versus noninvasive devices. But during clinical trials, the Ancure system may also not have been validated, as trained physicians perceived the Guidant Ancure system (Figure 8.26) as more difficult to use than a competing Medtronic AneuRx system. Validation would not have been met if ease of use was considered a clinical need. After market release, some Ancure systems, in the hands of physicians not specially trained for Ancure use, became lodged in patients. A Guidant sales representative assisted in developing the handle breaking technique so that the system could be removed from a patient's body after abdominal aortic graft deployment (Ryan, 2003).

When well-defined essential requirements are unavailable, substantial design iterations may occur *outside* each design control cycle, rather than just *within* one design control cycle. In the case of medical devices for CAD treatment, clinicians and engineers were unaware that host responses would cause restenosis after balloon angioplasty. They were also unaware during initial bare metal stent deployment that host responses leading to neointimal growth would also cause restenosis. This awareness and more recent observations have affected how each generation of drug-eluting stents is designed.

The Historic Devices section of each medical device chapter is not just a narrative of medical device precursors. Rather, each section is an illustration of the design iterations that occurred while each medical device was developed and refined. As seen in Chapter 10, early poor experiences with positive pressure caused mechanical ventilator engineers to use negative pressure for the iron lung. Only when positive pressure was demonstrated to reduce mortality during the polio epidemic of 1952 did positive pressure become reincorporated into mechanical ventilators. As seen in Chapter 11, early oximeters and pulse oximeters could not accurately estimate arterial saturation of oxygen. Only when engineer Scott Wilber discarded the Beer-Lambert law and introduced a calibration curve did this noninvasive measurement become accurate.

## DRUG-ELUTING STENT REQUIREMENTS

In the first generation of drug-eluting stents, the reduction of neointimal hyperplasia was the main goal. Stainless steel stents were coated with a polymer matrix, which targeted a precursor of neointimal hyperplasia. Boston Scientific's Taxus Express stent is its bare metal Express stent, coated with a mixture of paclitaxel and SIBS. Johnson & Johnson Cordis' Cypher stent is its Velocity stent, coated with a sirolimus/PEVA/PBMA basecoat and PBMA topcoat. Unlike paclitaxel, sirolimus directly inhibits smooth muscle cell migration and proliferation by inhibiting the transition from the G1 to the S phase of the cell cycle.

Clinical trials indicated that the Cypher and Taxus Express stents remained patent longer, compared to bare metal stents. However, a higher rate of subacute thromboses began to be reported after market release. After substantial clinical data analysis, it was determined that thrombosis was related to off-label use, such as implantation in patients with complex conditions or complex lesions. It was also determined that antiplatelet therapy of clopidogrel should be continued for at least 12 months, rather than the initially recommended 3 months, in patients who are at low risk for bleeding (Maisel, 2007).

For subsequent generations of drug-eluting stents, clinicians noted desirable design criteria based on deliverability, efficacy, and safety:

- A stent with less recoil so that the struts remain apposed to the vessel well.
- Thinner struts for less disruption of blood flow.

- A drug that reduces neointimal proliferation.
- A drug-coating technology that minimizes stent thrombosis (Ako et al., 2007).

## SYSTEM DESCRIPTIONS AND DIAGRAMS

In Chapter 8, we described a bare metal stent biocompatibility system diagram in which the stent reaction to the plaque is a function of strut design, strut spacing, radius of curvature, strut thickness, and stent-to-artery ratio. With the addition of a stent coating that elutes a drug, the original diagram is modified to include drug efficacy and coating thrombosis (Figure 19.8).

### First Generation

First-generation Cypher and Taxus Express stents met the requirement of keeping their target lesions patent, compared to bare metal counterparts. They were approved by the FDA in 2003 and 2004, respectively. In one study, a pooled analysis was performed, using data from four double-blinded trials in which 1748 patients were randomly assigned to receive either Cypher or Velocity (bare metal) stents and data from five double-blinded trials in which 3513 patients were randomly assigned to receive either Taxus Express or Express (bare metal) stents. At 4 yr, revascularization was markedly reduced in the Cypher (7.8% vs. 23.6%, $p<0.001$) and Taxus Express (10.1% vs. 20.0%, $p < 0.001$) stents (Stone et al., 2007).

An illustration of patency can be seen in histologic sections taken from a 65-yr-old man, who received both a Cypher and Velocity stent in his left anterior descending coronary artery before suffering a fatal head injury 15 months later. The Cypher stent struts are minimally covered by fibrin, whereas the Velocity struts have abundant overlying neointimal tissue (Joner et al., 2006) (Figure 19.9). This man was not part of the Cypher double-blinded trials.

In the pooled analysis study, the rates of death or myocardial infarction did not differ between drug-eluting and bare metal stents at 4 years. However, a small but significant difference in stent thrombosis was observed, beginning 1 yr after the procedure. Stent thrombosis increased in the Cypher (0.6% vs. 0%, $p = 0.025$) and Taxus Express (0.7% vs. 0.2%, $p = 0.028$) stents. It should be noted that the minimum administration of clopidogrel was 2 or 6 months, respectively (Stone et al., 2007).

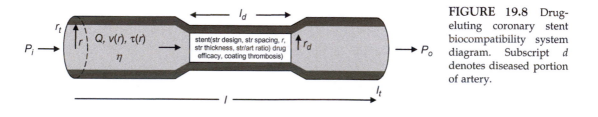

FIGURE 19.8 Drug-eluting coronary stent biocompatibility system diagram. Subscript $d$ denotes diseased portion of artery.

## Second Generation

Three second-generation stents addressed thrombosis by considering polymer coating interactions. The presence of a polymer drug coating may contribute to stent thrombosis as a result of delayed healing and possible hypersensitivity reactions. Hypersensitivity reactions may occur more than 4 months after implantation, long after the drug has been released (Kukreja et al., 2008). Polymer coating interactions were minimized by changing strut parameters, as well as by using more biocompatible polymer drug coatings. All three stents were approved in 2008.

The Taxus Liberte stent resembles its sister Taxus Express stent, except that the stainless steel geometry was modified from an open- to closed-cell geometry (Figure 19.10). This geometry change does not improve clinical outcomes (Turco et al., 2007; Yamamoto et al., 2010).

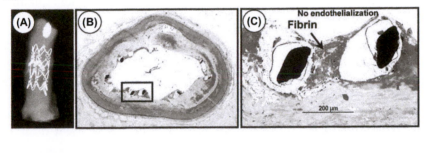

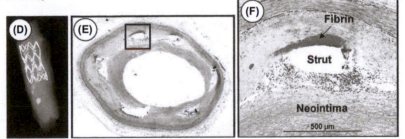

FIGURE 19.9  A 65-year-old man suffered traumatic injury to the head, resulting in death. This patient had received 2 stents in the left anterior descending coronary artery (LAD), a proximal Cypher and a distal Bx Velocity 15 months before death, which were found to be patent at autopsy. A (Cypher) and D (Bx Velocity): Radiographs of the stented LAD segment. All sections shown have been stained with Movat pentachrome. B: Histologic section of the Cypher-stented artery showing minimal coverage of the struts by fibrin. C: High-power magnification showing peristrut fibrin with rare endothelial cells but no luminal thrombus, whereas inflammation and smooth muscle cells are only rarely observed. E: Section of the Bx Velocity stent. There is an abundant neointimal tissue consisting of smooth muscle cells in a proteoglycan collagen matrix and an overlying endothelium. F: High-power section of the boxed area in E. Lymphocytes are present around the stent strut with minimal fibrin underneath the strut. The luminal surface of the stent is covered by smooth muscle cells in a proteoglycan/collagen rich matrix [Reproduced by permission from Joner et al. (2006)].

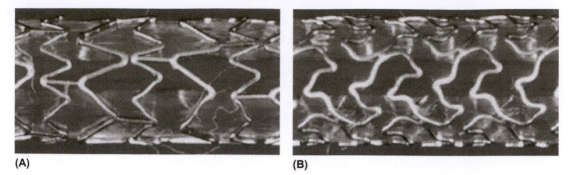

**(A)**                                                    **(B)**

FIGURE 19.10    Taxus stent geometries. **A**: Taxus Express, **B**: Taxus Liberte *[Reproduced by permission from Turco et al. (2007)].*

Two other second-generation stents, the Medtronic Endeavor and Abbott XIENCE V stents, use cobalt chromium (Co-Cr) rather than stainless steel stent platforms. Because Co-Cr is stronger than stainless steel, thinner struts can be used while meeting the same luminal support requirements (high pressure with less elastic recoil).

The Medtronic Endeavor stent also incorporated a sirolimus analogue, zotarolimus, and a phosphorylcholine (PC) polymer biomimetic coating in its design. The phosphorylcholine molecule is present in cell membranes. A biomimetic process is a human-made process that imitates nature. In a randomized, single-blinded clinical study of patients receiving either an Endeavor ($n = 190$) and Cypher ($n = 68$) stent, the mean neointimal thickness at 8 months was significantly higher for the Endeavor stent ($0.26 \pm 0.19$ mm vs. $0.04 \pm 0.04$ mm, $p < 0.01$) (Miyazawa et al., 2008).

The Abbott XIENCE V stent incorporated another sirolimus analogue, everolimus, and a polymer coating composed of PVDF-HFP. The XIENCE V coating is much thinner at 5.3 μm than that of the Cypher or Taxus (7.2 or 15.6 μm, respectively). In a randomized, single-blinded clinical study of patients receiving either a XIENCE V ($n = 642$) and Taxus Express ($n = 309$) stent, the target lesion revascularization rate at 2 yr was significantly lower for the XIENCE V stent (6.1% vs. 11.3%, $p = 0.006$). Major adverse cardiovascular events, which included death and myocardial infarction, were also significantly decreased (7.7% vs. 13.8%, $p = 0.005$). The rate of stent thrombosis did not change (Stone et al., 2009).

## Third Generation

To address the requirement for an optimal drug coating technology, Abbott has moved beyond coatings and is developing the ABSORB bioresorbable vascular scaffold (BVS). This is a new type of device that provides temporary support of the vessel lumen while it heals and that is then removed via biodegradation.

ABSORB is composed of a bioresorbable poly(L-lactide) (PLLA) backbone and a bioresorbable PDLLA coating containing everolimus (Figure 19.11). The semicrystalline PLLA replaces metallic alloys as the load-bearing material. According to Abbott, the device is referred to as a scaffold rather than as a stent, because the term stent implies permanence

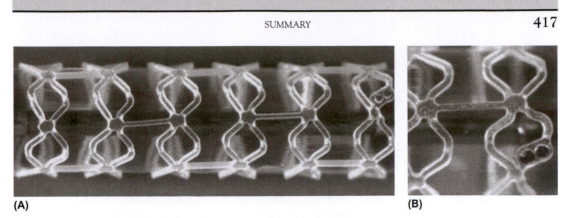

(A)                                                                                     (B)

**FIGURE 19.11**    The ABSORB BVS. **A**: Overview of the stent. **B**: Magnified image of the stent *[Reproduced by permission from Ormiston et al. (2008)].*

whereas a BVS provides structure temporarily before eventually being fully resorbed. It is hypothesized that the potential for late stent thrombosis is reduced for a device that disappears, and it has been suggested that the uncaging of the artery permits the restoration of natural vasoreactivity and late luminal enlargement (Ormiston et al., 2008; Serruys et al., 2009).

Although the ABSORB BVS has not yet been approved by the FDA, the early 2-yr study results for a small sample ($n = 28$) of the so-called Cohort A version of the device are impressive and included in this chapter. At 2 yr, no cardiac deaths or stent thrombosis was observed. The mean lumen diameters after the procedure, at 6 months and at 2 yr were $2.87 \pm 0.21$ mm, $2.41 \pm 0.31$ mm ($p = 0.03$), and $2.63 \pm 0.61$ mm ($p = 0.30$), respectively (Serruys et al., 2009), where the latter two time points support the hypothesized late luminal enlargement. Optical coherence tomography (OCT) cross-sectional images from the three time points of that study are reproduced in Figure 19.12, showing the evolution of the device and the vessel from implantation to full resorption.

For a discussion of other new drug-eluting stent concepts that have not been incorporated into FDA-approved stents, see Wessely (2010).

## KEY FEATURES FROM ENGINEERING STANDARDS

As of 2010, the FDA does not recognize any consensus standards that are specific to drug-eluting stents.

## SUMMARY

A drug-eluting stent is a bare metal stent medical device, coated with a polymer-drug formulation that controls the release rate of the drug. Similar to the steroid-eluting pacemaker lead, a drug is eluted from the stent over time to minimize host reactions so that the associated implanted artery does not restenose. Both drug-eluting devices are examples of combination products.

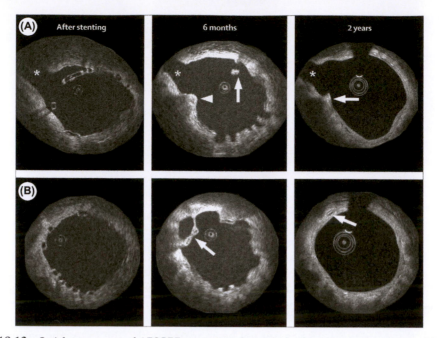

**FIGURE 19.12**   Serial assessment of ABSORB stent struts by optical coherence tomography. **A**: After stenting, incomplete apposition of struts (preserved box) in front of a side-branch ostium. At 6 months, persistent incomplete stent apposition (arrow) and resolved incomplete stent apposition (arrowhead), with open box appearance. At 2 yr, there is now a smooth appearance of the endoluminal lining without strut malapposition because struts have been absorbed. There is guidewire shadowing (at the top of the imaging), and a strut is still just discernible as a bright spot (arrow). **B**: Complete apposition of strut (box appearance) after the procedure. At 6 months, there is late acquired incomplete stent apposition of the struts (preserved box appearance) with tissue bridging connecting the struts (arrow). The endoluminal lining is corrugated. At 2 yr, there is a smooth endothelial lining with almost circular cross section. Generally, the struts are no longer discernible, although the bright reflection could indicate a strut (arrow). Asterisk indicates a side branch *[Reproduced by permission from Serruys et al. (2009)]*.

In the first generation of drug-eluting stents, reduction of neointimal hyperplasia was the main goal. Stainless steel stents were coated with a polymer matrix, which targeted neointimal hyperplasia or a precursor. For subsequent generations of drug-eluting stents, clinicians noted desirable design criteria, based on deliverability, efficacy, and safety: less recoil, thinner struts, a drug that reduces neotintimal proliferation, drug-coating technology that minimizes stent thrombosis.

The second-generation stent with the best clinical performance to date is the Abbott XIENCE V stent. It incorporates the antiproliferative, everolimus, in a PVDF-HFP coating. The everolimus-polymer matrix layer is much thinner than the thickness of the Cypher or Taxus layers. Abbott is also investigating a fully bioresorbable stent, which looks promising.

When well-defined essential requirements are unavailable, substantial design iterations may occur outside each design control cycle, rather than just within one design control cycle. In the case of medical devices for CAD treatment, clinicians and engineers were unaware that host responses would cause restenosis after balloon angioplasty. They were also unaware during initial bare metal stent deployment that host responses leading to neointimal growth would cause restenosis. This awareness and more recent observations have affected how each generation of drug-eluting stents is designed.

## Acknowledgment

The author thanks Drs. James P. Oberhauser and Shrirang Ranade for their detailed reviews. Dr. Oberhauser is Manager of BVS Polymer R&D at Abbott Vascular. Dr. Ranade is Manager of Research and Development at Acclarent, Inc., a Johnson & Johnson company.

## Exercises

**19.1** Use the MAUDE database to determine the total number of adverse events that were reported and documented for the Boston Scientific Contak Renewal 1 ICD: http://www.accessdata.fda.gov/scripts/cdrh/cfdocs/cfMAUDE/search.cfm. We discussed this ICD's recall in Chapter 5. How easy is the MAUDE database to use? What manufacturer, brand name, and date range did you use? What was the most commonly reported specific adverse event? Do you attribute the adverse events to lack of resources for product development, a lack of well-defined essential requirements, or a lack of well executed design verification and validation? Explain your reasoning.

**19.2** Describe the design iterations that occurred over several product development cycles for another medical device discussed in a previous chapter. What issues occurred that caused major redesigns after the first market release? Could these issues have been prevented during the original product development cycle? Explain your reasoning.

**19.3** Can the kinetic drug release of paclitaxel from SIBS (Figure 19.5) be described by a compartmental model? Explain your reasoning.

**19.4** Based on clinical study results discussed in this chapter, what stent platform parameters, drug, and drug carrier combination would you recommend? Explain your reasoning.

Read Schofer et al. (2003):

**19.5** What is neointimal hyperplasia? How does sirolimus combat it?

**19.6** Describe the coating. What is the purpose of Parylene C? Describe the implant protocol. Does the addition of a sirolimus coating statistically improve outcomes? Be specific.

Read Lansky et al. (2002):

**19.7** How was the lumen diameter measured in the E-SIRIUS trial? Describe the specific process.

**19.8** How accurate is this method of measurement?

## References

ABET (2009). *Criteria for accrediting engineering programs: Effective for evaluations during the 2010–2011 accreditation cycle*. Baltimore, MD: ABET.

AHA (2009). *Heart disease & stroke statistics: 2009 update at-a-glance*. Dallas, TX: AHA.

Ako, J., Bonneau, H. N., Honda, Y., & Fitzgerald, P. J. (2007). Design criteria for the ideal drug-eluting stent. *American Journal of Cardiology, 100*, 3M—9M.

Baura, G. D. (2013). *U.S. Medical device regulation: An introduction to biomedical product development, market release, and postmarket surveillance*, Manuscript in preparation.

Bauters, C., & Isner, J. M. (1998). The biology of restenosis. In E. J. Topol (Ed.), *Textbook of cardiovascular medicine* (2nd Ed.). Philadelphia: Lippincott-Raven.

CDRH (1997). *Design control guidance for medical device manufacturers*. Rockville, MD: FDA.

CDRH (2010). *MAUDE: Manufacturer and user facility device experience*. July, 2010 access. Silver Spring, MD: FDA.

Farb, A., Weber, D. K., Kolodgie, F. D., Burke, A. P., & Virmani, R. (2002). Morphological predictors of restenosis after coronary stenting in humans. *Circulation, 105*, 2974—2980.

Heller, J., & Hoffman, A. S. (2004). Drug delivery systems. In B. D. Ratner, A. S. Hoffman, F. J. Schoen, & J. E. Lemons (Eds.), *Biomaterials science* (2nd ed.). San Diego, CA: Elsevier.

Joner, M., Finn, A. V., Farb, A., Mont, E. K., Kolodgie, F. D., & Ladich, E., et al. (2006). Pathology of drug-eluting stents in humans: Delayed healing and late thrombotic risk. *Journal of the American College of Cardiology, 48*, 193—202.

Kohn, J., Abramson, S., & Langer, R. (2004). Bioresorbable and bioerodible materials. In B. D. Ratner, A. S. Hoffman, F. J. Schoen, & J. E. Lemons (Eds.), *Biomaterials science* (2nd ed.). San Diego, CA: Elsevier.

Kukreja, N., Onuma, Y., Daemen, J., & Serruys, P. W. (2008). The future of drug-eluting stents. *Pharmacological Research, 57*, 171—180.

Lansky, A. J., Desai, K., & Leon, M. B. (2002). Quantitative coronary angiography in regression trials: a review of methodologic considerations, endpoint selection, and limitations. *American Journal of Cardiology, 89*, 4B—9B.

Liechty, W. B., Kryscio, D. R., Slaughter, B. V., & Peppas, N. A. (2010). Polymers for drug delivery systems. *Annual Review of Chemical and Biomolecular Engineering, 1*, 149—173.

Maisel, W. H. (2007). Unanswered questions—drug-eluting stents and the risk of late thrombosis. *New England Journal of Medicine, 356*, 981—984.

Miyazawa, A., Ako, J., Hongo, Y., Hur, S. H., Tsujino, I., & Courtney, B. K., et al. (2008). Comparison of vascular response to zotarolimus-eluting stent versus sirolimus-eluting stent: intravascular ultrasound results from ENDEAVOR III. *American Heart Journal, 155*, 108—113.

Ormiston, J. A., Serruys, P. W., Regar, E., Dudek, D., Thuesen, L., & Webster, M. W., et al. (2008). A bioabsorbable everolimus-eluting coronary stent system for patients with single de-novo coronary artery lesions (ABSORB): a prospective open-label trial. *Lancet, 371*, 899—907.

Philp, A., Macdonald, A. L., & Watt, P. W. (2005). Lactate—a signal coordinating cell and systemic function. *Journal of Experimental Biology, 208*, 4561—4575.

Rajagopal, V., & Rockson, S. G. (2003). Coronary restenosis: a review of mechanisms and management. *American Journal of Medicine, 115*, 547—553.

Ranade, S. V., Miller, K. M., Richard, R. E., Chan, A. K., Allen, M. J., & Helmus, M. N. (2004). Physical characterization of controlled release of paclitaxel from the TAXUS Express2 drug-eluting stent. *Journal of Biomedical Materials Research A, 71*, 625—634.

Ryan, K. V. (2003). *United States of America v. Endovascular Technologies, Inc. Criminal Information, CR 02-0179 SI*. Department of Justice Northern District of California. San Francisco: United States District Court.

Schofer, J., Schluter, M., Gershlick, A. H., Wijns, W., Garcia, E., & Schampaert, E., et al. (2003). Sirolimus-eluting stents for treatment of patients with long atherosclerotic lesions in small coronary arteries: double-blind, randomised controlled trial (E-SIRIUS). *Lancet, 362*, 1093—1099.

Serruys, P. W., Ormiston, J. A., Onuma, Y., Regar, E., Gonzalo, N., & Garcia-Garcia, H. M., et al. (2009). A bioabsorbable everolimus-eluting coronary stent system (ABSORB): 2-year outcomes and results from multiple imaging methods. *Lancet, 373*, 897—910.

Stone, G. W., Midei, M., Newman, W., Sanz, M., Hermiller, J. B., & Williams, J., et al. (2009). Randomized comparison of everolimus-eluting and paclitaxel-eluting stents: two-year clinical follow-up from the clinical evaluation of the Xience V Everolimus Eluting Coronary Stent system in the treatment of patients with de novo native coronary artery lesions (SPIRIT) III trial. *Circulation, 119*, 680—686.

Stone, G. W., Moses, J. W., Ellis, S. G., Schofer, J., Dawkins, K. D., & Morice, M. C., et al. (2007). Safety and efficacy of sirolimus- and paclitaxel-eluting coronary stents. *New England Journal of Medicine, 356*, 998—1008.

Turco, M. A., Ormiston, J. A., Popma, J. J., Mandinov, L., O'Shaughnessy, C. D., & Mann, T., et al. (2007). Polymer-based, paclitaxel-eluting TAXUS Liberte stent in de novo lesions: the pivotal TAXUS ATLAS trial. *Journal of the American College of Cardiology, 49*, 1676–1683.

U.S. Code of Federal Regulations. (2010). *Combination product definition*. Title 21, Chapter 1, §3.2(e).

Wessely, R. (2010). New drug-eluting stent concepts. *Nature Reviews Cardiology, 7*, 194–203.

Yamamoto, M., Takano, M., Murakami, D., Okamatsu, K., Seino, Y., & Mizuno, K. (2010). Comparative angioscopic evaluation of neointimal coverage and thrombus between TAXUS-Express and TAXUS-Liberte stents: Is the stent platform type associated with the vascular response? *International Journal of Cardiology, 145*, 587–589.

# Artificial Pancreas

In this last medical device chapter, we detail another combination product that is a current research topic. Specifically, we discuss the appropriate physiology, history, and key features of medical devices that are precursors of a future artificial pancreas. An artificial pancreas would administer hormones as needed, to keep blood glucose concentration in the normal range during daily activities. Current artificial pancreas research approaches include a combination product consisting of an insulin pump/glucose/control algorithm and implanted beta cells that secrete insulin. This discussion is limited to the closed-loop insulin pump/glucose sensor approach (Figure 20.1). Specific closed-loop strategies for the intensive care unit are beyond the range of this discussion.

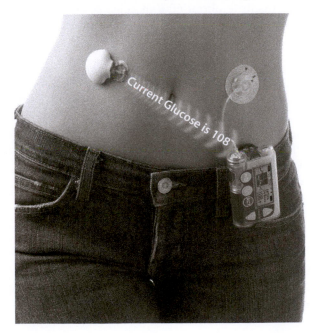

**FIGURE 20.1** The Medtronic Paradigm Real-Time system consists of an insulin infusion pump, glucose sensor, and transmitter (Medtronic Diabetes, Northridge, California).

Upon completion of this chapter, each student shall be able to:

1. Understand the mechanisms underlying type 1 and type 2 diabetes mellitus.
2. Identify biocompatibility issues that affect the realization of an artificial pancreas.
3. Describe five key features of medical devices related to an artificial pancreas.

## BLOOD GLUCOSE REGULATION

In previous chapters, we discussed how body fluid and core temperature are regulated for homeostasis. In this section, we discuss how blood glucose is regulated for homeostasis. Glucose is the only nutrient that supplies sufficient energy in the quantities required by the brain, retina, and gonads.

### Pancreatic Hormones

Glucose is regulated by the hormones insulin and glucagon, which are both secreted by the pancreas. This organ contains two major types of tissues: the *acini*, which secrete digestive juices into the duodenum, and the *islets of Langerhans*, which secrete glucagon and insulin into the blood.

The pancreas contains 1–2 million islets. Each islet is about 0.3 mm in diameter and is organized around small capillaries, into which hormones are secreted. Each islet contains alpha, beta, and delta cells. The alpha cells, which constitute about 25% of total cells in an islet, secrete glucagon. The beta cells, which constitute about 60% of cells, secrete insulin and amylin. The delta cells, which constitute about 10% of cells, secrete somatostatin (Guyton & Hall, 2006) (Figures 20.2 and 20.3).

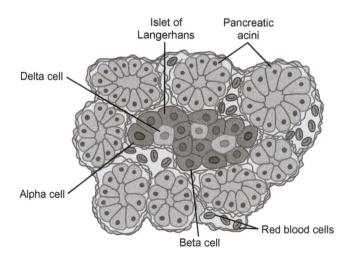

FIGURE 20.2 Physiologic anatomy of an islet of Langerhans in the pancreas *[Reproduced by permission from Guyton & Hall (2006)].*

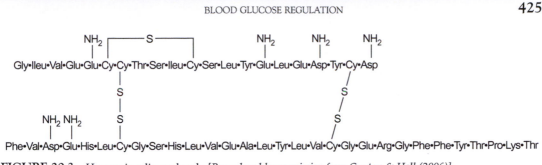

**FIGURE 20.3**  Human insulin molecule *[Reproduced by permission from Guyton & Hall (2006)].*

Each of these hormones is a *peptide*, that is, a short chain of amino acids. Peptides are normally administered intravenously during therapy, although some work is being conducted on peptide delivery through the lung. Peptides are not orally administered as pills because of a peptide's short half-life (the eigenvalue from a compartmental model) and rapid clearance from the liver (see Chapter 9). Only 10% of an initial peptide pill dosage may arrive at its target destination.

## Hormonal Control

Normally, blood glucose concentration is narrowly controlled. In this chapter, we refer to glucose concentrations in the blood, as well as in plasma. Blood glucose concentration is often measured from capillary blood obtained from a finger prick. Plasma glucose concentration is often measured in laboratory experiments. A fasting person before breakfast possesses a blood glucose concentration of 80–90 mg/dL of blood. During the first hour after a meal, the concentration increases to 120–140 mL/dL. Then feedback systems rapidly return the glucose concentration back to control levels, usually within 2 hr after the last absorption of carbohydrates. Conversely, during starvation, gluconeogenesis in the liver provides glucose.

When a high-carbohydrate meal is ingested, the glucose in the blood causes rapid secretion of insulin (Figure 20.4). Insulin causes rapid uptake of glucose by almost all tissues, especially the muscles, the liver, and adipose tissues. If the muscles are not exercising, the glucose is stored as muscle glycogen. As shown in Figure 20.5, insulin facilitates glucose transport through the muscle cell membrane. Glycogen can later be used for energy by the muscle. In the liver, glucose is stored immediately as glycogen. In adipose tissue, insulin promotes fatty acid synthesis.

Other effects of insulin include the promotion of protein synthesis and storage, synergetic interaction with growth hormone to promote growth (Guyton & Hall, 2006), and control of food intake and body weight (Porte et al., 2005; Baura et al., 1993).

When food is not ingested, the blood glucose concentration plummets. This hypoglycemia causes glucagon secretion (Figure 20.4). Glucagon causes the breakdown of liver glycogen (*glycogenolysis*), which in turn increases the blood glucose concentration within minutes. Glucagon also increases the rate of amino acid uptake by liver cells, as well as the conversion of many of these amino acids to glucose by gluconeogenesis.

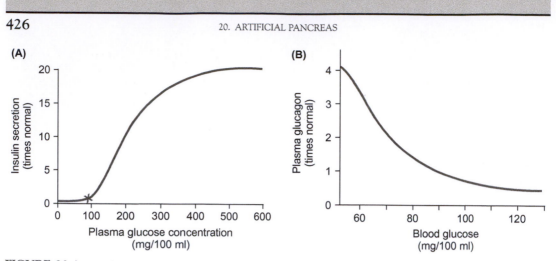

FIGURE 20.4   Insulin and glucagon levels as a function of plasma or blood glucose concentration. **A**: Insulin level. **B**: Glucagon level *[Reproduced by permission from Guyton & Hall (2006)].*

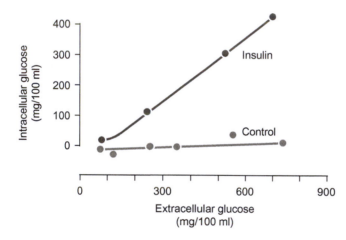

**FIGURE 20.5**   Effect of insulin in enhancing the concentration of glucose inside muscle cells. Note that, in the absence of insulin (control), the intracellular glucose concentration remains near zero, despite high extracellular glucose concentrations. *[Data from Eisenstein (1964). Figure reproduced by permission from Guyton & Hall (2006)].*

The principal role of somatostatin may be to extend the period of time over which food nutrients are assimilated into the blood (Guyton & Hall, 2006). The role of amylin, which is also known as *islet amyloid polypeptide (IAPP)*, is a current research topic.

## Diabetes Mellitus

Impaired insulin secretion or insulin sensitivity is referred to as *diabetes mellitus*. This syndrome is characterized by impaired carbohydrate, fat, and protein metabolism; it is classified as type 1, type 2, or gestational diabetes.

- With *type 1 diabetes*, which is also known as insulin-dependent diabetes mellitus (IDDM), insulin is not secreted. A patient's immune system has destroyed pancreatic beta cells.

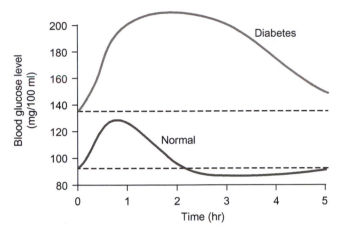

**FIGURE 20.6** Glucose tolerance curve in a normal person and in a person with diabetes *[Reproduced by permission from Guyton & Hall (2006)].*

- With *type 2 diabetes*, which is also known as non-insulin-dependent diabetes mellitus (NIDDM), insulin is secreted, but the body does not use it effectively. Due to insulin resistance, target tissues possess impaired carbohydrate utilization and storage, which eventually causes a compensatory increase in insulin secretion.
- *Gestational diabetes* is any degree of glucose intolerance that begins or is diagnosed during pregnancy.

Diabetes may be diagnosed by measuring the fasting blood glucose concentration in the early morning after overnight fasting, the fasting plasma insulin concentration, and the response to a glucose tolerance test. The upper limit of normal fasting blood glucose is 100 mg/dL. A fasting blood glucose level of 100–125 mL/dL is considered to be prediabetic and is referred to as *impaired fasting glucose (IFG)*. All diabetics possess a high fasting blood glucose concentration and an abnormal response to ingestion of glucose. As shown in Figure 20.6, diabetics respond with a higher maximum blood glucose level and an extended delay until blood glucose drops. A type 1 diabetic also possesses a low or undetectable fasting plasma insulin concentration.

Diabetes may also be diagnosed and monitored by measuring the percentage of glycosylated hemoglobin (glucose attached to hemoglobin, $HbA_{1c}$). This so-called *A1C test* reflects a patient's average blood glucose level for the past 2–3 months. Fasting is not required before the blood draw. An A1C percentage of 4.5–6% is considered normal; an A1C percentage of 6–6.5% is considered prediabetes. An uncontrolled diabetic may have an A1C percentage above 9%. For diagnosed diabetics, an A1C percentage of 7% is a common treatment target.

## COMPARTMENTAL MODELS

We can model the relationships between insulin and glucose transport in the body and between plasma and insulin glucose, using compartmental modeling. An important concept for compartmental models is model identifiability, which we discuss first.

## Model Identifiability

A postulated model may have several parameters, but each parameter may not be uniquely identifiable. Theoretical, or *a priori*, identifiability addresses the issue that a postulated model structure may be too complex for a particular set of ideal data. By using the transfer function matrix approach, the uniquely identifiable parameters can be determined.

First, compartmental model equations are rearranged into canonical form, which requires an explicit definition of the outputs:

$$\dot{\mathbf{c}}(t) = \underline{\mathbf{A}}(\mathbf{k})\mathbf{c}(t) + \underline{\mathbf{B}}(\mathbf{k})\mathbf{u}(t) \tag{20.1}$$

$$\mathbf{y}(t) = \underline{\mathbf{C}}(\mathbf{k})\mathbf{c}(t), \tag{20.2}$$

where $\mathbf{c}(t)$ is the concentration vector with dimension $(n-1) \times 1$, $\dot{\mathbf{c}}(t)$ is the concentration derivative vector, $\mathbf{u}(t)$ is the exogeneous input vector, $\mathbf{k}$ is the admissible parameter space of rate constants with dimension $p \times 1$, $\mathbf{y}(t)$ is the desired output vector, $\underline{\mathbf{A}}(\mathbf{k})$ is the feedback matrix containing rate constants, $\underline{\mathbf{B}}(\mathbf{k})$ is the feedforward matrix, and $\underline{\mathbf{C}}(\mathbf{k})$ is the relationship between the concentration and desired output vectors.

Next, Eqs. (20.1) and (20.2) are transformed into the Laplace domain, and the transfer function, $\underline{\mathbf{H}}(s, \mathbf{k})$, of dimension $r \times m$ is calculated as:

$$\underline{\mathbf{H}}(s,\mathbf{k}) = \underline{\mathbf{C}}(\mathbf{k})[s\underline{\mathbf{I}} - \underline{\mathbf{A}}(\mathbf{k})]^{-1}\underline{\mathbf{B}}(\mathbf{k}), \tag{20.3}$$

where $\underline{\mathbf{I}}(s)$ is the identity matrix. Each element, $\underline{\mathbf{H}}_{ij}(s, \mathbf{k})$, reflects an experiment performed on the system between input $j$ and output $i$. Each element can be further subdivided into the coefficients of the numerator polynomial, $\beta_1^{ij}(\mathbf{k}) \ldots \beta_{n-1}^{ij}(\mathbf{k})$, and the coefficients of the denominator polynomial, $\alpha_1^{ij}(\mathbf{k}) \ldots \alpha_n^{ij}(\mathbf{k})$. From these coefficients, the $(2n-1)rm \times p$ Jacobian matrix, $\underline{\mathbf{G}}(\mathbf{k})$, is formed, such that

$$\underline{\mathbf{G}}(\mathbf{k}) = \begin{bmatrix} \dfrac{\partial \beta_1^{11}}{\partial k_1} & \cdots & \dfrac{\partial \beta_1^{11}}{\partial k_p} \\ \vdots & & \vdots \\ \dfrac{\partial \alpha_n^{11}}{\partial k_1} & \cdots & \dfrac{\partial \alpha_n^{rm}}{\partial k_p} \\ \vdots & & \vdots \\ \dfrac{\partial \beta_1^{rm}}{\partial k_1} & \cdots & \dfrac{\partial \beta_1^{rm}}{\partial k_p} \\ \vdots & & \vdots \\ \dfrac{\partial \alpha_n^{rm}}{\partial k_1} & \cdots & \dfrac{\partial \alpha_n^{rm}}{\partial k_p} \end{bmatrix} \tag{20.4}$$

The model is identifiable if, and only if, the rank of the Jacobian matrix equals $p$ (Baura, 2002).

As an example, let us take the two compartmental model of urea kinetics from Chapter 9, and rearrange Eqs. (9.9) and (9.10) in canonical form:

$$\begin{bmatrix} \dot{c}_1(t) \\ \dot{c}_2(t) \end{bmatrix} = \begin{bmatrix} -(k_d + k_{21}) & k_{12} \\ k_{21} & -k_{21} \end{bmatrix} \begin{bmatrix} c_1(t) \\ c_2(t) \end{bmatrix} + \begin{bmatrix} 0 \\ 0 \end{bmatrix} u(t) \tag{20.5}$$

$$y(t) = \begin{bmatrix} 1 & 0 \end{bmatrix} \begin{bmatrix} c_1(t) \\ c_2(t) \end{bmatrix}. \tag{20.6}$$

Calculation of the transfer function yields:

$$\underline{\mathbf{H}}(s,\mathbf{k}) = \begin{bmatrix} 1 & 0 \end{bmatrix} [s\underline{\mathbf{I}} - \underline{\mathbf{A}}(\mathbf{k})]^{-1} \begin{bmatrix} 0 \\ 0 \end{bmatrix}, \tag{20.7}$$

$$\underline{\mathbf{H}}(s, \mathbf{k}) = 0. \tag{20.8}$$

Because the transfer function equals zero, the Jacobian Matrix equals zero, and the Jacobian Matrix's rank equals zero. The model is not uniquely identifiable. As stated in Chapter 9, not enough information is provided for modeling the three parameters in this system.

## Minimal Model of Insulin Sensitivity

The minimum model of insulin sensitivity was first proposed by physiologist Richard Bergman and electrical engineer Claudio Cobelli in 1979. This model was developed to quantify the influence of decreased peripheral insulin sensitivity on the impairment of a patient's ability to tolerate a standard glucose load. The model is based on insulin and glucose plasma samples obtained after an intravenous glucose injection, which is also known as an *intravenous glucose tolerance test* (IVGTT). It attempts to characterize the abrupt, multiphase insulin secretor response to a rapid intravenous injection of glucose, which is associated with a rapidly decreasing glucose concentration.

The blood glucose regulating system may be simplified by excluding glucagon effects and separating the system into two subsystems (Figure 20.7). In the first subsystem, the plasma glucose concentration, $g(t)$, is considered the input to the beta cells in the pancreas, which secrete, distribute, and metabolize insulin. In the second subsystem, the plasma insulin concentration, $i(t)$, is considered the input to the tissues that metabolize glucose. The minimal model of insulin sensitivity is based on the second subsystem, thus eliminating the difficulty of modeling beta cell function.

For this model, it is assumed that insulin as a forcing function enters a remote compartment, $r(t)$, with rate constant $k_2$, and is cleared from this compartment with rate constant $k_3$. It is also assumed that glucose uptake into tissues is directly dependent on the insulin in this remote compartment, $k_4 r(t) g(t)$. This bilinear relation is consistent with the existence of a remote compartment intimately involved in the action of insulin, increasing the mobility of glucose across the cell membrane.

To further simplify the model, insulin-independent tissues are separate from insulin dependent, glucose-utilizing tissues and are grouped as the "liver" and periphery, respectively. The rate of change of glucose is the difference between the net hepatic glucose

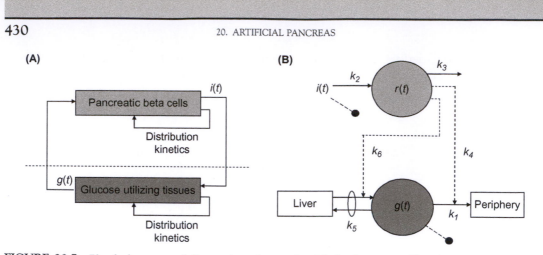

**FIGURE 20.7** Blood glucose regulating system. **A:** two simplified subsystems: $g(t)$ = plasma glucose concentration, $i(t)$ = plasma insulin concentration. **B:** Compartmental model of glucose and insulin in a remote compartment, $r(t)$, representing glucose subsystem. $k_i$ are rate constants.

balance, $b(t)$, and the disappearance of glucose into the peripheral tissues only, $u_p(t)$. In previous work, Bergman et al. (1979) showed that the hepatic glucose balance varies according to

$$b(t) = B_0 - [k_5 + k_6 r(t)]g(t), \tag{20.9}$$

where $B_0$ is the net balance expected when plasma glucose concentration is extrapolated to 0, $k_5$ is the net transport of glucose from liver to the glucose compartment, and $k_6$ is the rate constant representing the direct dependence of glucose production in liver on insulin in the remote compartment (Bergman & Bucolo, 1974). Glucose utilization is assumed to depend on insulin in the remote compartment, as just detailed, and on the glucose concentration:

$$u_p(t) = [k_1 + k_4 r(t)]g(t), \tag{20.10}$$

where $k_1$ is the rate constant associated with transport of glucose to the periphery.

The rate of change in glucose is then the difference between Eqs. (20.9) and (20.10):

$$\dot{g}(t) = b(t) - u_p(t), \tag{20.11}$$

$$\dot{g}(t) = B_0 - [k_5 + k_6 r(t)]g(t) - [k_1 + k_4 r(t)]g(t), \tag{20.12}$$

$$\dot{g}(t) = B_0 + [-(k_1 + k_5) - (k_4 + k_6)r(t)]g(t). \tag{20.13}$$

By modeling only IVGTT data, with $g(0) \neq 0$, the extremes of hypoglycemia and hyperglycemia are avoided and renal glucose loss may be ignored.

From Figure 20.7, the rate of change of insulin in the remote compartment is

$$\dot{r}(t) = k_2 i(t) - k_3 r(t). \tag{20.14}$$

Multiplying both sides of Eq. (20.14) by the sum $(k_4 + k_6)$ results in

$$(k_4 + k_6)\dot{r}(t) = (k_4 + k_6)k_2 i(t) - (k_4 + k_6)k_3 r(t). \tag{20.15}$$

If we define a new state variable, $x(t)$, such that

$$x(t) = (k_4 + k_6)r(t), \qquad (20.16)$$

then Eq. (20.15) may be rewritten as

$$\dot{x}(t) = p_2 x(t) + p_3 i(t), \qquad (20.17)$$

where $p_2 = -k_3$ and $p_3 = k_2(k_4 + k_6)$. Similarly, Eq. (20.13) may be rewritten as

$$\dot{g}(t) = [p_1 - x(t)]g(t) + p_4, \qquad (20.18)$$

where $p_1 = -(k_1 + k_5)$ and $p_4 = B_0$. We also assume that $p_5$ represents the initial glucose concentration, $g(0)$.

Eqs. (20.17) and (20.18) represent the minimal model. As demonstrated by the transfer function matrix approach, the **p** vector can be uniquely identified from this model:

$$\mathbf{p} = [p_1 \; p_2 \; p_3 \; p_4 \; p_5]^T, \qquad (20.19)$$

The minimal model is used to assess insulin sensitivity. Glucose effectiveness, $E(t)$, is defined as the quantitative enhancement of glucose disappearance due to an increase in the plasma glucose concentration:

$$E(t) = -\frac{\partial \dot{g}(t)}{\partial g(t)}, \qquad (20.20)$$

where $\dot{g}(t)$ is the sample rate of change of the plasma glucose concentration. Combining Eqs. (20.18) and (20.20), glucose effectiveness may be calculated from

$$E(t) = x(t) - p_1. \qquad (20.21)$$

Insulin sensitivity, $S_I$, is then defined, in terms of steady-state (SS) conditions, as the quantitative influence of insulin to increase the enhancement of glucose of its own disappearance:

$$S_I \equiv -\frac{\partial E_{SS}}{\partial i_{SS}}. \qquad (20.22)$$

Based on Eq. (20.17), at steady state, when $\dot{x}(t) = 0$,

$$x_{SS} = -\frac{p_3}{p_2} i_{SS}. \qquad (20.23)$$

Combining Eqs. (20.21) and (20.23) results in

$$E_{SS} = x_{SS} - p_1 \qquad (20.24)$$

$$E_{SS} = -\frac{p_3}{p_2} i_{SS} - p_1. \qquad (20.25)$$

From Eq. (20.22), insulin sensitivity is then (Bergman et al., 1979)

$$S_I = -\frac{p_3}{p_2}. \qquad (20.26)$$

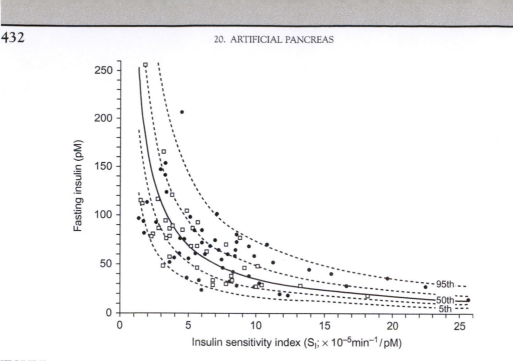

**FIGURE 20.8**    Relationship between $S_I$ and fasting insulin in 55 males (●) and 38 females (□). The best fit relationship is represented as a solid line. The 5th, 25th, 75th, and 95th percentiles are represented as dashed lines. The relationship is described by $S_I \times$ fasting insulin $= 3.518 \times 10^{-3}$ ($n = 93$) *[Reproduced by permission from Kahn et al. (1993)].*

In 93 human subjects, fasting insulin levels were determined to be related to insulin sensitivity, $S_I$, by a hyperbolic relationship (Figure 20.8). Greater degrees of insulin resistance (lower $S_I$) were associated with increased fasting insulin levels, whereas increasing insulin sensitivities (higher $S_I$) were associated with lower fasting insulin levels. Because of the hyperbolic relationship, when insulin sensitivity is high, large changes in insulin sensitivities would be expected to be associated with relatively small changes in fasting insulin. Similarly, when insulin sensitivity is low, small changes in insulin sensitivity would be expected to be associated with relatively large changes in fasting insulin (Kahn et al., 1993; Baura, 2002).

## Plasma Interstitial Glucose Equilibration Model

The minimal model approximates the relationship between plasma insulin and plasma (blood) glucose. However, subcutaneous glucose sensors monitor interstitial fluid (ISF) glucose, rather than plasma glucose. To determine this glucose relationship, Medtronic Minimed scientists modeled the interactions between these two compartments (Figure 20.9). In the figure, compartment 1 represents plasma glucose concentration, $g_1(t)$, and compartment 2 represents ISF glucose concentration, $g_2(t)$. Glucose moves between compartments with rate constants $k_{21}$ and $k_{12}$. Glucose is cleared from ISF by cellular glucose uptake, with rate constant $k_{02}$. The specific equations for this model are discussed in Exercise 20.3.

Plasma glucose can be measured accurately, but it is not known whether ISF glucose is measured accurately with a glucose sensor. Output current from this sensor, $I(t)$, was assumed to be proportional to the ISF glucose concentration, as

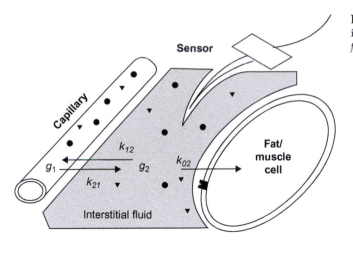

**FIGURE 20.9** Plasma ISF glucose equilibration model [*Reproduced by permission from Steil et al. (2005)*].

$$I(t) = c \cdot g_2(t). \tag{20.27}$$

This model was used to fit the mean glucose sensor current derived from ten nondiabetic subjects, who experienced a three-step euglycemic-hypoglycemic-hyperinsulinemic clamp. After an overnight fast, at 0 min, insulin (360 pmol·m$^{-2}$·min$^{-1}$) and glucose (20%, variable rate) infusions began. Plasma glucose was clamped at ~5 mmol/l until 90 min, at ~4.2 mmol/l until 180 min, and at 3.1 mmol/l until 270 min (1 mmol/l equals 18 mg/dL). Insulin and glucose infusions were then stopped, and plasma glucose was allowed to return to basal. The resulting glucose, insulin, and sensor current data are shown in Figure 20.10.

The sensor current was separately fitted for the current fall and recovery, with $r^2 = 0.899 \pm 0.0177$ and $0.928 \pm 0.0197$, respectively. Brief periods when the sensor current was systematically under- or overestimated were reportedly due to an increase in the model estimate of sensor sensitivity (Steil et al., 2005). Because the compartmental model is linear, a single model fit should have been possible for both fall and recovery. This lack of fit suggests that either the true model is more complicated, the sensor adds nonlinearity to the system, or both.

## CLINICAL NEED

According to the National Institute of Diabetes and Digestive and Kidney Diseases (NIDDK), 7.8% of the U.S. population, or 23.6 million people, had diabetes in 2007. 17.9 million people were diagnosed; 5.7 million were undiagnosed. About 186,300 people younger than 20 yr had type 1 or type 2 diabetes in 2007, representing 0.2% of people in this age group. In 1999–2000, 7.0% of U.S. adolescents ages 12–19 yr had IFG. These prevalence estimates were derived from various federal data systems.

Diabetes was the seventh leading cause of death listed on U.S. death certificates in 2006. Diabetic adults have heart disease death rates 2–4 times higher than nondiabetic adults. Diabetic retinopathy causes 12,000–24,000 new cases of blindness each year. Diabetes is

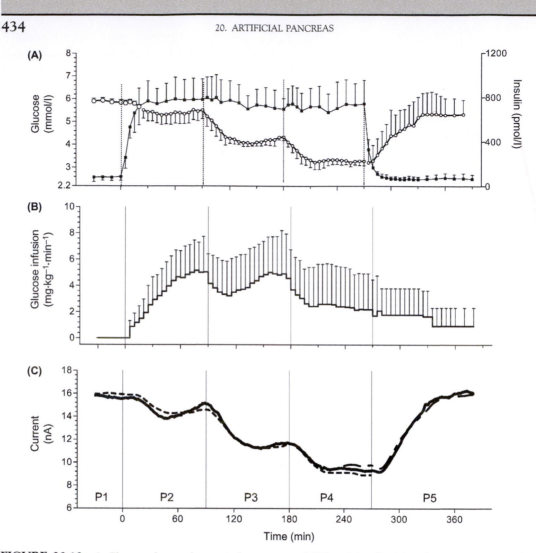

**FIGURE 20.10** **A**: Plasma glucose (open circles; mean and SD) and insulin (closed squares; mean and SD) concentration. **B**: Glucose infusion rate. **C**: Sensor current (mean), during insulin-induced hypoglycemia. Steady-state glucose was calculated during the last 30 min of each clamp phase (P1–P5). Sensor current (solid line) was separately fitted for the fall ($-30 < t < 270$ min; fit shown by dotted line) and recovery ($240 < t < 270$ min; fit shown by dashed line) using the model of Figure 20.9. Sensor current is shown with standard deviation bars removed for clarity [*Reproduced by permission from Steil et al. (2005)*].

the leading cause of kidney failure, accounting for 44% of new cases in 2005. Diabetic nerve disease contributes to lower-extremity amputations. The total direct and indirect costs in the United States for diabetes in 2007 were estimated to be $174 billion (NIDDK, 2008).

Improved glycemic control benefits patients with either type 1 or type 2 diabetes. During the Diabetes Control and Complications Trial (DCCT), 1441 patients with type 1 diabetes were randomly assigned to intensive (2% lower A1C) or conventional insulin therapy. Over 17 yr of follow-up, intensive insulin therapy reduced the risk of any cardiovascular disease event by 42% ($p = 0.02$), and the risk of nonfatal myocardial infarction, stroke, or death from cardiovascular disease decreased by 57% ($p = 0.02$) (Nathan et al., 2005). A recent meta-analysis of five randomized controlled trials followed 33,040 type 2 diabetics for about 163,000 person-years. Intensive glycemic control (0.9% lower A1C) resulted in a 17% reduction ($p$ not given) in events of nonfatal myocardial infarction and a 15% reduction ($p$ not given) in events of coronary heart disease (Ray et al., 2009).

Intensive insulin therapy, such as that performed in the DCCT, requires three or more daily subcutaneous injections of insulin or treatment with an external insulin pump, with dose adjustments based on at least four self-monitored glucose measurements per day (Nathan et al., 2005). Each glucose measurement requires a capillary blood sample obtained by pricking a finger with a lancet. Compliance with this regimen cannot be guaranteed with type 1 diabetics. Compliance for any insulin therapy regimen is lower for type 2 diabetics.

The use of an artificial pancreas would remove a patient from the decision process of insulin administration time and dose. It could also prevent hypoglycemic episodes due to accidental insulin administration overdoses, especially while sleeping (Hovorka et al., 2010).

## HISTORIC DEVICES

Clinical observation of diabetes can be traced 1550 BCE, when the Egyptian Ebers Papyrus mentioned remedies for polyuria. Physicians Frederick Banting and Charles Best isolated insulin from canine pancreatic extracts in 1921, and they treated their first human patient in 1922. The artificial pancreas, for automated administration of insulin, is a current research topic but has many impediments to true implementation.

### Ideal Artificial Pancreas System

An ideal artificial pancreas system is fully implantable. The patient is connected to a glucose sensor, which outputs accurate blood or plasma glucose readings. The sensor is isolated from system electronics. The continuous glucose readings are amplified and undergo data acquisition. The processor module receives the digitized glucose data. An internal control algorithm in the processor module calculates an appropriate insulin dose, which is fed to the pump module. The pump module administers this dose, completing the closed-loop system. The pump module may be periodically refilled with insulin. The entire system is battery operated. A telemetry transceiver enables system interrogation, data display, and parameter changes via a system programmer (Figure 20.11).

### Early Devices

In contrast to this ideal, the artificial pancreas was first realized by bioengineer A. Michael Albisser's group as a closed-loop system consisting of a glucose analyzer,

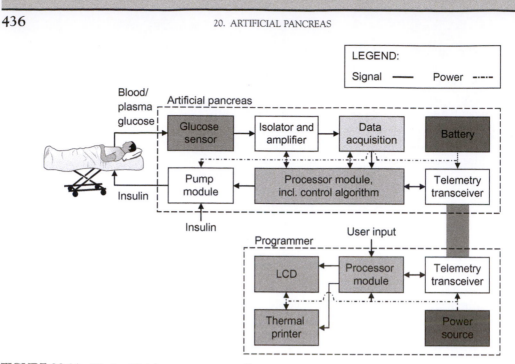

**FIGURE 20.11**   Ideal artificial pancreas system diagram.

minicomputer, and two infusion pumps for insulin and dextrose (Figure 20.12). The system was developed using pancreatized beagles (Albisser et al., 1974a) and clinically trialed in humans in 1974 (Albisser et al., 1974b). A few years later, in 1979, Miles Laboratories developed its Biostator Glucose Controlled Insulin Infusion System. This integrated system consisted of a blood pump, glucose analyzer, computer, infusion pump, and printer (Clemens, 1979). Both early systems were hampered by catheters in whole blood, oversimplified feedback control, large size, and constant need for clinical support.

More recently, separate components for a future closed-loop system have been developed. The Cygnus GlucoWatch Biographer (Figure 20.13) had a disposable sensor that noninvasively measured glucose three times per hour for 12 hr. Using reverse iontophoresis, current from a 0.42-V load was applied, which brought ISF in contact with the sensor. ISF glucose reacted with the enzyme glucose oxidase, GOx, to form hydrogen peroxide, $H_2O_2$, after an intermediate product reacted with oxygen:

$$FAD-GOx + glucose \rightarrow FADH2-GOx + gluconolactone, \quad (20.28)$$

$$FADH2-GOx + O_2 \rightarrow H_2O_2. \quad (20.29)$$

For GOx to act as a catalyst in Eq. (20.29), the cofactor flavin adenine dinucleotide (FAD) was required. Hydrogen peroxide was detected by a platinum/carbon electrode through the reaction (McGarraugh, 2009):

$$H_2O_2 \rightarrow 2H^+ + O_2 + 2e^-. \quad (20.30)$$

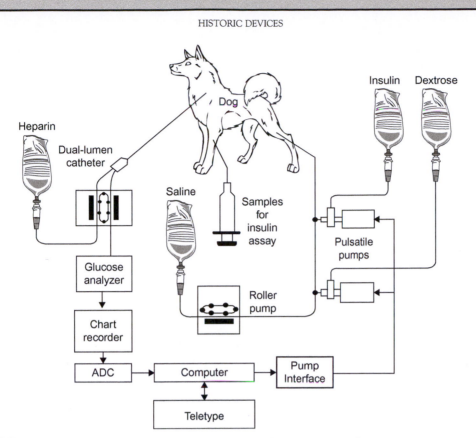

**FIGURE 20.12**    Albisser's artificial pancreas *[Adapted from Albisser et al. (1974a)]*.

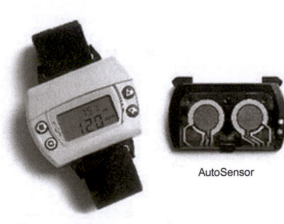

GlucoWatch® Biographer

**FIGURE 20.13**    Animas Glucowatch Biographer (Animas Corporation, West Chester, Pennsylvania).

Electrode current was passed to the Biographer circuitry, which translated the current to a blood glucose level for display. The sensor was initialized during a 3-hr warm-up period, followed by calibration with a traditional glucose meter. After 12 hr, a new sensor was required. In 420 glucose measurements submitted to the FDA, in which Biographer readings were compared to a standard glucose laboratory analyzer blood glucose readings, only 71% of glucose readings were within 20% of analyzer readings.

The Biographer was first approved by the FDA in 2001 (CDRH, 2001). In addition to inaccuracy, this sensor suffered from skipped readings, skin irritation, sweating precluding glucose measurements, and false alarms (Skyler, 2009). Its manufacture was discontinued in 2007 by Animas Corporation, which purchased Cygnus.

## ARTIFICIAL PANCREAS REQUIREMENTS

An ideal artificial pancreas would incorporate the following design criteria:

1. The components are integrated and fully implantable.
2. The glucose sensor is accurate, functions for at least several weeks, and does not require frequent recalibration every few hours.
3. The control algorithm prevents hypoglycemic or hyperglycemic excursions, so that blood glucose stays in the range of 80–180 mg/dL.

Current FDA-approved systems are not fully implantable because the glucose sensor lags behind the pump and processor. Microcontrolled implantable pumps already exist and are used to deliver drugs for chemotherapy and pain. An implantable insulin pump developed by Medtronic was investigated in clinical trials but was recently discontinued. This continuous intraperitoneal insulin infusion (CIPII) pump was implanted in the abdomen, with a catheter accessing the peritoneal cavity (Figure 20.14). Intraperitoneal insulin absorption, compared to subcutaneous insulin absorption, more closely mimics normal physiological action and results in improved hepatic uptake and lower peripheral plasma insulin levels (Logtenberg et al., 2009). The pump was filled with a specific U-400 insulin formulation, which includes a surface-reactive agent that prevents insulin aggregation promoted by body temperature and permanent contact with hydrophobic materials (Renard et al., 2006). We discuss this pump further in Exercise 20.2.

Currently marketed continuous glucose sensors that have been approved by the FDA are short-term subcutaneous amperometric electrodes, which continue to use the glucose oxidase reaction. The Medtronic MiniMed Paradigm and DexCom SEVEN sensors use $O_2/H_2O_2$ as the mediator. The Abbott FreeStyle Navigator sensor uses a vinyl pyridine polymer with pendant osmium complexes as the mediator (McGarraugh, 2009). The utility of each sensor is limited to 3, 7, or 5 days, respectively. Each sensor requires several recalibrations with a glucose meter during its lifetime. Sensor accuracy is similar to the Biographer accuracy.

Not surprisingly, subcutaneous sensor insertion initiates a wound-healing response, which includes the foreign body reaction (Figure 20.15). The sensor must undergo an initialization period of 2–10 hr, until the sensor signal stabilizes after insertion. This signal begins to drift due to biofouling (fluid exchange obstruction after nonspecific protein

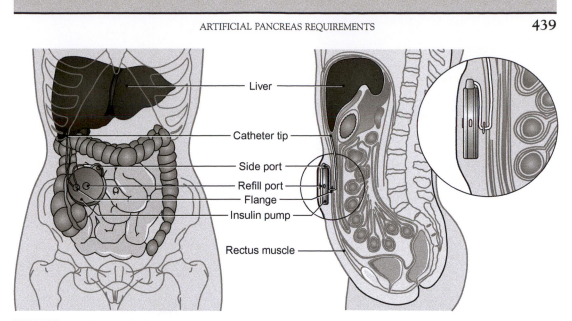

**FIGURE 20.14**    Positioning of the CIPII pump and catheter *in vivo [Adapted from Logtenberg et al. (2009)].*

adsorption), electrode passivation (signal attenuation through conductivity reduction), and degeneration (Boyne et al., 2003; Mang et al., 2005). The electrode surface may become covered with cells (Koschinsky & Heinemann, 2001); the sensor may be damaged by the effects of proteolytic enzymes and free radicals (Clark et al., 1988, Kyrolainen et al., 1995). Inflammatory cells may consume glucose around the implanted sensor. Later, more capillaries may supply glucose to the area (Gerritsen et al., 2001). The resulting signal drift is addressed by recalibration at fixed intervals (Cengiz & Tamborlane, 2009).

Sensor accuracy is also affected by sensor delay. Part of the delay is self-inflicted, due to the filtering of the sensor signal. For example, a three-point moving average on a signal obtained every 5 min will smooth the response but take 15 min to reach steady state (Voskanyan et al., 2007). But these subcutaneous sensors also measure ISF glucose, while the reference standard is blood (or plasma) glucose. Because plasma glucose and ISF glucose are separate compartments in the body, ISF glucose naturally lags behind blood glucose during steady state conditions. During nonsteady-state conditions, when glucose is decreasing, ISF glucose may fall in advance of plasma glucose (Sternberg et al., 1996) and reach minima that are lower than corresponding venous glucose levels (Caplin et al., 2003; Cengiz & Tamborlane, 2009).

An experimental intravenous glucose sensor developed by Medtronic also uses the glucose oxidase reaction. It was implanted in the vena cava by catheterization of the jugular or subclavian vein, eliminating the reference standard issue. It was connected to the Medtronic implanted insulin pump via a subcutaneous lead. A control algorithm was uploaded to the pump. Together, the sensor, pump, and control algorithm constituted the Medtronic Long Term Sensing System. System feasibility was demonstrated for 48 hr in three adult patients. In the fourth patient, significant drift between sensor glucose and

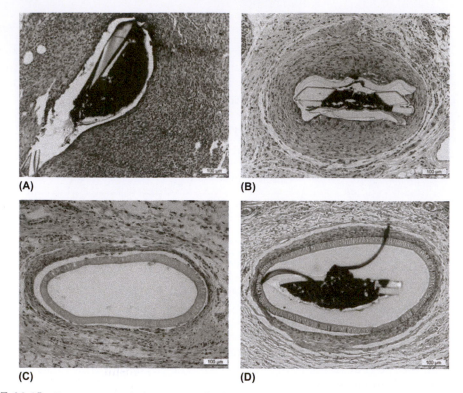

**FIGURE 20.15**　Representative light micrographs from subcutaneous tissue neighboring the glucose oxidase electrode of a glucose sensor, after an implantation period of 10 days in male Sprague-Dawley rats. An optional MPC (2-methacryloyloxyethyl phosphorylcholine) coating or polyamide hollow fiber membrane was used. **A**; Without MPC and without membrane results in a high-grade foreign body reaction. **B**: With MPC results in a moderate, foreign body reaction. **C**: With membrane results in a slight foreign body reaction. **D**; With MPC and with membrane, results in a slight foreign body reaction. *[Reproduced by permission from Mang et al. (2005)].*

actual blood glucose at 24 hr caused the protocol to be discontinued in this patient. During these 48 hr, blood glucose was kept within 80–240 mL/dL for 84% of the total patient time (Renard et al., 2006).

Without accurate and timely glucose and plasma insulin readings, it is difficult to develop and validate a closed-loop algorithm. The algorithms being developed can be described by two general classes. *Model-based algorithms,* such as model predictive control, calculate insulin infusion rates based on a mathematical model describing insulin and glucose dynamics, such as the minimal model of insulin sensitivity. Simpler *proportional-integral-derivative (PID)* algorithms and their subsets, such as PD, calculate insulin infusion rates according to a controller transfer function, $H_c(s)$, that contains a proportional gain, $K_p$, an integral term, $K_i/s$, and a derivative term, $K_d\,s$:

$$H_c(s) = K_p + \frac{K_i}{s} + K_d \cdot s. \tag{20.31}$$

# SYSTEM DESCRIPTION AND DIAGRAM

Medtronic, Dexcom, and Abbott all use a similar architecture for continuous glucose monitoring. This architecture, which is referred to as the *subQ-subQ approach* by the Juvenile Diabetes Research Foundation (JDRF), is the combination of a subcutaneous glucose sensor and subcutaneous insulin pump. JDRF launched its Artificial Pancreas Project in 2006 with $6 million of grant funding, "to accelerate the delivery of technologies that improve glycemic control in diabetes" (Kowalski, 2009).

As a subcutaneous glucose sensor measures ISF glucose, an attached transmitter periodically (every few minutes) sends the glucose current reading to an external insulin pump or a glucose monitor. Before the sensor current reading is transmitted, it is isolated, amplified, and undergoes data acquisition. The pump or monitor receives and displays the readings and other user information, such as a graph of glucose over time and an upward/ downward trend arrow. When the pump or monitor alarms because a preset hypoglycemic or hyperglycemic threshold is reached, the patient measures glucose with a home glucose monitor. Based on the reading, the patient may choose to change the insulin flow rate of the pump or inject more insulin with a syringe. The FDA does not approve these systems for directly making therapy adjustments. Sensor and glucose readings may be downloaded to a personal computer for further analysis. The transmitter, insulin pump, and glucose monitor are all battery operated. If an insulin pump is used, the patient refills it with insulin every few days (Figure 20.16).

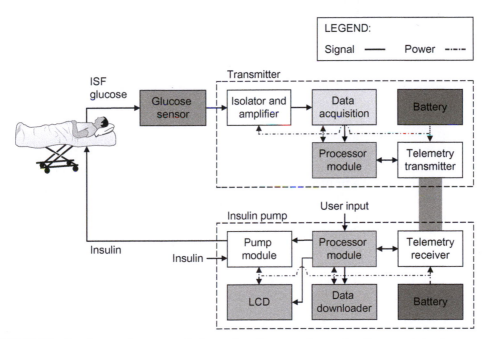

**FIGURE 20.16**    Continuous glucose monitoring system diagram, when an insulin pump is used.

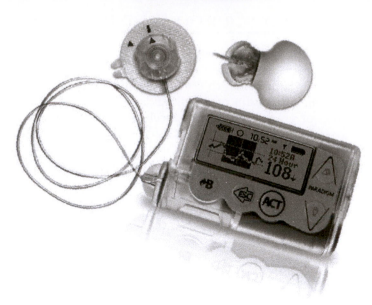

**FIGURE 20.17**    Medtronic Paradigm Real-Time Revel system (Medtronic Diabetes, Northridge, California).

Let us use the Medtronic Paradigm system as the basis of this system discussion because it is the only system in 2010 with an FDA-approved glucose sensor and FDA-cleared insulin pump that communicate with each other, for both adult (CDRH, 2006) and pediatric (CDRH, 2007) patients (Figures 20.1 and 20.17). The Paradigm pump is a bolus infusion pump. The pump mechanism is a piston that pushes against a reservoir, containing a 2- to 3-day supply of insulin, to infuse small boluses of insulin. The reservoir is connected to infusion tubing that terminates in a cannula. The cannula is inserted with an automatic insertion device and remains stationary with the cloth-covered adhesive above it. The pump is carried in a holster that attaches to a belt.

The pump is powered by one AAA alkaline battery. It receives and displays pump data, as well as transmitted glucose sensor data. It also downloads insulin and glucose data to a personal computer, using a universal serial bus link, for preformatted reports. Because it is not waterproof, it should be disconnected from the tubing before showering or swimming.

The glucose sensor is a three-electrode device, enabling the reactions in Eqs. (20.28)–(20.30) to occur. It has a 22-gauge, 10-mm cannula, which is inserted subcutaneously with an automatic insertion device. The sensor insertion site must be at least 3 in. from the pump insertion site. After a 2-hr initialization period, the sensor must be recalibrated at least once every 12 hr. Its lifetime is 72 hr.

A transmitter may be attached to the glucose sensor 10–15 min after subcutaneous insertion. The transmitter sends glucose values to the pump by radiofrequency (~900 MHz) every 5 min. The rechargeable battery in the transmitter lasts approximately fourteen days. Because the transmitter and sensor are waterproof, patients can shower or swim during sensing and transmission.

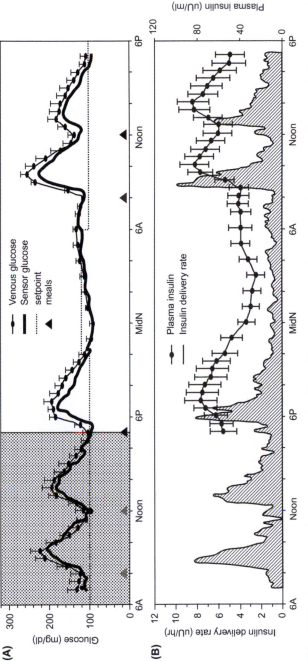

**FIGURE 20.18** Summary of closed-loop experiment in eight subjects undergoing glucose control. *x*-axes are aligned to facilitate comparison of panels. **A**: Sensor glucose levels (thick line) closely track venous blood glucose levels (thin line). The dotted area during the first day indicates period excluded from analysis. Daytime (6–10 a.m.) and nighttime (10 p.m.–6 a.m.) target glucose levels of 100 and 120 mL/dl, respectively, are indicated by the thin dashed line. Meals are marked by triangles. **B**: Insulin pump delivery rates (in units per hour) are indicated by the shaded region and left axis; plasma insulin levels sampled every 30–60 min during the last 24 hr of the closed-loop period are shown sampled line. Note persistent late postprandial hyperinsulinemia despite decreasing insulin delivery rates [*Reproduced by permission from Weinzimer et al. (2008)*].

In a clinical trial of 60 pediatric (ages 7–17 yr) subjects that was submitted for FDA approval, subjects wore the sensor 6–7 days. They were instructed to record reference OneTouch Ultra fingerstick glucose values seven times per day. In 2599 paired measurements, the sensor was within 20% of reference values 68.3% of the time (CDRH, 2006).

With the addition of an experimental PID algorithm to the system, 17 type 1 diabetes (ages 13–20 yr) were able to keep 85% of sensor glucose values within 70–180 mL/dL over 24 hr. Three hypoglycemic episodes occurred in three patients. These hypoglycemic episodes are discussed further in Exercises 20.7 and 20.8 (Weinzimer et al., 2008).

## KEY FEATURES FROM ENGINEERING STANDARDS

In May, 2010, the FDA added *CLSI POCT05-A Performance Metrics for Continuous Interstitial Glucose Monitoring; Approved Guideline* to its list of approved consensus standards. Clinical studies conducted before this addition refer to the accuracy requirement in an older *in vitro* standard, which is no longer acceptable by the FDA for continuous monitoring systems: *ISO 15197 In Vitro Diagnostic Systems—Requirement for in Vitro Whole Blood Glucose Monitoring Systems Intended for Use by Patients for Self-Testing in Management of Diabetes.*

In this section, we discuss requirements from more applicable *in vivo* CLSI POCT05-A. We also discuss some of the initial organizational work a new Glycemia Modeling Working Group is conducting toward closed-loop algorithm current practices. We do not discuss insulin bolus infusion pump requirements because the FDA does not recommend a consensus standard for this medical device.

### Sensor Point Accuracy

Point accuracy is tested in a clinical trial. Sensor measurements are compared to accepted reference measurements, which are taken at most 15 min apart. CLSI POCT05 does not specify an exact number of subjects or actual number of samples, but that both parameters should be determined by a power calculation. The device should be tested randomly during its life cycle. There should be "an adequate number of points in each testing range": <70 mg/dL, 70–180 mg/dL, >180 mg/dL. There should also be "a statistically significant sample" at the extreme values of <70 mg/dL and >400 mg/dL. No specific minimum point accuracy requirement is given (CLSI, 2008).

### Sensor Trend Accuracy

Using the same clinical trial data, trend accuracy is assessed. Accurate measurement of glucose dynamics is critical for real-time diabetes management, including decisions about self-administered insulin dose, closed-loop control, and hypoglycemia and hyperglycemia alarms. Paired samples should be collected for the categories in Table 20.1.

To obtain paired samples in each category, the clinical trial should include type 1 subjects. It may be necessary for competent investigators to actively manipulate their control by giving high–glycemic-index foods, alerting their insulin dose, or administering hypoglycemic agents.

**TABLE 20.1**   Trend accuracy categories

| Initial Glucose (mg/dL) | Rate of change (mg/dL/min) | | | | |
|---|---|---|---|---|---|
|  | $<-1.5$ | $<-1.5$ to $-0.5$ | $-0.5$ to $+0.5$ | $+0.5$ to $+1.5$ | $>+1.5$ |
| <70 | 0 | + | + + | + | + |
| 70–180 | + | + + | + + | + + | + |
| >180 | + | + + | + + | + + | + |

More samples should be collected from categories labeled " + + ," than from categories labeled " + ." To ensure patient safety, extreme negative slopes (<1.5 mg/dL/min) in hypoglycemic subjects do not have to be collected.
*Source: Adapted from CLSI (2008).*

The significance of observed differences can then be determined. A sufficient number of paired samples should be obtained at the extreme rates of change (<1.5 mL/dL/min and >1.5 mg/dL/min) to be able to determine the trend accuracy in each glucose zone at the 95% confidence level. No specific minimum trend accuracy requirement is given (CLSI, 2008).

CLSI POCT05 states that point and trend accuracy may also be determined by clinical agreement, as well as by numerical agreement, as described above. For clinical agreement analysis, sensor glucose estimates, as a function of reference glucose estimates, are divided into zones:

- Zone A = clinically accurate estimates.
- Zone B = benign errors that do not cause inaccurate clinical interpretation or that do not result in negative outcome.
- Zone C = overcorrective errors that may lead to overcorrection of hypoglycemia or hyperglycemia.
- Zone D = failure to detect errors, in which critical hypoglycemia (upper D-zone) or hyperglycemia (lower D-zone) are not detected.
- Zone E = erroneous readings, in which readings are opposite the reference glucose, which may lead to serious treatment mistakes (CLSI, 2008).

The zones were first proposed as the Clarke Error Grid Analysis (EGA), for testing clinical accuracy of self-monitoring devices. The Continuous Glucose-Error Grid Analysis (CG-ERA) builds on this well established tool. Paired data are input into the graphs in Figure 20.19. No specific minimum CG-EGA accuracy requirement is given (CLSI, 2008).

## Sensor Threshold Alarm Accuracy

Patients expect their hypoglycemic and hyperglycemic threshold alarms to be accurate. According to CLSI POCT05, paired clinical samples can be compared for agreement within 15 and at most 30 min. For hypoglycemic alarms, a true alert rate is the time percentage when sensor and reference values were at or below/above (hypoglycemic versus hyperglycemic) the alert level. A missed alert rate is the time percentage when the reference

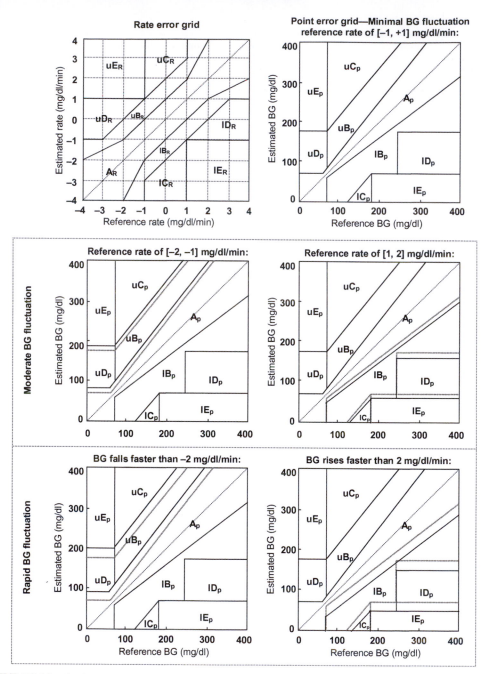

**FIGURE 20.19**  Performance metrics for continuous interstitial glucose monitoring [*From* Diabetes Care, *vol. 27, 2004; 1922–1928. Reprinted with permission from The American Diabetes Association. Figure reproduced with permission from CLSI (2008)*].

**TABLE 20.2** Summary table of percentage of concurrence of continuous glucose monitor values compared to reference values

| | % concurrence of CGM values compared to reference values | | | | | | | | |
| CGM value (mg/dL) | Reference value (mg/dL) | | | | | | | | |
| | <63 | 63–67 | 68–72 | 73–77 | 78–82 | 83–87 | 88–92 | >93 | >180 |
|---|---|---|---|---|---|---|---|---|---|
| <63 | | | | | | | | | |
| 63–67 | | | | | | | | | |
| 68–72 | | | | | | | | | |
| 73–77 | | | | | | | | | |
| 78–82 | | | | | | | | | |
| 83–87 | | | | | | | | | |
| 88–92 | | | | | | | | | |
| >93 | | | | | | | | | |
| >180 | | | | | | | | | |

*Source: Adapted from CLSI (2008).*

glucose value was at or below/above the alert setting, but the sensor value was not. A false alert rate is the time percentage when the sensor value was below/above the alert level, but the reference glucose value was above/below the alert level. A table of alert rates may be constructed, as in Table 20.2. No specific minimum true alert rate requirement is given (CLSI, 2008).

## Sensor Stability

According to CLSI POCT05, sensor stability is hypothesized to result when:

- The implanted sensor is mechanically stable within the subcutaneous tissue.
- The sensor-tissue interface has minimal blood and cell damage following sensor implantation.
- The sensor-tissue interface is not mechanically disturbed (e.g., by activity).
- The immune response at the sensor surface changes slowly.
- Changes in oxygen, pH, temperature, and blood flow minimally affect the chemical/physical mechanism of glucose transduction.
- Biofouling of the sensor's membranes, enzyme, electrolytes, and electrodes occurs slowly.
- The sensor's electronics, mechanics, and power source function in a stable fashion (CLSI, 2008).

An algorithm may determine whether the sensor data are physiologically possible and probable, based on prior glucose excursions and human physiology. It is suggested that the manufacturer consider withholding data deemed inaccurate from the patient and provide notification to the patient (CLSI, 2008).

## Insulin-Glucose Interaction Simulation Model

At the Eighth Annual Diabetes Technology Meeting in November 2008, a new Glycemia Modeling Working Group met to discuss current practices in mathematical modeling and to make recommendations for its use in developing automated insulin-delivery systems. Little progress was made toward the 2008 recommendation of creating a core data set for evaluating new simulation models when the Group met again in 2009 (Steil & Reifman, 2009). It was acknowledged that intellectual property royalties, licensing agreements, and a competitive edge when applying for extramural funding prevented free exchange of information. A new workshop in 2011 is proposed, during which Group members could present and discuss their models, based on public data sets (Steil et al., 2010).

Note that, in other patient monitoring specialties, common data sets are used to assess algorithm performance. For example, the American Heart Association database (discussed in Chapter 2) is used to assess arrhythmia detection performance.

## SUMMARY

According to NIDDK, 7.8% of the U.S. population, or 23.6 million people, had diabetes in 2007. 17.9 million people were diagnosed; 5.7 million were undiagnosed. Improved glycemic control benefits patients with either type 1 or type 2 diabetes. The use of an artificial pancreas would remove a patient from the decision process of insulin administration time and dose. It could also prevent hypoglycemic episodes due to accidental insulin administration overdoses, especially while sleeping.

An ideal artificial pancreas system is fully implantable. The patient is connected to a glucose sensor, which outputs accurate blood or plasma glucose readings. The processing module receives the digitized glucose data. An internal control algorithm in the processing module calculates an appropriate insulin dose, which is fed to the pump module. The pump module administers this dose, completing the closed-loop system.

Current FDA-approved systems are not fully implantable because the glucose sensor lags behind the pump and processor. Microcontrolled implantable pumps already exist and are used to deliver drugs for chemotherapy and pain. Continuous glucose sensors that have been approved by the FDA are short-term subcutaneous amperometric electrodes, based on the glucose oxidase reaction. Not surprisingly, subcutaneous sensor insertion initiates a wound-healing response, which includes the foreign body reaction. The sensor must undergo an initialization period of 2–10 hr, until the sensor signal stabilizes after insertion. This signal begins to drift due to biofouling, passivation of electrodes, and degeneration. Sensor accuracy is also affected by sensor delay, due to filtering of the sensor signal and the natural lag of ISF glucose behind blood glucose during steady state conditions.

Without accurate and timely glucose and plasma insulin readings, it is difficult to develop and validate a closed-loop algorithm. The algorithms being developed are model-based algorithms, such as model predictive control, and simpler proportional-integral-derivative algorithms and their subsets.

Key artificial pancreas features include sensor point accuracy, sensor trend accuracy, sensor threshold alarm accuracy, sensor stability, and the insulin-glucose interaction simulation model.

## Acknowledgment

The author thanks Stuart L. Gallant for his review of this chapter. Mr. Gallant is Vice President, Product Development at Glumetrics, Inc.

## Exercises

**20.1** The Pima Indians in Arizona (Figure 20.20) have the highest reported prevalence and incidence rate of type 2 diabetes in the world. NIDDK has a clinical research branch in Phoenix that studies the Pimas. Watch the short film from the *New York Times*: http://video.nytimes.com/video/2008/07/30/us/1194817478153/water-returns-to-the-pima.html. What is the Thrifty Genotype hypothesis? Why is it important to restore the Gila River to the Pimas? What is the implication of the Arizona Pima experience for the American public?

**FIGURE 20.20** Three Pima Indian farmers resting from work on irrigation ditches, ca. 1900 *(Digitally reproduced by the USC Digital Archive © 2004, California Historical Society: TICOR/Pierce, CHS-3608. Figure reproduced by permission from the University of Southern California, Los Angeles).*

**20.2** Knowing only that the implantable insulin pump in clinical trials was manufactured by Medtronic, predict the following: material used for the sealed case, type of battery, battery lifetime, methodology for communicating with the pump to change its flow rate, catheter material. What biocompatibility complications may occur with the catheter?

**20.3** Using the transfer function matrix approach, show that the **p** vector is uniquely identifiable for the minimal model of insulin sensitivity. Write the equations for the plasma ISF glucose equilibrium model, omitting the glucose sensor current. Using the transfer function matrix approach, determine which parameters are identifiable.

**20.4** What accuracy or frequency response requirements were discussed for ECG, blood pressure, temperature, and arterial saturation of oxygen? Are the glucose sensor point and trend accuracy requirements more or less stringent? What could be reasons for this difference in stringency?

**20.5** One of the themes of the medical device chapters (Chapters 2–20) is that biocompatibility affects implantable devices. We have seen how biocompatibility affects pacemaker leads, heart valves, intraocular lens implants, hip and knee prostheses, bare metal stents, and subcutaneous glucose sensors. Name two other device chapter themes and at least three devices affected by each theme. Be specific.

Read Albisser et al. (1974a):

**20.6** Describe how an insulin infusion rate was calculated, based on the current and desired glucose values. Based on the results summarized in Figures 4 and 5, what are the limitations of this control algorithm?

Read Weinzimer et al. (2008):

**20.7** Did the sensor glucose track the venous glucose? What was the point accuracy? What was the threshold value for hypoglycemia? What were the sensor-vs.-venous glucose values when hypoglycemia occurred? Did sensor glucose accuracy contribute to these hypoglycemic events?

**20.8** What is the time delay between peak ISF insulin administration and peak plasma insulin level? What causes this delay? During what period of the day did the hypoglycemic events occur? Did the time delay contribute to these hypoglycemic events? How did the algorithm attempt to prevent hypoglycemic events during this period of the day?

# References

Albisser, A. M., Leibel, B. S., Ewart, T. G., Davidovac, Z., Botz, C. K., & Zingg, W. (1974a). An artificial endocrine pancreas. *Diabetes, 23*, 389–396.

Albisser, A. M., Leibel, B. S., Ewart, T. G., Davidovac, Z., Botz, C. K., & Zingg, W., et al. (1974b). Clinical control of diabetes by the artificial pancreas. *Diabetes, 23*, 397–404.

Baura, G. D. (2002). *System theory and practical applications of biomedical signals.* Hoboken, NJ: Wiley-IEEE.

Baura, G. D., Foster, D. M., Porte, D., Jr., Kahn, S. E., Bergman, R. N., & Cobelli, C., et al. (1993). Saturable transport of insulin from plasma into the central nervous system of dogs in vivo. A mechanism for regulated insulin delivery to the brain. *Journal of Clinical Investigation, 92*, 1824–1830.

Bergman, R. N., & Bucolo, R. J. (1974). Interaction of insulin and glucose in the control of hepatic glucose balance. *American Journal of Physiology, 227*, 1314–1322.

Bergman, R. N., Ider, Y. Z., Bowden, C. R., & Cobelli, C. (1979). Quantitative estimation of insulin sensitivity. *American Journal of Physiology, 236,* E667–E677.

Boyne, M. S., Silver, D. M., Kaplan, J., & Saudek, C. D. (2003). Timing of changes in interstitial and venous blood glucose measured with a continuous subcutaneous glucose sensor. *Diabetes, 52,* 2790–2794.

Caplin, N. J., O'Leary, P., Bulsara, M., Davis, E. A., & Jones, T. W. (2003). Subcutaneous glucose sensor values closely parallel blood glucose during insulin-induced hypoglycaemia. *Diabetic Medicine, 20,* 238–241.

CDRH (2001). *Approval order P990026.* Rockville, MD: FDA.

CDRH (2006). *PMA supplement P980022/S013.* Rockville, MD: FDA.

CDRH (2007). *PMA supplement P980022/S015.* Rockville, MD: FDA.

Cengiz, E., & Tamborlane, W. V. (2009). A tale of two compartments: interstitial versus blood glucose monitoring. *Diabetes Technology and Therapeutics, 11*(Suppl 1), S11–S16.

Clark, L. C., Jr., Spokane, R. B., Homan, M. M., Sudan, R., & Miller, M. (1988). Long-term stability of electroenzymatic glucose sensors implanted in mice. An update. *ASAIO Transactions, 34,* 259–265.

Clemens, A. H. (1979). Feedback control dynamics for glucose controlled insulin infusion system. *Medical Progress Technology, 6,* 91–98.

CLSI (2008). *POCT05-A performance metrics for continuous interstitial glucose monitoring: Approved guideline.* Wayne, PA: CLSI.

Eisenstein, A. B. (1964). *The Biochemical Aspects of Hormone Action.* Boston: Little, Brown.

Gerritsen, M., Jansen, J. A., Kros, A., Vreizema, D. M., Sommerdijk, N. A., & Nolte, R. J., et al. (2001). Influence of inflammatory cells and serum on the performance of implantable glucose sensors. *Journal of Biomedical Materials Research, 54,* 69–75.

Guyton, A. C., & Hall, J. E. (2006). *Textbook of Medical Physiology* (11th Ed.). Philadelphia: Elsevier Saunders.

Hovorka, R., Allen, J. M., Eelleri, D., Chassin, L. J., Harris, J., & Xing, D., et al. (2010). Manual closed-loop insulin delivery in children and adolescents with type 1 diabetes: A phase 2 randomised crossover trial. *Lancet, 375,* 743–751.

Kahn, S. E., Prigeon, R. L., McCulloch, D. K., Boyko, E. J., Bergman, R. N., & Schwartz, M. W., et al. (1993). Quantification of the relationship between insulin sensitivity and beta-cell function in human subjects. Evidence for a hyperbolic function. *Diabetes, 42,* 1663–1672.

Koschinsky, T., & Heinemann, L. (2001). Sensors for glucose monitoring: Technical and clinical aspects. *Diabetes and Metabolism Research Review, 17,* 113–123.

Kowalski, A. J. (2009). Can we really close the loop and how soon? Accelerating the availability of an artificial pancreas: a roadmap to better diabetes outcomes. *Diabetes Technology and Therapeutics, 11*(Suppl 1), S113–S119.

Kyrolainen, M., Rigsby, P., Eddy, S., & Vadgama, P. (1995). Bio-/haemocompatibility: Implications and outcomes for sensors? *Acta Anaesthesiologica Scandinavica. Supplementum, 104,* 55–60.

Logtenberg, S. J., Kleefstra, N., Houweling, S. T., Groenier, K. H., Gans, R. O., & Van Ballegooie, E., et al. (2009). Improved glycemic control with intraperitoneal versus subcutaneous insulin in type 1 diabetes: A randomized controlled trial. *Diabetes Care, 32,* 1372–1377.

Mang, A., Pill, J., Gretz, N., Kranzlin, B., Buck, H., & Schoemaker, M., et al. (2005). Biocompatibility of an electrochemical sensor for continuous glucose monitoring in subcutaneous tissue. *Diabetes Technology and Therapeutics, 7,* 163–173.

McGarraugh, G. (2009). The chemistry of commercial continuous glucose monitors. *Diabetes Technology and Therapeutics, 11*(Suppl 1), S17–S24.

Nathan, D. M., Cleary, P. A., Backlund, J. Y., Genuth, S. M., Lachin, J. M., & Orchard, T. J., et al. (2005). Intensive diabetes treatment and cardiovascular disease in patients with type 1 diabetes. *New England Journal of Medicine, 353,* 2643–2653.

NIDDK (2008). *National diabetes statistics, 2007: NIH publication no. 08-3892.* Bethesda, MD: NIH.

Porte, D., Jr., Baskin, D. G., & Schwartz, M. W. (2005). Insulin signaling in the central nervous system: A critical role in metabolic homeostasis and disease from *C. elegans* to humans. *Diabetes, 54,* 1264–1276.

Ray, K. K., Seshasai, S. R., Wijesuriya, S., Sivakumaran, R., Nethercott, S., & Preiss, D., et al. (2009). Effect of intensive control of glucose on cardiovascular outcomes and death in patients with diabetes mellitus: A meta-analysis of randomised controlled trials. *Lancet, 373,* 1765–1772.

Renard, E., Costalat, G., Chevassus, H., & Bringer, J. (2006). Closed loop insulin delivery using implanted insulin pumps and sensors in type 1 diabetic patients. *Diabetes Research and Clinical Practice, 74,* S173–S177.

Skyler, J. S. (2009). Continuous glucose monitoring: An overview of its development. *Diabetes Technology and Therapeutics, 11*(Suppl 1), S5–S10.

Steil, G. M., Hipszer, B., & Reifman, J. (2010). Update on mathematical modeling research to support the development of automated insulin delivery systems. *Journal of Diabetes Science and Technology, 4,* 759–769.

Steil, G. M., Rebrin, K., Hariri, F., Jinagonda, S., Tadros, S., & Darwin, C., et al. (2005). Interstitial fluid glucose dynamics during insulin-induced hypoglycaemia. *Diabetologia, 48,* 1833–1840.

Steil, G. M., & Reifman, J. (2009). Mathematical modeling research to support the development of automated insulin-delivery systems. *Journal of Diabetes Science and Technology, 3,* 388–395.

Sternberg, F., Meyerhoff, C., Mennel, F. J., Mayer, H., Bischof, F., & Pfeiffer, E. F. (1996). Does fall in tissue glucose precede fall in blood glucose? *Diabetologia, 39,* 609–712.

Voskanyan, G., Barry Keenan, D., Mastrototaro, J. J., & Steil, G. M. (2007). Putative delays in interstitial fluid (ISF) glucose kinetics can be attributed to the glucose sensing systems used to measure them rather than the delay in ISF glucose itself. *Journal of Diabetes Science and Technology, 1,* 639–644.

Weinzimer, S. A., Steil, G. M., Swan, K. L., Dziura, J., Kurtz, N., & Tamborlane, W. V. (2008). Fully automated closed-loop insulin delivery versus semiautomated hybrid control in pediatric patients with type 1 diabetes using an artificial pancreas. *Diabetes Care, 31,* 934–939.

# LAB EXPERIMENTS

In Part II, we present lab experiments that are intended to supplement course lectures. As a undergraduate student intern, the author became interested in medical devices when she calibrated hospital devices for three years. In this spirit, we provide these lab experiments in Part II as a way for students to *play* with medical devices. From the instructor's standpoint, these lab experiments provide problem-based learning.

Because the author firmly believes that students should personally experience real medical devices, many of these lab experiments are based on market-released devices. Specifically, the pacemaker programming, echocardiography, patient monitoring artifact, and thermometry accuracy labs use market-released devices. But in consideration of cost, the thermometry accuracy, electrocardiograph design, electrocardiograph filtering, surface characterization, and entrepreneurship labs use lower-cost parts. This second group of labs assumes that standard laboratory instruments are available for student use. In some of the labs, software, such as executable LabVIEW subroutines, must be downloaded by the instructor before the lab occurs.

The labs are designed for groups of three students. Often, one student acts as the patient (which requires access to the thorax); so at least one male student should be included in each group. With the exception of the entrepreneurship lab, it is expected that lab reports summarizing experimental results contain the following sections: Introduction, Materials and Methods, Results, Discussion, Conclusion.

CHAPTER

# 21

# Electrocardiograph Design Lab

In this chapter, we describe the experimental steps for building an electrocardiograph, including consideration of the Nyquist sampling theorem. This lab is intended to supplement the electrocardiograph lecture based on Chapter 2 and the patient and operator safety lecture based on Chapter 1. It requires no specialized piece of equipment, other than a National Instruments data acquisition board. It is helpful to use a Fluke 217A patient simulator for testing the circuit, before the circuit is connected to a student.

Upon completion of this chapter, each student shall be able to:

1. Understand the foundation of data acquisition.
2. Monitor and calculate heart rate.
3. Practice patient isolation.

## STRATEGIC PLANNING

During this experiment, you will be building an electronic circuit that will be used to measure the electrocardiogram (ECG) of a fellow student. This ECG will be filtered with frequency-selective and wavelet filters in Chapter 22. In preparation for this lab, please review the circuit in Figure 21.1 carefully. To minimize the probability of short circuits, as you build the circuit, spread the components out along the breadboard.

This lab should take approximately 2 hr to conduct.

## MATERIALS AND METHODS

This experiment requires the following electrical components:

- 2 9-V batteries with battery strap
- 1 100-$\Omega$ resistor
- 1 680-k$\Omega$ resistor
- 1 4.7-k$\Omega$ resistor

*Medical Device Technologies*                    455

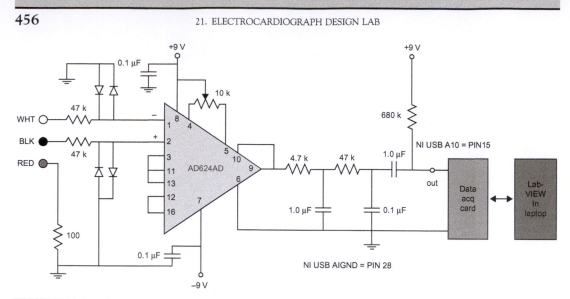

**FIGURE 21.1**    Electrocardiograph Design Lab test setup. This circuit was modified from Pico Technology's circuit *(Pico Technologies, 2010).*

- 3 47-kΩ resistors
- 1 10-kΩ potentiometer
- 3 0.1-μF capacitors
- 2 1.0-μF capacitors
- 4 diodes
- 1 AD624AD instrumentation amplifier

  This experiment also requires these parts:

- 1 breadboard
- 1 black ECG lead wire, with distal connector removed so lead wire can be inserted into breadboard
- 1 red lead wire, with distal connector removed
- 1 white lead wire, with distal connector removed
- 3 ECG electrodes
- 1 laptop, running on battery
- 2 LabVIEW executable vis for data acquisition and analysis (downloaded by your instructor)
- 1 National Instruments NI USB-6210 data acquisition card, with cable
- 1 screwdriver
- 1 Fluke 217A patient simulator, optional
- jumper wire, as needed

At its core, a cardiograph consists of circuitry that monitors cardiac biopotentials and then displays them on a strip chart (Figure 21.1). We assemble the monitoring circuitry onto a breadboard and then monitor the heartbeats using a virtual strip chart.

## Procedure

First, identify your parts. Remove the batteries from their straps, and connect the complete circuit in Figure 21.1. When you are finished, have your instructor check the circuit by observation. If a Fluke 217A patient simulator is available, connect the simulator to the circuit and batteries, and verify that an ECG is observed.

The analog output voltage we obtain from the ECG circuit is digitized and displayed, using National Instruments' data acquisition card, USB-6210, and its LabVIEW software. The card conducts *antialiasing* analog filtering and analog-to-digital conversion. The LabVIEW executable, *CardiographLab.vi*, that is loaded on the laptop displays the digitized data in 1.5 s segments.

Connect the BLK lead wire and ECG electrode to left arm (LA), WHT lead wire and electrode to right arm (RA), and RED lead wire and electrode to left leg (LL) of your test subject. Remember that we are assuming the body is purely resistive; so position all electrodes on the edge of the thorax. Enter the desired sampling rate (default = 1000 Hz), and run *CardiographLab.vi* by clicking the white arrow in the upper left-hand corner. You may have to adjust the potentiometer with a screwdriver. Each time this vi begins to run, it asks for the file name you wish to save. Stop *CardiographLab.vi* with the STOP button. Rerun *CardiographLab.vi* several times at lower sampling rates. You can review your saved files by running *Read_Cardiograph_DataLab.vi*. Use the graph palette beneath the waveform graph to zoom in on data. You can also download a file you have read into this vi as an ascii file, for analysis in a program such as Matlab.

What is the minimum sampling rate at which you observe good resolution cardiac beats? In your lab report, plot 3 continuous beats of a data file sampled at 1000 Hz and a data file sampled at this minimum sampling rate. Within *Read_Cardiograph_DataLab.vi*, you can copy/paste the waveform graph to MS Word. Just remember to remove the Autoscale function from the *x*-axis, if you want to zoom/cut/paste 3 beats.

- What is the mean heart rate, based on the 3 beats, for each data file?
- What ECG lead is being measured?

## RESULTS AND ANALYSIS

Answers to the questions in the preceding written procedure should be recorded in the Results section of your lab report.

## DISCUSSION

Answer the following questions in the Discussion section of your lab report:

- What was your process for determining the minimum sampling rate at which you observe good resolution cardiac beats? Include intermediate calculations in your description.
- How was patient safety practiced during this lab experiment?

## References

Pico Technologies (2010). *Electrocardiogram (ECG) project for DrDaq*. Cambridgeshire, UK: Pico Technologies.

# Electrocardiograph Filtering Lab

In Chapter 21, we built a simple cardiograph and determined the minimum sampling rate required for acquiring electrocardiograms (ECGs). In this chapter, we use the same cardiograph to acquire ECG data, separate the data into ECG and respiration waveforms, and process these waveforms using frequency-selective and wavelet filters.

This lab is intended to supplement the electrocardiograph lecture based on Chapter 2 and respiration monitoring lecture based on Chapter 11. It requires the equipment from the Electrocardiograph Design Lab in Chapter 21 and National Instruments' LabVIEW software. LabVIEW is standard software for many industries, such as the medical device industry. If your school does not have a LabVIEW license, you can download LabVIEW for a free 30-day trial (http://www.ni.com/trylabview/).

Because frequency-selective filters, such as Butterworth filters, are not specifically discussed in this textbook, students are advised to read introductory material about frequency-selective filters, such as Chapter 7 in (Oppenheim & Schafer, 1989). If students are not familiar with LabVIEW programming, they are advised to review National Instruments' *LabVIEW Fundamentals Manual* (http://www.ni.com/pdf/manuals/374029a.pdf).

Upon completion of this chapter, each student shall be able to:

1. Understand how ECG and respiration waveforms are acquired.
2. Separate ECG and respiration waveforms with Butterworth filters.
3. Detect QRS complexes with wavelets in the midst of noise.

## STRATEGIC PLANNING

During this experiment, you will be acquiring the electrocardiogram of a fellow student and filtering it with frequency-selective and wavelet filters. In preparation for this lab, review the frequency-selective filter chapter and/or *LabVIEW Fundamentals Manual* given by your instructor. You need to be familiar with LabVIEW because the LabVIEW code you write during the lab will be turned in at the end of the lab session.

Before you acquire the data file you use for analysis, have your test subject practice small random movements and large random movements. For small random movements, such as head or arm movement, there should be some movement of the baseline signal, but the ECG should still be prominent. For large random movements, such as jumping, the ECG should be somewhat discernable in the midst of large noise.

This lab should take approximately 3 hr to conduct.

## INTRODUCTION TO WAVELET FILTERS

By definition, a wavelet, $\psi(t)$, is a function with a mean of zero, such that

$$\int \psi(t) = 0. \tag{22.1}$$

We can use a wavelet transform, which is based on a chosen wavelet, for a variety of tasks. Wavelet transforms are based on different powers and may be continuous or discrete. The dyadic discrete wavelet transform, $WT(k, 2^j)$, which is based on powers of 2, is widely used. All wavelet transforms are functions of time and scale, $2^j$. As $2^j$ decreases, wavelets are compressed and analyze high frequencies. As $2^j$ increases, wavelets are stretched, and analyze low frequencies.

The dyadic discrete wavelet transform is calculated as

$$WT(k, 2^j) = \frac{1}{\sqrt{2}} \sum_{i=0}^{N-1} u(i) \psi^* \left( \frac{i-k}{2^j} \right), \tag{22.2}$$

where $u(k)$ is the input signal. This calculation is essentially filtering through convolution and downsampling. It is often used for image compression. We use the wavelet transform to isolate and detect low- or high-frequency features.

A wavelet is associated with wavelet and scaling filters. When a scaling filter is used, convolution results in smoothing or lowpass filtering. The resulting wavelet transform is composed of approximation coefficients. When a wavelet filter is used, convolution results in emphasis of discontinuities or highpass filtering. The resulting wavelet transform is composed of detail coefficients. Several common scaling and wavelet filters are shown in Figure 22.1.

The effects of Haar scaling and wavelet filters are shown in Figure 22.2. The original ECG was acquired with the laboratory setup of Chapter 21. The Haar wavelet is illustrated because it is excellent at detecting discontinuities. As the scale is increased, the Haar approximation coefficients overemphasize the T wave in the ECG. As the scale is increased, the Haar detail coefficients overemphasize the R wave. Even though the scale technically is $2^j$, $j$ is commonly referred to as the scale for dyadic wavelet transforms. If noise were added to this ECG, much of the noise would be reduced through the wavelet transforms.

For a more thorough discussion of wavelets, see Mallat (1998) and Baura (2002).

Wavelet | Scaling filter | Wavelet filter

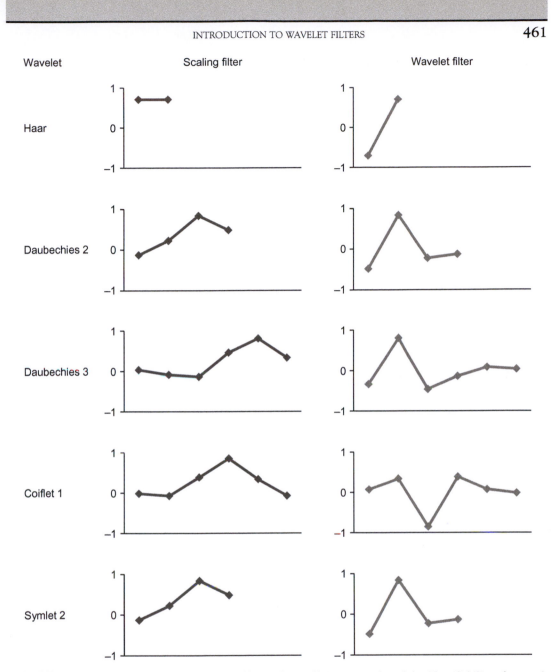

**FIGURE 22.1** Common scaling and wavelet filters. The coefficients are plotted for identifiability of numeric values, rather than in the classic wavelet style.

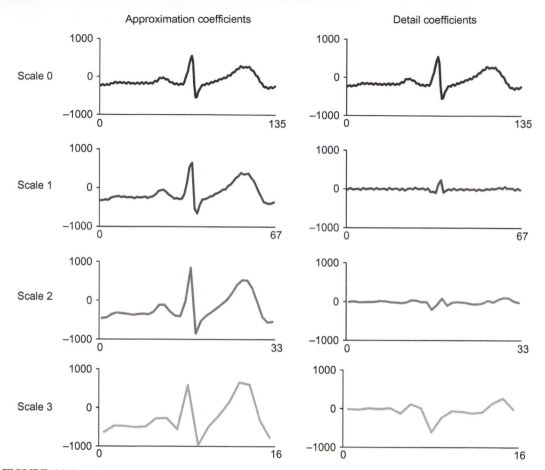

**FIGURE 22.2**  Effects of Haar scaling and wavelet filters for an ECG at 3 scales. The original sampling frequency is 1000 Hz.

## MATERIALS AND METHODS

This experiment requires the following equipment:

- 1 working cardiograph, with all hardware and software from the Electrocardiograph Design Lab
- 1 licensed or evaluation LabVIEW program, loaded onto the same laptop used for data acquisition
- 3 ECG electrodes
- 1 LabVIEW.llb library for analysis (downloaded by your instructor)

## Procedure

Your analysis will be based on an ECG file that you acquire today. Connect your test subject to the ECG leads. Acquire one 90-s file using the executable, *CardiographLab.vi*, with the default 1000-Hz sampling rate. During these 90 s, your subject should be quiet and sitting for the first 30 s, incorporate random small movements (i.e., moving head or arms) for the next 15 s, quiet and sitting for 15 s, incorporate random large movements (i.e., jumping) for the next 15 s, and quiet and sitting for 15 s. Record the random movements your subject conducted.

Open *BlockProcessing.vi* within the *ECGDetection.llb* library. Observe Sequences 0, 1, and 2 within the block diagram. What downsampling occurs? Why is downsampling implemented?

Observe the entire front panel of *BlockProcessing.vi*, noting available tools for analysis. Use *BlockProcessing.vi* to determine the best lowpass cutoff frequency to isolate respiration and the best highpass cutoff frequency to isolate ECG. Each implemented Butterworth filter is a tenth-order filter. Describe your methodology for selecting these cutoff frequencies. Merge these results, and select a single frequency for both highpass and lowpass filtering.

Open *PacketProcessing.vi* in the *ECGDetection.llb* library. When this vi runs, data is input in packets of 2560 samples, after a 15-s so-called learning period in Sequence 0. This vi could be used after an analog-to-digital converter to detect QRS complexes in real time. Observe the importance of shift registers in Sequence 1. Note that the output of this main vi is the most recent packet of ECG and a corresponding detection array. Currently this detection array is filled with 0 values.

Input your merged cutoff frequency on the front panel. As you run *PacketProcessing.vi*, you will see that the Haar detail coefficients at scales 1, 2, and 3 are calculated for the current packet. Find a wavelet transform parameter during the learning period that you can use as the basis of QRS complex detection in Sequence 2. Although your stated goal is to detect QRS complexes, your real goal is to detect R waves, because this high-frequency feature is the easiest part of a QRS complex to detect. You may choose your parameter from scale 1, 2, or 3. For this lab, you are *not allowed* to choose an absolute threshold for detection. Absolute threshold detection (i.e., *ThresholdPeakDetector.vi*) may not generalize well for a wide range of cardiac patients. Program QRS complex detection in Sequence 1, making sure to change the detection array output so that detected QRS complexes can be seen in the main graph.

## RESULTS AND ANALYSIS

Record the answers to the questions in the written procedure in the Results section of your lab report.

For your data file, determine the sensitivity and specificity of your detection algorithm. Count a detected QRS complex as a true positive (TP) when it is within ±2 downsampled samples of a true R wave. Count a detected QRS complex as a false positive (FP) when no

R wave is present within 2 downsampled samples. Assume true negative (TN) is the (number of true R waves $+1-$ FP). Count each undetected true R wave as a false negative (FN). Sensitivity and specificity are then calculated as:

$$\text{sensitivity} = TP/(TP + FN) \tag{22.1}$$

$$\text{specificity} = TN/(TN + FP) \tag{22.2}$$

## DISCUSSION

Answer the following questions in the Discussion section of your lab report:

- Why was a respiratory waveform acquired when you attempted to acquire only an electrocardiogram?
- Why does Butterworth highpass filtering not remove most of the random movement in the ECG?
- Is your detection algorithm affected by random movement?

Turn in a copy of your data file and LabVIEW.llb to your instructor at the end of the lab session. Your lab report will be turned in separately later.

### References

Baura, G. D. (2002). *System theory and practical applications of biomedical signals*. Hoboken, NJ: Wiley-IEEE Press.
Mallat, S. (1998). *A wavelet tour of signal processing*. San Diego, CA: Elsevier Academic Press.
Oppenheim, A. V., & Schafer, R. W. (1989). *Discrete-time signal processing*. Englewood Cliffs, NJ: Prentice Hall.

# Pacemaker Programming Lab

In this chapter, we describe the steps for a pacemaker programming laboratory experiment. This lab is intended to supplement the pacemaker lecture based on Chapter 3. It is written for execution using St. Jude Medical equipment but can be customized for pacemakers from other manufacturers.

Upon completion of this chapter, each student shall be able to:

1. Conduct right atrial and right ventricular capture tests.
2. Identify various arrhythmias.
3. Program a pacemaker appropriately to treat an arrhythmia.

## STRATEGIC PLANNING

During this experiment you will be working with three simulated patients. In preparation for this lab, please review your lecture notes regarding various arrhythmias. You will be expected to identify and treat three arrhythmias with pacemaker therapy. Note that all your printouts are a required part of your lab report.

This lab should take approximately 2 hr to conduct.

## MATERIALS AND METHODS

This experiment requires the following equipment:

- 1 St. Jude Medical Model 3510 programmer, with sufficient strip chart paper
- 1 St. Jude Medical telemetry wand
- 1 St. Jude Medical Frontier II pulse generator, or similar pulse generator, with at least RA, RV pacing capability, that communicates with St. Jude Medical Model 3510 programmer
- 1 St. Jude Medical Model 3510 programmer manual
- 1 Rivertek RSIM-1500 cardiac simulator

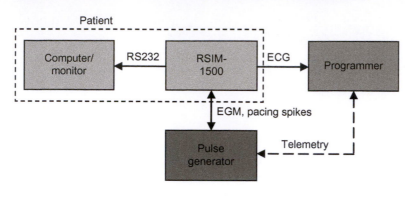

**FIGURE 23.1** Pacemaker
Programming Lab test setup.

- 1 Rivertek PG interface cable
- 1 Rivertek programmer interface cable
- 1 RS232 null modem cable
- 1 desktop or laptop computer, running Rivertek software
- 3 simulated patients for the Rivertek software (downloaded by your instructor)
- 1 computer monitor (if desktop computer)
- St. Jude Medical pacemaker screwdriver

The equipment you are using is standard in the medical device industry. As discussed in Chapter 3, a pacemaker system consists of three components: the programmer, pulse generator, and pacing leads. You are using the St. Jude Medical Model 3510 programmer, which communicates with a variety of pulse generators using a telemetry wand. One such pulse generator is the Frontier II pulse generator. In this experiment, we replace the pacing leads with a Rivertek PG interface cable, which interfaces to the Rivertek RSIM-1500 cardiac simulator.

Each of the simulated patients consists of the Rivertek simulator and its null modem cable-connected desktop or laptop computer, with a computer monitor if necessary. The simulator requires a host computer to execute its software and provide display capability. Through the Rivertek PG interface cable, electrogram (EGM) data are transmitted from the patient to the pulse generator, and pacing spikes are transmitted from the pulse generator to the patient. To replace a surface electrocardiogram (ECG), the Rivertek simulator outputs an ECG through its programmer interface cable to the Model 3510 programmer. The test setup is given in Figure 23.1.

For this experiment, we treat three patients. For Patient 1, we identify the arrhythmia, conduct right atrium (RA) and right ventricle (RV) capture threshold tests, and perform pacemaker therapy. For Patients 2 and 3, we assume RA and RV capture thresholds are 5.0 V, identify the arrhythmias, and perform pacemaker therapy.

## Patient 1

Your instructor has preloaded the first patient waveform. Power on the programmer. Immediately position the telemetry wand over the pulse generator. Using the touch screen, select INTERROGATE. Wait for interrogation to occur. Select PRINT to print your initial programmer settings. Never tear off the sheets; just let them accumulate till the end of the experiment. On pp. 3-3–3-5 of the manual provided, read how to interpret the programmer waveform display.

Observe and identify the arrhythmia being created by RSIM. Analyzing the arrhythmia with a paper copy is easier than observing the arrhythmia on the computer screen. Print the arrhythmia with these directions:

1. Select FREEZE.
2. Select PRINT.
3. Select ADVANCE PAPER.
4. Select CLOSE.

Before we can program the pulse generator, we need to run capture tests to determine the minimal RA and RV stimulation thresholds. During each test, you will observe the programmer, running in temporary mode, step down in stimulation voltage, from 4 V to 0.25 V. For the atrial test, you will observe the P wave. Because the programmer screen is small, observation may be easier on the computer monitor. For the ventricular test, you will observe the QRS complex. When the quality of the P-wave or QRS complex changes, you have seen the capture limit.

Here are the specific directions for the capture tests:

5. Connect the RA and RV RSIM leads to the pulse generator using the screwdriver. Keep the telemetry wand over the pulse generator. Why is this necessary?
6. Select TESTS.
7. Select ATRIAL CAPTURE.
8. Select START TEST.
9. When you see the quality of the P wave change, select END TEST. The programmer will summarize results. You can always rerun the test by selecting CLOSE and using steps 8–9.
10. Select PRINT. After printing, select RECORD RESULTS. Select CLOSE.
11. Select RIGHT VENT. CAPTURE.
12. Select START TEST.
13. When you see the quality of the QRS complex change, select END TEST. The programmer will summarize results. You can always rerun the test by selecting CLOSE and using steps 12–13.
14. Select PRINT. After printing, select RECORD RESULTS. Select CLOSE.

For extra safety margins, we will program the pulse amplitudes to $2\times$ the detected thresholds:

15. Select PROGRAMMED PARAMETERS.
16. Select A. PULSE AMPLITUDE.
17. Select the value of ($2\times$ detected RA threshold).
18. Select RV PULSE AMPLITUDE.
19. Select the value of ($2\times$ detected RV threshold).
20. Note that the new parameter values are identified in blue, with "#." Select PERMANENT PROGRAM. After a few seconds and an audible tone, these new parameters are programmed. The parameters are back in black text, with a check mark next to them.

Observe the patient ECG. Is the patient better? Was RA only, RV only, or both RA and RV pacing required for therapy? Select FREEZE. Select PRINT. Select CLOSE.

Permanently program the RV pulse amplitude to the difference value of (RV threshold − 0.25). What happens to the patient? Select FREEZE. Select PRINT. Select CLOSE.

Permanently program the RV pulse amplitude to the value of (2 × detected RV threshold). Disconnect the RA and RV RSIM leads to the pulse generator.

## Patient 2

Advance the simulator to Patient 2 by the following steps:

21. On the simulator display, hit SELECT.
22. Select the file named *patient2.scn*.
23. Select OK.
24. Select START.

Now that the patient ECG has been downloaded, follow these programmer steps:

25. Select A. PULSE AMPLITUDE.
26. Select the value of 5.0 V.
27. Select RV PULSE AMPLITUDE.
28. Select the value of 5.0 V.
29. Select PERMANENT PROGRAM.
30. Observe and identify the arrhythmia being created by RSIM using steps 1−4.
31. Connect the RA and RV RSIM leads to the pulse generator. What pacing takes place?
32. Select FREEZE. Select PRINT. Select CLOSE.
33. Disconnect the RA and RV RSIM leads to the pulse generator.

## Patient 3

Advance the simulator to Patient 3 by the following steps:

34. On the simulator display, hit SELECT.
35. Select the file named *patient3.scn*.
36. Select OK.
37. Select START.

Now that the patient ECG has been downloaded, follow these programmer steps:

38. Observe and identify the arrhythmia using steps 1−4.
39. Apply therapy using steps 31−33. What pacing takes place?

It is always good practice to reprogram the initial settings when warranted. Looking at your first printout, reprogram the initial RA and RV pulse amplitudes.

## RESULTS AND ANALYSIS

If you correctly conducted the capture tests, connecting the simulated leads to the pulse generator should improve the conditions of your patients. Record the answers to the questions in the written procedure in the Results section of your lab report.

## DISCUSSION

Answer the following questions in the Discussion section of your lab report:

- What do the acts of observing your patient without and then with the leads attached represent in the electrophysiology clinic?
- What data did you acquire from the telemetry wand? Are these data alone enough information for you to diagnose your patients? Provide justification.

# Echocardiography Lab

In this chapter, we describe the steps for an echocardiography laboratory experiment. This lab is intended to supplement the heart valve lecture based on Chapter 6. Because ultrasound is not discussed in detail in this textbook, students are advised to read introductory material about ultrasound, such as the ultrasound chapter in Webb (2003). This lab is written for execution using Sonosite equipment, but it can be customized for ultrasound monitors from other manufacturers.

Upon completion of this chapter, each student shall be able to:

1. Identify the left ventricular outflow tract (LVOT) in the left parasternal long axis view, and the left ventricle in the apical two chamber view.
2. Understand the anatomy associated with these views.
3. Measure LVOT diameter, velocity time integral, and left ventricular volume; calculate stroke volume and ejection fraction from these measurements.

## STRATEGIC PLANNING

During this experiment, you will be measuring hemodynamic parameters from one of your fellow students. In preparation for this lab, review the ultrasound chapter given by your instructor. Echocardiography is usually performed by skilled operators called sonographers, who have studied echocardiography for many years. Be patient because it takes time to obtain the views you are seeking. For the parasternal view, find the mitral valve leaflets first, and then make slight adjustments to the transducer to image the other landmarks. For the apical view, try to position the transducer toward the subject's right shoulder. Note that all your printouts are a required part of your lab report.

This lab should take approximately 2.5 hr to conduct.

## MATERIALS AND METHODS

This experiment requires the following equipment:

- 1 Sonosite MicroMaxx ultrasound system with P17 transducer
- 1 Sonosite ECG trunk cable, with leadwires
- 1 compact flash
- 1 Sonosite MicroMaxx manual
- 1 compatible Sony printer
- 3 ECG electrodes
- 1 Aquasonic ultrasound transmission gel
- 1 cot or camping pad
- Alcohol wipes and paper towels, as needed

### Background

This experiment uses two echo views to calculate left ventricular (LV) stroke volume (SV) and ejection fraction (EF). LV SV is the volume of blood ejected from the left ventricle with each cardiac beat. LV EF is the percentage of blood ejected from the left ventricle with each cardiac beat. The transducer positions required for parasternal and apical views are shown in Figure 24.1.

Your patient lies in the left lateral decubitus position: lying on left side, head propped up by elbow, legs together and bent at the knees. The parasternal long-axis view is usually obtained by placing the transducer in the left parasternal area in the second or third intercostal space (under the second or third rib from the top). This view is illustrated in Figure 24.2. The apical view is usually obtained lateral (left) and inferior (beneath) to the nipple. This view is illustrated in Figure 24.3.

While the transducer is positioned along the parasternal long-axis, we can make a pulsed Doppler recording of left ventricular outflow tract velocity, which is shown in Figure 24.4.

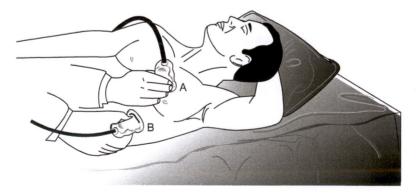

FIGURE 24.1 The parasternal (**A**) and apical (**B**) transducer positions, with the patient in the left lateral decubitus position [*Adapted from Oh et al. (1999)*].

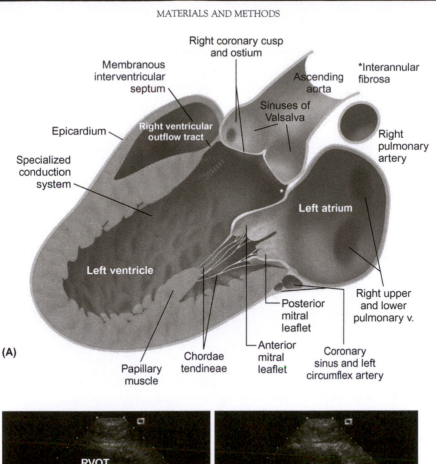

(A)

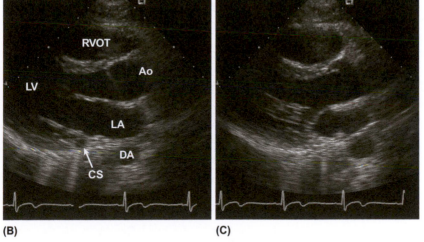

(B)  (C)

**FIGURE 24.2**  Anatomy (**A**) observed during left parasternal long axis view. Echo images were recorded at end-diastole (**B**) and end-systole (**C**). RVOT = right ventricular outflow tract, LV = left ventricle, Ao = aorta, LA = left atrium, CD = coronary sinus, DA = descending thoracic aorta [*The anatomical drawing from Otto (2009a). Figures reproduced by permission from Otto (2009b)*].

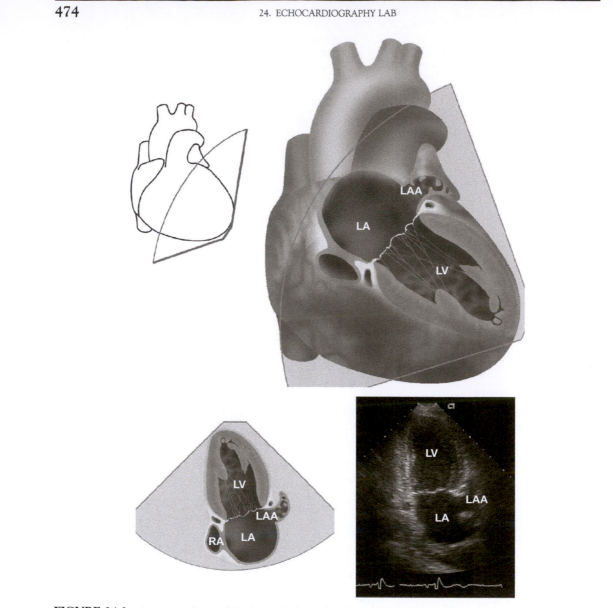

**FIGURE 24.3** Anatomy observed during apical two chamber view. LA = left atrium, LAA = left atrial append-age, LV = left ventricle, RA = right atrium *[Reproduced by permission from Otto (2009b)]*.

## Measurements

Stroke volume is estimated from measurements of LVOT diameter and the velocity time integral (VTI). This method was first validated by bioengineer Lee Huntsman in 1983 (Huntsman et al., 1983; Otto, 2009b). The amount of blood flow through a fixed orifice is directly proportional to the product of the orifice cross-sectional area and flow velocity. We

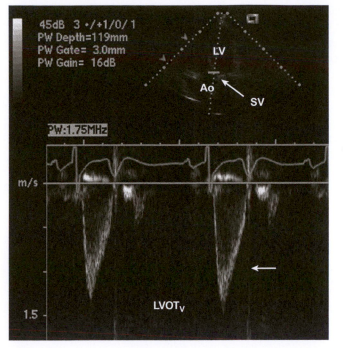

**FIGURE 24.4** Pulse Doppler recording of left ventricular (LV) outflow tract velocity (LVOT$_v$) just proximal to the aortic (Ao) valve from an apical approach. The Ao valve closing click (arrow) of the outflow tract velocity recording ensures that the sample volume (SV) location is immediately adjacent to the valve, corresponding with the site of outflow tract diameter measurement [*Reproduced by permission from Otto (2009b)*].

obtain this cross-sectional area by measuring the LVOT diameter and assuming a circular cross-section. We obtain the flow velocity by measuring the area under the velocity curve (VTI) because this is the distance blood flow travels with one stroke (Huntsman et al., 1983):

$$SV(cm^3) = \left[\frac{LVOT\_diam(cm)}{2}\right]^2 \cdot \pi \cdot VTI \ (cm) \tag{24.1}$$

Ejection fraction (EF) is estimated as the percentage difference between the end diastolic volume (EDV) and the end systolic volume (ESV) of the left ventricle. Using Simpson's rule, each ventricular volume is calculated as the sum of a series of parallel slices from apex to base. This method was first validated by cardiologist Louis Teichholz in 1974 (Teichholz, et al., 1974; Otto, 2009b). EF is then the percentage change in volume:

$$EF \ (\%) = \left[\frac{EDV - ESV}{EDV}\right] \cdot 100\% \tag{24.2}$$

## Procedure

Attach the ECG electrodes in the proper locations (LA = black, RA = white, LL = red), and adjust the ECG amplitude so that the QRS complex is visible. Measure LVOT diameter through the following steps:

1. Apply gel to the MicroMaxx transducer.
2. Select 2D.

3. Position the transducer on the patient so that the image replicates Figure 24.2. The length of the transducer should be parallel to the length of the intercostal space. Work to find the mitral valve and inferior wall first. Then make small angular adjustments to image the aortic valve also. You may need to adjust the NEAR, FAR, and GAIN knobs. You may have to adjust the DEPTH keys.
4. Leave the transducer in the proper position for 15 s so that the MicroMaxx buffer fills with data.
5. Select FREEZE. Do not hit FREEZE again till you are finished with step 10.
6. Use the FREEZE left and right arrows to position to the first (or next) systolic LVOT diameter. To determine the systolic LVOT diameter, check your ECG and that the diameter is at its widest length.
7. Select CALIPER. Select SELECT, which will enable you to position over one side of the LVOT diameter.
8. Select SELECT, which will enable you to position over the other side of the diameter, to make the measurement.
9. Select SAVE on the blue panel. Manually record the LVOT diameter.
10. Repeat steps 6–9 to obtain two other estimates of LVOT diameter.
11. Select FREEZE to unfreeze.

Measure VTI through the following steps:

12. Position the transducer on the patient so that the image replicates Figure 24.2.
13. Select DOPPLER.
14. Position sample volume so that it is in the center of LVOT.
15. Select UPDATE.
16. Leave the transducer in the proper position (look for a V-shaped velocity profile just past the QRS complex with high-frequency sound, per Figure 24.4) for 15 s so that the buffer fills with data.
17. Select FREEZE. Do not hit FREEZE again till you are finished with step 25.
18. Use the FREEZE left and right arrows to position to the first (next) systolic Doppler spectrum.
19. Select CALCS.
20. Move cursor to select AV, then LVOT VTI.
21. Select SELECT. This enables you to start tracing the systolic Doppler spectrum. Begin to trace from 0 m/s (the $x$-axis) as the valve opens, around the V, and back to 0 m/s as the valve closes.
22. When you have finished tracing, select SELECT.
23. Select SET.
24. Select SAVE on the blue panel. Manually record LVOT VTI.
25. Repeat steps 18–24 to obtain two other estimates of VTI.
26. Select FREEZE to unfreeze.

You can view your six measurements by:

27. Select REVIEW.
28. Select each saved measurement.

**29.** Select DONE.

Show these measurements to your instructor. Print your images on the Sony printer interfaced to your MicroMaxx. When the measurements are approved, you can manually calculate the mean LVOT diameter and mean VTI, and calculate SV.

Measure EF through the following steps:

**30.** Apply gel to the MicroMaxx transducer.
**31.** Select 2D.
**32.** Position the transducer on the patient so that the image replicates Figure 24.3. Again, the length of the transducer should be parallel to the length of the intercostal space. Work to image four chambers first. When these are in view, turn the transducer 90° to isolate only the left atrium and left ventricle in the image. You may need to adjust the NEAR, FAR, and GAIN knobs. You may have to adjust the DEPTH keys.
**33.** Leave the transducer in the proper position for 15 s so that the MicroMaxx buffer fills with data.
**34.** Select FREEZE. Do not hit FREEZE again till you are finished with step 42.
**35.** Use the FREEZE left and right arrows to position to the first (or next) left ventricular end diastolic volume. To determine diastolic volume, check your ECG and that the volume is at its largest area.
**36.** Select CALCS.
**37.** Move cursor to select A2Cd.
**38.** Select SELECT to start cursor. Trace around volume.
**39.** Select SELECT to end cursor.
**40.** Select SAVE on the blue panel. Manually record value. Select CALCS to remove volume display.
**41.** Repeat steps 35–40 to obtain two other estimates of end diastolic volume.
**42.** Repeat steps 35–41 to obtain three estimates of end systolic volume. Make sure you use the same beats as were selected for end diastolic volume. In these iterations, use the label A2Cs, instead of A2Cd.
**43.** Select FREEZE to unfreeze.

You can view your six measurements by:

**44.** Select REVIEW.
**45.** Select each saved measurement.
**46.** Select DONE.

You can view your calculated EF by:

**47.** Select REPORT.
**48.** View calculated EF.
**49.** Select DONE.

Show these measurements to your instructor. Print your images on the Sony printer interfaced to your MicroMaxx.

## RESULTS AND ANALYSIS

Record the answers to the questions in the written procedure in the Results section of your lab report.

## DISCUSSION

Answer the following questions in the Discussion section of your lab report:

- Why is the VTI transducer position tuned by using sound?
- How accurate do you believe your four measurements (LVOT diameter, VTI, EDV, ESV) are? Give reasons.
- Which do you believe is the least accurate measurement? Give reasons.

## References

Huntsman, L. L., Stewart, D. K., Barnes, S. R., Franklin, S. B., Colocousis, J. S., & Hessel, E. A. (1983). Noninvasive Doppler determination of cardiac output in man. Clinical validation. *Circulation*, *67*, 593–602.

Oh, J. K., Seward, J. B., & Tajik, A. J. (1999). *The echo manual* (2nd ed.). Philadelphia: Lippincott Williams & Wilkins.

Otto, C. M. (2009a). Echocardiographic evaluation of valvular heart disease. In: C. M. Otto, & R. Bonow (Eds.), *Valvular heart disease: A companion to Braunwald's heart disease*. Philadelphia: Elsevier/Saunders.

Otto, C. M. (2009b). *Textbook of clinical echocardiography* (4th ed.). Philadelphia: Elsevier Saunders.

Teichholz, L. E., Cohen, M. V., Sonnenblick, E. H., & Gorlin, R. (1974). Study of left ventricular geometry and function by B-scan ultrasonography in patients with and without asynergy. *New England Journal of Medicine*, *291*, 1220–1226.

Webb, A. (2003). *Introduction to biomedical imaging*. Hoboken, NJ: John Wiley/IEEE Press.

# Patient Monitoring Lab

In this chapter, we describe the steps for a patient monitoring laboratory experiment. This lab is intended to supplement the electrocardiograph, pulse oximeter, respiration monitor, and bispectral index material in Chapters 2, 11, and 13. This lab is written for execution using Philips patient monitoring equipment, but it can be customized for patient monitors from other manufacturers.

Upon completion of this chapter, each student shall be able to:

1. Monitor electrocardiogram (ECG), respiration, pulse oximetry, and the bispectral index (BIS).
2. Understand the physiologic sources of these signals.
3. Develop a clinical protocol to assess motion artifact.
4. Observe the occurrence of false alarms during motion artifact.

## STRATEGIC PLANNING

During this experiment, you will be monitoring parameters from one of your fellow students. The other two students should observe the monitor waveforms, parameters, and alarms. One of these two students should also act as the scribe for recording observations.

The lab procedure will intentionally cause nuisance alarms. At times, alarms are generated when a parameter momentarily goes out of range. To stop this type of intermittent alarm quickly, put the monitor quickly in and out of standby mode. Note that all your printouts are a required part of your lab report.

This lab should take approximately 2 hr to conduct.

## PATIENT MONITORING

Patient monitoring is the continuous observation of repeating events of physiologic function to guide therapy or to monitor the effectiveness of interventions. Historically, patient monitoring devices were most prevalent in the operating room and intensive care

479

unit (ICU). Over time, these devices migrated to other hospital units, clinics, and physician offices because of their great utility (Baura, 2002).

Vital signs monitoring may be conducted with patient monitoring devices to assess an individual's level of physiologic function. The classic vital signs measurements are heart rate, respiration, blood pressure, and temperature. The Veteran's Health Administration first added pain to their facilities' checklist of indicators in February 1999. However, medical device industry workers often identify pulse oximetry as the fifth vital sign. A typical patient monitor display is illustrated in Figure 25.1.

## Motion Artifact and False Alarms

Patient monitoring is typically conducted in the ICU, simultaneously for several patients. Nurses sit at the central station, which is centered between several individual ICU rooms within a single ICU module. At the central station, high and low alarms for each physiologic signal are typically set. In theory, these alarms allow a nurse to immediately know when patient status is compromised. In practice, these threshold alarms are often triggered when no physiologic change occurs, due to noise sources such as motion artifact. These false alarms lead to continuous stress for both patients and staff. It has also been reported that loud alarms lead to patient sleep deprivation (Aaron et al., 1996).

Motion artifact may trigger a false alarm because it can occur within the same frequency range as the original signal. For example, electrocardiogram alarm levels are typically set at 60 and 100 bpm, or 1 and about 2 Hz. Because motion artifact is not restricted to a single frequency, it can easily occur within this 1 to 2 Hz range. During drug infusion, motion artifact has been traced to eating with a catheterized arm, coughing, and wheelchair

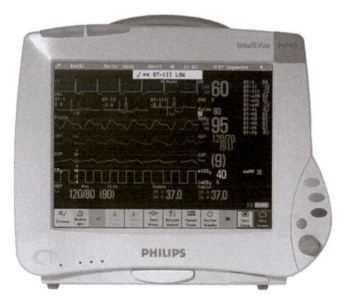

FIGURE 25.1   Philips Intellivue MP40 Patient Monitor (Philips Medical Systems, Andover, Massachusetts).

self-ambulation. Wheelchair self-ambulation is wheelchair movement at one rate, while an infusion pump on a pole behind the patient moves at a second rate (Baura, 2002).

One solution implemented by device manufacturers is individualized alarm sounds. But the "cuckoo" sound of a supposedly occluded infusion pump may be even more annoying than the originally implemented alarm. A chorus of individualized alarm sounds in unison for one patient increases the stress for staff. A second solution implemented by some device manufacturers is true minimization of the occurrence of motion artifact—based false alarms by mathematics, such as Masimo Signal Extraction Technology (SET). In today's lab experiment, we will personally experience false alarms in unison.

## MATERIALS AND METHODS

This experiment requires the following equipment:

- 1 Philips IntelliVue MP50 monitor
- 1 Philips M3001A measurement server with Masimo SET
- 1 Philips BIS module
- 1 Philips recorder module
- 1 Philips ECG lead set and trunk cable
- 3 ECG electrodes
- 1 Masimo pulse oximetry cable and reusable sensor
- 1 Aspect Medical BIS cable
- 1 Aspect Medical BIS sensor
- 1 cot or camping pad
- Strip chart paper, paper towels, alcohol wipes, as needed

## Procedure

Measurements will be made using the Philips IntelliVue MP50 monitor and modules. Have your patient sit in a chair quietly. When the monitor is turned on, you will note that the signals are set to normal ranges, except for BIS:

- ECG: 60—80 bpm
- Respiration: 12—18 bpm
- $SpO_2$: 93—100%
- BIS: 20—70 (unitless)

Follow these directions to monitor your patient in steady state:

1. Using the diagrams on the BIS sensor package, ECG cable, and pulse oximeter sensor, connect your subject to all three sensor/cables.
2. Power on the monitor.
3. Depress 1, 2, 3, and 4 on the BIS sensor for 5 s to initiate BIS monitoring.

4. If an alarm is triggered by the patient in this quiet state, adjust the alarm threshold to turn off the alarm. Specifically, touch the offending numeric and follow the menu to change the offending threshold.
5. Place the monitor in standby and then touch the main screen to stop any alarms.
   - During the quiet period, what sounds do you hear?
   - To which signals are the sounds tied?
   - How would you describe the differences between sounds?
   - Is one sound modulating in tone?

Print the waveforms generated during steady state through the following steps:

6. Touch RECORDINGS.
7. Touch RT A RECORDING.
8. As soon as you see the waveforms printing after parameters, wait 5 s.
9. Touch RT B RECORDING.
10. As soon as you see the waveforms, print after parameters, wait 5 s.
11. Touch STOP ALL RECORDINGS.
    - What is the source of the ECG signal? What is the source of the respiration signal? What is the source of the $SpO_2$ signal? What is the source of the BIS signal?
    - Which signal(s) is/are affected by holding your breath? Consider both the waveform and the numeric.
    - Which signal(s) is/are affected by tapping the finger with the $SpO_2$ sensor gently against a desk randomly at a rate that is not the heart rate? Consider both the waveform and the numeric.
    - Which signal(s) is/are affected by quickly moving from a sitting to standing position? Consider both the waveform and the numeric.
    - Try to generate three simultaneous alarms through any "repeatable simulated means." Describe the repeatable simulated means in your report. How does the MP50 allow you to distinguish three alarms from each other? Print the waveforms generated through your "repeatable simulated means" using steps 6–11.

Use the touch screen to turn off all alarms and tones. Conduct BIS monitoring through the following steps:

12. Have the patient lie on a cot or camping pad for 10 min in the dark.
13. Your instructor will observe your patient's final BIS value at 10 min.
14. Record the patient's initial and final BIS values. The initial value was printed during the first strip chart recording.

As the patient removes the BIS and ECG sensors, reflect on their construction.

## RESULTS AND ANALYSIS

Record the answers to the questions in the written procedure in the Results section of your lab report.

# DISCUSSION

Please answer the following questions in the Discussion section of your lab report:

- If one sound is modulating in tone, what is the source of modulation?
- Why are only certain signals affected by holding your breath?
- Why are only certain signals affected by tapping the finger with the $SpO_2$ sensor gently against a desk at a rate that is not the heart rate?
- Why are only certain signals affected by quickly moving from a sitting to standing position?
- Which signal is the most difficult to alarm?
- How do you account for this BIS change after 10 min?
- Why is the BIS sensor so wet? Why is the ECG adhesive so sticky?
- How difficult is it to generate false alarms during this patient monitoring lab?

## References

Aaron, J. N., Carlisle, C. C., Carskadon, M. A., Meyer, T. J., Hill, N. S., & Millman, R. P. (1996). Environmental noise as a cause of sleep disruption in an intermediate respiratory care unit. *Sleep, 19*, 707–710.

Baura, G. D. (2002). *System theory and practical applications of biomedical signals*. Hoboken, NJ: Wiley-IEEE.

# Thermometry Accuracy Lab

In this chapter, we describe the steps for a thermometry laboratory experiment. This lab is intended to supplement the thermometry lecture based on Chapter 12. It is written for execution using a hospital-grade electronic thermometer and consumer infrared thermometer.

Upon completion of this chapter, each student shall be able to:

1. Execute portions of two FDA-recognized consensus standards.
2. Understand the sources of measurement error in oral and ear thermometers.

## STRATEGIC PLANNING

During this experiment, you will be testing an electronic and infrared thermometer under laboratory conditions to determine which thermometer is more accurate. The more accurate thermometer will be used for reference measurements made with your group members to assess clinical accuracy.

This lab is based on the following FDA-recognized consensus standards:

- *ASTM E1112-00 Standard Specification for Electronic Thermometer for Intermittent Determination of Patient Temperature* (ASTM, 2006).
- *ASTM E1965-98 Standard Specification for Infrared Thermometer for Intermittent Determination of Patient Temperature* (ASTM, 2009).

Review these standards before your lab. If there is only one copy of each standard, make sure that you check the standards out for review before your scheduled lab time. Make sure you understand how the blackbody in Figure 26.1 was constructed.

This lab should take approximately 3 hr to conduct.

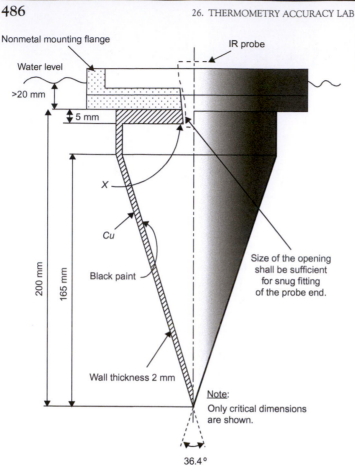

**FIGURE 26.1** A blackbody immersed in water [*Reproduced by permission from ASTM (2009)*].

## MATERIALS AND METHODS

This experiment requires the following calibrated equipment:

- 1 room thermometer
- 1 relative humidity meter
- 1 stopwatch
- 1 Welch Allyn Sure Temp thermometer
- 1 consumer ear thermometer
- 42 Welch Allyn Sure Temp thermometer probe covers
- 42 consumer ear thermometer probe covers
- 3 new AA batteries for Sure Temp thermometer
- new batteries for the consumer ear thermometer
- 1 water bath for electronic thermometer measurements: $\geq$ 1-L volume

- 1 water bath for blackbody measurements: $\geq 2$-L volume, $\pm 0.02°$C temperature stability
- 2 water reference thermometers (contact imbedded or immersed), with uncertainty $\leq \pm 0.03°$C
- 1 blackbody, constructed according to ASTM E1965-98

## Procedure

At your station, you will see two water baths, each holding a water reference thermometer. The water baths have been set to approximately 37.5°C. The water bath for ear thermometer measurements contains the blackbody.

To begin this protocol, record the room temperature and room relative humidity. Insert new batteries into each thermometer.

We will measure temperature at four temperature settings. If this experiment were conducted in industry, the order of temperature settings would be randomized. But in consideration of total lab time, we will conduct measurements according to increasing temperature settings. Follow these directions to measure temperature at the following temperature settings, in this order: 37.5, 38.0, 39.0, 39.5°C:

1. Adjust both water baths so that your water reference thermometer reads the desired setting $\pm 0.5°$C for 5 min.
2. In random order, immerse the electronic thermometer 6 times and ear thermometer 6 times. Before each measurement, use a new disposable probe cover. Before each measurement, record the water reference temperature.
3. For each immersion, record the displayed temperature reading.

Determine your reference thermometer through the following steps:

4. Calculate the mean error and mean absolute error for each thermometer, based on 24 readings.
5. Designate the more accurate thermometer as your clinical reference thermometer.

Record the room temperature and room relative humidity. Construct a table for recording ear and oral thermometry measurements from each person in the group, in random order. You will measure ear and oral temperatures simultaneously, with 6 replications, for a total number of 18 paired measurements. Follow these directions to make 18 clinical measurements:

6. Use a new disposable oral probe; use a new disposable ear probe.
7. Insert both probes simultaneously into the subject, following the best practices we discussed during lecture.
8. For each thermometer, record the displayed temperature readings.
9. Calculate the mean error and mean absolute error, based on 18 paired readings and your designation of clinical reference thermometer.

Record the room temperature and room relative humidity. Did these measurements change during the course of the lab?

## RESULTS AND ANALYSIS

Record the answers to the questions in the written procedure in the Results section of your lab report.

## DISCUSSION

Answer the following questions in the Discussion section of your lab report:

- Why is it necessary to record the room temperature and relative humidity? If the readings changed between the beginning and end of the lab session, how did this affect your lab measurements?
- Why is it necessary to randomize temperature measurements?
- Explain the mechanisms behind the water bath accuracy differences in the thermometers.
- What is the specified blackbody accuracy requirement for the ear thermometer? Do your measurements meet this requirement? Does a blackbody replicate the ear canal?
- What are the specified water bath accuracy requirements for the oral thermometer? Do your measurements meet this requirement? Does a water bath accurately replicate the sublingual pocket?
- For your clinical measurements, what are the specified clinical requirements? Do your measurements meet these requirements?
- Looking at your calculations, which is a more appropriate calculation for error: mean error or mean absolute error? Explain your reasoning.
- If you had a fever, which thermometer would you want to use to determine whether you should go to the hospital?

### References

ASTM (2006). *ASTM E1112-00 standard specification for electronic thermometer for intermittent determination of patient temperature*. West Conshohocken, PA: ASTM.

ASTM (2009). *ASTM E1965-98 standard specification for infrared thermometers for intermittent determination of patient temperature*. West Conshohocken, PA: ASTM.

# Surface Characterization Lab

In this chapter, we describe the steps for a surface characterization laboratory experiment. This lab is intended to supplement the drug-eluting stent material in Chapter 19. Because scanning electron microscopes (SEMs) are not discussed in this textbook, students are advised to read introductory material about SEMs, such as Chapters 2 and 4 of Goldstein et al. (2003). The lab is written for execution using a LEO DSM 982 Scanning Electron Microscope (SEM) and SurModics coating samples, but it can be customized for an SEM and coating samples from other manufacturers.

Upon completion of this chapter, each student shall be able to:

1. Mount and load samples for examination.
2. Acquire, refine, and capture an image using an SEM.
3. Distinguish between coated and uncoated catheter samples.

## STRATEGIC PLANNING

During this experiment, you will use an SEM to distinguish between coated and uncoated catheter samples. One student should load the samples, one should take notes, and one should put the samples away. Each student should take a turn at adjusting the SEM focus and magnification.

This lab should take approximately 2 hr to conduct.

## MEDICAL DEVICE COATINGS

Although medical device coatings are not a new technology, they gained much greater attention upon FDA approval of two drug-eluting stents: Johnson & Johnson Cordis's Cypher stent and Boston Scientific's Taxus Express stent. In both cases, stainless steel stents were coated with a polymer matrix containing a drug, which disrupted or inhibited smooth muscle cell migration and proliferation. The Taxus Express coating consists of a thermoplastic elastomer with discrete paclitaxel nanoparticles embedded in the elastomer matrix (Figure 27.1).

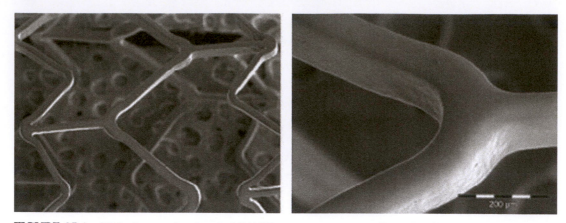

**FIGURE 27.1**   SEM micrographs of typical TAXUS stents postexpansion *[Reproduced by permission from Ranade et al. (2004)].*

## Surface Modification

The coatings we are examining today are manufactured by SurModics, who provided the coating for the Cypher stent. The SurModics APPLAUSE coating is a heparin coating; the SurModics HARMONY coating is a hydrophilic coating.

Surface modification is the process of changing the existing characteristics of a material surface to a more desirable characteristic. Using SurModics' PhotoLink technology, coating molecules (containing a diradical aromatic ketone group and molecule of interest) are photochemically coupled to a polymer surface. A diradical aromatic ketone group such as a benzophenone derivative is illuminated with blue or ultraviolet light to initially form a singlet excited state. This short-lived species undergoes a rapid intersystem crossing to generate the longer-lived triplet state. "This highly reactive intermediate is then capable of insertion into carbon-hydrogen bonds by abstraction of a hydrogen atom from the polymer surface, followed by the collapse of the resulting radical pair to form a new carbon-carbon bond. The high energy of the triplet state makes the photochemical coupling process relatively independent of the chemical composition of the surface, with the efficiency of the process being determined by the relative stability of the free radicals formed on the surface of the polymer" (Anderson et al., 2003) (Figure 27.2).

## MATERIALS AND METHODS

This experiment requires the following equipment:

- 1 LEO DSM 982 Scanning Electron Microscope station
- 2 SurModics white samples A and B, coated with a 100 Å layer of gold
- 2 SurModics blue samples C and D, coated with a 100 Å layer of gold

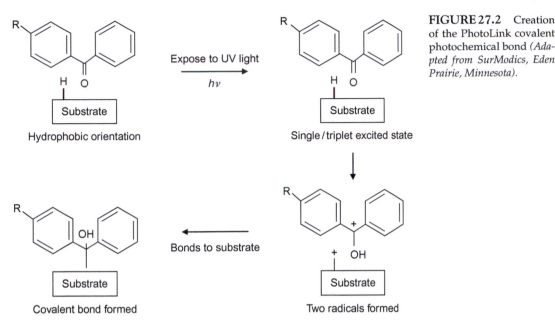

**FIGURE 27.2** Creation of the PhotoLink covalent photochemical bond *(Adapted from SurModics, Eden Prairie, Minnesota).*

## Procedure

During this lab experiment, you will image samples A, B, C, and D and determine how each sample is coated.

To start, prepare sample A:

1. Clean and dry sample. Mount it on the pin stub.
2. Use nitrile gloves and tweezers to handle anything going in the SEM.
3. If sample has never been pumped out, prepump to below $10^{-5}$ Torr in the auxiliary pumping system adjacent to the SEM area.

Vent the sample:

4. Turn power on for the 3 CRT monitors (move joystick to activate the fourth monitor).
5. Check all samples and the stage are well below the lens (cone) on the Sony monitor.
6. Check that the HIGH VOLTAGE OFF button is lit green.
7. Check that the CURRENT display reads a value between 25 and 150 $\mu$A.
8. Check the gun vacuum (the red LED on the vacuum panel is <2–8).
9. Check that the nitrogen cylinder is open, and record the tank pressure on a yellow pad.
10. Press the VENT button, and wait for a hiss from the SEM door ($\sim$3–5 min).

Load the sample:

11. Using gloves, mount sample stub, using black Allen driver to gently tighten.
12. Close the SEM door; watch the Sony monitor to see the sample clears the lens.
13. Hold door closed and press the EVAC button on vacuum panel; monitor vacuum system.
14. Wait $\sim$10 min until the HIGH VOLTAGE ON button lights up.

Adjust the settings:

15. Set kV −6 kV with knob in the HIGH VOLTAGE region.
16. Choose detector (SEInLens is better for flat samples) and #3 aperture.
17. If using SEInLens detector, make sure the SE detector on side console is turned ON.
18. Press the TV button in the SCANNING MODE section of console, and type "L."
19. Press the HIGH VOLTAGE ON button to begin imaging.

Focus and align aperture:

20. Set course FOCUS to a working distance ∼12 mm. Then use medium FOCUS.
21. Increase MAGNIFICATION gradually to ∼10,000× using medium FOCUS.
22. Press FOCUS WOBBLE, and align the aperture with APERTURE ALIGN X and Y.
23. If imaging above 10,000×, optimize focus then STIGMATE X and Y.
24. Optionally use SMALL mode to optimize at high magnifications.

Store an image:

25. Use SLOW scan (set knob to S3) or TV integration (type "I" in TV mode) to get a high-quality image.
26. In the Memory menu, set PATH.
27. Save images to the external drive.
28. Repeat steps 1−27 for the other three samples.

Finish using the SEM:

29. Reduce MAGNIFICATION to the lowest value and turn HIGH VOLTAGE OFF.
30. Lower the stage, and confirm that it will clear the lens cone and press VENT key.
31. Use gloves to gently loosen the set screw and remove the stub from stage.
32. Hold the door closed, and press EVAC button.
33. Place the stub back inside the plastic case.
34. Record the tank pressure.
35. Transfer images from the external drive to the personal computer to send over the network if desired.

## RESULTS AND ANALYSIS

Include one image each of samples A, B, C, and D in the Results section, and answer the following questions:

- Why would you coat a device using a heparin coating?
- Why would you coat a device with a hydrophilic coating?

## DISCUSSION

Please answer the following questions in the Discussion section of your lab report:

- Which of the white samples (A or B) is coated? Explain your reasoning.
- Which of the blue samples (C or D) is coated? Explain your reasoning.

- Which of the coated samples has a heparin coating? Which of the coated samples has a hydrophilic coating? Explain your reasoning.

## Acknowledgment

The author thanks Jennifer Wong, who contributed the SEM procedure as part of her Independent Studies course in 2007.

## References

Anderson, A. B., Dallmier, A. W., Chudzik, S. J., Duran, L. W., Guire, P. E., & Hergenrother, R. W., et al. (2003). Technologies for the surface modification of biomaterials. In: M. J. Yaszemski, D. J. Trantolo, K. Lewandrowski, V. Hasirci, D. E. Altobelli, & D. L. Wise (Eds.), *Biomaterials in orthopedics* (2nd ed.). New York: Marcel Dekker.

Goldstein, J., Newbury, D., Joy, D., Lyman, C., Echlin, P., & Lifshin, E., et al. *Scanning electron microscopy and X-ray microanalysis* (3rd ed.). New York: Springer.

Ranade, S. V., Miller, K. M., Richard, R. E., Chan, A. K., Allen, M. J., & Helmus, M. N. (2004). Physical characterization of controlled release of paclitaxel from the TAXUS Express2 drug-eluting stent. *Journal of Biomedical Materials Research Applied Biomaterials, 71*, 625–634.

# 28

# Entrepreneurship Lab

In this chapter, we describe the steps for a semester-long entrepreneurship project. This lab is intended as a technology and marketing exercise for a medical device course.

Upon completion of this chapter, each student shall be able to:

1. Understand how both technology and marketing are necessary for a start-up company's success.
2. Conduct a preliminary technology analysis.
3. Conduct a preliminary marketing analysis.

## STRATEGIC PLANNING

During this project, your group will be investigating the technology and marketing of a medical device company that failed to cross the chasm. For the purpose of this project, we specifically define this type of company as one that has been unable to overcome the gap, or chasm, between $50 million and $100 million in annual sales.

Your instructor will reveal the names of companies to be investigated at the end of a predetermined lecture. Some of the companies no longer exist; some are existing public companies. Because it is much more difficult to investigate failed private companies versus existing public companies, your group will be asked to e-mail its company choice after the lecture ends. To make this a fair process, you will be assigned companies in the order that the e-mails are received. Balance the time you research companies against the time required to get the company you request. In the author's past experience, all the companies may be assigned in as little as one hour.

If your company has more than one major product, you should choose the product with the largest annual sales for technology assessment. If your group believes that more than one product should be analyzed, confirm this need with your instructor first.

At times, students perform their due diligence by contacting their companies directly, but by no means is company contact required. However, if you contact a company, do not state that your instructor has assigned this company for investigation because it did not cross the chasm. A better method for encouraging information sharing is to state that your instructor assigned this company for its exemplary technology and marketing.

This project will require several weeks to complete. Do not attempt to complete this project within one week, such as during spring break. Your level of effort will be very obvious to your instructor in your report.

## Required Readings

The following readings are required for this project:

- Moore, *Crossing the Chasm*, provides the foundation for this project (2002).
- *Marketing's Four P's: First Steps for New Entrepreneurs* provides a foundation for your marketing analysis (Ehmke et al., 2005, http://www.extension.purdue.edu/extmedia/ EC/EC-730.pdf).
- Food and Drug Administration (FDA) regulation may be considered part of marketing. Without a 510(k) or Premarket Approval, new products cannot typically be marketed in the United States (http://www.fda.gov/MedicalDevices/DeviceRegulationand Guidance/HowtoMarketYourDevice/default.htm). From this URL, go to Step 3, 510(k) information and PMA information.
- Medicare reimbursement may also be considered part of marketing: Read Mannen (2004, http://www.mddionline.com/article/changing-perspectives-reimbursement-roller-coaster).
- The FDA considers human factors to be a critical component for medical devices. Human factors will be part of your technology analysis: Read Chapters 1 and 2 of Shneiderman and Plaisant (2005).

## CROSSING THE CHASM BACKGROUND

In October, 1981, start-up Nellcor was founded as a manufacturer of pulse oximeters. As discussed in Chapter 11, Nellcor did not originate the key technology of the calibrate curve; Biox did. However, because one of the founders was an anesthesiologist, Nellcor realized that pulse oximetry should be targeted toward operating rooms for safety, not pulmonary function labs for research. Nellcor did invent the disposable pulse oximetry probe, which sold well during the early days of the AIDS crisis.

Less than 6 yr later, with $46.2 million in revenue in the last 12 months and annual growth of 250.9%, Nellcor's initial public offering raised $38.4 million, at $16 per share. To medical device venture capitalists (VCs), Nellcor is the ideal medical device start-up (Baura, 2008).

Most start-ups are not so fortunate. According to research based on interviews with 200 individuals in 150 companies, 60% of high-tech companies that succeed in obtaining venture capital end up bankrupt. Only 10% of VC-funded start-ups succeed. These successful start-ups compensate for the other 90% of poorer performing companies in a VC's investment portfolio (Nesheim, 2000).

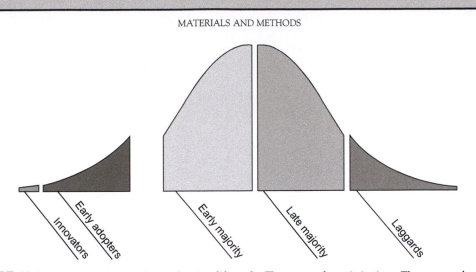

**FIGURE 28.1** The revised technology adoption life cycle. The assumed *x*-axis is time. The assumed y-axis is number of product purchasers *[Reproduced by permission from Moore (2002)]*.

When companies achieve initial success by selling their first product, they are directing their marketing efforts toward two initial groups of customers. Innovators purchase new technology products first, followed by early adopters. Though similar groups of customers, early adopters are not technologists. As shown in the technology adoption life cycle (Figure 28.1), the majority of customers are the early majority and late majority. The early majority likes new technology and purchases established technology. The late majority purchases standardized technology. The laggards purchase a technology without awareness of its existence when it is incorporated into a larger product (Moore, 2002).

Geoffrey Moore added a chasm between the purchase habits of the early adopters and early majority to signify a change in purchasing behavior. The two groups are dissociated, with the early majority not accepting a new product if it is marketed with the same methods used for marketing to the early adopters. This dissociation requires that a company change its marketing and sales strategies when different groups purchase the same product. This strategy change is critical for a company that wants to cross the chasm from $50 million in annual sales to $100 million (Moore, 2002).

## MATERIALS AND METHODS

Your group will conduct technology and marketing assessment to understand why a company failed to cross the chasm. The majority of assigned companies are start-up companies. However, a few companies have tried to cross the chasm unsuccessfully for many years.

Company material is available from many sources. For technology assessment, you may access patents, journal articles, reference textbooks (if available), clinicians, white papers, and online PowerPoint presentations. For marketing assessment, you may also access annual reports, marketing blogs, and press releases. Remember that any material may be presented in an overly positive manner, if it has been written by the company itself without peer review.

Allocate your time so that each section of your report is adequately researched, with consideration of potential points, graded on a 100-point scale.

## Report Structure and Grading

Because this is a technology course, 40% of the grade will be based on your technology assessment:

- Clinical need: 5 points
- Previous solutions: 5 points
- Sensor description (if electrical device)/biocompatibility (if mechanical device): 10 points
- System diagram: 10 points
- Human factors analysis: 5 points
- System limitations: 5 points

Twenty percent of the grade will be based on your marketing and sales assessment:

- Market definition/size: 2 points.
- Direct/indirect sales force: 3 points
- Annual sales revenue: 3 points
- Competitors: 3 points
- Regulation (FDA approval or Medicare reimbursement): 3 points
- Market strategy: 3 points
- Assessment summary: 3 points

Thirty percent of the grade will be based on your crossing the chasm analysis:

- High-tech parable similarities: 10 points. Construct a parable in two paragraphs of your company's growth, in a style similar to pages 23–25 of Moore's book.
- Alternative crossing strategies: 20 points

Ten percent of the grade will be based on the Executive Summary, Introduction, and Conclusion.

In consideration of the instructor's time for grading, your report, excluding title page and reference pages, is limited to this standard: 15 pages, single-spaced Times Roman, 1-inch margins, minimum 10-point font. All figures are to be directly incorporated into the text, with a figure legend below each figure. Each figure must be legible on a paper copy of your report Your instructor may require that your report be submitted using TurnItIn software.

The group with the best project report will receive 15 bonus points out of a total of 100 points. This group will be asked to present 10 slides summarizing their report during the Crossing the Chasm Discussion Lecture.

# CROSSING THE CHASM DISCUSSION LECTURE

A few weeks after the reports are due, they will be graded and the best project will be announced. A few lectures later, you will have a lecture and discussion completely dedicated to Crossing the Chasm.

As an introduction to marketing, the lecture will begin with an abbreviated screening of 15 min of the film *The Hudsucker Proxy* (Coen, 1994). For this film clip, you will be asked to identify elements of the technology adoption life cycle and The Four Ps: price, place, promotion, and product. Advanced Tissue Sciences' quest for FDA approval and Medicare reimbursement for their Dermagraft product will be used to illustrate the impact of regulation on sales (Baura, 2012).

The group with the best project will present their project summary. As a class, you will then discuss individual issues that prevented your companies from crossing the chasm.

# References

Baura, G. D. (2008). *A biosystems approach to industrial patient monitoring and diagnostic devices*. San Rafael, CA: Morgan Claypool.

Baura, G. D. (2012). Corporate considerations on biomaterials. In: B. Ratner, F. Schoen, A. Hoffman, & J. Lemons, (Eds.), *Biomaterials science: An introduction to materials in medicine* (3rd ed.). Burlington, MA: Elsevier Academic Press.

Coen, J. (1994). *The hudsucker proxy*. Warner Brothers Pictures.

Ehmke, C., Fulton, J., & Lusk, J. (2005). *Marketing's four P's: First steps for new entrepreneurs EC-730*, West Lafayette, IN: Purdue Extension.

Mannen, T. (2004). Changing perspectives: The reimbursement roller coaster. *Medical Device and Diagnostic Industry*, August.

Moore, G. A. (2002). *Crossing the chasm* (Rev. ed.). New York: Collins Business Essentials.

Nesheim, J. L. (2000). *High tech start up* (Rev. ed.). New York: Free Press.

Shneiderman, B., & Plaisant, C. (2005). *Designing the user interface* (4th ed.). Boston: Pearson.

# Index